The Sugar Code

Edited by
Hans-Joachim Gabius

Further Reading

Demchenko, A. V. (ed.)

Handbook of Chemical Glycosylation

Advances in Stereoselectivity and Therapeutic Relevance

524 pages in 1 volumes with 492 figures and 14 tables
2008
Hardcover
ISBN: 978-3-527-31780-6

Wong, C.-H. (ed.)

Carbohydrate-based Drug Discovery

980 pages in 2 volumes with 296 figures and 42 tables
2003
Hardcover
ISBN: 978-3-527-30632-9

Ernst, B., Hart, G. W., Sinaÿ, P. (eds.)

Carbohydrates in Chemistry and Biology

2578 pages in 4 volumes with 949 figures and 111 tables
2000
Hardcover
ISBN: 978-3-527-29511-1

The Sugar Code

Fundamentals of Glycosciences

Edited by
Hans-Joachim Gabius

WILEY-VCH Verlag GmbH & Co. KGaA

The Editor

Prof. Dr. Hans-Joachim Gabius
Ludwig-Maximilians-University Munich
Faculty of Veterinary Medicine
Department of Veterinary Sciences
Chair of Physiological Chemistry
Veterinärstrasse 13
80539 München
Germany

Cover
Staining for α2,6-sialylated N-glycans of a bovine blastocyst (together with DNA and F-actin staining; please see Fig. 24.4 for details) and for α2,8-linked polysialic acid of a rat embryo (please see Fig. 6.1 and Chapter 30.7 for details) is exemplarily illustrated to document the importance of glycosylation and protein-carbohydrate recognition, shown in the center (please see Fig. 13.1 for details), from fertilization and different stages of embryogenesis to reach the adult and enter the new cycle for progeny.

Library of Congress Card No.: applied for

British Library Cataloguing-in-Publication Data
A catalogue record for this book is available from the British Library.

Bibliographic information published by the Deutsche Nationalbibliothek
The Deutsche Nationalbibliothek lists this publication in the Deutsche Nationalbibliografie; detailed bibliographic data are available on the Internet at http://dnb.d-nb.de.

Cover Adam Design, Weinheim
Typesetting SNP Best-set Typesetter Ltd., Hong Kong
Printing and Binding Strauss GmbH, Mörlenbach

Printed in the Federal Republic of Germany
Printed on acid-free paper

ISBN: 978-3-527-32089-9

Foreword

The book of life is written with a *molecular* alphabet that is not limited to the four letters of the genetic code. This fact is being increasingly recognized not only by chemists but also by biologists and physicians, who have termed emerging fields '*molecular* biology, *molecular* genetics, *molecular* medicine, *molecular* cardiology, etc'. For glycosciences, this is also the 'age of the molecule'. Modern views largely accept that the co- and post-translational glycosylation of proteins has a function in molecular recognition. Protein-carbohydrate and carbohydrate-carbohydrate interactions control salient aspects of intra- and intercellular communication and trafficking, and are at the basis of a variety of essential biological phenomena, such as clearance of glycoproteins from the circulatory system, adhesion of infectious agents to host cells, and cell adhesion in the immune system, malignancy and metastasis.

To encode this complex biomolecular recognition ability, Nature has at its disposal a very powerful information tool, **the sugar code**. The role of reading the sugar-encoded messages is mainly played by a class of carbohydrate-binding proteins called lectins, which, together with their synthetic analogues and the complementary glycomimetics, have attracted the attention of many scientists with a different cultural background and training, coming from organic, medicinal and pharmaceutical chemistry, biochemistry, biology, medicine and even material science.

The modern concepts of glycosciences are not covered in the currently available organic chemistry, biochemistry and medicine textbooks, the former dealing mainly with the synthetic and conformational properties of carbohydrates and the last two with the biosynthesis of polysaccharides such as glycogen and the role of sugars as biochemical fuel in energy metabolism. Therefore, the textbook *The Sugar Code*, edited by Hans-Joachim Gabius, arrives at the right time and is highly welcome. Thanks to the Editor's efforts, the 30 chapters are consistently structured around **the sugar code** concept, convincingly attaining the aim of teaching fundamentals and revealing the molecular relations between apparently different phenomena. All the most important topics of the field of glycosciences are elegantly treated in an easily digestible manner, from the classic structure and analysis of carbohydrates to glycoproteins and glycolipids, up to the modern aspects of lectinology. Of particular relevance for the interdisciplinary nature of this textbook

The Sugar Code. Edited by Hans-Joachim Gabius

ISBN: 978-3-527-32089-9

are the interesting chapters on multivalency, versatility of protein-carbohydrate and carbohydrate-carbohydrate interactions, mammalian mini-lectins, and emerging biomedical relevance of the sugar code as well as the valuable frequent cross-referencing between chapters. Tables and figures in all cases are didactically well-tailored for teaching purposes.

The textbook will play a major role in helping teachers to introduce undergraduate and graduate students in chemistry, biochemistry, biology, pharmacy, medicine and bio- and nanotechnology to this fascinating and rapidly expanding field. Moreover, other scientists in academia and industry, even those having a loose contact with glycosciences, will also benefit from this outstanding book. The view that glycans are more complex and difficult to study than proteins and nucleic acids is no longer true: *The Sugar Code* textbook teaches us they are sweet and easy.

Parma, July 2009 *Rocco Ungaro*

Foreword

There are three major classes of biological macromolecules, all of which encode information essential for the life of the organism. Nucleic acids and proteins are linear molecules in which the individual subunits (nucleotides and amino acids, respectively) are linked by identical phosphate and peptide bonds, respectively. Glycans are significantly more complex. The linkages between individual subunits (monosaccharides such as hexoses, hexosamines, pentoses, etc.) are not identical; the anomeric carbon atom can be linked either by an alpha or beta bond to one of several carbon atoms on the adjoining monosaccharide. The resulting macromolecule may be linear or branched and may be covalently attached to a protein or lipid. Glycans are therefore highly efficient vehicles for information storage – more value for the money.

The apparatus required for glycan assembly is unique. Whereas the biosynthesis of nucleic acids and proteins is carried out by relatively simple template mechanisms, glycans require a complicated non-template assembly line (very similar to an automobile assembly line) where many different workers and machines (membranous organelles and vesicles, enzymes, transporters, structural proteins) function in harmony to manufacture the final product from a complex collection of primary building materials.

Glycosciences have received far less attention than they deserve in undergraduate, graduate and post-graduate life sciences courses. This is primarily because of the complexity of glycan structures and the diverse functions that have been attributed to them. There has also been a relative lack of suitable textbooks. The editor and authors of this book have fully succeeded to present the many aspects of glycosciences including their roots in a clear and efficient manner ready to enter classrooms. It is also gratifying that 'Info boxes' are used in this book to cover interesting 'side' topics relevant to the general discussion. An extensive 'Glossary' allows rapid access to unfamiliar terms. Glycobiology is a huge field that affects every known biological organism: prokaryotes, protozoa, fungi, plants, invertebrates and vertebrates. Since it is impossible to deal with everything, the highly relevant and important topics are covered in this book, and an excellent compromise between total knowledge and teaching requirements has been achieved.

The Sugar Code. Edited by Hans-Joachim Gabius

ISBN: 978-3-527-32089-9

The original concept of the Central Dogma by Francis Crick ignored post-translational modifications (such as protein glycosylation) that greatly magnify the functions of a single protein encoded by a particular gene. The above brief discussion should be sufficient to convince the reader that glycans certainly play pivotal functional roles in biology. This book provides the detailed evidence to teach this lesson.

Toronto, July 2009 *Harry Schachter*

Contents

The Sugar Code. Edited by Hans-Joachim Gabius

ISBN: 978-3-527-32089-9

Part Four Protein–Carbohydrate Interactions

Part Five Biomedical Aspects and Case Studies

Preface

Why this book? Carbohydrates are much, much more than simple biochemical fuels or – as polymers – the molecular concrete to convey stability to plants or insects. As obvious sign for a wide physiological role, glycan chains are also frequently presented by proteins and lipids. Their significance 'is to impart a discrete recognitional role' on the carrier (P. J. Winterburn and C. F. Phelps, 1972), the essence of the concept of the sugar code. The alphabet of its components (letters) and the sophisticated transport and enzymatic machinery for oligosaccharide (code word) formation provide the basis for the generation of these enormously versatile biochemical signals. They are handy for all aspects of bioregulation, because sugars are second to no other class of biomolecules in the capacity of information coding using oligomers. As instrumental as this feature is for far-reaching functionality *in vivo* the unsurpassed structural complexity automatically turns the task of breaking the sugar code into a highly demanding interdisciplinary challenge. With chemistry, biology and medicine being involved, there is the need for a coherent guide to the fundamentals of glycosciences designed for students and experts with interest in interdisciplinary work. As L. Stryer so elegantly and timelessly wrote in 1974, a textbook should give readers command of concepts and language and prepare a fertile ground for future discoveries. These are the aims of this multi-authored introduction to glycosciences.

The book is formatted to allow it to serve as a source for self-study and teaching, eventually enabling the reader to embrace the breadth of glycosciences at a comprehensible and professional level. To do so, all chapters are deliberately adjusted to a common standard and style and presented in a logical order from structure to biology and medicine. Illustrations and tables are designed to be ready for direct use in classrooms for lectures or even complete courses. Written by renowned experts and enthusiastic teachers, the chapters – with their concise summary boxes and entertaining info boxes – can even be read separately, to allow students and colleagues to be selective. Frequent cross-referencing between chapters is intended to minimize the need to turn to specialized literature to lay solid foundations. Consequently, aided by editorial encouragement, it was possible to limit the number of references. Complaints in this regard should be sent to me at gabius@lectins.de. Naturally, any advice, criticism or suggestions on what to improve in

The Sugar Code. Edited by Hans-Joachim Gabius

ISBN: 978-3-527-32089-9

order to best achieve our aims, that is to provide proper guidance and effective material for teaching as well as to fascinate and to inspire, are very welcome. So we invite you to contact us!

What you now hold in your hands would not have been possible without the enthusiasm and expertise of all members of the team of authors, without the invaluable advice, encouragement and help given by colleagues, Jürgen Kopitz and Harold Rüdiger deserve to be mentioned for being special friends, coworkers, Ruth Ohl and Herbert Kaltner deserve to be mentioned for being preciously dedicated and profoundly professional, and especially my wife Sigrun Ortrud along the way from project plan to product – and without an essential factor in teamwork, which quickly made its presence felt: the team spirit! It is a great pleasure and privilege for me to say 'Thank you', also for your interest. In the name of each and everyone involved we hope that you, our readers, enjoy the book and consider it helpful for your contribution to breaking the sugar code!

Munich, July 2009 *Hans-Joachim Gabius*

List of Contributors

Sabine André
Ludwig-Maximilians-University
Munich
Faculty of Veterinary Medicine
Department of Veterinary
Sciences
Chair of Physiological Chemistry
Veterinärstrasse 13
80539 München
Germany

Iwona Bucior
University of California
San Francisco
Department of Medicine
513 Parnassus Avenue
San Francisco, CA 94143
USA

Eckhart Buddecke
University of Muenster
Institute of Physiological
Chemistry
and Pathobiochemistry and
Leibniz-Institute of
Arteriosclerosis Research
Domagkstrasse 3
48149 Münster
Germany

Max M. Burger
Friedrich Miescher Institute for
Biomedical Research
Novartis Research Foundation
Novartis International AG
4002 Basel
Switzerland

Yoann M. Chabre
University of Québec at Montréal
Department of Chemistry
P.O. Box 8888, succ. Centre-Ville
Montréal (Québec) H3C 3P8
Canada

Anthony Corfield
Bristol Royal Infirmary
Division of Clinical Sciences South
Bristol
Mucin Research Group
Marlborough Street
Bristol BS2 8HW
UK

Françoise Debierre-Grockiego
Philipps-University Marburg
Medical School
Medical Center for Hygiene and
Medical Microbiology
Institute for Virology
Laboratory of Parasitology
Robert-Koch-Strasse 17
35043 Marburg
Germany

The Sugar Code. Edited by Hans-Joachim Gabius

ISBN: 978-3-527-32089-9

Hervé Falet
Harvard Medical School
Brigham and Women's Hospital
Department of Medicine
75 Francis Street
Boston, MA 02115
USA

Xavier Fernàndez-Busquets
University of Barcelona
Institute for Bioengineering of Catalonia
Nanoscience and Nanotechnology
Baldiri Reixac 10-12
08028 Barcelona
Spain

Hans-Joachim Gabius
Ludwig-Maximilians-University Munich
Faculty of Veterinary Medicine
Department of Veterinary Sciences
Chair of Physiological Chemistry
Veterinärstrasse 13
80539 München
Germany

Stefan Gaunitz
Karolinska Institute
Karolinska University Hospital in Huddinge
Division of Clinical Immunology
141 86 Stockholm
Sweden

Jill E. Gready
Australian National University
John Curtin School of Medical Research
College of Medicine, Biology and Environment
Computational Proteomics and Therapy Design Group
Genome Biology Program
GPO Box 334
Canberra ACT 2601
Australia

Anki Gustafsson
Recopharma AB
Hälsovägen 7
141 57 Huddinge
Sweden

Felix A. Habermann
Ludwig-Maximilians-University Munich
Faculty of Veterinary Medicine
Department of Veterinary Sciences
Chair of Veterinary Anatomy (Histology, Embryology)
Veterinärstrasse 13
80539 München
Germany

Thierry Hennet
University of Zurich
Institute of Physiology
Winterthurerstrasse 190
8057 Zürich
Switzerland

Jun Hirabayashi
National Institute of Advanced Industrial Science and Technology
Research Center for Medical Glycoscience
Lectin Application and Analysis Team
AIST Tsukuba Central 2 1-1-1, Umezono
Tsukuba, Ibaraki 305-8568
Japan

Karin M. Hoffmeister
Harvard Medical School
Brigham and Women's Hospital
Department of Medicine
75 Francis Street
Boston, MA 02115
USA

Jan Holgersson
Karolinska Institute
Karolinska University Hospital in Huddinge
Division of Clinical Immunology
141 86 Stockholm
Sweden

Koichi Honke
Kochi University Medical School
Department of Biochemistry
Kochi System Glycobiology Center
Kohasu, Oko-cho, Nankoku
Kochi 783-8505
Japan

Jesús Jiménez-Barbero
Consejo Superior de Investigaciones Científicas
Centro de Investigaciones Biológicas
Biología Físico-Química
Ramiro de Maeztu 9
28040 Madrid
Spain

Herbert Kaltner
Ludwig-Maximilians-University Munich
Faculty of Veterinary Medicine
Department of Veterinary Sciences
Chair of Physiological Chemistry
Veterinärstrasse 13
80539 München
Germany

Jürgen Kopitz
Ruprecht-Karls-University Heidelberg
Medical School
Institute of Pathology
Applied Tumor Biology
Im Neuenheimer Feld 220
69120 Heidelberg
Germany

Tibor Kožár
Slovak Academy of Sciences
Department of Biophysics
Institute of Experimental Physics
Watsonova 47
04001 Košice
Slovak Republic

Atsushi Kuno
National Institute of Advanced Industrial Science and Technology
Research Center for Medical Glycoscience
Lectin Application and Analysis Team
AIST Tsukuba Central 2 1-1-1, Umezono
Tsukuba, Ibaraki 305-8568
Japan

Robert W. Ledeen
University of Medicine & Dentistry of New Jersey
New Jersey Medical School
Department of Neurology & Neurosciences
185 South Orange Avenue
Newark, NJ 07103-2714
USA

Robert I. Lehrer
University of California Los Angeles
David Geffen School of Medicine
Department of Medicine
10833 Le Conte Avenue
Los Angeles, CA 90095
USA

Margarita Menéndez
Consejo Superior de Investigaciones Científicas
Instituto de Química Física Rocasolano
Serrano 119
28006 Madrid
Spain

Hans Merzendorfer
University of Osnabrueck
Department of Biology/Chemistry
Division of Animal Physiology
Barbarastrasse 11
49069 Osnabrück
Germany

Hiroaki Nakagawa
Hokkaido University
Graduate School of Advanced Life Science and
Frontier Research Center for Post-Genomic Science and Technology
Sapporo 001-0021
Japan
(current address: Hitachi High-Technologies Co.
Naka Application Center
Hitachinaka 312-0057
Japan)

Helen M. I. Osborn
University of Reading
School of Pharmacy
Whiteknights
Reading RG6 6AD
UK

Stefan Oscarson
University College Dublin
UCD School of Chemistry and Chemical Biology
Centre for Synthesis and Chemical Biology
Belfield, Dublin 4
Ireland

Katharina Paschinger
University of Natural Resources and Applied Life Sciences
Department of Chemistry
Muthgasse 18
1190 Wien
Austria

Georgios Patsos
Bristol Royal Infirmary
Division of Clinical Sciences South Bristol
Mucin Research Group
Marlborough Street
Bristol BS2 8HW
UK

Dubravko Rendić
University of Natural Resources and Applied Life Sciences
Department of Chemistry
Muthgasse 18
1190 Wien
Austria

Antonio Romero
Consejo Superior de Investigaciones Científicas
Centro de Investigaciones Biológicas
Departamento de Ciencia de Proteínas
Ramiro de Maeztu 9
28040 Madrid
Spain

Jürgen Roth
University of Zurich
Department of Pathology
Division of Cell and Molecular Pathology
Schmelzbergstrasse 12
8091 Zürich
Switzerland
(current adress: Yonsei University
Graduate School
Department of Biomedical Science
WCU Program
262 Seongsanro, Seodaemun-gu
Seoul 120-749
Korea)

René Roy
University of Québec at Montréal
Department of Chemistry
P.O. Box 8888, succ. Centre-Ville
Montréal (Québec) H3C 3P8
Canada

Harold Rüdiger
Julius-Maximilians-University
Wuerzburg
Institute of Pharmacy and Food
Chemistry
Am Hubland
97074 Würzburg
Germany

Reinhard Schwartz-Albiez
German Cancer Research Center
Department of Translational
Immunology
Im Neuenheimer Feld 580
69120 Heidelberg
Germany

Ralph T. Schwarz
Philipps-University Marburg
Medical School
Medical Center for Hygiene and
Medical Microbiology
Institute for Virology
Laboratory of Parasitology
Robert-Koch-Strasse 17
35043 Marburg
Germany

Hosam Shams-Eldin
Philipps-University Marburg
Medical School
Medical Center for Hygiene and
Medical Microbiology
Institute for Virology
Laboratory of Parasitology
Robert-Koch-Strasse 17
35043 Marburg
Germany

Fred Sinowatz
Ludwig-Maximilians-University
Munich
Faculty of Veterinary Medicine
Department of Veterinary Sciences
Chair of Veterinary Anatomy
(Histology, Embryology)
Veterinärstrasse 13
80539 München
Germany

Dolores Solís
Consejo Superior de Investigaciones
Científicas
Instituto de Química Física Rocasolano
Serrano 119
28006 Madrid
Spain

Hiroaki Tateno
National Institute of Advanced
Industrial Science and Technology
Research Center for Medical
Glycoscience
Lectin Application and Analysis Team
AIST Tsukuba Central 2 1-1-1,
Umezono
Tsukuba, Ibaraki 305-8568
Japan

Naoyuki Taniguchi
Osaka University
Institute of Scientific and Industrial
Research
Department of Disease Glycomics
Mihogaoka 8-1, Ibaraki
Osaka 567-0047
Japan

Andrea Turkson
University of Reading
School of Pharmacy
Whiteknights
Reading
RG6 6AD
UK

Jozef Uličný
Pavol Jozef Šafárik University
Faculty of Science
Department of Biophysics
Jesenná 5
04145 Košice
Slovak Republic

Antonio Villalobo
Consejo Superior de Investigaciones Científicas
Instituto de Investigaciones Biomédicas
Arturo Duperier 4
28029 Madrid
Spain

Iain B. H. Wilson
University of Natural Resources and Applied Life Sciences
Department of Chemistry
Muthgasse 18
1190 Wien
Austria

Gusheng Wu
University of Medicine & Dentistry of New Jersey
New Jersey Medical School
Department of Neurology & Neurosciences
185 South Orange Avenue
Newark, NJ 07103-2714
USA

Alex N. Zelensky
Erasmus Medical Center
Department of Genetics
Dr. Molewaterplein 50
3015 GE Rotterdam
The Netherlands

Christian Zuber
University of Zurich
Department of Pathology
Division of Cell and Molecular Pathology
Schmelzbergstrasse 12
8091 Zürich
Switzerland

Part One
Chemical Basis

The Sugar Code. Edited by Hans-Joachim Gabius

ISBN: 978-3-527-32089-9

1
The Biochemical Basis and Coding Capacity of the Sugar Code

Harold Rüdiger and Hans-Joachim Gabius

Teaching the biochemistry of carbohydrates is not simply an exercise in terminology. It has much more to offer than commonly touched upon in basic courses, if we deliberately pay attention to the far-reaching potential of sugars beyond energy metabolism and cell wall stability. In fact, then there is no reason why complex carbohydrates should shy at competition with nucleic acids and proteins for the top spot in high-density biocoding. On the contrary, sugars have ideal properties for this purpose, as will be concluded at the end of this chapter. In this sense, an obvious explanation why research in glycosciences (structural and functional glycomics and lectinomics) has lagged behind the fields of genomics and proteomics, also in the public eye, is 'that glycoconjugates are much more complex, variegated, and difficult to study than proteins and nucleic acids' [1]. What is a boon for decorating cell surfaces with a maximum number of molecular messages at the same time has been and still is a demanding challenge for analytical and synthetic chemistry (please see Chapters 3–5 for details on how to address it properly). That the sugar code can give up its secrets rather easily depends on a solid understanding of the unique aspects of its biochemical basis. With hindsight, E. Chargaff's rule on the regularities of nucleotide composition, derived from biochemical analysis of DNA in 1949/1950, was conspicuously more than a descriptive parameter. Its ratio of unity signified base complementarity. It is thus worthwhile to carefully examine basic carbohydrate biochemistry to define what makes sugars genuinely special in chemical terms. In doing so, we guide readers from the etymological roots of frequently used terms to raising awareness for what common structural depictions tell us about information coding by sugars.

1.1
Etymological Roots

Elementary analysis of hexoses revealed presence of carbon (Latin '*carbo*' = 'coal') and water (Greek '*hydor*' = 'ΰδωρ') in a stoichiometric proportion, that is, $C_n(H_2O)_m$, with $n \geq m$. This result explains the origin of the term 'carbohydrates'. Its synonym

The Sugar Code. Edited by Hans-Joachim Gabius

ISBN: 978-3-527-32089-9

'sugar' goes back to the Sanskrit word '*sarkar*', which means sugar cane and its product. This word has entered many languages (for example Persian, Greek, Latin and Arabic). The Greek variant '*saccharon*' ('σάκχαρον') and its Latin form '*saccharum*' have led us to designate sugars as 'saccharides'. When using 'glycoside', the Greek term '*glykys*' ('γλυκύς'), translated into 'sweet', is the etymological root.

The most abundant hexose is glucose (Glc) or grape sugar. Its name is derived from the Latin loanword 'glucus'. This sugar and its 2-deoxy-2-acetamido derivative *N*-acetylglucosamine (GlcNAc) are the building blocks for rigid cell walls, that is, the polymers cellulose and chitin (please see Chapter 12 for details). GlcNAc is ubiquitously present in glycosaminoglycan chains and peptidoglycans as well as in glycoconjugates such as glycoproteins with *N*- and *O*-glycans (please see Chapters 6–11 and 29). De-*N*-acetylation without proper sulfation results in the occurrence of small amounts of glucosamine (GlcN) in glycosaminoglycan chains (please see Chapter 11). Historically, the scientific detection of the natural presence of GlcN and its derivative GlcNAc dates back to a fortunate menu selection in 1875 (see Info Box 1 and also Info Box 1 in Chapter 12). An answer why glucose and its derivatives are so abundant in Nature is given in the next paragraph by closely looking at the different projection formulas.

Info Box 1

'In 1875 a young physician named Georg Ledderhose was working during the summer semester in the laboratory of Friedrich Wöhler in Göttingen when Ledderhose's uncle, Felix Hoppe-Seyler, a noted physiological chemist, invited him to dinner. At his uncle's suggestion he took the remains of the lobster they had eaten back to the laboratory, where he found that the claws and the shell dissolved in hot concentrated hydrochloric acid and that on evaporation the solution yielded characteristic crystals. He soon identified the crystalline compound as a new nitrogen-containing sugar, which he named *glycosamin*' [N. Sharon. Carbohydrates. *Sci Am* 1980; **243**, 80–97].

Note: F. Hoppe-Seyler was the founder of the *Zeitschrift für physiologische Chemie* in June 1877, which today is known as *Biological Chemistry*.

1.2
What Projection Formulas Tell Us

The classical chain structure introduced by E. Fischer figures how the $(H_2O)_m$ molecules are distributed over the carbon backbone. A series of hydroxy groups is present together with an aldehyde function at C1 (Figure 1.1a). Of note, their high density establishes a platform with exceptional properties (please see below). When looking at experimental numbers in Figure 1.2, it becomes obvious that the chain structure will not at all be the predominant form of a carbohydrate.

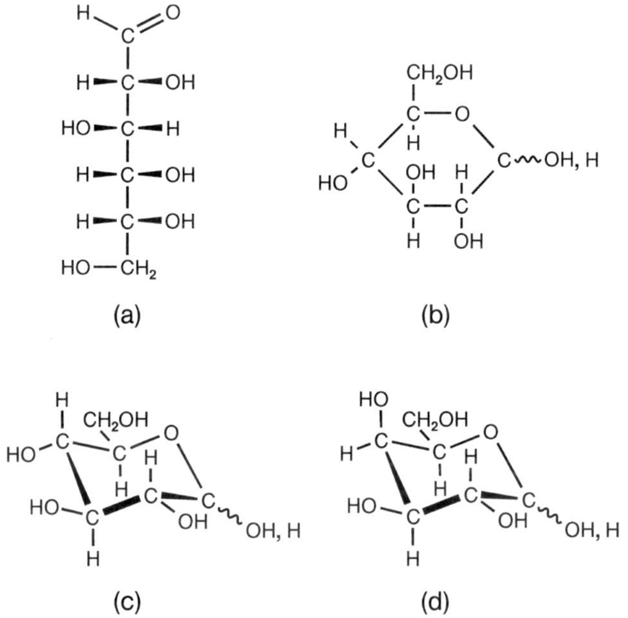

(a) (b)

(c) (d)

Figure 1.1 Illustration of the two types of projection formulas and the chair-like conformation of D-glucose. The open-chain (Fischer) (a) and hexopyranose (Haworth) projection formulas (b) as well as the $^4C_1$ low-energy chair-like pyranose conformation (c) are presented. Structural variability at the anomeric center (α or β) is symbolized by a wavy line. For further information on assignment of anomeric positions and contributions of pyranose and open-chain forms to the equilibrium, please see Figure 1.2 and its legend. Epimer formation from D-glucose (c) to D-galactose (d) leads to the axial positioning of the 4-hydroxy group in D-galactose and changes in the topological signature of hydroxy and polarized C—H groups.

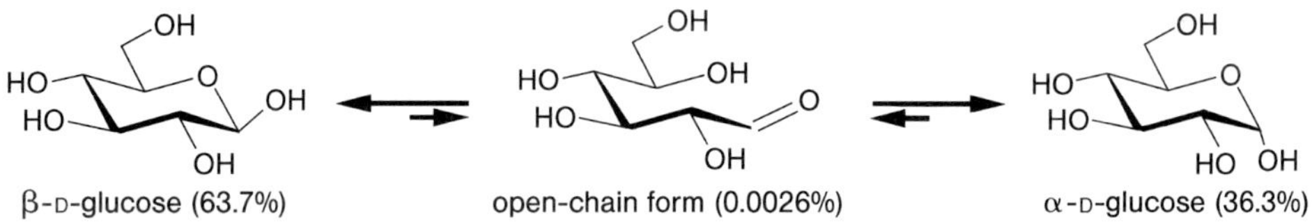

β-D-glucose (63.7%) open-chain form (0.0026%) α-D-glucose (36.3%)

Figure 1.2 Illustration of the equilibrium including the two anomeric forms of D-glucose. The percentages of presence of the two anomeric hexopyranose and the open-chain forms in equilibrium are given in the bottom line.

As early as 1883, B. Tollens demonstrated that monosaccharides fail to react in common tests for free aldehydes. This observation intimates an intramolecular reaction. Its product is depicted in the Haworth projection in Figure 1.1b. This structure, in chemical terms a semialdehyde, is a derivative of the heterocyclus pyran, which explains the term 'pyranose'. Therefore, a monosaccharide symbol can be accompanied by an italic '*p*'. However, it is often omitted, since most of the physiologically important hexoses occur as pyranoses. Like cyclohexane, the pyranose ring is not planar, but adopts a low-energy chair-like conformation (Figure 1.1c). Each substituent can assume either an axial or an equatorial position. Thus, the chemically equivalent groups are subdivided into geometrically different constellations and these are energetically not identical in the chair conformation. Owing to the free rotation around a single bond, an axial substituent can 'collide' with other axial groups in its vicinity (1,3-diaxial 'clashes'). Evidently, the lowest energy level is attained for a pyranose if all substituents larger than a hydrogen atom reside in an equatorial position and this formula represents glucose, the most abundant sugar in Nature (Figure 1.1c). Placing a plane through carbon atoms C2, C3, C5 and the ring oxygen atom readily explains why this conformation is referred to as $^4C_1$ (C = chair) (for further information on conformational flexibility of the pyranose ring, please see Figure 2.1). It harbors a characteristic topological signature of hydroxy groups, ready for directional hydrogen bonds in protein/carbohydrate–carbohydrate recognition or coordination bonds with Ca^{2+} (please see Chapters 13, 16 and 21). The remaining ambiguity at the C1 position, the anomeric center, is clarified with the reaction mechanism given in Figure 1.2.

All D-sugars can present this hydroxy group in either axial (α) or equatorial (β) positions. Mutarotation (Lat.: '*mutare*' = 'to alter'; alteration of optical rotation using polarized light) is the consequence of the equilibrium between α- and β-anomers, when starting measurements with a pure anomer. This phenomenon was first observed in 1846 by A.P. Dubrunfaut, who discovered D-fructose one year later. The geometry at the C1-atom of a hexopyranose has not only a bearing on optical properties. It also determines the shape of disaccharides and thus enhances total coding capacity (please see below). What happens if a hydroxy group other than at the anomeric center changes position? Is this only a subtle change?

This process yields epimers. Epimerization will entail emergence of the mentioned 1,3-diaxial clashes. Being a source of structural destabilization, their number will most likely be restricted to a minimum in natural glycans, that is to one in the 4-epimer galactose (and also the 2-epimer mannose) as shown in Figure 1.1d. By the way, 3-epimerization of D-glucose to D-allose leads to two clashes, whereas the 5-epimer L-idose can avoid them by adopting the $^1C_4$ conformation [please see below: case study of D-glucuronic acid (GlcA) versus its natural 5-epimer L-iduronic acid (IdoA)]. Beyond 1,3-diaxial contacts and, of course, the alteration of the signature of hydroxy group presentation, a further parameter is automatically affected in epimers: the spatial distribution of the positively polarized C—H bonds. When comparing the presentation of C—H bonds in D-glucose/D-galactose one should pay attention to the constellation between the C3 and C5 atoms (Figure 1.1c and d). The 4-epimer has a contiguous stretch of these C—H bonds, inviting polar contact with π-electrons. This patch is ideally suited for C—H/π interaction. On these grounds the presence of an aromatic residue in proteins, preferably of tryptophan, is predicted in binding sites of proteins specific for D-galactose (please see Figure 13.1 for the answer to this question). In sum, epimerization – together with derivative formation – accounts for the origin of many constituents of natural glycans.

A final point arising from inspecting the structures in Figure 1.1 is versatility for oligomer formation. In comparison to the phosphodiester and peptide bonds, strictly constrained in this respect, a large panel of glycosidic linkages can be formed. This unique property is visualized by arrows in Figure 1.3. That this statement is not of solely academic value is underscored by the natural occurrence in honey and other sources of all theoretically possible diglucosides (Table 1.1). In forming a glycosidic bond, the anomeric center of one partner is always involved. Even if the linkage points (for example 1–4) are identical, the decision on the anomeric position will markedly bear upon the biochemical properties of the

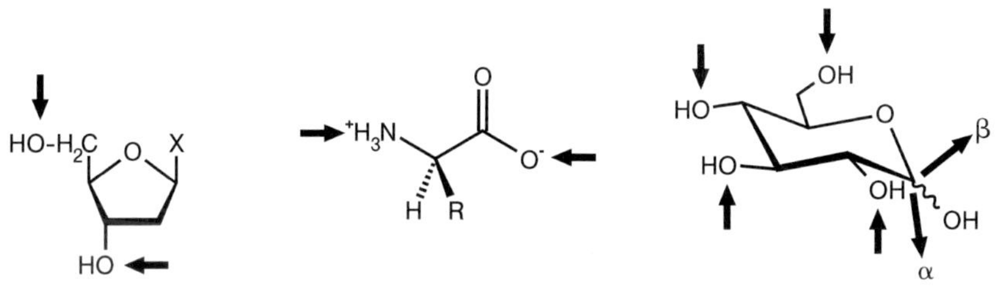

Figure 1.3 Illustration of the linkage points for oligomer formation in biomolecules by arrows. The phosphodiester bond in nucleic acid biosynthesis (a) and the peptide bond in protein biosynthesis (b) yield linear oligomers. In contrast, the glycosidic linkage in oligosaccharides can involve any hydroxy group, opening the way to linear and also branched structures (c) (for an example of branching, please see Figure 1.5). Using D-glucose (please see Figure 1.1c) as an example, its active form UDP-Glc allows conjugation of this sugar to carbohydrate acceptors to any hydroxy group, as symbolized by arrows directed towards the hydroxy groups (for a list of resulting diglucosides, please see Table 1.1). The anomeric position in chain elongation can vary, as symbolized by two bold arrows pointing away from the molecule (for structures of the anomers and the two 1–4-linked diglucosides with different anomeric positions, please see Figures 1.2 and 1.4).

Table 1.1 Naturally occurring disaccharides formed from two glucose units.

Type of linkage	Common name
α1–2	Kojibiose
β1–2	Sophorose
α1–3	Nigerose
β1–3	Laminaribiose
α1–4	Maltose
β1–4	Cellobiose
α1–6	Isomaltose
β1–6	Gentiobiose
α1–1'α	Trehalose

All disaccharides are conversion or degradation products of natural polysaccharides and glycosides, except for trehalose, which is present in bacteria, fungi and insects.

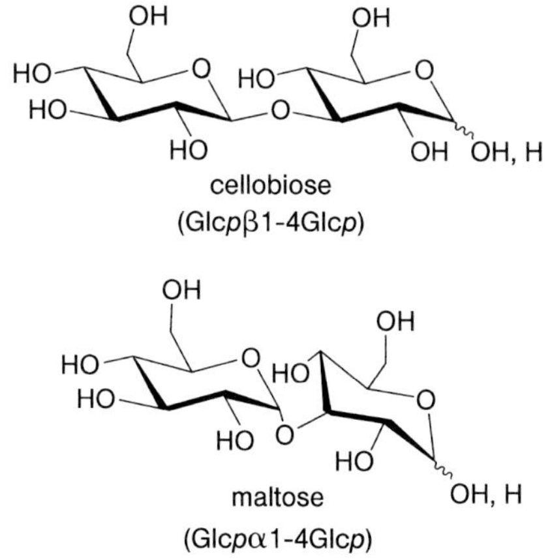

cellobiose
(Glcpβ1-4Glcp)

maltose
(Glcpα1-4Glcp)

Figure 1.4 Illustration of the two 1–4-linked diglucosides cellobiose and maltose. The α- and β-anomers of D-glucose (please see Figure 1.2 for structures) produce diglucosides with different shapes, underscoring spatial consequences of anomer selection.

product. Cellulose and glycogen/starch are telling examples, and Figure 1.4 illustrates the structures of the respective α/β1–4-linked diglucopyranosides. Taken together, the projection formulas tell us that (i) hydroxy groups in a hexapyranose are presented in high density with implications for hydrogen and C—H/π bonding as well as oligomer formation, and (ii) distinct hexoses will be energetically favored (glucose and its epimers mannose and galactose). What are the actual consequences of these structural aspects for the coding capacity of oligosaccharides?

1.3 The Coding Capacity of the Sugar Code

A quantitative measure of this characteristic is the total number of 'words' (isomers) built from a set of 'letters' (monomers). As deduced from Figure 1.3, calculations to define the range of structural permutations are simple in the cases of nucleotides and amino acids. It is only the linear sequence that counts to master this task. In order to completely define a disaccharide structurally, however, it is not sufficient to determine just the sequence, as already explained above. Beyond this parameter, the linkage points, the anomeric position and also the ring size (pyranose/furanose) must be known. These attributes distinguish carbohydrates from nucleotides and amino acids, and there is more.

Compactness of structural units is achieved by branching. Toward this end, a monosaccharide will then engage more than two hydroxy groups for glycosidic linkages. Figure 1.5 shows a classical example for branched oligosaccharides, that is, the ABH(0) histo-blood group epitopes. Delineation of their biochemical nature was aided by application of an eel lectin (see Info Box 2; its crystal structure is shown in Figure 16.1g), their structures are given in Figure 1.5. Branching is also

Info Box 2

'An early observation of Landsteiner had established that for artificial antigens a simple substance with a structure closely related to, or identical with, the immunologically determinant (haptenic) group of the antigen can combine with the antibody and thereby competitively inhibit the reactions between antigen and antibody. Although this principle is employed today in many forms of inhibition tests, in the 1950s it had not been used to find the determinants in naturally occurring antigens. By 1952 we had accumulated a large selection of anti-H reagents of human and animal origin and we decided, with no great expectation of the outcome, to screen them for inhibition of the agglutination of 0 cells with the component sugars present in the blood-group active substances. Somewhat to our surprise one of the many reagents, that from the eel, *Anguilla anguilla,* was quite strongly inhibited by L-fucose and to a greater extent by α-methyl L-fucoside and not by the other monosaccharides. Our conclusions were somewhat tentative at first because this was an isolated result with a rather exotic reagent, but the inference that L-fucose in α-linkage is more important than the other sugars for H specificity was reinforced when we were given some plant agglutinins (later called lectins)...' [W.M. Watkins. A half century of blood-group antigen research: some personal recollections. *Trends Glycosci Glycotechnol* 1999; **11**, 391–411; for illustration of the folding of the eel agglutinins, please see Figure 16.1g].

a common feature of *N*- and *O*-glycan structures (please see Chapters 6–8), and has a bearing on bioaffinity in receptor (lectin) binding [2]. Consideration of the factors of sequence, of linkage-point and ring-size permutations as well as branching sets glycans far apart from nucleic acids and proteins in terms of coding capacity. In actual numbers, only 4096 (4^6) hexanucleotides are possible with the four letters in the DNA language and still 6.4×10^7 (20^6) hexapeptides from 20 proteinogenic amino acids, but the staggering number of 1.44×10^{15} hexasaccharides from 20 monosaccharides [3]. Even though not every combination is realized, since oligomer synthesis is confined to using exclusively the anomeric center of the activated donors (please see also Table 1.1), the case for an enormous potential

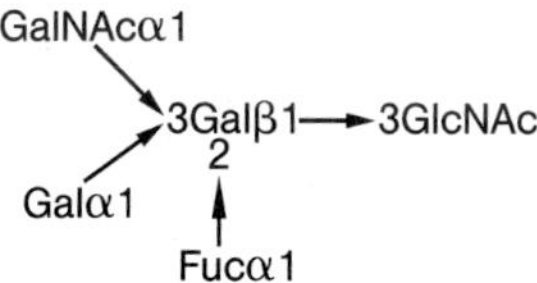

Figure 1.5 Illustration of the linkage pattern in ABH(0) histo-blood group tri- and tetrasaccharides. The core H(0)-trisaccharide (type I: α1–2-fucosylated Gal*p*β1–3GlcNAc*p*), whose L-fucose part is freely accessible to the eel lectin (please see Info Box 2), can be extended in α1,3-linkage by either *N*-acetylgalactosamine (A epitope) or galactose (B epitope). A branched structure is generated, as intimated by arrows in Figure 1.3. For structures of the individual 'letters' of the ABH(0) 'words', please see Figure 1.6.

of glycans as bioinformatic toolbox is nonetheless convincing. The basic alphabet of the sugar language is comprised by a set of 'letters'. The structures, symbols and acceptor characteristics of common monosaccharides are compiled in Figure 1.6. In addition to building up glycans of cellular glycoconjugates (glycoproteins, glycolipids and proteoglycans) and also cell walls (please see Chapters 6–12 for details), saccharides can also enter low-molecular-weight molecules. An instructive example from basic biochemistry is the conjugation of two GlcA molecules to bilirubin. This glycosylation enhances solubility and in consequence promotes excretion of this otherwise hardly soluble compound. Glycosylation processes of this type are also fairly common in plants, for example, concerning alkaloids, cyanohydrins, phenolics or terpenoids. By doing so, a wide variety of plant glycosides is generated such as the famous *Digitalis* compounds, opening a fertile field for synthetic glyco-randomization with the therapeutic/biotechnological aims of glyco-optimization [4]. All in all, the monosaccharides in Figure 1.6 establish the third alphabet of life.

The modification of nucleotides and amino acids is a powerful means to enlarge the size of a basic alphabet. Akin to posttranslational protein phosphorylation or sulfation glycan epitopes, too, are subjected to such reactions in order to convey particular properties to specific sites [5, 6]. Examples of how the cores for routing signals to lysosomes and to endothelial cells in the liver or for a determinant of cell communication in the nervous system look like are given in Figure 1.7 (top panel) (for further information on lectins binding these epitopes, please see Chapters 19.3 and 30.7 as well as Figure 16.1j). Beyond mammalian biochemistry, sulfation of egg jelly glycans at fucose residues is relevant for fertilization in invertebrates, this topic explained in Chapter 24.

Illustration of the anticoagulant pentasaccharide of heparin serves a second purpose besides documenting presence of sulfations, especially the rare 3-*O*-sulfation in the structure's center (Figure 1.7, bottom panel). It gives an impression of the ingenious combinatorial way to turn a seemingly dull repetition of the basic disaccharide $[\text{-4GlcNAc}\alpha\text{1–4GlcA}\beta\text{1-}]_n$ into an amazing structural (heparanomic) complexity. In fact, 48 different dimers are possible. As alluded to above, the epimerization from D-GlcA to L-IdoA (for structures, please see Figure 1.6) has significant consequences, because IdoA is an ideal hinge for shape rearrangements [7]. The conformational flexibility of the $^{1}C_{4}$ form to adopt a $^{2}S_{O}$ skew-boat structure (Figure 1.6) underlies this property (for further details on designation of ring geometry, please see next chapter, and on proteoglycans and drug design with heparin, please see Chapters 11 and 28.5). As current research is revealing, these substitutions are by no means a rare or random event. Thus, the enzymatic machinery for glycan assembly, processing and remodeling as well as site-specific substitution is elaborate and well developed, as expected for professional tasks in information handling. Whereas only two sulfotransferases are assigned to protein substitution in the human Golgi region, a total of 35 enzymes adds sulfate groups to carbohydrates [8]. The overall investment in genomic coding pays off by producing the glycomic complexity. It is not template derived and can be adjusted dynamically by the availability of enzymes, substrates and acceptors in space and

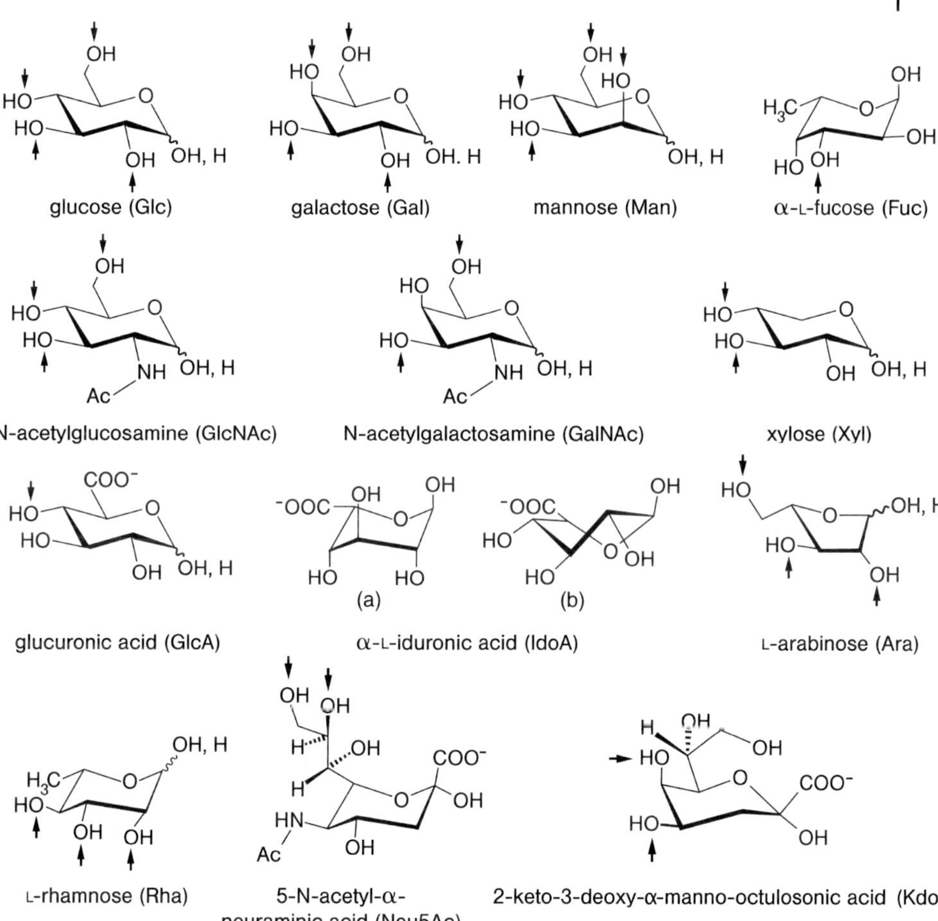

Figure 1.6 Illustration of the alphabet of the sugar language. Structural representation, name and symbol as well as the set of known acceptor positions (arrows) in glycoconjugates are given for each letter. Four sugars have L-configuration: fucose (6-deoxy-L-galactose), rhamnose (6-deoxy-L-mannose) and arabinose are introduced during chain elongation, whereas L-iduronic acid (IdoA) results from postsynthetic epimerization of GlcA at C5. The $^{1}C_{4}$ conformation of IdoA (a) is in equilibrium with the $^{2}S_{O}$ form (b) in glycosaminoglycan chains where this uronic acid can be 2-sulfated (please see Figure 1.7d). All other 'letters' are D-sugars. Neu5Ac, one of the more than 50 sialic acids, often terminates sugar chains in animal glycoconjugates. Kdo is a constituent of lipopolysaccharides in the cell walls of Gram-negative bacteria, and is also found in cell wall polysaccharides of green algae and higher plants. Foreign to mammalian glycobiochemistry, microbial polysaccharides contain the furanose ring form of D-galactose and also D/L-arabinose indicated by an italic '*f*' derived from the heterocyclus furan. The α-anomer is prevalent for the pentose arabinose, for example, in mycobacterial cell wall arabinogalactan and lipoarabinomannan. β1–5/6-Linked galactofuranoside is present in the arabinogalactan and the β1–3/6 linkage in lipopolysaccharides.

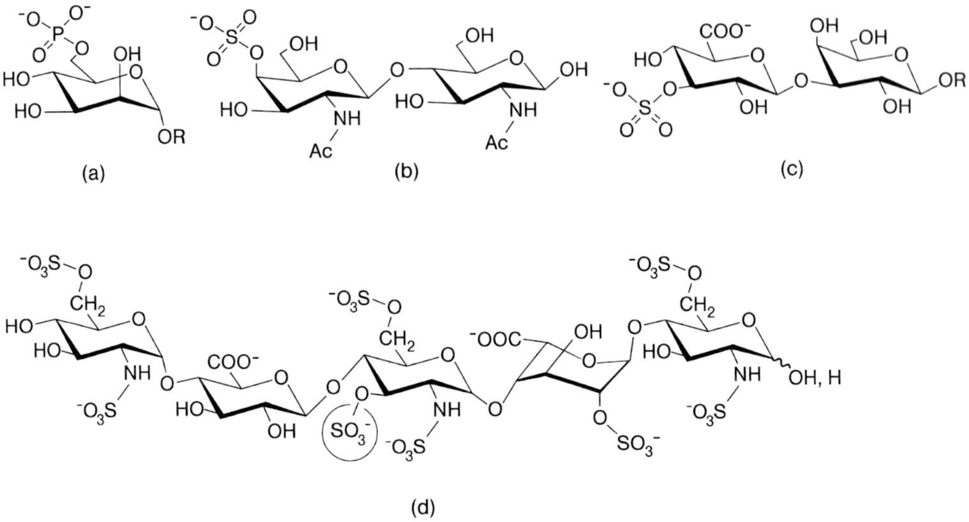

Figure 1.7 Illustration of phosphorylated (phosphated) and sulfated (sulfurylated) glycan 'words'. 6-Phosphorylation of a mannose moiety (in the context of a mannose-rich pentasaccharide) is the key section of a routing signal in lysosomal enzymes (a), 4-sulfation of the GalNAcβ1–4GlcNAc (LacdiNAc) epitope forms the 'postal code' for clearance from circulation by hepatic endothelial cells of pituitary glycoprotein hormones labeled in such a manner (b), the HNK (human natural killer)-1 epitope (3-sulfated GlcAβ1–3Galβ1–4GlcNAc) is involved in cell adhesion/migration in the nervous system (c) and the encircled 3-*O*-sulfation in the pentasaccharide's center is essential for heparin's anticoagulant activity (d). All sugars are in their pyranose form. Please note that the central GlcN unit has *N*,*O*-trisulfation and that the 2-sulfated IdoA, given in the ${}^{1}C_{4}$ conformation, can also adopt the hinge-like ${}^{2}S_{O}$ skew-boat structure (please see Figure 1.6; about 60% or more for the ${}^{2}S_{O}$ form in equilibrium depending on the structural context) when present within glycosaminoglycan chains of the proteoglycan heparin. 2-Sulfation of IdoA serves two purposes: favoring the hinge-like ${}^{2}S_{O}$ conformation and precluding reconversion to GlcA.

time [5, 9], and its structural and functional aspects are the topic of the following chapters. That said, we have built the evidence for the following conclusions.

1.4 Conclusions

'Carbohydrates are ideal for generating compact units with explicit informational properties, since the permutations on linkages are larger than can be achieved by amino acids, and, uniquely in biological polymers, branching is possible. Moreover, the oligosaccharide units are not flexible but exhibit highly specific structures with only limited degrees of freedom' [10]. This statement highlights that the sugar code has a third dimension. What this means is explained in the next chapter and its relevance for protein–carbohydrate interactions is outlined in Chapter 13.

Summary Box

Carbohydrates form the third alphabet of life. Compared to amino acids and nucleotides their versatility for isomer formation (code words) is unsurpassed. The resulting high-density coding capacity of oligosaccharides is established by variability in (i) anomeric status, (ii) linkage positions, (iii) ring size, (iv) by branching and (v) introduction of site-specific substitutions.

Note: Nomenclature rules for carbohydrates are presented in detail in *Carbohydr Res* 1997; **297**, 1–92.

References

1 Roseman S. Reflections on glycobiology. *J Biol Chem* 2001;276:41527–42.

2 André S *et al.* Substitutions in the *N*-glycan core as regulators of biorecognition: the case of core-fucose and bisecting GlcNAc moieties. *Biochemistry* 2007;46: 6984–95.

3 Laine RA. The information-storing potential of the sugar code. In: *Glycosciences: Status and Perspectives* (Eds.: Gabius H-J, Gabius S), pp. 1–14. Chapman & Hall, London, 1997.

4 Griffith BR *et al.* 'Sweetening' natural products via glycorandomization. *Curr Opin Biotechnol* 2005;16:622–30.

5 Reuter G, Gabius H-J. Eukaryotic glycosylation – whim of nature or multipurpose tool? *Cell Mol Life Sci* 1999;55:368–422.

6 Gabius H-J. Cell surface glycans: the why and how of their functionality as biochemical signals in lectin-mediated information transfer. *Crit Rev Immunol* 2006;26:43–80.

7 Casu B *et al.* Conformational flexibility: a new concept for explaining binding and biological properties of iduronic acid-containing glycosaminoglycans. *Trends Biochem Sci* 1988;13:221–5.

8 Hemmerich S *et al.* Strategies for drug discovery by targeting sulfation pathways. *Drug Discov Today* 2004;9:967–75.

9 Gabius H-J *et al.* The sugar code: functional lectinomics. *Biochim Biophys Acta* 2002; 1572:165–77.

10 Winterburn PJ, Phelps CF. The significance of glycosylated proteins. *Nature* 1972;236: 147–51.

2
Three-Dimensional Aspects of the Sugar Code

Tibor Kožár, Sabine André, Jozef Uličný, and Hans-Joachim Gabius

The different building blocks of glycans (the alphabet of the sugar code) and the way they can be linked to form oligo- and polysaccharides (words, messages) have been introduced in Chapter 1. In addition to describing these two-dimensional aspects of glycan assembly, initial consideration of three-dimensional structures was given when discussing the different conformations of the pyranose ring using L-iduronic acid (please see Figure 1.6) – a component of glycosaminoglycans (see Chapter 11). This chapter builds on this introduction. We initially describe glycan flexibility at the monosaccharide level, and then present insights into conformational analysis of oligo- and polysaccharides, highlighting biologically active epitopes at branch ends and *N*-glycans (for details on their biosynthesis and functionality, please see Chapters 6–8, 22 and 23). Naturally, the glycans' shape will have implications for their recognition by proteins (for basic principles of protein–carbohydrate interactions, please see Chapter 13) and for drug design (please see Chapter 28, for examples). In the following we describe experimental techniques employed to obtain information about carbohydrate conformation.

2.1
How to Obtain Information about Carbohydrate Conformation

The experimental methods most commonly used to elucidate carbohydrate structures are based on X-ray crystallography and nuclear magnetic resonance (NMR) (for further details, please see Chapters 13 and 16). A limitation associated with crystallographic studies is their inability to reflect the conformational dynamics. However, measurements of the chemical shifts, coupling constants and nuclear Overhauser effects (NOEs; please see Chapter 13.3 for further information on this process which serves as a molecular ruler to determine interproton distances) represent key steps towards determining these dynamics. Combined with crystallographic data, NMR analysis often yields a definitive description of glycan structures. Respective compilations (X-ray coordinates) are available (see Cambridge

The Sugar Code. Edited by Hans-Joachim Gabius

ISBN: 978-3-527-32089-9

Structural Database; www.ccdc.cam.ac.uk). Data on complex glycoconjugates, various polysaccharides and carbohydrates in complex with proteins (X-ray and NMR structures) are also available in the public domain (see Brookhaven Protein Database; www.rcsb.org; for example on illustrations of protein–sugar complexes, please see Figure 16.1). Information on carbohydrate–protein complexes is accessible in the receptor–ligand database (Relibase; http://www.ccdc.cam.ac.uk/free_services/relibase_free). Other experimental methods important for structural studies such as mass spectrometry (see Chapter 5 on structural aspects) or circular dichroism do not provide direct information on atomic coordinates and essentially require correlations with NMR data or molecular modeling. Therefore, to gain a full understanding of the three-dimensional attributes of carbohydrates in solution (i.e., the shape and the dynamics of its alterations) a strategic combination of experimental methods with molecular modeling is required; *in silico* protocols having thus become an indispensable tool [1].

Computer modeling of carbohydrates started around 40 years ago on a simple platform based on a trouble-free 'allow' and 'reject' approach for conformations judged to be possible owing to interatomic distances. Further progress in this field resulted in the development of powerful computational tools, built on contemporary 'state-of-the-art' hardware (cluster and grid computing) and software (quantum and molecular mechanics, molecular dynamics, Monte Carlo, and so on) applicable to glycosciences. Various custom-made carbohydrate-modeling tools are publicly available at www.glycosciences.de. They facilitate *in silico* building of oligosaccharide structures and calculations of their conformational features. In addition, conformational energy maps for selected oligosaccharides are also accessible at this resource. To guide the reader to an understanding of the conformational behavior of oligosaccharides we start our tour through the realm of computational glycochemistry by identifying the levels of complexity.

2.2 Complexity of Carbohydrate Flexibility

The following parameters are tied to the spatial appearance of glycans: the conformation of the monosaccharide rings (either five-membered furanose or six-membered pyranose rings; please see Chapter 1), the mutual orientation of the rings specified in terms of flexibility around a glycosidic bond, and the positions of the side chains attached to these rings. All of these factors influence the molecular shape, partial charge distribution, formation of intra- and inter-residual hydrogen bonds, solvent accessibility, and bonding ability to proteins and other biomacromolecules [2]. Due to the apparent complexity of this issue, we start by defining in further detail the flexibility at the level of the building blocks (i.e., the monosaccharides).

2.3
How to Describe the Shape of Monosaccharides

Intramolecular hemiacetal formation of monosaccharides yields two different ring structures. In comparison, the flexibility of the furanose ring is higher than that of the pyranose ring. The level of flexibility of the pyranose ring is schematically depicted in Figure 2.1 (please see Chapter 1.2, where the chair/boat forms were introduced), showing the pseudorotational pathway in the ring interconversion. Polar coordinates q, θ and ϕ describe the ring conformation in a rather simple manner in comparison to the traditional ring-torsion angles. The polar q, θ and ϕ coordinates correspond to the ring-puckering parameters introduced by Cremer and Pople [3] and their visualizations [4]. The more stable chair conformations are localized on the poles of the sphere ($^4C_1$ with $\theta \sim 0°$ and $^1C_4$ with $\theta \sim 180°$; the $^4C_1/^1C_4$ terminology is explained in Chapter 1.2); the less-stable boat (starting from $^{3,O}B$ at $\phi \sim 0°$) and skew (beginning with $^3S_1$ at $\phi \sim 30°$) conformations reside around the equator ($\theta \sim 90°$). The half-chair and half-boat or half-skew forms lie somewhere between the pole and the equator on the meridian of the sphere. The $^4C_1$ form is typical for D-sugars, whereas L-sugars prefer the $^1C_4$ conformation. The energy difference, based on *ab initio* quantum chemical calculations, is only 1.2 kcal/mol in favor for the $^4C_1$ form over the $^1C_4$ conformation of methoxytetrahydropyran shown in Figure 2.1. Actual structural parameters also depend on the presence of ring substitutions. In addition to steric factors, the conformational

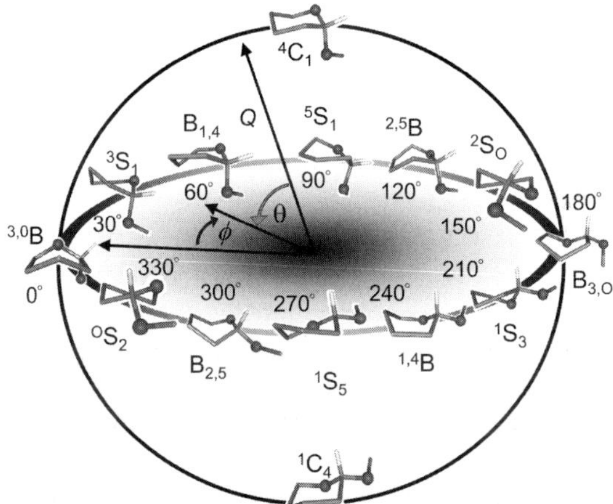

Figure 2.1 Conformational flexibility of the pyranose ring presented in the spherical visualization according to the Cremer–Pople ring-puckering parameters. The chair conformations occupy the poles of the sphere, the twist (skew) and boat forms are located around the equator. The *S* chirality of the anomeric C1 carbon is maintained along the pseudo-rotational pathway. The half-boat/half-chair forms (not shown) are located around the meridians. The intra-ring torsion angles of all 14 forms are presented in the on-line version of Supplementary Material.

equilibrium around the glycosidic bond is driven by two electronic effects, anomeric and exoanomeric [5] (see Info Box 1 for details). These effects are important for the conformational preference of the –C–O–C–O–C– segment present in *O*-glycosides and for the –C–O–C–N–C– segment in *N*-glycosides, with impact also on the corresponding *O*/*N* carbohydrate–protein linkage (please see Chapters 6 and 7 for such linkages). Having analyzed the monosaccharide (a letter), we can now move on to inspect the di- and oligosaccharides (encoded words).

Info Box 1

The anomeric effect refers to the preference for the axial orientation of the electronegative substituent at the anomeric C1 carbon over the equatorial one in the $^4C_1$ chair conformation of the pyranose ring. The *gauche* (*synclinal, sc*) conformational preference over the *trans* (*antiperiplanar, ap*) position for the rotation around the anomeric C1–O1 bond is referred to as the exoanomeric effect. As a consequence of the generalized anomeric and exoanomeric effects, the *gauche* conformation (*synclinal*) is preferred over the *trans* (*antiperiplanar*) form in all acetals and related compounds having an X–C–Y–C segment, where X and Y are heteroatoms (such as O, N and F) that possess free electron pairs. High-level *ab initio* studies are required to properly describe the stereoelectronic effects and orbital interactions resulting in the anomeric and exoanomeric effects.

2.4
How to Describe the Shape of Di- and Oligosaccharides

In any disaccharide, the mutual orientation of the two pyranose rings can be easily characterized by two (three in the case of the 1–6 bond) torsion angles of the glycosidic bond, Φ, Ψ and ω (Figure 2.2). In analogy to Ramachandran's two-dimensional Φ and Ψ conformational maps for dipeptides, similar diagrams can be established for disaccharides. As illustrated in Figure 2.3 with examples, they depict a simplified projection of the potential energy hypersurface as the function of the given coordinates Φ, Ψ or/and ω. To describe glycans of increasing structural complexity, equivalent maps are calculated for each disaccharide portion of the complete molecule. In the case of a tetrasaccharide, for example, three different conformational energy maps need to be generated – one for each glycosidic bond present in the molecule. This procedure embodies a significant simplification, understanding that the conformational behavior of the selected disaccharide fragment might be tangibly influenced by the shape of the rest of the molecule. Of note, backfolding of an antenna of a core-substituted biantennary *N*-glycan is a typical example where such a situation occurs (please see below). The minima in such maps define energetically privileged conformations of the oligosaccharide.

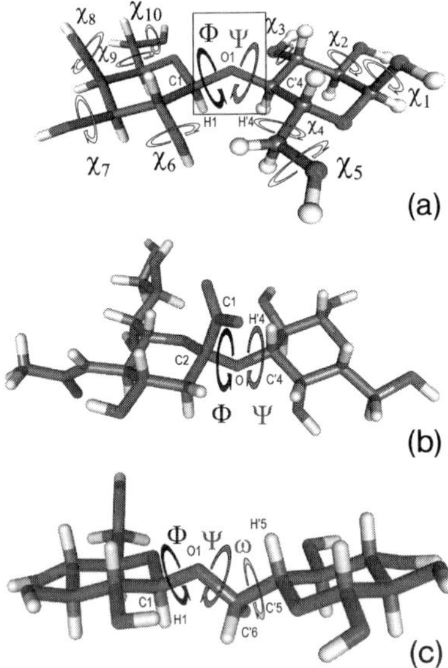

Figure 2.2 (a) Definition of the torsion angles in α-lactose at Φ = Ψ = 0°. The most common definition of the angles is based on the following convention: Φ = H1—C1—O1—C′4, Ψ = C1—O1—C′4—H′4. The reducing end is shown as a ball-and-stick model. (b) Different numbering and scheme apply for oligosaccharides with sialic acid. (c) A 1,6-glycosidic bond includes a third linkage with implications for the conformational space, as illustrated for Glcβ1,6Glc.

The transition from one to another conformation ('energetic valley') is thus an intramolecular rearrangement that does not require breaking or creation of a new chemical bond. These structures can be likened to molecular keys so that a glycan is able to present itself in different shapes. Changes in hydrogen bonding or non-bonded stacking interactions are typical for conformational transitions in equilibrium. In comparison to peptides only few conformers (keys) are energetically privileged in glycans [6]. Due to ensemble averaging during the NMR measurements – every conformation will contribute its characteristic signals to the measured spectrum; in other words, it is not a single (virtual) conformation that is the source of the signals [1, 6] – and lack of other experimental techniques to provide detailed information on the conformational dynamics, computational methods are required to help assign signals to structures. Both the quantum chemical (QC) and the molecular mechanics (MM) methods (with properly constructed force-field parameters) allow one to address this issue. Molecular dynamics (MD) (please see Info Box 2) is valuable to predict the most likely paths and frequency of conformational changes. What do we need to know to predict the molecular shape of a disaccharide?

Molecular conformations reside in the minima within the valleys of an *n*-dimensional energy hypersurface of the molecules. Originally, very simple empirical functions were used to estimate the potential energy surface with

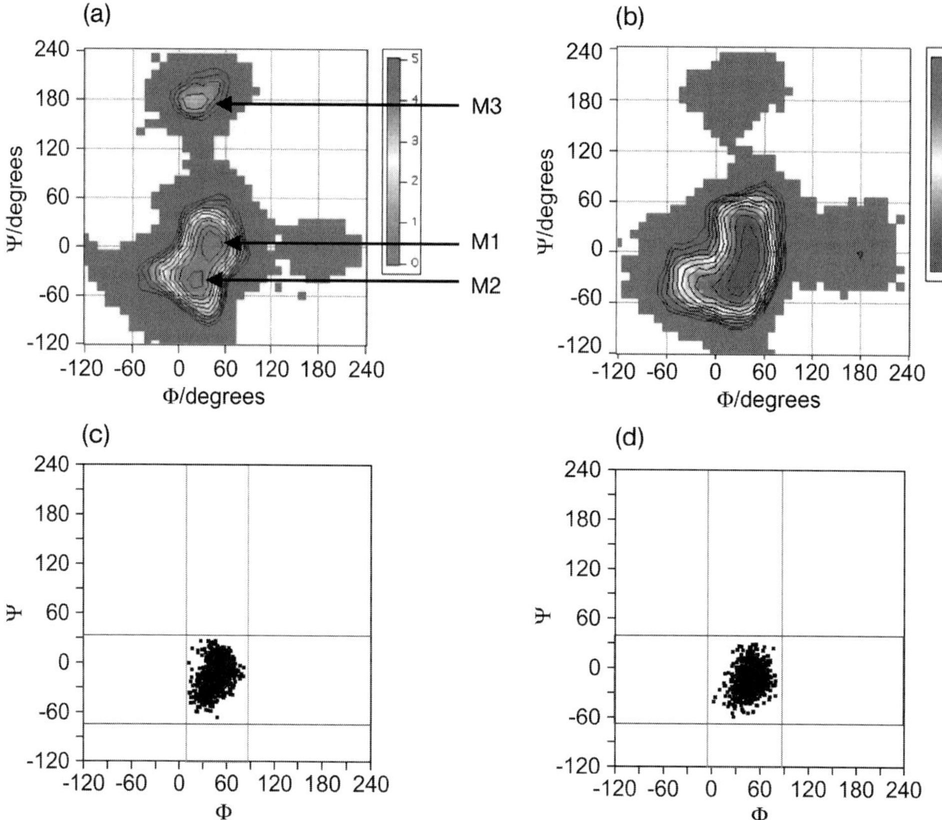

Figure 2.3 Conformational energy maps of Galβ1,4Glc (β-lactose) and Galβ1,4GlcNAc (β-LacNAc). The first two maps (A and B) also showing the energy profiles originate from www.glycosciences.de; the two additional maps were obtained from 1-ns water simulations (see Figure 2.5 for details) and illustrate the Φ, Ψ sampling during the simulations. The conformations, corresponding to energy minima located on the map, are marked M1, M2 and M3.

significant reduction of the dimensionality to two, considering only the Φ, Ψ dihedral angles for rotation around the glycosidic bond. Consequently, the quality of these calculations was improved by expanding the energy terms and also by using improved parameterization of energy functions. The energy contours were drawn into the Φ, Ψ maps, indicating the minima for the most probable conformations. Although simple potential energy functions were used in the first calculations, these conformational maps were already instrumental to delineate the molecular shape of the studied disaccharides.

Together with the progress in development of more sophisticated force field methods, significant efforts were also devoted towards detailed energy calculation using quantum mechanical (QM) methods. The first-ever published semiempirical quantum chemical study of a disaccharide appeared in the early 1970s [7].

Info Box 2

MD simulations can be used as a tool for the exploration of the actually populated conformational space as a measure of time. In principle, the energy of a particular configuration of atoms that form molecules and molecular complexes can be treated as electronic energy of the corresponding QM problem. This energy is a complicated function of atomic coordinates, its spatial distribution being described by a potential energy surface. In almost all practical cases, the dynamics of molecules can be described sufficiently well as the dynamics of nuclei, moving on a potential energy surface according to classical Newtonian equations of motion. Although the origin of the potential energy is quantum mechanical, the final property of interest, energy $E(\vec{R})$, where $\vec{R}$ stands for coordinates of nuclei treated as classical point masses, is a simple smooth hypersurface defined in a coordinate space. It is thus possible to approximate the shape of this surface by a purely empirical way, avoiding the need for generally very costly QM calculations. In general, we are not interested in all possible configurations, since only a very minor part of them is sampled under real experimental conditions with significant probability. Minima on a potential energy surface correspond to stable molecular conformations, while the saddle points define the transition states. They are pivotal characteristics of the reaction path. The reaction path is an idealized route in the configurational space describing the topological course of the reaction–the transition from one set of stable configurations (reactants) to another set of stable configurations (products). The empirical description of the potential energy surface is written as a sum of terms expressing various physical contributions to the total energy of the molecular system. The most common functional form can be given as:

$$
\begin{aligned}
E(\vec{R}) = & \sum_{bonds} K_B(r_i - r_{i0})^2 && \text{(bond stretching terms)} \\
& + \sum_{angles} K_A(\vartheta_i - \vartheta_{i0})^2 && \text{(angular bending terms)} \\
& + \sum_{torsions} \sum_{n} K_{\tau n}(1 + \cos(n\omega - \gamma)) && \text{(torsion and wagging terms)} \\
& + \sum_{i<j} \left\{ 4\varepsilon_{ij} \left[\left(\frac{\sigma_{ij}}{r_{ij}} \right)^{12} - \left(\frac{\sigma_{ij}}{r_{ij}} \right)^{6} \right] + \frac{q_i q_j}{4\pi\varepsilon_0 r_{ij}} \right\} && \text{(van der Waals and Coulomb terms)}
\end{aligned}
$$

This expression is often called force field (FF), because the forces acting on nuclei can be easily calculated according $\vec{F} = -grad\, E(\vec{R})$ parameters of FF, i.e. K_B, r_{i0}, K_A, ϑ_{i0}, $K_{\tau n}$, γ, ε_{ij}, σ_{ij}, are called *force constants*. The actual configuration of nuclei $\vec{R}$ is expressed in internal coordinates, represented by r_i, ϑ_i, and ω_i. Most biological molecules can access several nuclear configurations and perform transitions between them without breaking chemical bonds. Thus, there is a dynamic equilibrium between these key-like structures, with not yet well-defined consequences for biological functions.

Today, as refinement in relation to early simple procedures, intricate hybrid MM/QC methods are applied [8]. The recent progress in methodology is exceptional and opens new horizons for further advances in carbohydrate modeling [9]. As to the shape, it is not only the inherent flexibility around the glycosidic bond that matters, as explained in the next part.

2.5
Additional Factors Influencing the Shape of Oligo- and Polysaccharides

The shape of the carbohydrates and oligosaccharides is also influenced by factors resulting from the carbohydrate pendant groups/side chains. For example, there are six flexible groups (OH and CH_2OH) present in glucose. Each such group has three possible conformational states, resulting in $3^6 = 729$ conformations for a given α or β anomer. This number is dramatically increased when moving from 3^{10} in a disaccharide (see Figure 2.2a for lactose χ_i) to 3^n in the case of oligosaccharides (n describes the number of flexible side chains). However, not all of the 3^n possible orientations are populated in complex glycans (or even a disaccharide) due to preferences resulting from both intramolecular forces and interactions with surrounding molecules (proteins, solvent and other molecules present in the biological environment). Figure 2.4 illustrates the superimposed small subset of 100 low-energy conformations of β-lactose obtained by running the Macromodel program. The Φ, Ψ values were restrained and only the conformational changes of side chains were calculated. The figure gives an impression of the challenges inherent to conformational analysis of glycans, aimed at finding the most probable shape of the molecule. Modeling and predicting the conformational properties of glycans are evidently a complex task, encompassing characterization of ring conformations as well as conformations around the glycosidic bond and of the side chains.

In addition to the internal factors influencing carbohydrate shape already discussed, one should be aware of the additional influence coming from the environment. Taking into consideration the presence of water in living cells, it is clear that a solvent effect has to be included in modeling studies in order to properly describe glycan conformations. The most commonly encountered solvent effect is due to water molecules, although solvents other than water may be used in the course of NMR measurements, in order to pick up solvent-exchangeable hydroxyl protons in an aprotic solvent [6]. In the simplest models the solvent is treated as a bulk dielectric continuum, with solvation effects represented by electrostatic, dispersion and cavity terms. Computationally more demanding are the discrete models of solvents, especially those, where several hundreds of solvent molecules form a solvation box using periodic boundary conditions. To give an impression on the molecules processed in this procedure, a hydrated lactose molecule within a water box of $30 \times 30 \times 30\,Å^3$ is presented in Figure 2.5 as an example for the discrete solvation model. Having provided the fundamentals of modeling including the solvent, we can now present information on actual case studies.

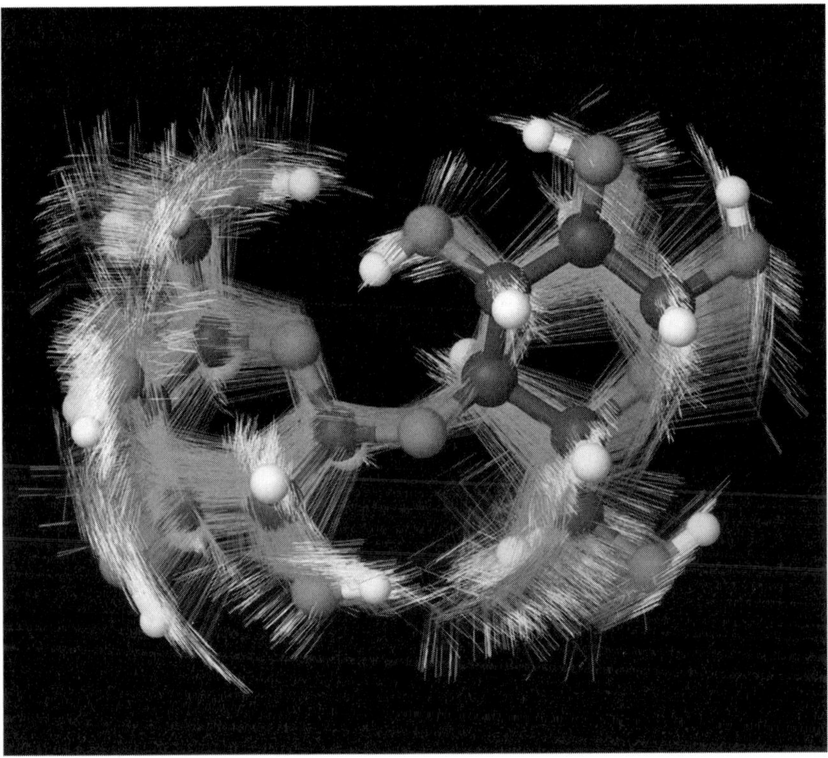

Figure 2.4 Illustration of a conformational search for side-chain orientations in β-lactose (minimum at $\Phi = 49°$, $\Psi = 7°$) as calculated with Macromodel. The global minimum is shown in ball-and-stick presentation.

2.6 Examples of Di- and Oligosaccharide Conformations

Alterations of glycosylation are associated with diseases, as described in detail in Chapters 22 and 23, indicating the need to define structure–activity profiles. Indeed, biological assays gave reasons to predict different conformations for substituted *N*-glycans to explain their differences in bioactivity, e.g., the interaction with the asialoglycoprotein receptor (that affects clearance of glycoproteins from serum; please see Chapter 19 for details on the lectin and Chapter 15 for the historical background of its detection) [10]. Below we provide instructive examples in a stepwise manner, starting from simple disaccharides and then presenting more complex structures.

Lactose (Galβ1,4Glc) is the most important carbohydrate component of milk and also one of the major disaccharides utilized in the food and pharmaceutical industries. *N*-Acetyllactosamine (Galβ1,4GlcNAc; LacNAc) is a modified version

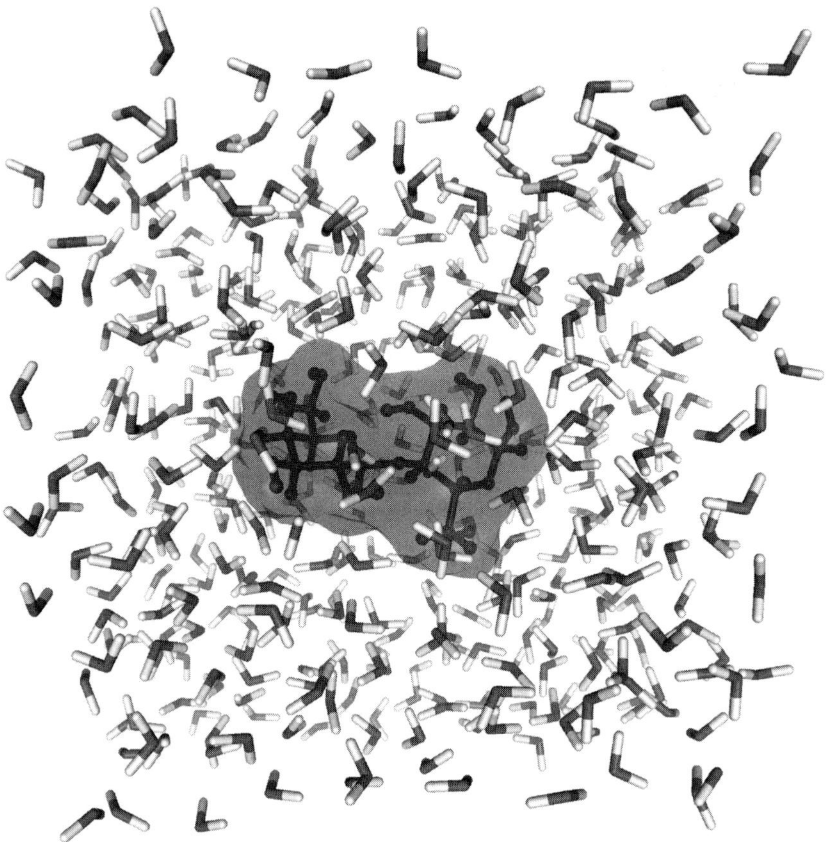

Figure 2.5 β-Lactose in a $30 \times 30 \times 30 Å^3$ solvent box. The Discovery Studio Visualizer (Accelrys) was used for visualization. The semi-transparent solvent-accessible surface of lactose is also visualized in order to readily spot the lactose position in the middle of the water box.

with *N*-acetylated glucose in position 2 at the reducing end moiety, an important part of ABH and Lewis blood group determinants and docking point for lectins (for structures of the epitopes, please see Figure 1.5 and Table 7.4). Understanding the conformational behavior of these disaccharides is essential to identify the relationship between the glycans' shape and their related biochemical and physiological functions, e.g., in inflammation (please see Chapter 27).

The conformational maps for these two natural disaccharides are illustrated in Figure 2.3. This information can be analyzed with respect to NMR data to observe the presence of conformers in solution [1, 6] (please see Chapter 13). Figure 2.3 clearly demonstrates that a relatively small structural change such as *N*-acetylation can have influence on the appearance of the conformational map, here the disappearance of the conformational area located in the *trans* region (minimum M3

Table 2.1 Relative energies of selected conformations of lactose and LacNAc calculated with the Jaguar program (Schrödinger).

Molecule	Φ	Ψ	ΔE[a] (kcal/mol)	ΔE[b] (kcal/mol)
Lactose	49	7	0	0
	27	−27	2.4	1.4
	32	174	1.6	1.9
LacNAc	48	−11	0	0
	23	−29	6.7	2.2
	28	176	5.1	2.9

[a] DFT (B3LYP/6-31G**).
[b] PBF solvation model.

on Figure 2.3a for β-lactose) of the Ψ glycosidic angle located around 180° in the case of LacNAc. During the MD simulation of the solvated molecule (Figure 2.5) both conformational regions around minima M1 and M2 were sampled. During the simulation (at 300K) there were no conformational transitions into the region of the M3 conformation.

The choice of the particular force field (see Info Box 2 for examples of the energy functions) is critical for the predictive value of results from such calculations. The force field parameterization results either from experimental data or from high-level *ab initio* calculations. MD simulations with the MM3 force field were used in the above example. The geometry of the conformations, located in the minima of the maps, can be further optimized using high-level QC methods. Atomic charges, dipole moments, chemical shifts and other molecular parameters can then be delineated. Accordingly, the relative energies of the selected conformations when reoptimized using density functional theory (DFT) with the 6–31G** basis set are summarized in Table 2.1, together with the corresponding Φ, Ψ values. The M3 *trans* conformation of LacNAc is a high-energy form, whereas the corresponding lactose M3 topology is only 1.6kcal/mol above the global minimum. For LacNAc the predicted value for M3 is 5.1kcal/mol. The second minimum M2 for LacNAc harbors a relatively high energy. When the solvation energies of lactose conformers are recalculated within the Poisson–Boltzmann finite-element (PBF) solvation model (DFT B3LYP with 6-31G** basis set), the relative energy value for M3 slightly increases, whereas the M2 energy decreases by 1kcal/mol. In the case of LacNAc the energy decrease is even more significant, to 2.2kcal/mol for M2 and to 2.9kcal/mol for M3 (for the interaction of lactose with a lectin, please see Figure 13.7). Having analyzed a disaccharide, we can now proceed to oligosaccharides.

The histo-blood ABH determinants (see Figure 1.5 for molecular composition) are typical branch-end epitopes where the question on relationships between structure, conformation and biological activity is of wide interest (for their crucial role in the history of lectin research, please see Chapter 15). To demonstrate the influence of solvent on their conformational behavior, all corresponding tri- and tetrasaccharides were constructed, the structures minimized, searched for side-chain orientations and solvated within a $30 \times 30 \times 30\,\text{Å}^3$ water box. Finally, 1-ns MD simulations were carried out. Figure 2.6 illustrates the resulting three-dimensional representations of these molecules. The fucosylated trisaccharide (see Figure 2.6 caption for letter abbreviations) exhibits a smaller degree of conformational flexibility around the glycosidic bond than the trisaccharides A and C. The GalNAc*p*-α1,3-Gal*p* linkage is predicted to be flexible for the trisaccharide C, exhibiting two regions located between $-60°$ and $+60°$ for Ψ with the Φ angle around $-60°$, in accord with the exoanomeric effect. This conformational flexibility is significantly reduced in the presence of the α1,2-fucosylation at the central Gal*p* unit in the corresponding tetrasaccharide E (compare C-2 and E-2 Φ, Ψ maps in Supplementary Material available in electronic source). The di- and oligosaccharide examples given illustrate the potential of computational methods for elucidation of the conformational flexibility of carbohydrates, with potential to affect the biological activity of these molecules. In fact, as the data in Table 25.2 attest, glycosylation changes can alter protein functionality. Also, the shape is a key to the ligand properties of glycans.

2.7
Carbohydrate–Protein Intermolecular Interactions and Reaction Mechanisms

Chapters 13 and 14 deal in detail with the principles of carbohydrate–protein interactions and experimental analysis techniques for specificity. For modeling of the structural complementarities in interactions between carbohydrates and proteins, a different computational strategy needs to be applied (although MD simulations are applicable also for protein–ligand–water systems). Docking of the carbohydrate ligands into the binding site of a lectin is a straightforward task using some of the well-established calculation procedures (for detailed information on lectins, please see Chapters 16–19, including the gallery of lectin structures in the figure of Chapter 16). Several programs are assigned to this task (for example Dock, AutoDock, FlexX, Glide, Gold, and so on.). The docking methodologies differ in the search algorithm and the parameterization of the scoring functions (empirical or force field functions to calculate the protein–ligand interaction energies). The efficiency of the docking runs can be influenced by the user-selectable input parameters of the mentioned programs. In general, the ligand molecules are treated as flexible entities undergoing conformational changes during the docking procedure, whereas the protein's conformational flexibility is usually not taken into consideration, treating the protein as a rigid body. Experimental input on the interplay between lectin or enzyme and ligand will help refine this situation, revealing that the interplay in solution can trigger responses on the level of protein

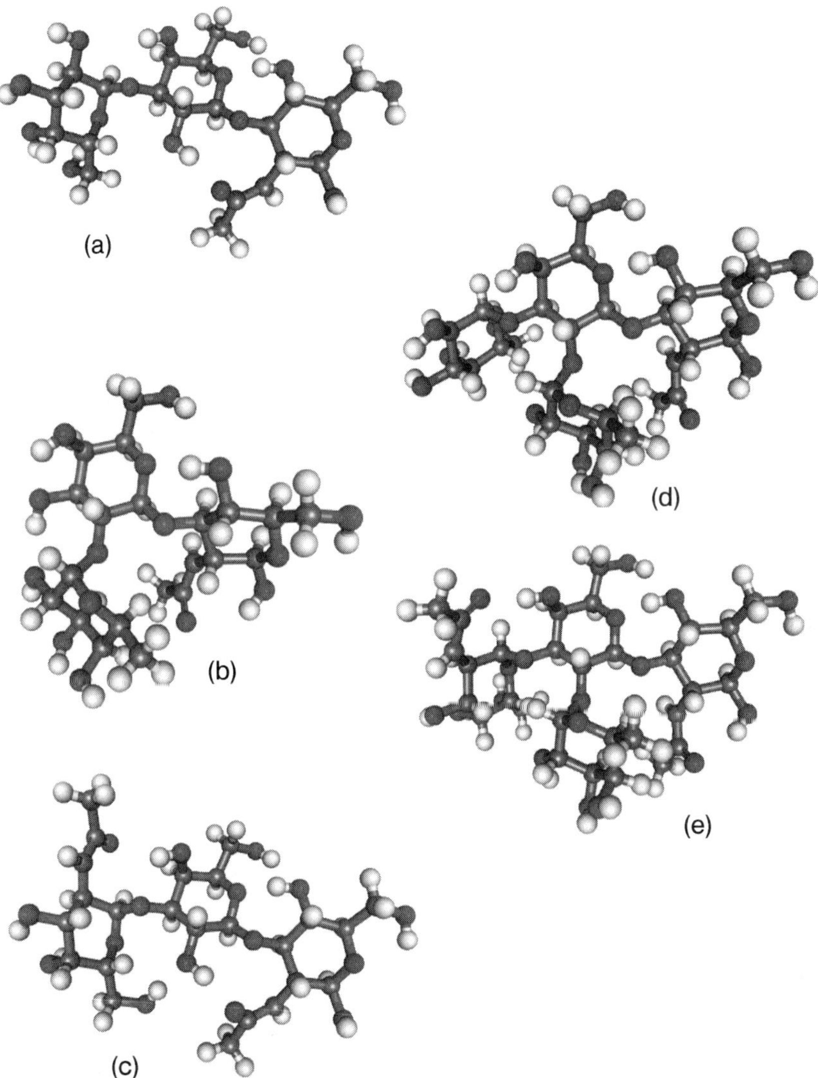

Figure 2.6 Stable conformations of ABH histo-blood group tri- and tetrasaccharides (type 1). (a)Gal*p*-α1,3Gal*p*-β1,3GlcNAc*p*;(b)Fuc*p*-α1,2Gal*p*-β1,3GlcNAc*p*;(c)GalNAc*p*-α1,3Gal*p*β1,3GlcNAc*p*; (d) Gal*p*-α1,3(Fuc*p*-α1,2)Gal*p*-β1,3GlcNAc*p*; (e) GalNAc*p*-α1,3(Fuc*p*-α1,2)Gal*p*-β1,3GlcNAc*p*.

[11]. Naturally, the flexibility of the polar hydroxyl groups of the ligand should also be taken into account, considering thus their involvement in hydrogen bonds with the protein's amino acids in the binding site.

In addition to force field-based conformational and docking calculations, computational chemistry applied to mechanisms of enzymatic reactions is a fertile field for drug design. This is especially important when the goal is to design inhibitors for carbohydrate-processing enzymes (for examples please see Figure 28.2). Bond breaking and creation are best studied at a QM level. These enzymes are rather large molecules, thus requiring combined MM and QC methods to model reaction mechanisms. In these so-called hybrid QM/MM calculations the carbohydrate is treated on the QM level together with the enzyme's immediate active-site architecture, whereas the rest of the protein is treated on an MM or MD level. Modeling the catalytic mechanism of the inverting Golgi enzyme *N*-acetylglucosaminyltransferase-I starting branch elaboration of *N*-glycans [8] is an example for this approach (please see Chapter 6.7 and Figure 23.5).

Although the current status of molecular modeling allows us to infer the conformational behavior of complex *O*- and *N*-glycans, experimental data are required to reveal the importance of conformational features for biological activity. For example, respective changes can be induced by substitutions in the core such as α1,6-fucosylation, which can therefore act as molecular switches (for structural aspects of the glycan chains, please see Chapters 6–8) [10]. Such studies combining synthetic, modeling and biological approaches will be instrumental for understanding sugar coding. As long-chain glycans such as glycosaminoglycans (please see Chapter 11.1 for details) also act as important regulators, the final section is devoted to the analysis of these compounds.

2.8
How to Perform Molecular Modeling of Large Glycans

The development in coarse-grained (CG) and mesoscopic modeling opens new routes for carbohydrate modeling. Briefly, in CG modeling, small groups of atoms – the pyranose ring – are treated as a single particle. This approach, compared to detailed atomistic simulations, will speed up dynamics studies by several orders of magnitude (on both time and/or spatial scales) on the same hardware. Thus, the possibility now arises of performing simulations on long-chain glycans such as chondroitin sulfate and hyaluronic acid, applicable to proteoglycans (Chapter 11) or chitin (Chapter 12) [12]. Computer modeling of large systems such as membrane fragments with glycoconjugates and lectins or modeling of erythrocyte surfaces interacting with antibodies thus becomes a viable perspective for the area of *in silico* work.

2.9
Conclusions

Carbohydrates are ideal for information coding. Understanding this talent and its modes of operation requires insights into their three-dimensional (conformational) behavior and the rules driving these properties. The importance and interconnection between the conformations of the pyranose ring, side chains and the glycosidic bond have been highlighted. The predictive quality of computational approaches is strategically combined with experimental data, as illustrated for core substitutions of *N*-glycans as molecular switches [10, 13]. Advances in synthetic carbohydrate chemistry, outlined in the next chapter, noticeably broaden the range of testable glycans. This fruitful combination of *in silico* work on shape and the inherent dynamics with the growing insights on functional relevance of the glycans present in the living cell (please see the chapters in the second part of the book) will help us to identify and decode the information stored in these molecules and facilitate the ability to read and interpret the sugar code.

Summary Box

There is a hierarchy of theoretical methods that allow estimations of the conformational energies of glycans. The modeling of conformational energetics and dynamics starting from simple, quick and imprecise methods to the high-level *ab initio* calculations is a key step guiding us to an understanding of the biological activities of glycan chains – a central issue in the concept of the sugar code. Modeling includes docking studies with relevance for rational drug design.

References

1 von der Lieth C-W *et al.* Lectin ligands: new insights into their conformations and their dynamic behavior and the discovery of conformer selection by lectins. *Acta Anat (Basel)* 1998;161:91–109.

2 Rao VSR *et al. Conformation of Carbohydrates.* Harwood Academic, Amsterdam, 1998.

3 Cremer D, Pople JA. A general definition of ring puckering coordinates. *J Am Chem Soc* 1975;97:1354–8.

4 Jeffrey GA, Yates JH. Stereographic representation of the Cremer–Pople ring-puckering parameters for pyranoid rings. *Carbohydr Res* 1979;74:319–22.

5 Kirby AJ. *The Anomeric Effect and Related Stereoelectronic Effects at Oxygen.* Springer, Berlin, 1983.

6 Gabius H-J *et al.* Chemical biology of the sugar code. *ChemBioChem* 2004;5:740–64.

7 Giacomini M *et al.* Recherches quantiques sur la conformation des disaccharides. *Theor Chim Acta* 1970;19:347–64.

8 Kozmon S, Tvaroska I. Catalytic mechanism of glycosyltransferases: hybrid quantum mechanical/molecular mechanical study of the inverting *N*-acetylglucosaminyltransferase I. *J Am Chem Soc* 2006;128:16921–7.

9 Gould IR *et al.* Correlated *ab initio* quantum chemical calculations of di- and trisaccharide conformations. *J Comput Chem* 2007;28:1965–73.

10 André S *et al.* Substitutions in the *N*-glycan core as regulators of biorecognition: the case of core-fucose and bisecting GlcNAc moieties. *Biochemistry* 2007;46:6984–95.

11 Wu AM *et al.* Activity–structure correlations in divergent lectin evolution: fine-specificity of chicken galectin CG-14 and computational analysis of flexible ligand docking for CG-14 and the closely related CG-16. *Glycobiology* 2007;17:165–84.

12 Bathe M *et al.* A coarse-grained molecular model for glycosaminoglycans: application to chondroitin, chondroitin sulfate, and hyaluronic acid. *Biophys J* 2005;88:3870–87.

13 André S *et al.* From structural to functional glycomics: core substitutions as molecular switches for shape and lectin affinity of *N*-glycans. *Biol Chem* 2009;390:557–65.

3
The Chemist's Way to Synthesize Glycosides

Stefan Oscarson

The preceding chapters have introduced the alphabet of the sugar code and the oligomers (words) (first and second dimension of the sugar code) as well as their conformational flexibility (third dimension of the sugar code). Evidently, the high-density coding capacity of oligosaccharides, mastered by Nature's intricate set of glycosyltransferases (for information of such enzymes in glycan synthesis of glycoproteins, please see Chapters 6–8) is a challenge to chemists. In contrast to nucleic acids and proteins it is not only just the simple sequence that needs to be accomplished, but also the correct connection points of, as well as the correct stereochemistry in, the glycosidic linkages (please see Chapter 1). When successfully completed, the synthetic products can enter biomedical experiments.

This chapter will discuss methodologies used (by chemists) to synthesize oligosaccharides of various types and complexity. To succeed in these syntheses, different aspects have to be considered and mastered:

(i) Strategy. After identification of the target oligosaccharide structure, a strategy for the synthesis has to be planned and decided upon so as to identify suitable building blocks and the order in which the different glycosidic bonds should be constructed.

(ii) Glycosidic bond formation. Once the overall strategy has been decided upon, the synthesis of the glycosidic bonds within the building blocks as well as between them has to be planned and performed with regio- and stereoselectivity.

(iii) Protecting group manipulation. At the start, all the designed, properly protected precursors have to be synthesized.

After a discussion and survey of principles involved, and methodologies developed and available, the application of these on an authentic example, the synthesis of the blood group Lewis^b (Le^b) hexasaccharide, with an anomeric spacer to allow labeling and conjugation to proteins, dendrimeric scaffolds and solid phases, will be discussed (for role of Lewis epitopes as ligands for bacterial and human lectins, please see Chapters 17, 19, 24 and 27).

The Sugar Code. Edited by Hans-Joachim Gabius

ISBN: 978-3-527-32089-9

One-step mechanism (with inversion of anomeric configuration)

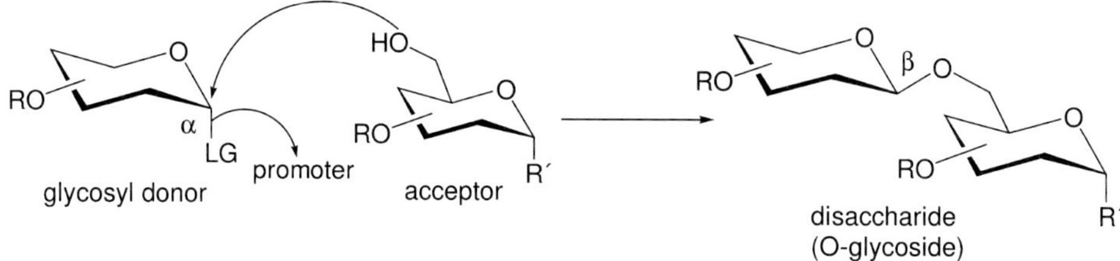

Two-step mechanism (via a positively charged intermediate)

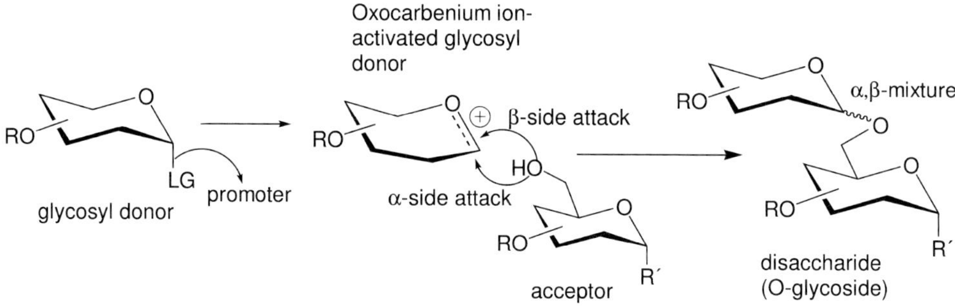

Figure 3.2 General glycosylation reaction mechanisms (LG = Leaving Group).

the aid of the glycosyltransferases we have to use other means to activate the glycosyl donor and to get the right regio- and stereoselectivity in the reaction with the acceptor and formation of the glycosidic bond.

Two different mechanisms can be envisaged – either a direct displacement of the leaving group by the acceptor (which would lead to inversion at the anomeric centre) or a two (or more)-step mechanism where the leaving group aided by the promoter leaves first to create a positively charged oxocarbenium intermediate, which then is attacked by the acceptor. Since the acceptor can attack the intermediate either from the α or the β side, this will result in an anomeric mixture. The latter mechanism is by far the most common in chemical glycosylations, which makes the problem of performing stereoselective glycosylations a major one. Methods that allow us to control the stereoselectivity in the attack of the acceptor on the activated glycosyl donor intermediate are required.

Thus, when performing a glycosylation reaction in the laboratory two main questions have to be asked and answered:

(i) What type of glycosyl donor/promoter system should be used?
(ii) How should the stereoselectivity in the reaction be controlled?

A summary of commonly used glycosyl donors and a survey of methodology developed to control the stereoselectivity in glycosylation reactions follows below.

3.3 Fischer Glycosylations

The glycosidic bond is part of a functional group called an acetal. An acetal is the product of an aldehyde (or ketone) with two alcohol groups and, thus, for a chemist, the obvious way to make a glycoside would be to mix a monosaccharide (aldehyde, cyclic hemiacetal) with an alcohol under acidic conditions. This was first done more than a century ago by one of the 'grandfathers' of carbohydrate chemistry, Emil Fischer (see Info Box 1), and is therefore called Fischer glycosylation (Figure 3.3). This is an excellent way to quickly (no protecting group manipulations needed, the free sugar is used as the glycosyl donor) produce large quantities of glycosides of simple alcohols. However, it requires that the acceptor (mono)alcohol can be used in large excess (preferably as the solvent) and that of the many products possible (α,β-furanosides and α,β-pyranosides, please see Chapter 1) there is one that is the major one, usually this is the α-pyranoside. Thus, with more complex acceptors other glycosylation techniques, involving other types of glycosyl donors, have to be used.

Info Box 1

Emil Fischer (1852–1919) was the founder of modern carbohydrate chemistry. His name and mission are remembered, for example, through the Fischer glycosylation and the Fischer projection (please see Chapter 1). In spite of the limited methods available at the time, he elegantly managed to determine the structure of the monosaccharides, arbitrarily assigning their absolute configuration (D-sugar with the hydroxy group of the last chiral center to the right in their Fischer projection). It took more than 60 years before X-ray crystallography in 1951 showed that he had made the correct choice. For his work on carbohydrates he received the Nobel Prize in Chemistry in 1902. In spite of his great contributions he still struggled with the complexity of carbohydrates. He wrote in a letter to a colleague: 'The investigations on sugars are proceeding very gradually. It will perhaps interest you that mannose is the geometrical isomer of grape sugar. Unfortunately, the experimental difficulties in this group are so great, that a single experiment takes more time in weeks than other classes of compounds take in hours, so only very rarely a student is found who can be used for this work. Thus, nowadays, I often face difficulties in trying to find themes for the doctoral theses'. Fortunately, today we have in our hands methodology to deal better with the complexity of sugars to make it possible to unravel their huge biomedical importance, and this is why this is now an exciting field for all students and scientists.

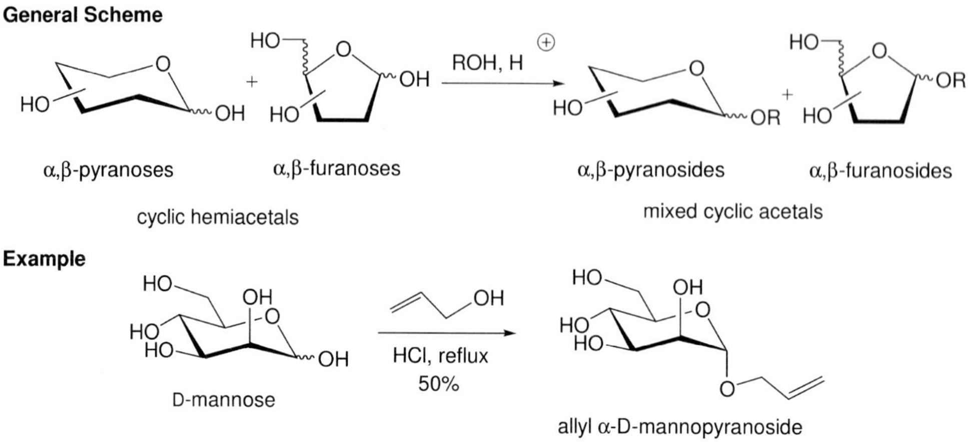

Figure 3.3 Fischer glycosylation.

3.4 Glycosyl Donors

Numerous glycosyl donors with different types of anomeric leaving group have been developed, but only a few of these are more extensively used [1]. The natural nucleotide sugar donors are not used in the chemical laboratory, instead mainly halide sugars, thioglycosides and trichloroacetimidates are used (Figure 3.4). The

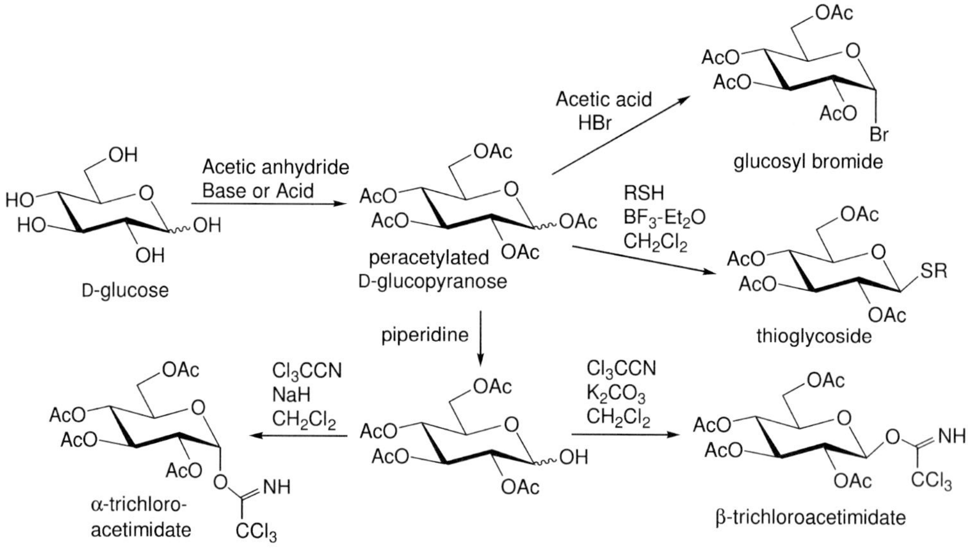

Figure 3.4 Common glycosyl donors and their preparation.

requirements of a glycosyl donor, and especially of its anomeric leaving group, are that it should be easy to prepare and be stable until it is intentionally activated by a promoter.

The traditional donors are the halide sugars, especially the bromides and chlorides, which have been used for more than a century.

Advantages
(i) They are easily prepared in large amounts.
(ii) They are efficiently activated by a number of promoters, mainly heavy atom (Hg^{2+}, Ag^{+}) salts.

Disadvantages
(i) The conditions under which they are prepared are usually quite strongly acidic (HBr/HOAc). This allows more or less only acyl protecting groups (acetates and benzoates) and smaller donors, since glycosidic linkages might be cleaved.

(ii) Stoichiometric amounts of promoter are needed, which might be expensive (silver salts) and not environmentally friendly (mercury salts).

(iii) They are quite labile and have to be prepared just before the glycosylation reaction is performed.

The trichloroacetimidate donors have some of the drawbacks of the halide sugars, but also a number of advantages in comparison.

Advantages
(i) The preparation of the trichloroacetimidates is performed under mild basic conditions, allowing almost any type of protecting group pattern and glycosidic linkages in the donor molecule.

(ii) The activation, using simple hard acids, typically BF_3-etherate or TMS-triflate (trimethylsilyl trifluoromethylsulfonate), requires only catalytic amount of the promoter – a quite unique feature among frequently used glycosyl donors.

(iii) They can, under optimized conditions, be activated to allow the glycosylation to proceed via a direct displacement mechanism (Figure 3.2, first mechanism). Since they can be prepared either as the α- or the β-imidate, by using different basic conditions, this allows one way to control stereoselectivity.

Disadvantages
(i) They are still rather labile and have to be prepared more or less immediately before the glycosylation reaction is performed, which, thus, require some temporary protection of the anomeric position up until the introduction of the trichloroacetimidate function.

Thioglycosides (and similarly *n*-pentenyl glycoside donors) have the following advantages and disadvantages.

Advantages
(i) They are easily prepared even at a large scale.

(ii) They are stable towards almost any type of reaction conditions. Hence, in contrast to the other donors discussed above, the thioglycoside functionality can be installed early in the synthesis, whereafter elaborate protecting group manipulations and also orthogonal glycosylations can be performed to build up a donor block of desired size.

(iii) When desired the addition of a thiophilic promoter facilitates their chemoselective activation and use as a glycosyl donor. A variety of such promoters with different reactivity has been found and developed, usually iodonium (I^+), carbonium (C^+) or sulfonium (S^+) compounds with triflate (trifluoromethylsulfonate, $CF_3SO_3^-$) as counteranion, for example NIS/triflic acid or silver triflate, dimethyl(thiomethyl)sulfonium triflate (DMTST) and methyl triflate.

(iv) They can under mild conditions be converted to other donors (for example halide sugars), allowing the synthesis of even quite labile such donors.

Disadvantages
(i) The thiols used for their preparation are usually smelly.
(ii) In spite of their stability, oxidations, debenzylations and some orthogonal glycosylations might still cause problems.

With this knowledge about available glycosyl donors and their strengths and weaknesses, we can now address the second question – how to perform stereoselective glycosylations.

3.5
Anomeric Configuration: Stereoselectivity

Without the aid of glycosyltransferases other ways to control the stereoselectivity of the glycosylation reaction, that is to get α- or β-selectivity, have to be developed. When creating a glycosidic linkage what is most interesting is if the desired product is a 1,2-*cis*- or a 1,2-*trans*-linkage, that is if the group in the anomeric position and the substituent in the 2-position are pointing in the same direction (*cis*) or the opposite direction (*trans*) in reference to the plane of the pyranoside ring (Figure 3.5).

As will be discussed below, there are good general methods to produce 1,2-*trans*-linkages, hence making β-D-glucosides, β-D-galactosides and α-D-mannosides is fairly easy. Formation of 1,2-*cis*-linkages, however, is still a problem, α-D-glucosides and α-D-galactosides a bit less so, since they are the thermodynamically more stable products, but construction of β-D-mannosides is one of the classical problems in carbohydrate chemistry. This challenge has biomedical relevance, because a β-mannoside linkage is common to the *N*-glycan core pentasaccharide (please see Chapters 6 and 8 for details).

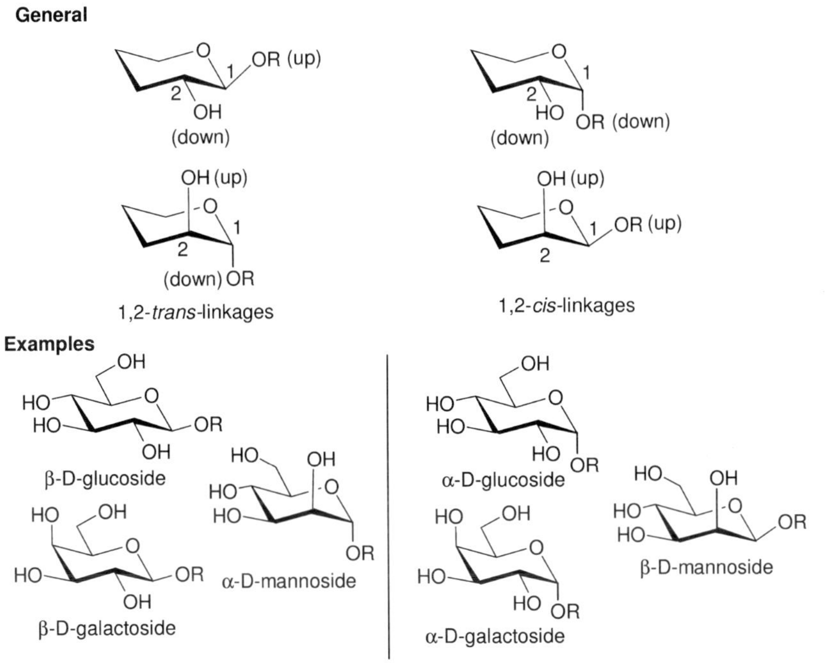

Figure 3.5 1,2-*Trans*- and *cis*-linkages.

3.5.1
Formation of 1,2-*Trans*-Linkages

To create a 1,2-*trans*-linkage a participating group in the 2-position, usually an acyl protection group (for example acetate or benzoate), is utilized. When the donor is activated and the oxocarbenium ion is starting to form, the free electron pair of the carbonyl oxygen participate and form a bond to the anomeric center to form a cyclic acyloxonium ion intermediate (Figure 3.6). When the acceptor now approaches, the *cis*-side of the anomeric position is protected by the cyclic ion and this is why the attack has to come from the *trans*-side to form a 1,2-*trans*-linkage. Another option for the acceptor is to attack the positively charged carbon, which will lead to formation of an orthoester. Under basic conditions this will be the main product, but under acidic conditions (read glycosidic conditions) the orthoester is not stable but will be in equilibrium with the acyloxonium ion to finally give the glycosidic product (*trans*-linked!). There is no other prerequisite for this methodology than having a participating group at the 2-position of the donor, and this is why this is a simple, general and reliable way to make 1,2-*trans*-linkages.

Saccharides with a 2-amino group (most often in the form of an acetamido group, for example GlcNAc and GalNAc) are abundant moieties in natural carbohydrate structures (please see Chapter 1). Also with this type of donor 1,2-*trans*-

linkages can be constructed using a participating group in the 2-position. Ideally, the natural acetamido function could be used, but with this type of donor, quite often the oxazoline, formed by the loss of a proton from the initial intermediate, is found to be the product instead of the glycoside (Figure 3.7). The oxazoline is more stable than the corresponding orthoester, but can be transformed into glycosides using strong acids and excess of a reactive acceptor. However, with other types of acceptors alternative participating *N*-protection groups have to be used. A variety of such groups have been developed, and used with high yields and complete β-selectivity in the glycosylation reaction [2], the most commonly used ones being the phthalimido (Phth) and the trichloroethylcarbamate (Troc) groups, but all still with the drawback of additional protection, deprotection and acetylation reaction steps required to give the target acetamido structures.

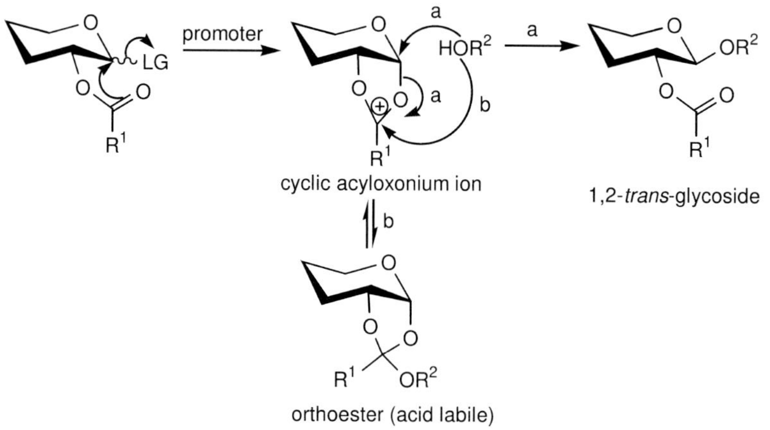

promoter
a
HOR^2
a
b
R^1
cyclic acyloxonium ion
b
R^1 OR^2
orthoester (acid labile)
OR^2
R^1
1,2-*trans*-glycoside

Figure 3.6 Synthesis of 1,2-*trans*-linkages using neighboring group participation.

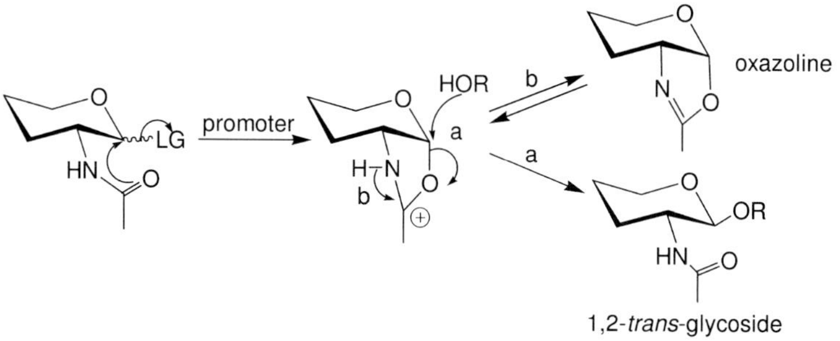

LG
HN
promoter
HOR
a
b
b
a
oxazoline
OR
HN
1,2-*trans*-glycoside

Figure 3.7 Competing formation between an oxazoline or a 1,2-*trans*-glycoside from a 2-acetamido-2-deoxy donor.

3.5.2
Formation of 1,2-*Cis*-Linkages

As discussed above, if no special technique is used, other than trying to optimize the reaction conditions (solvents, promoter, temperature), a glycosylation using a glycosyl donor with a 2-nonparticipating group (often a *O*-benzyl) will usually give the α-linked product in excess (due to the anomeric effect; please see Info Box 1 in Chapter 2). Some trends can be observed – a sterically more hindered acceptor will increase the α/β ratio, as will the use of some solvents and promoters, whereas the temperature dependence is more uncertain. Mannosides can often be obtained purely α-linked, but since this is a *trans*-linkage, α-mannosides can also be obtained in a more controlled way by using participating groups. Galactosyl donors give usually better α/β ratios than the corresponding glucosyl donors, probably due to steric hindrance by the axial 4-*O*-protecting group to β-side attack of the acceptor. Also, 6-*O*-acyl groups can participate to increase α-selectivity. Mainly because there is still a lack of a good general alternatives for the formation of α-D-glucosides and -galactosides (for example found as key epitopes in histo-blood group B determinants, please see Figure 1.5) this is frequently the approach taken: optimizing the reaction conditions and acceptor and donor properties to get as good α/β ratio as possible and then separating the obtained anomers (which is most often possible nowadays using silica gel high-performance liquid chromatography (HPLC)) before continuing the synthesis with the desired α-anomer. For the formation of the corresponding α-linked 2-acetamido glycosides, α-D-GlcNAc and α-D-GalNAc (blood group A determinant, please see Figure 1.5), the same approach is also generally employed. Most frequently the azido group (N_3) is then utilized as the 2-non-participating group. This group can efficiently be transformed into an acetamido group by reduction followed by acetylation.

The drawbacks of this approach to 1,2-*cis*-glycosides are obvious – optimization takes time, requires substantial amounts of the donors and acceptors, and the selectivity obtained is rarely complete, so material is lost both during chromatography and due to the β-anomer being formed. Consequently, there is a continuous quest for new methodologies to construct 1,2-*cis*-linkages in a stereospecific way [3] (please see also Info Box 2).

During glycosylations using halide donors with non-participating protecting groups it was observed that added halide salts increased the ratio of α-product formed and this finding was developed into a methodology called 'halide-assisted' (or '*in situ* anomerization') glycosylations. In this approach no 'real' promoter is added to the glycosylation reaction, only a halide salt, usually a tetraalkyl ammonium bromide. The role of the halide salt is to establish a quick equilibrium between the α- and the β-bromide of the donor, but not to remove the bromide as a normal promoter does. This would give the undesired oxocarbenium intermediate and thereby loss of stereoselectivity. Of the two anomers, the α-bromide is by far the most stable and in large excess, but under these conditions there is always some β-bromide present as well. The rationale is now that the α-bromide is too stable to react with the acceptor. The β-bromide, however, is highly reactive and can be directly displaced by the acceptor to afford the α-linked glycoside (Figure 3.8). Once the (small amount of) β-bromide has reacted to give the product, new β-bromide will be formed due to the equilibrium and finally all the halide donor has been turned into the desired α-product. This procedure utilizes easily available

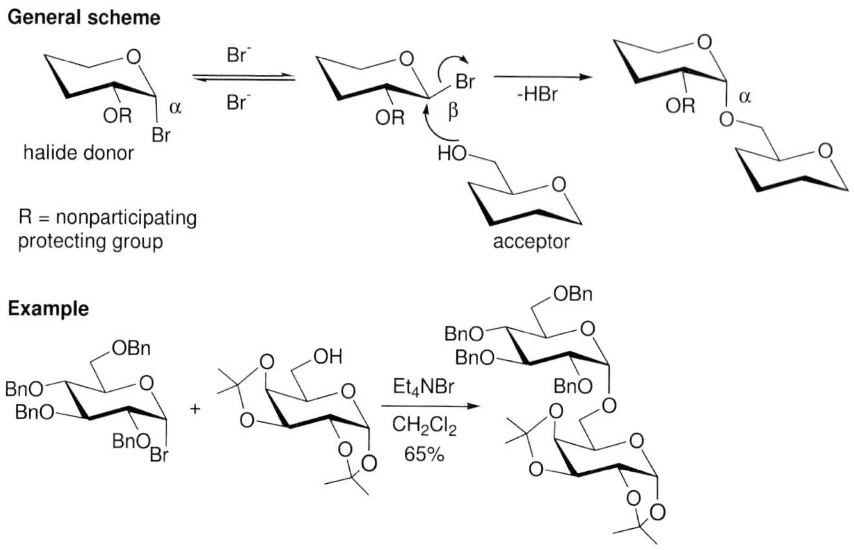

Figure 3.8 Halide-assisted glycosylation.

donors and, if correctly performed, gives excellent α-selectivity. The drawback is that, due to the lack of promoter, it is a very mild glycosylation methodology, requiring extended reaction time and sometimes heating, and yields with less reactive acceptors (for example many secondary hydroxyl groups) are usually not very high.

3.6 Neuraminic Acid and Kdo-Glycoside Synthesis

Another difficult form of glycosides to synthesize in a stereoselective manner is glycosides with no substituent at all in the position next to the anomeric center–2-deoxy-glycosides. Such structures found in Nature are Kdo-glycosides, found in bacterial polysaccharides, and sialic acid glycosides (for structures, please see Figure 1.6; for relevance of sialic acids in biorecognition, please see Chapters 18, 19, 25 and 27; for relevance of sialic acids in drug design, please see Chapter 28). Nomenclature-wise these are 3-deoxy sugars, since the carboxyl function is C1 and the anomeric center is C2, but the problem is the same–how to perform a stereoselective glycosylation when you have no functional group next to the anomeric center to use as a handle.

With NeuAc and Kdo the naturally occurring glycosides are both α-linked. However, due to the different preferred conformations of the pyranose ring of the two sugars this means that in neuraminic acid glycosides the carboxyl group is in the axial and the glycosidic linkage in the equatorial position, whereas in Kdo it is the opposite.

Info Box 2

Sucrose, normal table sugar, which plants can make tons of, is very difficult to synthesize chemically. When looking at the structure (below) one can realize why – sucrose is a nonreducing sugar where the anomeric positions of an α-glucopyranosyl and a β-fructofuranosyl are linked together. Thus, when performing the glycosylation the stereoconfigurations in both the anomeric positions have to be controlled and they are both 1,2-*cis*-linkages! The first chemical synthesis of sucrose was reported by R.U. Lemieux and G. Huber in 1953. Their synthesis gave a mixture of all anomers and the yield of sucrose in the coupling was only 5%. The solution to a stereoselective synthesis of sucrose came almost 50 years later, when a β-directing fructofuranosyl donor was developed by Oscarson and Seghelmeble. This thioglycoside donor contained a bridge, a silyl acetal, between the 1- and 4-OH group. Since the 4-OH group is on the α-side, this locked the anomeric CH_2OH group in the α-position allowing incoming acceptors to attack only from the β-side. As acceptor tetra-*O*-benzyl-α-D-glycopyranose was selected, the pure α-anomer obtainable by crystallization. The DMTST-promoted glycosylation afforded protected sucrose in 80% yield, which could be deprotected to give sucrose.

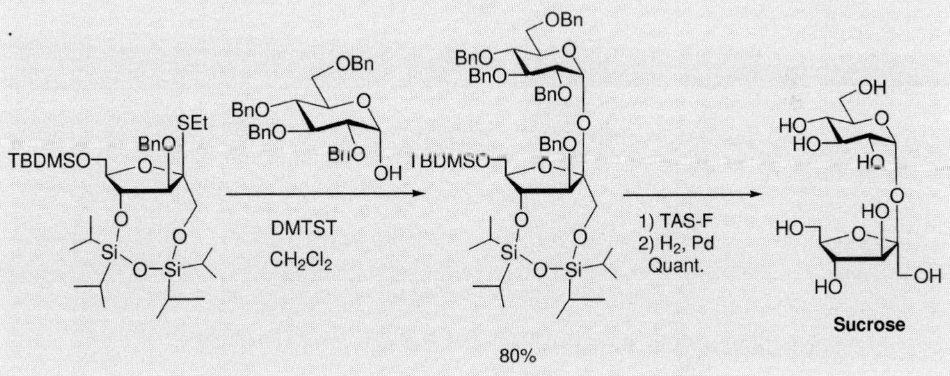

The general way to synthesize these types of glycosides is the same as discussed for α-D-glucopyranosides and galactopyranosides – an optimization of the glycosylation conditions to obtain as high a α/β ratio as possible and then separate the anomers (Figure 3.9). The type of donors is also the same, halide sugars and thioglycosides being most frequently used. For NeuAc glycosides conditions often can be found which afford high yields and stereoselectivity [4], but with Kdo this is often still a problem [5]. Therefore, methods have been developed to try to deal with this problem.

These methodologies involve the use of a temporary participating group to control the stereoselectivity in the glycosylation – a group which is then removed to give the natural deoxy function. This temporary participating group is generally a halide (bromine or iodine) or a selenium or thio ether and can be introduced *in situ* if the glycal is used as donor (Figure 3.10). The participating group is preferably introduced on the β-side, allowing acceptor attack only from the α-side,

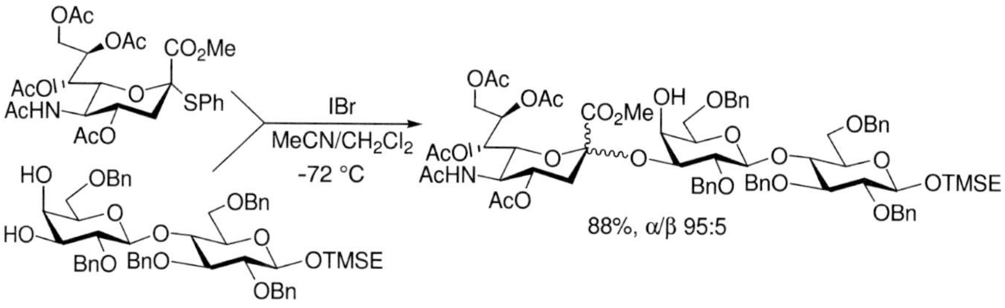

Figure 3.9 Example of Neu5Ac glycoside synthesis.

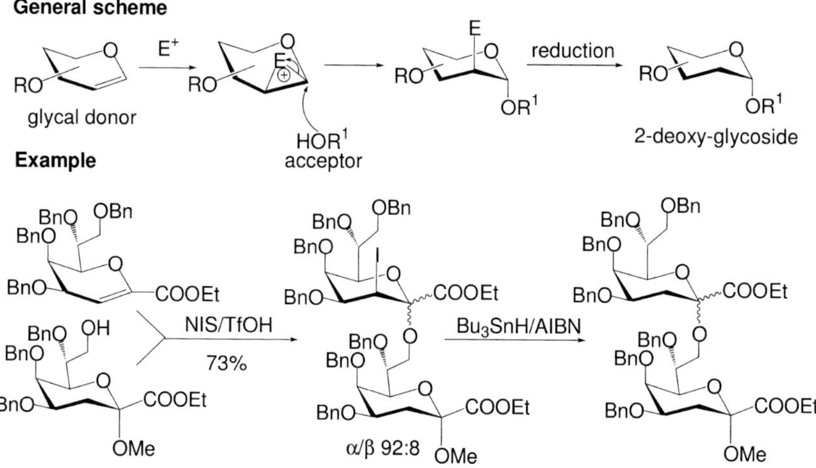

Figure 3.10 Synthesis of Kdo derivatives from glycal donors.

resulting in the formation of the desired α-(*trans*)-glycoside. Removal of the auxiliary group, usually by Bu_3SnH reduction, then affords the target 3-deoxy-α-glycoside. For NeuAc there are also good commercial glycosyltransferases that can be used in chemoenzymatic synthetic approaches to oligosaccharides.

Now we know which the normal glycosyl donors are and also a variety of glycosylation methods. What we need to learn to be able to build larger oligosaccharide blocks is which donors are stable, and which are activated using the various promoters and glycosylation methods discussed.

3.7
Formation of Building Blocks: Orthogonal Glycosylations

If a block synthesis is chosen, building blocks have to be designed and synthesized. Only a few oligosaccharides (for example lactose and cellobiose) are commercially available from natural sources so monosaccharides will generally be the initial starting materials used. Human carbohydrate structures contain mainly easily available monosaccharides (D-Glc, D-Gal, D-Man, D-GlcA, D-GlcNAc, D-GalNAc, L-Fuc, and Neu5Ac, please see Chapter 1; L-IdoA, however, found in heparin and heparan sulfate is not; for details of these proteoglycans, please see Chapter 11; for biological relevance of IdoA, please see Chapter 1.3). Bacterial structures, in contrast, often include rare monosaccharides (for example Kdo) that have to be synthesized from commercially available precursors before being used in oligosaccharide synthesis, which severely complicates these syntheses. When designing the building blocks a decision has to be made about what type of donors should be targeted. If a trichloroacetimidate donor is chosen, then the reducing end anomeric position has to be protected with a protecting group that eventually can be removed selectively to free this anomeric position and allow the formation of the trichloroacetamidate. If a thioglycoside or pentenyl donor block is chosen, then this function is usually introduced at the anomeric position at the very beginning of the block synthesis, since it is stable to most protecting group manipulations as well as to many glycosylation conditions. Hence, a thioglycoside, containing a free hydroxyl group, can function as an acceptor if proper donors and reaction conditions are used. Since the thioglycoside itself is also a potential donor this is often described as orthogonal coupling (only one of the two possible donors is activated). Promoters used for the activation of halide donors (silver or mercury salts) and trichloroacetimidate donors (TMS-triflate or BF_3-etherate) normally do not activate thioglycosides, and can therefore be used in the building up of thioglycoside blocks.

Protecting groups not only protect but confer different reactivity to the protected derivative. Both in a donor and in an acceptor, electron-withdrawing protecting groups (that is, acyl groups, acetates and benzoates) decrease the reactivity, whereas electron-donating groups (that is, ether groups like benzyls) increase the reactivity. This introduced reactivity difference can be large enough to facilitate glycosylation reactions between the same type of glycoside donors, but differently protected (Figure 3.11).

Hence, a benzylated thioglycoside can be activated using a rather weak promoter and used as a donor, whereas a benzoylated thioglycoside can be inert under these conditions and used as an acceptor, allowing the construction of a thioglycoside disaccharide, which in turn subsequently can be activated by a stronger promoter and used as a donor. This is quite elegant, but it also puts a number of restrictions

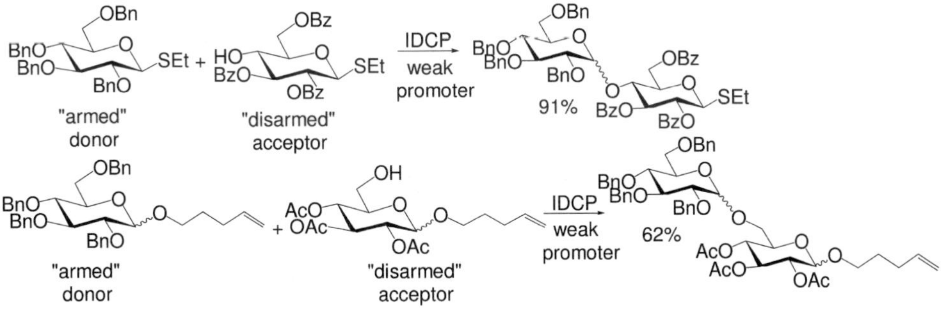

Figure 3.11 Use of the 'armed–disarmed' approach in orthogonal glycoside synthesis.

on the protecting group pattern allowed, so most often it is easier to use different types of donors in these orthogonal glycosylations and in the build-up of the different blocks. Still, thioglycosides can be used as the sole precursors since they are easily transformed into halide donors or trichloroacetimidate donors under mild conditions.

3.8 Protecting Group Manipulations

Without the help of the glycosyltransferases, apart from the problem with stereoselectivity we also have to address the problem of regioselectivity. How do we allow only one of the many hydroxyl groups present in the acceptor and in the donor saccharides to react with the anomeric position of the donor to form the new glycosidic bond? This problem is solved by protecting all the other hydroxyl groups. A major part of carbohydrate chemistry is therefore involved with protective group manipulations – both the introduction of protective groups as well as their removal [6].

In spite of the fact that it is more or less only hydroxyl protecting groups that have to be considered, often several different types of protecting groups have to be utilized. This is partly due to the fact that we want different protecting group patterns to get different reactivities, as discussed above, but also because if larger structures than a disaccharide are to be synthesized, two types of protecting groups are required – persistent (permanent) ones that will remain until the very last deprotection step(s) and temporary ones that will be removed somewhere along the synthetic pathway to expose a new hydroxyl group allowing another glycosylation reaction (please compare Figure 3.1). Obviously, these types of protecting groups have to be what is called orthogonal, that is, that they can be removed in the presence of each other (compare definition of orthogonal glycosylations in Section 3.7).

By far the most used hydroxyl protecting groups, which can be used both as temporary and permanent ones, are acetyl, benzoyl and benzyl groups, all of which are easy to introduce and also to remove even at multiple positions. In addition, there are various other groups that can be used as temporary protecting groups, being efficiently introduced at one or a few positions and also selectively cleaved when desired (examples are silyl, chloroacetyl, *p*-methoxybenzyl and various acetal groups).

3.9 An Example

With all this knowledge in hand, let us now apply them to a real case – the synthesis of the Le^b hexasaccharide (Figure 3.12; please see Chapter 17.1.2.1 for information on its role as docking points for infections with *Helicobacter pylori* causing gastric ulcers; its availability has a therapeutic perspective, especially when presented as clusters: please see next chapter for design of glycoclusters) [7]. Of the six glyco-

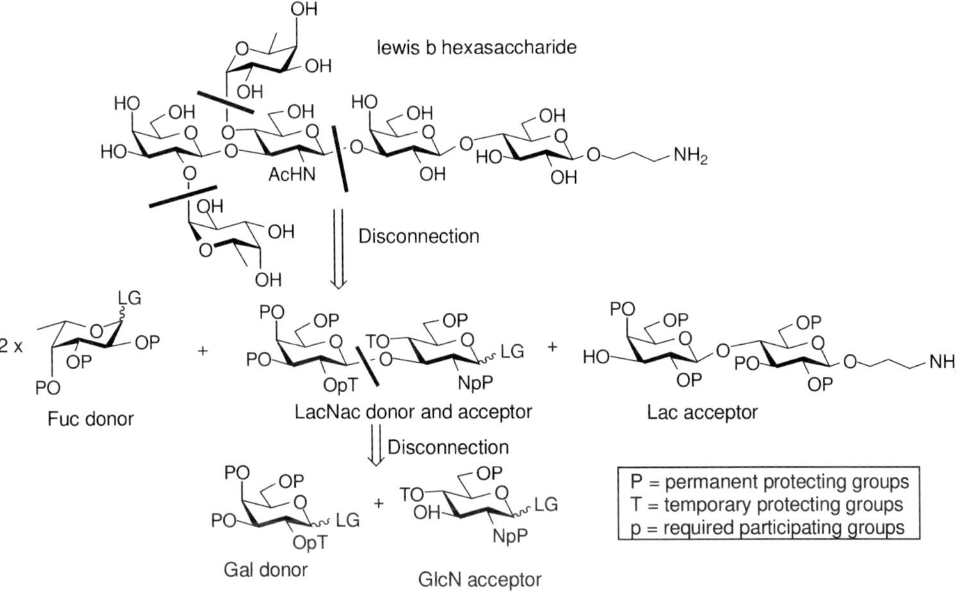

Figure 3.12 Retrosynthesis of the Le^b hexasaccharide.

sidic linkages, two are 1,2-*cis*-linkages, the two α-L-fucopyranosides, and four are 1,2-*trans*-linkages. Two identical motifs are recognized, the two α-L-fucopyranosides, and also a lactose disaccharide moiety, β-D-Gal-(1 → 4)-D-Glc, which is commercially available (and cheap). Thus, a straightforward retrosynthetic strategy is to perform the disconnections according to Figure 3.12 identifying three building blocks:

(i) A monosaccharide fucopyranosyl donor.
(ii) A disaccharide LacNAc (Type 1) block.
(iii) A disaccharide Lac (Type 2) block.

Fucopyranosyl donor
Requirements
(i) Function as an α-directing donor in the glycosylation of the LacNac part.

Since *cis*-linkages are to be formed, nonparticipating benzyl protecting groups are chosen. The halide-assisted glycosylation is known to work well in the formation of α-L-fucosyl linkages and, thus, the glycosyl bromide is selected.

Lactosamine building block
Requirements
(i) Function as a 4″, 2‴ acceptor for the fucosyl donor.
(ii) Function as a β-directing disaccharide donor in the glycosylation of the lactose acceptor.

This block has to function both as a donor, to be linked to the lactose acceptor, and as an acceptor, in the glycosylation with the fucosyl donors. The two glycosidic bonds to be formed are both *trans*-linkages, so a 2-participating group is needed both in the galactose part (acetate chosen) as well as in the glucosamine (phthalimido chosen) part. Temporary protecting groups are required in the 4- and 2′-positions to allow introduction of the two fucopyranosyl moieties. Benzyl groups are chosen as permanent protecting groups and acetates as temporary protecting groups. The block is synthesized as its thioglycoside, which allows the use of the galactosyl bromide in the orthogonal glycosylation to form the desired disaccharide (Figure 3.13).

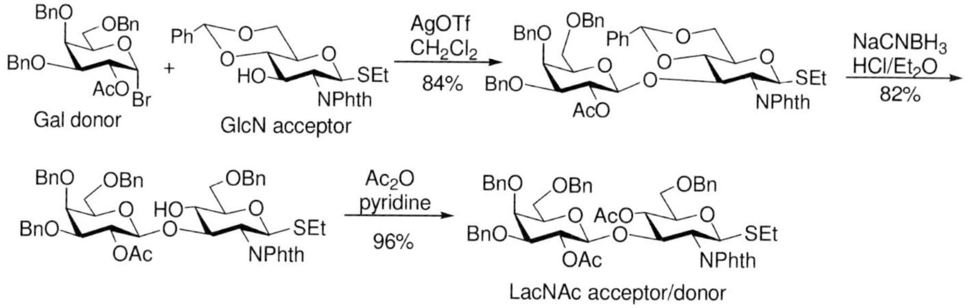

Figure 3.13 Synthesis of LacNAc acceptor/donor.

Lactose disaccharide acceptor block

Requirements

(i) A β-linked spacer is desired at the reducing end to allow subsequent conjugation reactions.

(ii) Function as a 3′-acceptor for the LacNAc donor.

To introduce the β-linked spacer, participating groups are needed (Figure 3.14). After the introduction of the spacer, the protecting groups are changed to permanent benzyl protecting groups in a regioselective way to afford the hydroxy acceptor. The protecting group scheme used actually affords the 3′,4′-diol acceptor, but it was known that the 4′-hydroxyl group is of low reactivity, why a 3′-regioselective glycosylation can be performed (please compare Figure 3.9).

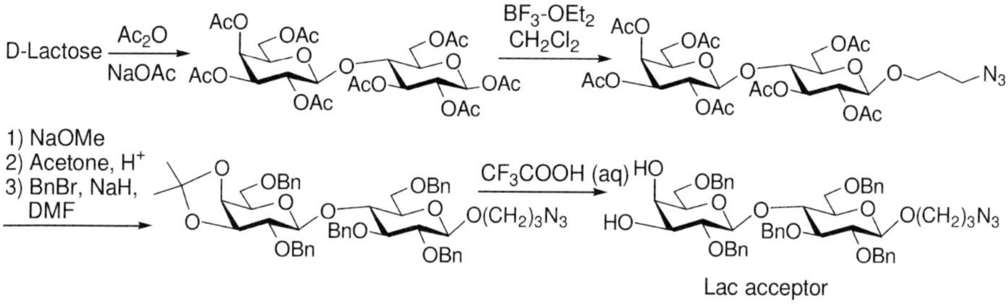

Figure 3.14 Synthesis of Lac acceptor.

Assembly of the hexasaccharide

With the building blocks available, their connection can now be accomplished. Two ways are possible:

(i) The fucosyl linkages are formed first (possible since halide-assisted glycosylation conditions are orthogonal to thioglycosides) to give a tetrasaccharide donor, which is then coupled to the lactose acceptor

(ii) The LacNAc donor is first coupled to the lactose acceptor to create a tetrasaccharide acceptor into which the fucosyl moieties are then introduced.

Since neither of theses glycosylations would be considered as 'difficult', both these potential approaches to the hexasaccharide are similar, but practically it was found that the latter approach is the preferred one. Hence, activation of the LacNAc thioglycoside with NIS/TfOH in the presence of the lactose acceptor afforded the (1-3)-β-linked tetrasaccharide in high yield (Figure 3.15). Protecting group manipulations then changed the *N*-phtalimido group into the target *N*-acetamido group. Concomitant removal of the temporary acetyl protecting groups gave the trihydroxyl acceptor, which was α-difucosylated using the fucosyl donor and halide-assisted conditions to yield the target hexasaccharide (Figure 3.16), again taking advantage of the low reactivity of the axial 4′-hydroxyl group to perform a regioselective glycosylation. Final deprotection, consisting of only one step (hydrogenolysis), removed the benzyl protecting groups as well as reduced the azido group of the spacer to yield an amino group ready for conjugation, afforded the target hexasaccharide.

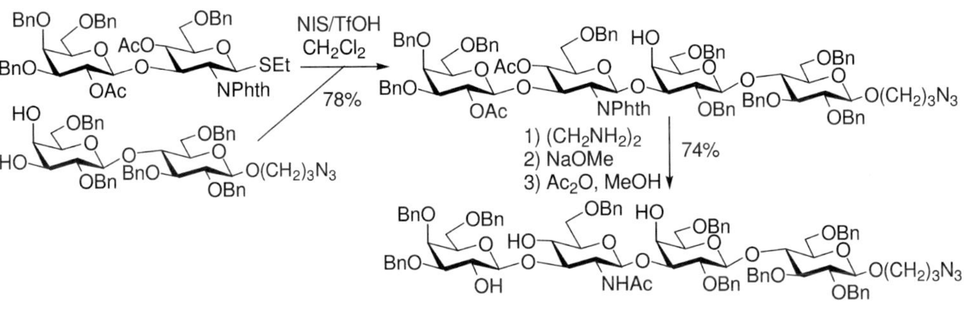

Figure 3.15 Synthesis of *N*-lactotetraose acceptor.

3.10 Conclusions

Regio- (polyhydroxy compounds) and stereo-(α/β-configuration) selective issues are encountered in oligosaccharide synthesis. The regioselectivity problem is addressed by protecting all hydroxyl groups except the one in the acceptor used to

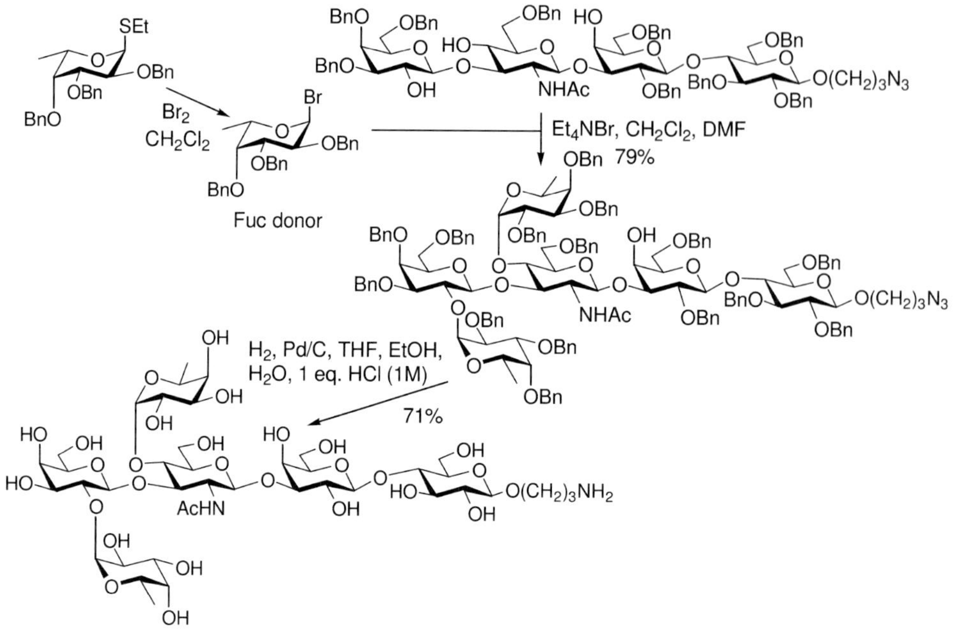

Figure 3.16 Synthesis of Le^b hexasaccharide.

form the new glycosidic linkage. As to the stereoselectivity, 1,2-*trans*-linkages (β-D-*gluco* and -*galacto*, α-D-*manno*) are constructed efficiently by using 2-participating groups. 1,2-*Cis*-linkages (α-D-*gluco* and -*galacto*, β-D-*manno*), however, have no general solution and an optimization for each coupling has to be performed. There is no universal glycosyl donor, but a variety (for example halide sugars, trichloroacetimidates and thioglycosides) to choose from with different properties as stability and reactivity. Larger oligosaccharides are preferentially synthesized using a convergent approach where building blocks are first constructed and then coupled together to give the target oligosaccharide. This minimizes manipulations on large structures and allows freedom in the glycosylation sequence. With the techniques described to master synthesis of glycans, the next step to gain access to potent ligands for lectins (please see Chapters 15–19 for surveys on this protein superfamily) is to optimize their presentation by achieving multivalency. How to do that is outlined in the following chapter.

Summary Box

Chemical synthesis can be utilized to get access to required oligosaccharide structures. Developed methodologies using protecting groups to get regioselectivity and also stereoselectivity in the glycosylation reactions in combination with efficient glycosyl donor/promoter system allows for the construction of complex (also labeled and nonnatural) structures.

References

1 Toshima K *et al.* Recent progress in *O*-glycosylation methods and its application to natural products synthesis. *Chem Rev* 1993;93:1503–31.
2 Banoub J *et al.* Synthesis of oligosaccharides of 2-amino-2-deoxysugars. *Chem Rev* 1992;92:1167–95.
3 Demchenko AV. 1,2-*Cis*-*O*-glycosylation: methods, strategies, principles. *Curr Org Chem* 2003;7:35–79.
4 Boons G-J, Demchenko AV. Recent advances in *O*-sialylation. *Chem Rev* 2000; 100:4539–65.
5 Oscarson S, Hansson J. Synthesis of bacterial carbohydrate surface structures containing Kdo- and *glycero*-D-*manno*-heptose residues or anomeric phosphodiester linkages. *Curr Org Chem* 2000;4:535–64.
6 Oscarson S. Protective group strategies. In: *The Organic Chemistry of Sugars* (Eds.: Levy DE, Fugedi P), pp. 53–87. Taylor & Francis, Boca Raton, FL, 2006.
7 Lahmann M *et al.* Synthesis of the Lewis^{b} hexasaccharide and HSA-conjugates thereof. *Glycoconj J* 2004;21:251–64.

4
The Chemist's Way to Prepare Multivalency

Yoann M. Chabre and René Roy

Carbohydrates have distinct characteristics to enable high-density coding, among which positional branching constitutes unique features (see Summary Box of Chapter 1). The branching generates multivalency. The valency of molecular structures bearing carbohydrate ligands at the periphery is defined as the number of identical units that can each contribute to binding contacts with suitable receptors (for example lectins) [1]. Thus, a glycoconjugate that has two copies of a binding carbohydrate moiety is said to be divalent. Similarly, multivalent interactions are defined as specific associations of multiple carbohydrate ligands present on a molecular surface that bind multiple receptors expressed on proteins or cell surfaces. The importance of these multivalent carbohydrate–protein interactions is critical to several biological phenomena discussed throughout this book. In fact, mammals only express nine different carbohydrates that are arranged in variable linkages (the 'sugar coding') (please see Figure 1.6 for monosaccharide structures and Chapter 3 for synthetic aspects leading to branched oligosaccharides). This family of natural as well as synthetic species has been designed to compensate for the usually low binding affinity (K_a) of carbohydrate ligands (for further details, please see Chapter 13). The main goal of this chapter is to highlight synthetic strategies that have been created to enhance carbohydrate–protein interactions at the molecular level by architectures resembling 'carbohydrate Velcro'. [Velcro, used to attach shoelaces, is made of several tiny hooks and loops each forming weak bindings. It was invented by a Swiss engineer George de Mistral in 1945 after his close observation of burdock (*Arctium lappa*, Figure 4.1c) sticking to his clothes during his daily summer walks in the Alps. The word velcro is derived from the French '*velours*' (velvet) and '*crochet*' (hook)]. These novel structures are assembled by strategies analogous to the constructions of 'Lego toys'. Chemists are therefore playing essential roles in modern glycobiology.

Nature's multivalency is often expressed by 'dendritic' architectures that represent the most pervasive topologies observed in plant and animal kingdoms. Numerous examples of these patterns may be found at different dimensional length scales (meters to microns) and typical examples can be observed in abiotic systems (for example snow crystals, fractal erosions, manganese dendrites on

The Sugar Code. Edited by Hans-Joachim Gabius

ISBN: 978-3-527-32089-9

Figure 4.1 Nature's way to provide multivalency. (a) Dendritic network of a tree's branches and roots. (b) Dendritic red blood cell. (c) Dendritic representation of tiny hooks on burdock. (d) Scanning electron micrograph of a human dendritic cell. (e) Manganese dendrites on rock. (f) Fractal snow crystal. (g) Dendritic network under a gecko's foot.

rock) or in the biological world (for example tree branching/roots, respiratory system, neurons, gecko adhesion) (Figure 4.1). The reasons for such extensive usage of these dendritic topologies at virtually all dimensional length scales are not entirely clear. However, one might speculate that these are optimized architectures that have evolved over the past several billion years to provide structures manifesting the most favorable interfaces and efficacy.

A nanoscale example of glycodendritic architecture in biological systems is found in the network of proteoglycans (for further details, please see Chapter 11). These macromolecules appear to provide energy-absorbing, cushioning properties and determine the viscoelastic properties of connective tissues. Another striking example is found on yeast and mammalian cell surfaces, also found on viral glycoproteins. Their carbohydrate structures play critical roles in multiple key cellular events such as cellular adhesion and recognition, physiological function regulations, and pathogenic infections. For HIV-1 gp120 (please see Chapter 17.2.3), several hypothesis were formulated concerning their role in infection processes and it has been speculated that the *N*-linked glycans (for further details on *N*-glycans, please see Chapters 6, 8, 22–25 and 27–30) exposed at the exterior of envelope glycoprotein gp120 ($Man_9GlcNAc_2$, Figure 4.2) help viruses to escape neutralizing antibodies and/or use the mannoside binding properties of dendritic cells to invade lymphocytes (please see Chapter 19 for further information on the relevant lectin DC-SIGN and Figure 16.1i for its crystal structure).

4.1 Blocking Viral/Bacterial Adhesion

Although carbohydrate structures have long been regarded only as space-filling matrices or posttranscriptional accessory elements in glycoproteins, which serve to protect them from premature degradation, it has become apparent that glyco-

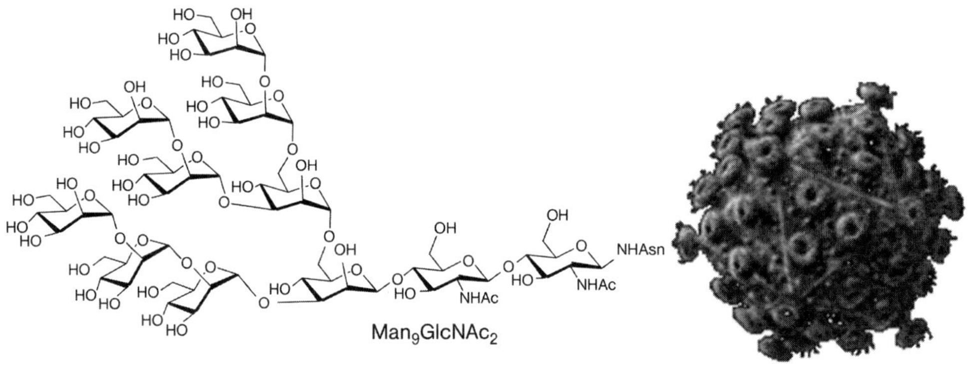

Figure 4.2 The *N*-linked glycan ($Man_9GlcNAc_2$) on HIV-1 gp120.

conjugates exhibit a broad variety of additional biological functions such as presentation of target structures for microorganisms, toxins and antibodies, control of protein half-life, modulation of protein function or provision of ligands for specific binding events. Multivalent interactions are now understood to be a ubiquitous strategy that has evolved in nature for a wide range of functions, that provides numerous and unique benefits that are not achievable with monovalent interactions. This observation is also present at the molecular level with multiple carbohydrate–protein interactions that are responsible for several biological events (for further details on mechanism of carbohydrate–protein interactions, please see Chapter 13). Indeed, saccharides are expressed on the majority of mammalian cell surfaces and are bound to proteins ('glycoproteins') or lipids ('glycolipids') that are entangled in the cell membranes and clustered in multiantennary configurations [2] (for further details, please see also Chapters 6 and 10). Surprisingly, although these multiple protein–carbohydrate interactions are responsible for several crucial biological events, it has been shown that, on a per-saccharide basis, these interactions are characterized by rather weak association constant (milli- to micromolar) with limited specificity and selectivity (for further details, please see Chapter 14). However, these interactions are transformed into very potent attractive forces, dramatically and naturally reinforced, when multiple ligand copies are presented to similarly clustered receptors. This phenomena, resulting in a synergic and cooperative effect, is known as the 'glycoside cluster or dendritic effect' and was initially observed with asialoglycoprotein receptors found on hepatocytes (please see survey figure in Chapter 19 showing the oligomer, and Chapter 15 for the history on Ashwell's discovery of the C-type lectin) [3]. In its widespread version, it is usually assumed that this effect has its source in the enhanced affinity of a given multivalent glycoside toward a carbohydrate recognition domain (CRD) by fully occupying one active site at a time (for further details, please see Chapter 16 providing X-ray illustrations on CRDs of various Ca^{2+}-dependent lectins). The phenomenon is now widely accepted as having its basis on stabilization by macroscopic 'cross-linking glycocluster effects'. With all these observed advantages in mind, research on synthetic multivalent macromolecules has intensified, giving rise to several original glycoconjugate structures that constitute high-affinity multivalent ligands that target surface receptors (that is enzymes, lectins, toxins, viruses and bacteria). These multivalent glycoconjugates have shown great potential as effective inhibitors of cell–pathogen interactions (Figure 4.3) (please see Chapter 17 for further details on bacterial/viral lectins). Notable examples include multivalent sialic acid and mannoside molecules that bind to the hemagglutinin receptor on the influenza virus surface or the FimH at the tip of pilliated *Escherichia coli (E. coli)*, where the multivalent enhancement factor ranges from 10 to 10^6, depending on the degree of valency and type of scaffolds utilized [4]. Here, again, the chemist's creativity has provided the impetus for access to highly elegant and effective architectures [5].

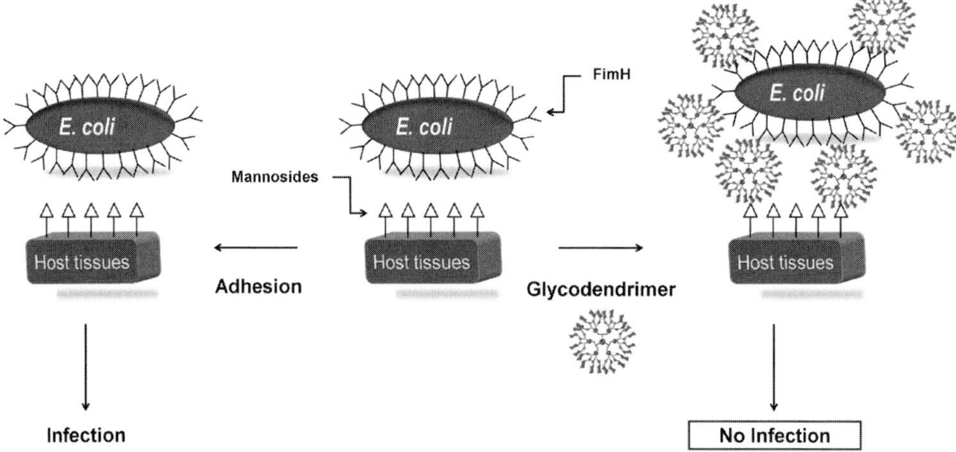

Figure 4.3 Blocking bacterial/viral infections by competitive binding to host tissues with glycodendrimers.

4.2
How to Prepare Multivalent Carbohydrates?

A thorough understanding of multivalent carbohydrate–protein interactions suffers from the natural carbohydrate complexities that are also the result of incomplete biosynthesis or subtle attachment of other functionalities at specific positions along the oligosaccharide sequences (sulfate, phosphate, acetate, and so on; for substituted sugars such as mannose-6-phosphate on the HNK-1 epitope, please see Chapter 1, especially Figure 1.7). In order to study, characterize, understand and manipulate these critical interactions, striking advances in isolation, purification, structural analyses and partial or selective degradation processes have been achieved, albeit with limited amounts of the resulting glycoconjugates (for further details, please see Chapter 5). Hence, chemical or chemoenzymatic synthesis of multivalent carbohydrate ligands is likely to remain the method of choice to afford various well-defined architectures that have been developed as effectors or inhibitors of biological mechanisms. Thus, synthetic analogs (neoglycoconjugates) in which carbohydrate residues are chemically or enzymatically attached to unnatural carriers present numerous advantages. Moreover, using structure–activity relationships, it is also possible to identify many of the essential carbohydrate residues (epitopes) responsible for the biological activity of interest (for further details, please see Chapter 14). One can consider that neoglycoconjugate syntheses are facilitated by simpler carbohydrate targets (for further details, please see Chapter 3). Progress encountered in terms of their synthetic accessibility and efficiency has allowed their modulation and their activity optimization remains necessary to investigate how these multivalent structures can influence the binding activity. These investigations are critical to highlight the potential of multivalent carbohydrate inhibitors as high-affinity ligands or as effectors capable of clustering cell-surface receptors and may lead to generate structures with tailored biological activities.

This section will be dedicated to the description of the multivalent neoglycoconjugate family, emphasizing the preparation of glycomimetics of prominent components at the surfaces of mammalian cells (that is neoglycoproteins, neoglycolipids), and the synthesis of glycopolymers and glycodendrimers, briefly mentioning their respective uses in biomedical applications (for further information, please see Chapters 24–30).

4.3 Neoglycoproteins

Neoglycoproteins represent the first class of synthetic neoglycoconjugates since they were initially synthesized as capsular polysaccharide protein conjugate vaccines related to pneumococcal infection [5]. Nowadays, several reasons and methods to generate neoglycoproteins exist, and the need for neoglycoproteins stems from the lack of carbohydrate homogeneity in natural glycoproteins. It has also been observed that certain threshold carbohydrate contents were necessary for the expression of desired biological activities, and one way to remedy this situation was to artificially introduce more and even unnatural carbohydrate moieties. Therapeutic glycoproteins such as antibodies recently need to be 'glycoengineered' for medical approval. Naturally occurring carbohydrates are linked to proteins via an *N*-glycosidic bond to the side-chain of asparagine or an *O*-glycosidic bond to hydroxylated amino acids serine and threonine (for further details on *N*- and *O*-glycan attachment, please see Chapters 6–8). Table 4.1 lists a few of the possible amino acid modifications by carbohydrates derivatives for neoglycoprotein synthesis.

Combined with site-directed mutagenesis, this approach allows control of both the site of attachment and choice of saccharide introduced. Furthermore, a mixed strategy of an initial linear assembly coupled with convergent build-up, involving

Table 4.1 The most commonly used conjugation chemistries in neoglycoprotein synthesis.

Amino acid	Carbohydrate derivative	Reaction type	Linkage
Lysine (ε-NH_2, N-terminal)	Carboxylic acid[a]	Amidation	Amide
	Aldehyde	Reductive amination	Amine
	Ketone	Reductive amination	Amine
	N-acryloyl	Conjugate addition	Amine
	Isocyanate	Ureidation	Urea
	Isothiocyanate	Thioureidation	Thiourea
	Imidate	Amidination	Amidine
	Pseudo-thiourea	Guanidination	Guanidine
	Imidazoylurethane	Carbamoylation	Carbamate
	Cyanate ester	Isoureidation	Isourea
	Imidocarbonate	Imidocarbonation/ carbamoylation	*N*-imidocarbonate/ carbamate
Aspartic, glutamic acid, C-terminal	Amine	Amidation	Amide
Tyrosine	Diazonium	Diazotation	Diazo
Cysteine (thiol)	Thiol	Oxidation/exchange	Disulfide
	Bromoacetyl	Substitution	Thioether
	Maleimide	Conjugate addition	Thioether
	N-acryloyl	Conjugate addition	Thioether

[a] Carboxylic derivatives: acyl halide, azide, hydrazide, anhydride, active ester, thiolactone.

extension of the glycan part with, for example, enzymatic-mediated glycosylation through chemoselective ligation, has also been described. These two distinct sets of strategies have been employed for glycoprotein remodeling and to remove unwanted glycoforms or to link saccharide residues to proteins (for further details on natural glycosylation, please see Chapters 6–8). Neoglycoproteins are particularly useful as antibacterial vaccines (for further applications, please see Chapter 18.3 and Table 25.1).

4.4 Neoglycolipids and Liposomes

A generic and elegant methodology for the design of carbohydrate biosensors has been the construction of neoglycolipids (for further details on glycolipids, please see Chapters 10 and 30). These molecular composites, based on synthetic oligosaccharides coupled to lipid residues, constitute interesting tools for deciphering carbohydrate sequences and structures using microarrays. Glycolipids are found at the surface of every cell and several hundred glycolipid structures have been characterized on mammalian cell surfaces. Their structures differ greatly from that of lipid A, which is present as part of the lipopolysaccharides of Gram-negative bacteria cell walls. Based on their hydrophobic portion, mammalian glycolipids can be divided into two classes – glycosphingolipids (or gangliosides when sialic acid is present) and glycoglycerolipids (Scheme 4.1). The facile synthesis of lipidic glucosylamides having potent immunomodulatory properties deprived of mitogenic activity reflects the high commercial potential of neoglycolipid mimetics and further exploits the strength of organic chemistry.

Concerning synthetic strategies, enzymatic degradation of natural glycosphingolipids by ceramide glycanase represents a powerful method. Thus, it has been possible to effect '*trans*-lipidation' reactions of natural glycolipids. The complete chemical syntheses are much more tedious. The aforementioned 'cluster effect' has been studied through these glycolipids that can form liposomes (Figure 4.4). These self-assemblies, obtained by suspending suitable lipids in aqueous media, can offer structural versatility in terms of chemical composition and molecular fluidity, and allow very special carbohydrate domains at the surface. However, in spite of their simplicity, one of the major drawbacks of liposomes results from the fact that they can be unspecifically incorporated into various cell membranes. Moreover, the micellar concentrations necessary to obtain micellar species are not necessarily retained in *in vivo* experiments.

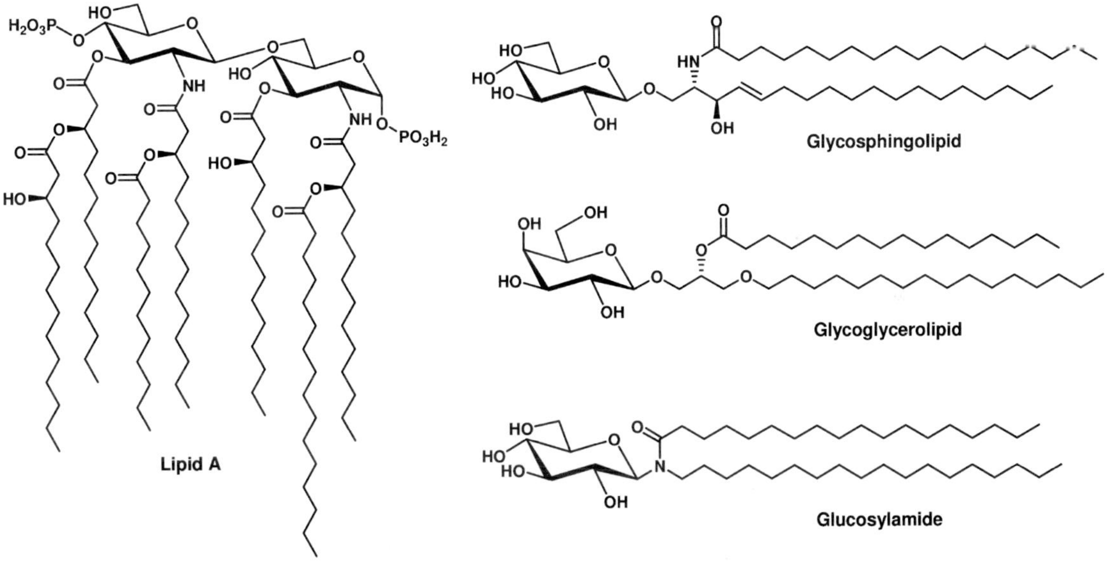

Lipid A

Glycosphingolipid

Glycoglycerolipid

Glucosylamide

Scheme 4.1 Structures of natural and synthetic glycolipids.

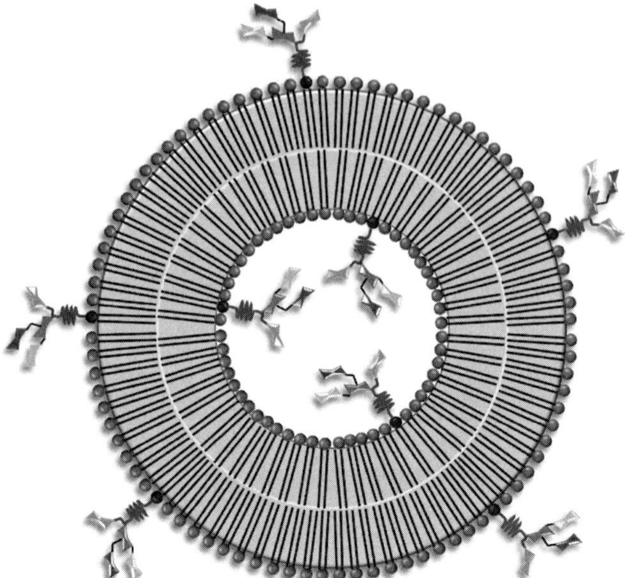

Figure 4.4 Schematic representation of glycoliposomes wherein the carbohydrate parts are randomly distributed over the lipid bilayer surface.

4.5 Glycopolymers

Another class of synthetic multivalent carbohydrates is represented by glycopolymers that can be defined as artificial macromolecules featuring pendant carbohydrate moieties. It consists of a relatively novel and rapidly expanding family of neoglycoconjugates, receiving increasing interest due to their numerous applications [6–10]. By virtue of their many carbohydrate moieties, they constitute potent carbohydrate scaffolds insuring multivalency, added to several practical and financial advantages over other forms of neoglycoconjugates. Indeed, they can be constructed with an almost infinite number of molecular weights, and progress in the field has allowed control over the glycosylation and functional density, and position on the polymeric backbones. Moreover, the use of comonomers (such as acrylamide, methacrylamide, and so on) has led to copolymer production in an inexpensive way. In addition, they possess homogeneous glycan structures and generally constitute better tools for carbohydrate–protein interaction studies. Finally, many polymer carriers have also been shown to be nontoxic, poorly or non-immunogenic and have increased stability towards external stimuli such as pH variations.

A number of elegant methods exist for their preparation that may vary significantly depending on the targeted applications. As mentioned before, recent progresses allowed better control over size, shape, valency and functional group incorporation. Thus, simple to very complex polymeric oligosaccharide sequences can be synthesized. By analogy with other neoglycoconjugate constructions, chemoenzymatic glycosylations on preformed monosaccharide glycopolymers have been envisaged to furnish multiantennary glycopolymers. Of course, glycopolymer synthetic strategies like homo/heteropolymerization or polymer postglycosylation need carbohydrate derivatives containing tailored chemical functionalities to be either 'polymerizable' or grafted (Figure 4.5).

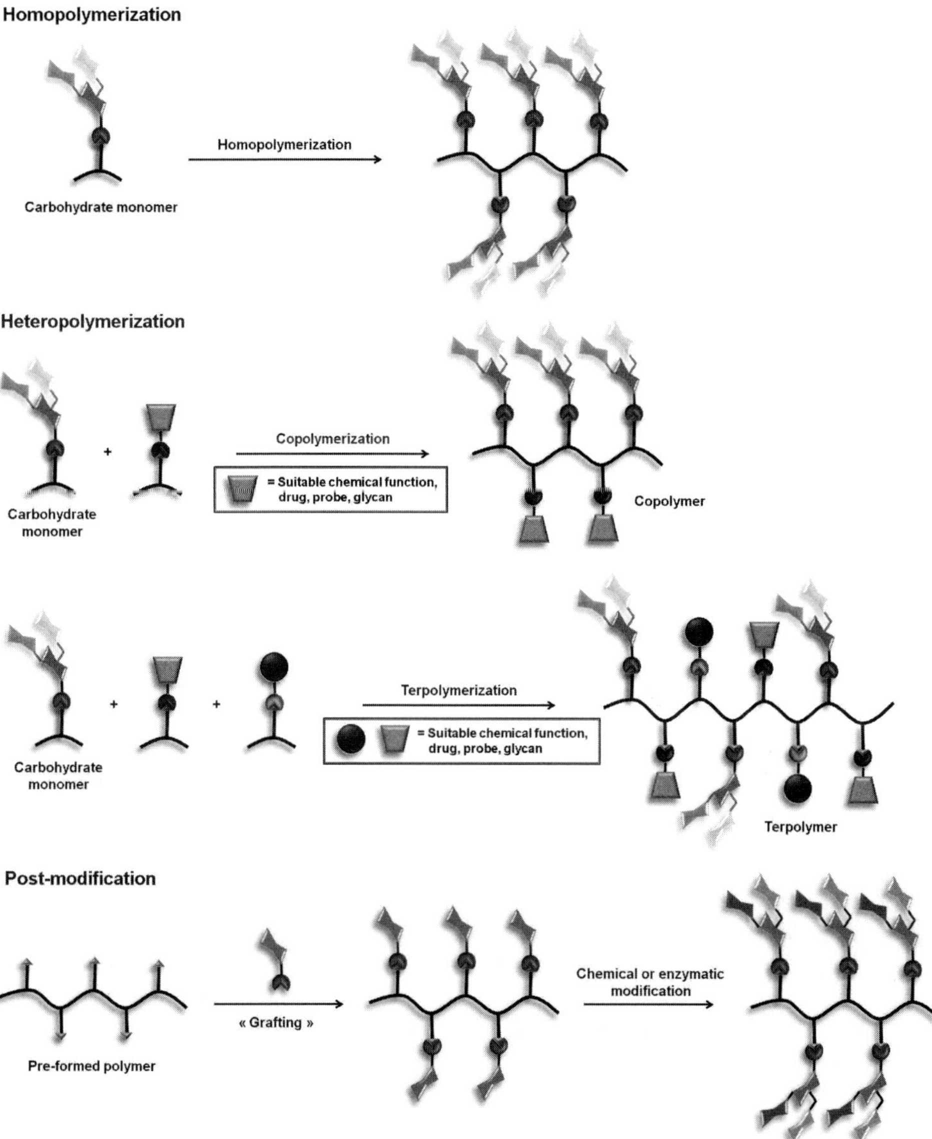

Figure 4.5 Schematic representations of glycopolymer syntheses by homo/heteropolymerization from carbohydrate monomers or by post-modification.

Homopolymers are the result of self-condensation of monomeric carbohydrate derivatives in the absence of any other added molecules. The polymerizable functionality can be placed at the anomeric center (for further details on strategies, please see Chapter 3) or at any other position on the carbohydrate ring. They can be prepared from fully protected carbohydrate monomers or undeprotected derivatives and polymerization processes are initiated or triggered by suitable catalysts or by heating, and their isolation remains easy by precipitation or dialysis. This strategy has allowed the successful synthesis of a variety of glycopolymers with polyvinyl, polystyrene or poly(iminomethylene) backbones with potential biological applications as biocompatible glycopolymers, drug delivery carriers and immunodiagnostics [8].

Heteropolymers constitute the most abundant and most useful form of glycopolymers, since they are built with added noncarbohydrate comonomers which confer special physical and biophysical properties. Copolymers can be synthesized by direct copolymerization of two different monomers or, alternatively, using the grafting strategy in a postpolymerization modification approach. The advantages result from the possibility to control the incorporation of the desired ratios of the two monomers and thus the saccharide density. In order to increase the number and complexity of the functionalities on the polymer backbone to better fit some of the desired properties, custom-designed hybrid glycopolymers have been synthesized by copolymerization of three distinct components (terpolymerization) [9–11]. Typically, reaction mixtures are composed of carbohydrate haptens, probes (or effectors molecules), drugs, solubility enhancers containing suitable polymerizable function and monomers (acrylamide, methacrylamide, and so on) that can be incorporated as polymer backbones.

Although a wide variety of methods are available for the construction of multivalent structures based on proteins, lipids and polymers, serious limitations arise with these neoglycoconjugates in relation to the heterogeneity of their structures or molecular weight. For this reason, strategies and procedures to obtain structurally well-defined and smaller multivalent glycoconjugate systems have become an important area of research activity in recent years. This has led to the newly designed family of glycodendrimers that can better mimic multiantennary glycans.

4.6 Glycodendrimers

Dendrimers are synthetic highly branched monodisperse polyfunctional macromolecules that are constituted by repetitive units that are chemically bound to each other by an arborescent process around a multifunctional central core (please see Figure 4.1 for natural examples). Thus, as opposed to traditional polymers, which often have poorly defined molecular structures that clearly represent an important disadvantage for medical application in terms of reproducibility, dendrimers are structurally well defined and can be synthesized by a fully controlled iterative approach (please see Info Box for details on historic development). Although dif-

Info Box

The concept of repetitive and controlled synthetic growth with branching was first introduced by Fritz Vögtle in 1978 to achieve the construction of rather low-molecular-weight 'cascade' polyamines. The introductory sentence of this article ['For the construction of large molecular cavities and pseudocavities that are capable of binding ionic guests and molecules (as complex or inclusion compounds) in a host–guest interaction, synthetic pathways allowing a frequent repetition of similar steps would be advantageous.'] opened the field of cascade and dendritic chemistry as an entirely novel field. [E. Buhleier *et al.* 'Cascade' – and 'nonskid-chain-like' syntheses of molecular cavity topologies. *Synthesis* 1978; 155–158]. However, it was not until 1985 that George Newkome and Donald A. Tomalia independently published two new iterative synthetic protocols for the preparation of large tree-like macromolecules named 'arborols' and 'dendrimers', respectively. Concerning glycodendrimers, they first appeared in 1993 [R. Roy *et al.* Solid phase synthesis of dendritic sialoside inhibitors of influenza A virus haemagglutinin. *J Chem Soc Chem Commun* 1993; 1869–1872].

ferences exist in terms of rigidity and compaction, dendrimers are often compared to 'artificial proteins' with their globular structures, mostly with a high density of peripheral functionalities and a small molecular 'volume' [12, 13].

Tomalia *et al.* first introduced the term 'dendrimer' that arises from the Greek '*dendron*' = 'tree' or 'branch' and '*meros*' = 'part', developing at the same time efficient original methodology that still constitutes the preferred commercial route to the trademarked Starburst dendrimer family with molecular weights ranging from several hundred to over 1 million Daltons (that is generations 1–13) [13].

Until the mid 1990s, the synthetic challenge of such aesthetic structures stimulated numerous research groups that had intensively investigated several synthetic approaches. Those efforts gave rise to original dendritic architectures emerging from two main chemical strategies that were used to construct perfectly branched dendrimers – the divergent and the convergent approaches. Over time, an accelerated version of the convergent strategy has been developed in order to increase its throughput and efficacy by using clever adaptations of cores or dendrons. Undoubtedly, biology and nano-medicine, more particularly biomedical and therapeutic applications, represent the domains that have generated the highest interest in these architectures.

The use of dendrimers in biological systems and systematic studies of the most common dendritic scaffolds to determine their biocompatibility, such as *in vitro* and *in vivo* cytotoxicity, their biostability or immunogenicity have been reviewed extensively [14]. One typical example concerns the use of dendrimers as 'glycocarriers' for the control of multimeric presentation of biologically relevant carbohydrate moieties that are useful for targeting modified tissue in malignant diseases for diagnostic and therapeutic purposes. In such molecules, called 'glycodendrimers', saccharide portions are conjugated according to the principles of dendritic growth or they are ligated to preexisting highly functionalized and repetitive dendritic scaffolds having varied molecular weights and structures. Since they first appeared in the literature, glycodendrimers and related glycoclusters have stimulated wide interests within the community [15–17].

Similarly to conventional dendritic structures, they can be obtained as dendrons, spherical or globular compounds or 'hybrid dendronized polymers' according to divergent, convergent or accelerated approaches. All these new clusters were synthesized in such a way that their valencies, shapes and carbohydrate contents could be varied at will (Figure 4.6) with controlled integration of dendritic building blocks [16, 17].

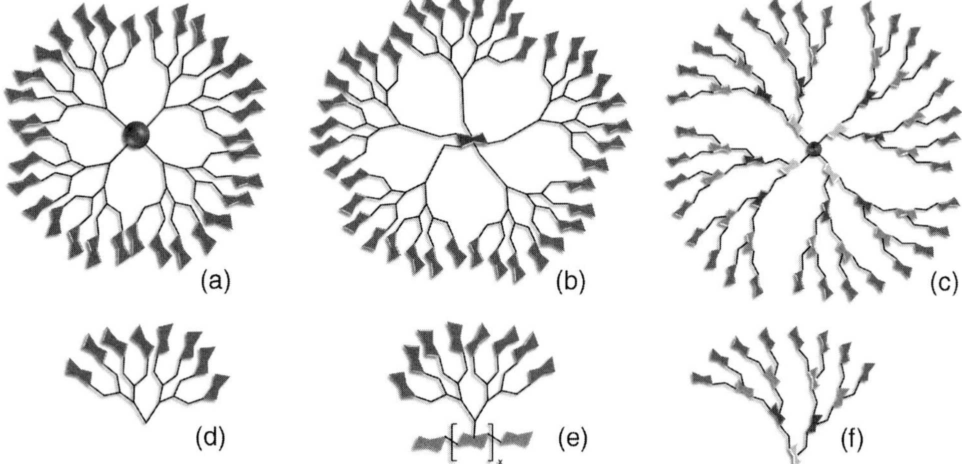

Figure 4.6 Schematic representation of glycodendrimers and glycodendrons varying by their synthetic construction. (a) Dendrimers with non-carbohydrate scaffolds. (b) Dendrimers with carbohydrate scaffold and aliphatic linkers. (c) All-carbohydrate-based glycodendrimers. (d) Glycodendrons with noncarbohydrate scaffolds. (e) Hybrid 'dendronized glycopolymers'. (f) All-carbohydrate-based glycodendrons.

4.7
Glycodendrimer Syntheses

Historically, glycodendrimers have been built in an iterative way out from a central core, layer by layer, requiring activation/addition steps to afford the desired dendritic structure: the focal and multifunctional molecules are systematically expanded outward using various and adapted chemical linkages. Thus, the first-generation glycodendrimers were simply formed by attaching branching units to simple multifunctional cores that constitute the 'seeding molecules' (heart of the dendrimers). The second-generation dendrimers were formed by treating the peripheral functional groups with complementary chemical functions present on the branching building blocks. Chemical transformations of the newly formed surface groups with, for example, sugars resulted in the desired second-generation glycodendrimers. The generation growth can quickly allow exponential multiplication of active terminal functions and the process is repeated until the required degree of branching (multivalency) is obtained.

Several dendrimers having various surface functionalities and building blocks are commercially available: poly(amidoamine) (PAMAM) (Starburst; Dendritic Nanotechnologies), poly(propylene)imine (Astramol; DSM Fine Chemicals), polyglycerols and Boltorn dendrimers are most commonly used as multibranched dendritic core or glycodendron precursors, most of them being known to be nontoxic and nonimmunogenic. One of the most striking examples is represented by the largest glycodendrimer built so far containing 256 mannoside residues that was prepared on generation G6-PAMAM dendrimer using thiourea linkages.

The convergent strategy was first reported in 1990 by Fréchet [13], using the symmetrical nature of these structures to its advantage, in order to overcome some of the synthetic and purification problems associated with the divergent methodology. It involves preliminary synthesis of peripheral branched dendritic arms named 'dendrons' or 'glycodendrons' from the 'outside-in' (Figure 4.7). The advantages of the convergent strategy lie with the reduced number of reactions carried out at each step. Moreover, purification of the desired dendrimer becomes easier than that in the divergent case, since the final products are structurally different from the corresponding smaller precursors.

Glycodendrimers may constitute an arsenal of novel principles for the treatment of infectious agents by the inhibition of attachment of the infecting

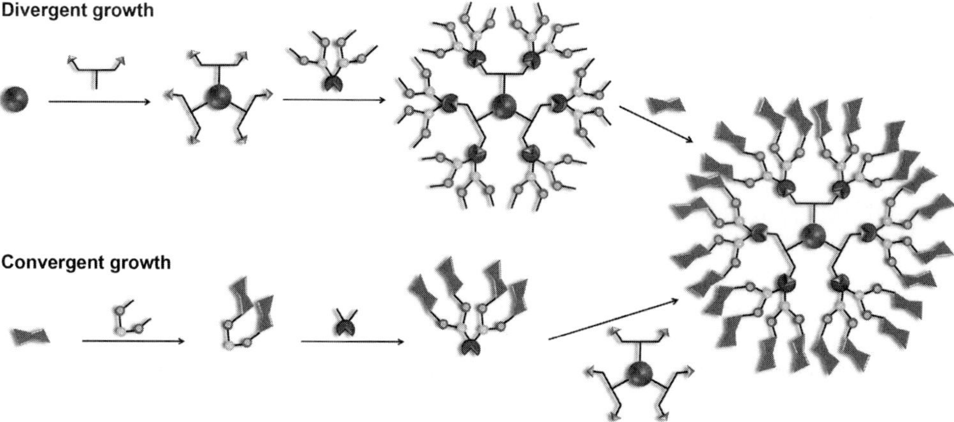

Figure 4.7 Schematic representation of the two major strategies for glycodendrimer growth.

pathogens to the host's cell surfaces (Figure 4.3). The attachment is often a prerequisite for the later stages of infection, colonization and invasion of specific human tissues. Due to their highly structured nature, their monodispersity and their enhanced affinities, glycodendrimers are therefore suitably designed for inhibition of bacterial or viral adhesions to host tissues. Among the most interesting cases, one can mention the use of mannosylated dendrimers as potential ligands for fimbriated *E. coli*, the major causative agent for urinary tract infections, and that it is responsible for serious nosocomial infections as it is the most common cause of Gram-negative cystitis and pyelonephritis. This uropathogenic bacterium possesses proteins at the tip of its fimbriae (FimH) that recognize and bind to mannosides of host human epithelial tissues as the premise for bacterial infections. Many mannosylated dendritic structures with different scaffolds and valencies having strong avidity for *E. coli* FimH have been proposed [15]. Early work in this field has shown that mannosylated dendrons built on L-lysine scaffolds (Figure 4.8) were rather potent inhibitors of attachment of *E. coli* to human erythrocytes [4]. In 2007, the best ligands known to date were synthesized and presented sub-nanomolar affinities [18]. These tetrameric mannosides were 1000 times more potent than mannose for their capacity to inhibit the binding of *E. coli* to erythrocytes *in vitro* (for further examples of design of antibacterial compounds, please see Chapter 17.3).

In addition, a wide range of immunodominant T-antigen-containing glycodendrimers have been synthesized, using nonimmunogenic scaffolds such as PAMAM (Scheme 4.2) [19]. T-Antigen disaccharide (β-D-Gal-(1–3)-α-D-GalNAc) is characteristic of certain cancer carcinomas, usually cryptic on healthy tissues and greatly increased on breast and colorectal cancer cells as a result of aberrant glycosylation of the *O*-linked sugars of mucins. The idea was to use such architectures described in this chapter [neoglycoprotein (vaccine), glycopolymer (diagnostic) and glycodendrimer (therapeutic)] for interaction with carcinoma-related T-antigen-binding receptors, among them galectins (please see Chapters 19 and 25) or to use them for generation of T-specific antibodies for diagnostic and therapeutic purposes. Results indicated a real dendritic effect going from G1 to G2 PAMAM dendrimer (tetramer and 8-mer respectively); on a per-saccharide basis, the inhibitory efficiencies of the dendritic structures were all similar and approximately 100 times better than the monomer. Also, calix[*n*]arenes have proven effective as scaffold [20].

In conclusion, numerous biological studies have shown that glycodendrimers represent potent synthetic mimics of natural glycoconjugates and will interact efficiently with natural carbohydrate receptors, in many cases to an extent that allows competition with natural binding substance as shown in this section.

4.8 Conclusions

Advances encountered in the field of macromolecular chemistry in terms of isolation, purification and structural analyses have allowed the efficient syntheses of

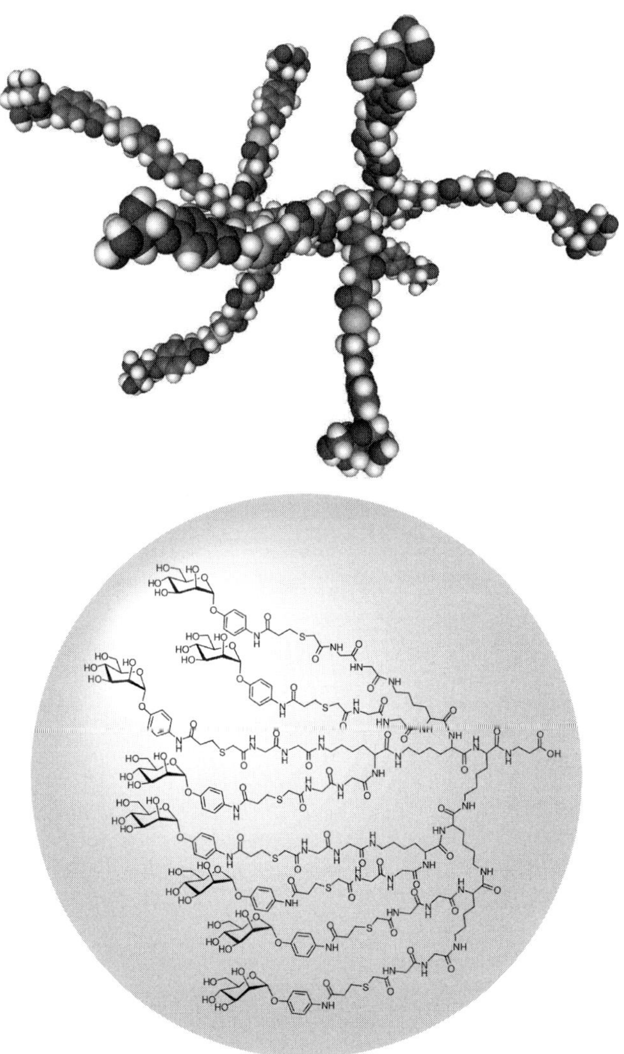

Figure 4.8 Arylated glycondendron bearing eight mannoside residues using a polylysine scaffold and showing an IC_{50} of 14 μM against the FimH of *E. coli* K12 [4].

several neoglycoconjugates in which carbohydrate residues are chemically or enzymatically attached to unnatural carriers. Organic chemistry is therefore at the forefront of glycobiology by allowing access to novel architectures that can provide a better and deeper understanding of biological processes involving carbohydrates. This relatively new field is open-ended since the chemists' imagination, coupled with appropriate biological guidance, will offer new targets in their design and applications.

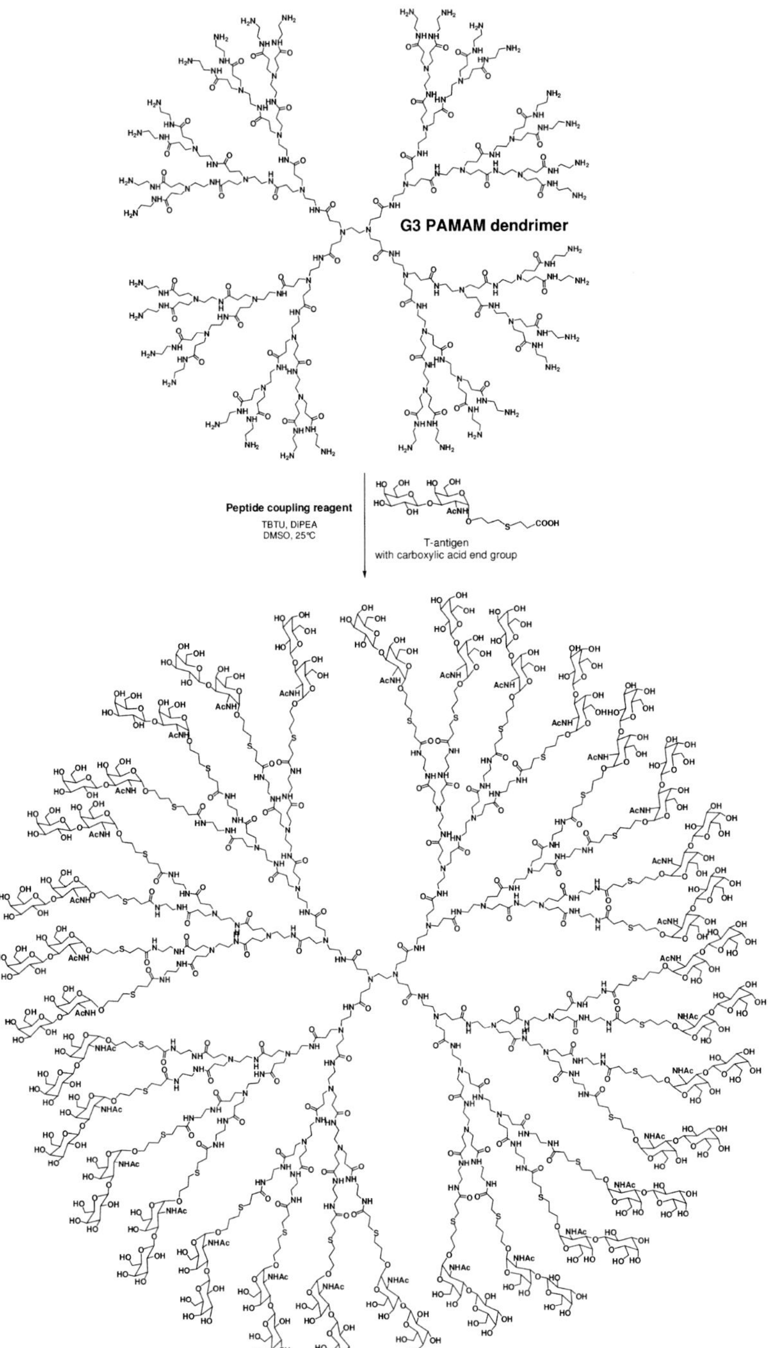

Scheme 4.2 Glycodendrimer bearing 32 disaccharides representing the cancer marker T-antigen built on a commercially available PAMAM using the divergent growth and final peptide coupling chemistry [19].

Summary Box

Multivalent carbohydrate–protein interactions are responsible, at the molecular level, for several crucial biological events, and mediate many important physiological and pathophysiological processes. In spite of the weaknesses of these interactions in terms of affinity and selectivity on a per-saccharide basis, these attractive forces are dramatically and naturally reinforced by a phenomenon known as the 'glycocluster or dendritic effect'. In order to study, characterize, understand and manipulate these critical multivalent interactions, multivalent neoglycoconjugates based on several scaffolds such as peptides or proteins, lipids, polymers or dendrimers have been synthesized using a large arsenal of simple but efficient chemical methods.

References

1 Mammen M *et al.* Polyvalent interactions in biological systems: implications for design and use of multivalent ligands and inhibitors. *Angew Chem Int Ed Engl* 1998; 37:2754–94.

2 Gabius HJ. Glycans: bioactive signals decoded by lectins. *Biochem Soc Trans* 2008;36:1491–6.

3 Lee YC, Lee RT. Carbohydrate–protein interactions: basis of glycobiology. *Acc Chem Res* 1995;28:321–7.

4 Nagahori N *et al.* Inhibition of adhesion of type 1 fimbriated *Escherichia coli* to highly mannosylated ligands. *ChemBioChem* 2002; 3:836–44.

5 Roy R. The chemistry of neoglycoconjugates. In: *Carbohydrate Chemistry* (Ed.: Boons GJ), pp. 243–321. Blackie Academic, London, 1998.

6 Bovin NV, Gabius H-J. Polymer immobilized carbohydrate ligands versatile chemical tools for biochemistry and medical science. *Chem Soc Rev* 1995;24:413–21.

7 Spain SG *et al.* Recent advances in the synthesis of well-defined glycopolymers. *J Polym Sci, Part A: Polym Chem* 2007;45: 2059–72.

8 Kobayashi K *et al.* Carbohydrate-containing polystyrenes. In: *Neoglycoconjugates: Preparation and Applications* (Eds.: Lee YC, Lee RT), pp. 261–84. Academic Press, San Diego, CA, 1994.

9 Roy R. Blue-prints, syntheses and applications of glycopolymers. *Trends Glycosci Glycotechnol* 1996;8:79–99.

10 Roy R. Design and synthesis of glycoconjugates. In: *Modern Methods in Carbohydrate Synthesis* (Eds.: Khan SH, O'Neil RA), pp. 378–402. Harwood Academic, Amsterdam, 1996.

11 Ouchi T, Ohya Y. Drug delivery systems using carbohydrate recognition. In: *Neoglycoconjugates: Preparation and Applications* (Eds.: Lee YC, Lee RT), pp. 465–98. Academic Press, San Diego, CA, 1994.

12 Newkome GR *et al. Dendrimers and Dendrons: Concepts, Synthesis, Applications.* Wiley-VCH, Weinheim, 2001.

13 Fréchet JMJ, Tomalia D. *Dendrimers and Other Dendritic Polymers.* John Wiley & Sons, New York, 2001.

14 Duncan R, Izzo L. Dendrimer biocompatibility and toxicity. *Adv Drug Deliv Rev* 2005; 57:2215–37.

15 Touaibia M, Roy R. Application of multivalent mannosylated dendrimers in glycobiology. In: *Comprehensive Glycoscience* (Ed.: Kamerling JP), pp. 821–70. Elsevier, Amsterdam, 2007.

16 Roy R. A decade of glycodendrimer chemistry. *Trends Glycosci Glycotechnol* 2003;15:291–310.

17 Turnbull WB, Stoddart JF. Design and synthesis of glycodendrimers. *Rev Mol Biotechnol* 2002;90:231–55.

18 Touaibia M *et al.* Mannosylated G(0) dendrimers with nanomolar affinities to *Escherichia coli* FimH. *ChemMedChem* 2007;2:1190–201.

19 Roy R, Baek MG. Glycodendrimers: novel glycotope isosteres unmasking sugar coding. Case study with T-antigen markers from breast cancer MUC1 glycoprotein. *Rev Mol Biotechnol* 2002;90:291–309.

20 André *et al.* Calix[*n*]arene-based glycoclusters: bioactivity of thiourea-linked galactose/lactose moieties as inhibitors of binding of medically relevant lectins to a glycoprotein and cell-surface glycoconjugates and selectivity among human adhesion/growth-regulatory galectins. *ChemBioChem* 2008;9:1649–61.

5
Analytical Aspects: Analysis of Protein-Bound Glycans

Hiroaki Nakagawa

The preceding chapters have introduced the biochemical basis and coding capacity of the sugar code and given us an idea of the three-dimensional structure and the chemical strategies for preparing glycans. Now we turn to analytical methods to delineate glycan structures. As described in Chapter 1, the structural variability poses a considerable challenge. For the analytical chemist the enormous coding capacity means that not only are there a lot of possible structures (please see Table 1.1 for a telling example), but also that most samples include several kinds of structures. In this chapter, analytical methods are described for detailed structural analysis and glycomics (Figure 5.1 and Table 5.1). The processing results in the complete structural description and glycomic profiling of glycan mixtures.

5.1
Detection of Glycans on Glycoproteins

The presence of sugar chains attached to proteins can be detected using the periodic acid-Schiff reaction (PAS) or lectin staining following sodium dodecylsulfate–polyacrylamide gel electrophoresis (SDS–PAGE). PAS staining was developed by R. A. Kapitany and E. J. Zebrowski in 1973. It is based on the reaction between an aldehyde, formed by the periodic acid, and Schiff's reagent. The original Schiff's reagent (magenta) has been improved and several staining kits are now commercially available. PAS is a general stain for carbohydrates and so is widely used in histochemical studies. Lectin staining can distinguish distinct glycan structures on glycoproteins. A list of lectin-reactive epitopes is given in Table 18.1. The presence of an *N*-glycan can also be detected by shifts in molecular weight after treatment with *N*-glycosidases, which remove the *N*-glycan from proteins. The molecular weight changes can be detected using SDS–PAGE, high-performance liquid chromatography (HPLC) or mass spectrometry (MS).

Amino sugars, *N*-acetylglucosamine (GlcNAc) and *N*-acetylgalactosamine (GalNAc) can be demonstrated by amino acid composition analysis. The presence

The Sugar Code. Edited by Hans-Joachim Gabius

ISBN: 978-3-527-32089-9

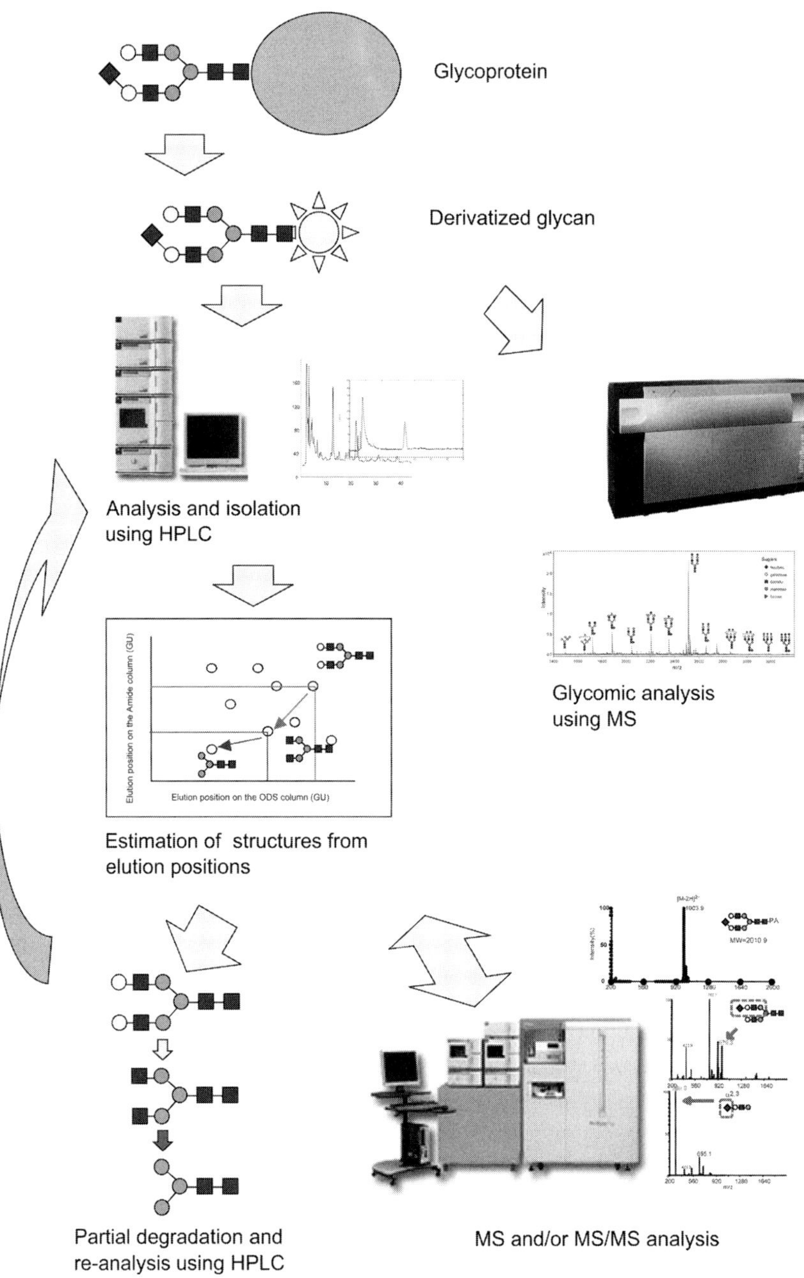

Figure 5.1 Flow chart of *N*-glycan structural analysis. Blue arrows show the strategy for detailed structural analysis and yellow arrows indicate the pathway for glycomic analysis. *N*-Glycans are released from the glycoprotein and derivatized to enhance the sensitivity of chromatography or MS. Derivatized glycans are isolated and their elution positions analyzed on two or three types of column using HPLC. Glycan structures are determined through comparison with the database of known glycans. Determined structures are confirmed by a combination of partial degradations and/or MS analysis. MS analysis provides information on sugar components and allows precise differentiation between glycan structures. The fragmentation pattern obtained using MS/MS or MS^3 analysis often provides detailed structure analysis. For a glycomic analysis (yellow arrows) a mixture of derivatized glycans is often processed using MALDI-TOF/MS. This strategy is suitable for high-throughput analysis and has been applied to the search for biomarkers. Photographs of the equipment were kindly provided by Bruker Daltonics and Hitachi High-Technologies.

Table 5.1 Equipment and technologies in each step of the experiments described in this chapter.

Step	Equipment and techniques
Sample preparation	
detect glycan	PAS, lectin and other glycostaining methods with SDS–PAGE
	N-glycosidase treatment
	amino acid analysis
release glycan	*N*-glycosidase (*N*-glycan)
	hydrazinolysis (*N*-glycan, *O*-glycan)
	alkali degradation (*O*-glycan)
purification	column chromatography or cartridge extraction (gel filtration, ion exchange, reversed phase, cellulose, active carbon, lectins)
	sugar-specific binding materials (hydrazide or oxime derivatives)
derivatization	amine, hydrazine and oxime derivatives
	methylation
Structural analysis	
detailed structure	elution positions on HPLC
	sensitivity for glycosidases
	fragmentation on MS
glycomics	MS
other methods	CE, PAGE, HPLC
	NMR, methylation analysis, lectin specificity
glycopeptides	MS

of an *O*-linked glycan is likely if GalNAc is detected, since an *N*-glycan rarely contains GalNAc (for an example, please see Figure 1.7b). In contrast, the detection of mannose by monosaccharide composition analysis indicates the presence of an *N*-glycan, with the exception of *O*-mannosylation (see Chapters 7.3.3 and 22.2.2). To make the strategy of this procedure clear, examples of detailed structural analysis are instructive (see below).

5.2
Release of Glycans from Glycoproteins

N-Glycosidase is eminently useful for examining oligosaccharides (see Info Box) and it is also commercially available. *N*-Glycosidase F from *Flavobacterium* spp. does not cleave *N*-glycans that possess core α1,3-fucose residues, known to exist in plant and insect glycoproteins (please see Figures 8.3a and 8.4a) [1]. *N*-Glycosidase A from almonds, however, does cleave these structures and is used commonly. *N*-Glycosidase F is smaller than A and is sometimes used for natural proteins. For reproducibility, it is strongly recommended with both these enzymes that the substrates are denatured before analysis (please see Info Box for further information).

To remove *N*-glycans, the established chemical procedure uses hydrazine [2]. This method removes the risk of the glycan remaining uncleaved due to the steric hindrance from the protein region or the loss of enzyme activity. However, this anhydrous reaction requires careful handling and equipment. In this chemical method, *N*-glycolylneuraminic acid can also be converted to *N*-acetylneuraminic acid.

As yet, there is no single enzyme for the removal of all types of *O*-glycans. Alkali degradation and hydrazine degradation are used to release *O*-glycans [3, 4]. However, these procedures can easily result in the formation of fragmented products. Endoglycoceramidase, which releases glycans from glycosphingolipids (introduced by M. Ito and T. Yamagata in 1986), can be used to analyze the structure of glycans in sphingolipids.

Info Box

N-Glycosidase, formally peptide-N^4-(*N*-acetyl-β-glucosaminyl) asparagine amidase (EC 3.5.1.52), is named as glycopeptidase, glycoamidase, glycanase and peptide-*N*-glycosidase. This enzyme was originally isolated from almonds by N. Takahashi in 1977. In 1984, T. H. Plummer *et al.* isolated another glycopeptidase from *Flavobacterium* spp. These enzymes cleave the N–C bond in the Asn branch, which binds the *N*-glycan, liberating a glycan that includes the -NH_2, and changing Asn to Asp (Figure 5.2). There are differences in the optimal pH of the two enzymes. The enzyme from almonds reacts under weak acidic conditions, so the amine of the reducing end of the glycan is rapidly removed. On the other hand, the enzyme from *Flavobacterium* reacts at pH 7–8 and the amine on the reducing end is not removed. The complexity in the reaction mechanisms and competition between suppliers has led to the use of several names for these enzymes. In addition, 1 U of the commercial enzyme from *Flavobacterium* corresponds to 1 mU of the enzyme from almonds. There are also differences in the active ingredient between manufacturers, so it is important to take particular care when using this enzyme.

5.3 Glycan Purification

Glycans can be purified, as for other biomaterials, on gel-filtration, ion-exchange or reversed-phase columns. In addition, cellulose and activated carbons, in particular, are known to interact with carbohydrates [5]. A sophisticated method for glycan isolation involves the use of reactive materials including oxime or hydrazide functional groups to bind to aldehyde on the reducing end of the carbohydrates (Figure 5.3) [6]. Sialic acids, which often form the non-reducing ends of glycans, are

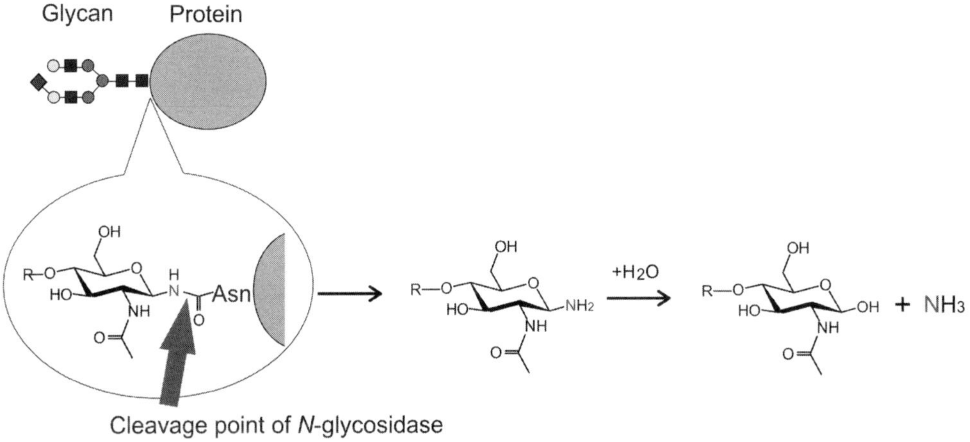

Figure 5.2 The *N*-glycosidase reaction mechanism. *N*-Glycosidase cleaves the bond between the N and C of the Asn to which the *N*-glycan is bound. Thus, the released *N*-glycan has an NH_2 at the reducing end and the Asn changes to Asp. The NH_2 can then be non-enzymatically removed as ammonia.

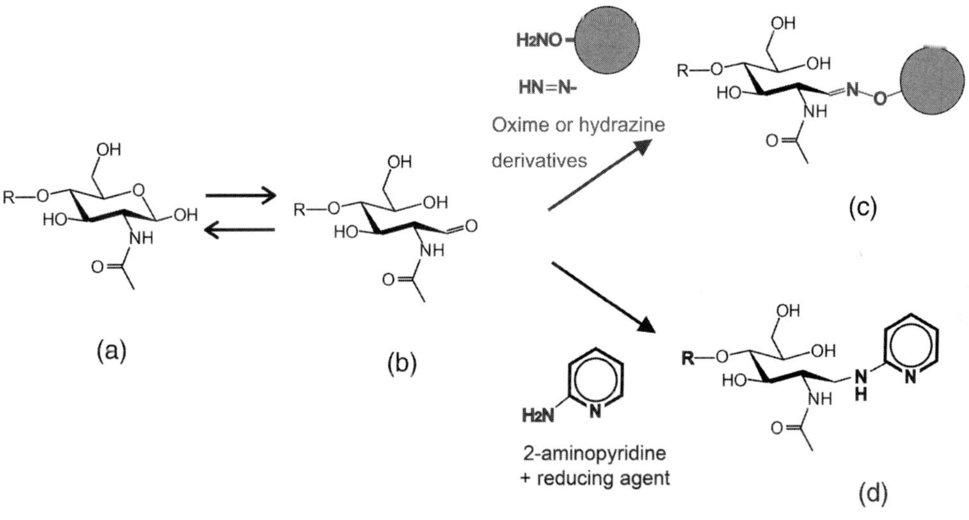

Figure 5.3 Modification of the reducing end of glycans. (a) and (b) The reducing end of an *N*-glycan, which is the binding site for proteins, has a sugar residue that is always *N*-GlcNAc, and this is changed to aldehyde. Few aldehyde structures exist in biomaterials that do not include a saccharide group, and this saccharide at the reducing end provides a good target for labeling and, in particular, for purification. (c) Oxime or hydrazide structures react with aldehyde. Materials possessing an oxime or hydrazide functional group can be used to capture saccharides. (d) The aldehyde also reacts with amine reagents (for example, 2-aminopyridine) and the structure is then protected by reducing agents.

removed under acidic conditions. Weak acid lysis at pH 2.0 and 90 °C for 60 min is used to release the sialic acid. Weaker acidic conditions in solution or on ion-exchange resins can result in the loss of sialic acids. Sialic acid residues are also displaced naturally from the glycan at a low pH by its own acidity in concentrated sialyl-glycan solutions. Purification methods are selected on the basis of the particular character of the glycoprotein and the purpose of the study.

Carbohydrates do not exhibit ultraviolet absorption, which is another major analytical problem. Glycans are detected by refractive index, pulsed amperometric detection and UV detector at 210 nm using HPLC. However, these methods also detect other biomaterials so that they require tight restrictions when used for analysis, and are often not sufficiently sensitive. Various derivatization methods have been developed for the detection of oligosaccharides [7] and most of them are based on the reaction of the reducing ends with aldehyde (Figure 5.3). Pyridylamination, developed by S. Hase in 1978, was the first labeling technique to use reductive amination of the reducing end, and it is still widely used in HPLC and MS. The labeling of glycans not only increases the sensitivity of HPLC through its fluorescence, but also enhances the separation obtained using reversed-phase columns. Improved separation of oligosaccharides including isomers through labeling is of great advantage, as the sensitivity of MS is dependent on the purity of the sample glycans. Derivatization for sialic acid or permethylation is also used, not only to resolve problems of sensitivity, but also to increase the stability of sialic acids during MS analysis [8].

5.4
Detailed Structural Analysis Using HPLC

A widely applied approach to analyze oligosaccharide structure uses the elution positions on two- or three-mode columns. Two-dimensional mapping is a strategy used for neutral *N*-glycans employing pyridylamino (PA) labeling and HPLC on two types of column, octadecylsilica (ODS) and amide [9]. To increase reproducibility, elution position is then converted to glucose units, which are relative elution positions compared to a glucose oligomer. Several PA-glycans are commercially available, and they help to increase reliability of the elution positions and subsequent enzymatic treatment (for further application of PA-glycans, please see Chapter 14.3). To analyze sialyl-glycans and clarify the number of sialic acid residues, a third dimension using a diethylaminoethyl column is applied [10]. Isomers are separated on the ODS column and interactions on the amide column are correlated with the size of the glycans. The *N*-glycan structure is indicated by the elution position on the two or three columns when compared with elution positions of about 500 known oligosaccharides (http://www.glycoanalysis.info/ENG/index.html). We can therefore derive the structure of the sample *N*-glycan corresponding to these elution positions using a database. The fluorescence intensity of the PA-*N*-glycan does not depend on the *N*-glycan structure and quantitative analysis is done using UV absorption. The amount and molar ratio of the *N*-glycan can be subsequently calculated.

To confirm the suggested structure, partial degradation is performed. The elution positions of the sample oligosaccharide treated serially with α- or β-galactosidase, β-*N*-acetylhexosaminidase, α-fucosidase, α-mannosidase, neuraminidase and weak acid lysis are then analyzed using HPLC. Using the database, we can predict the structure of the treated products from the estimated structure. After treatment we confirm that the elution positions of the treated sample coincides with those of the predicted product. After several steps of treatment, we can be certain of the oligosaccharide structure. Of help, several enzymes have substrate specificity for distinct linkages and so that we can distinguish between isomers (for example α2,3/6-sialylation and β1,3/4-galactosylation). An example of the analysis of isomers is shown in Figure 5.4. Four monosialyl-biantennary oligosaccharide isomers, 1A1-200.4, 1A2-200.4, 1A3-200.4 and 1A4-200.4, are separated on ODS and amide columns. The elution positions of 1A3-200.4 and 1A4-200.4 are close to each other; these could be further differentiated if necessary. They each

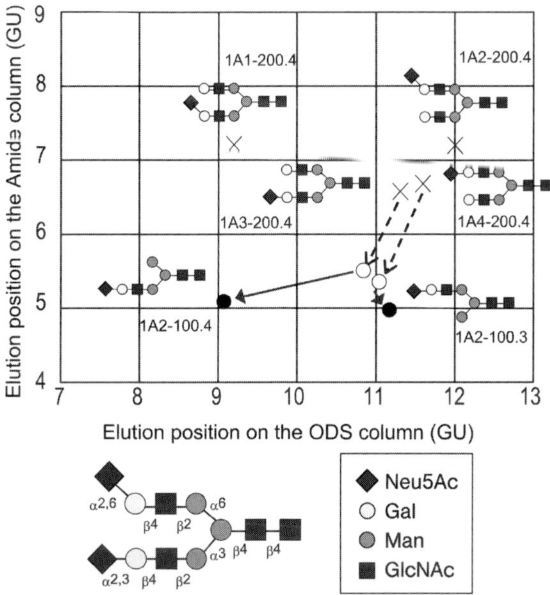

Figure 5.4 Estimation of glycan structure from its elution position. The *x*-axis represents the elution position on the ODS column, and the *y*-axis, the position on the amide column. Broken arrows represent β-galactosidase treatment, and solid arrows, treatment with β-*N*-acetylhexosaminidase, which releases β-GlcNAc and β-GalNAc. Open circles represent the products of β-galactosidase treatment and the closed circles, the products after further treatment with β-*N*-acetylhexosaminidase. The four isomers of monosialyl-biantennary glycans have different elution positions and estimated structures. If there are two or more candidates with similar elution positions, partial degradation is effective. As an example, the elution positions of 1A3-200.4 and 1A4-200.4 are very near to each other, but the elution positions of the products obtained after treatment with β-galactosidase and β-*N*-acetylhexosaminidase, 1A2-100.4 and 1A2-100.3, can be easily distinguished.

have one sialic acid residue in their non-reducing end, but on a different branch; 1A3-200.4 is sialylated on the α3Man branch, whereas 1A4-200.4 is so on the α6Man branch. Removal of the sialic acid produces the same product from both isomers, and we can confirm that both isomers have a sialic acid on the same component, but cannot distinguish the branching pattern. The non-sialylated branches can be digested by β-galactosidase. However, their elution positions still remain very close to each other. Further treatment next with β-*N*-acetylhexosaminidase removes the GlcNAc residue and the resultant products have clearly distinguishable elution positions. In this way, the estimated structure can be ascertained through a combination of several degradation pathways.

5.5 Detailed Structural Analysis Using MS

Analysis using partial treatment is time consuming, and planning partial degradation requires detailed knowledge of enzyme substrate specificities and sample specificities. MS analysis has also recently been applied for detailed structure analysis [11–13]. As an example, negative-mode MS analysis with electrospray ionization (ESI) of monosialyl-biantennary oligosaccharides shows a peak at $m/z = 1003.9$ (Figure 5.5a) [14]. This corresponds to the ionized PA-glycan after removal of the two protons and charged −2, and is calculated as: $m/z = 2010.9$ (molecular weight of the PA-glycan) − 2 (molecular weight of the two protons removed)/2 (number of charge). The four isomers described above show similar spectra and cannot be distinguished. Tandem MS (MS/MS) is MS analysis in which isolated peaks in the initial mass spectrum, in this case the ion at $m/z = 1003.9$, are broken by collision with helium gas. Each of the glycans shows different fragmentation patterns (Figure 5.5b). As an example, the α3Man branch cleaved at the Manα1,3Man linkage, whereas the α6 branch fragment included Manα1,6Man. We can now, therefore, determine which branch is sialylated. The linkage position of sialic acid is indicated by MS^3 analysis, because the α2,3-linkage is cleaved more easily than the α2,6-linkage. The ion of sialic acid alone is clearly seen at $m/z = 290$ in the spectrum of the fragment from the glycan that includes α2,3-sialylation (Figure 5.5c).

5.6 Glycomic Analysis Using MS

MS is highly sensitive and rapid, and can be applied to glycomic analysis. This approach to glycomics is used to determine biomarkers of disease. Sialic acid residues are often removed during MS analysis and the sensitivity of MS can vary greatly depending on the charge of the sialic acid. Therefore, methylation (or another method of capping the COOH on sialic acid) or permethylation (methylation of all OH groups on an oligosaccharide) is used in the MS analysis of sialyl

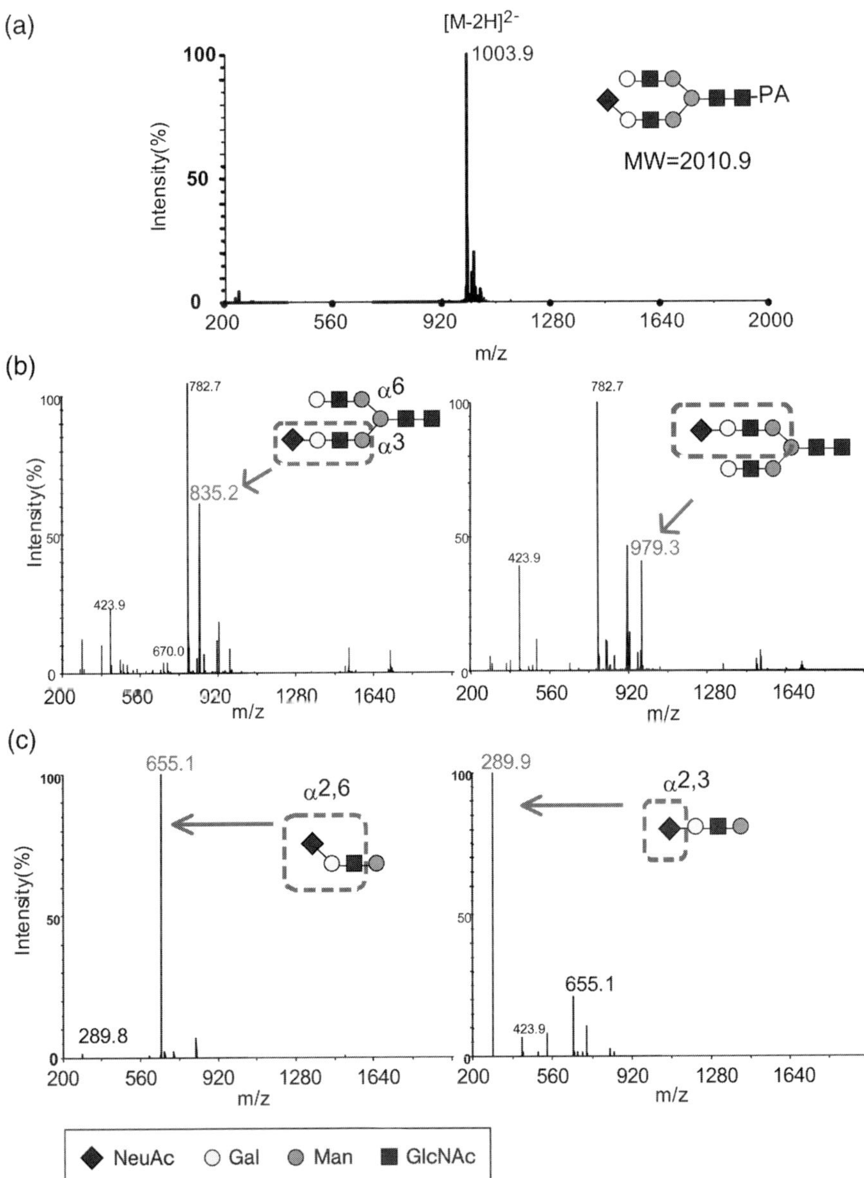

Figure 5.5 MS analysis of glycans. (a) Mass spectrum of pyridylaminated monosialyl-biantennary glycan using ESI-MS in negative mode, which is effective for the analysis of sialyl-glycans. Mass number provides information on sugar composition, but the isomers present the same spectra. (b) MS/MS analysis of isomers of monosialyl-biantennary glycans. The peak of at 1003.9 in (a) was degraded and analyzed again. The α3Man branch was cut at the Manα1,3Man linkage and formed a NANA-Gal-GlcNAc-Man fragment, detected at $m/z = 835.2$. The α6Man branch was cleaved at the βMan residue and formed a NANA-Gal-GlcNAc-Man-Man fragment, detected at $m/z = 979.3$. In this way we can determine which branch was NANA linked. (c) MS^3 analysis of the $m/z = 835.2$ fragment in (b). NANAα2,3Gal is cleaved more easily than is NANAα2,6Gal, thus we can determine the type of NANA linkage. Data provided by Dr K. Deguchi.

glycans. Figure 5.6 shows an example of glycomic analysis [15]. *N*-Glycans are enzymatically released from human serum and purified using hydrazine resin. The glycans are then labeled using an oxime derivative and analyzed by matrix-assisted laser desorption/ionization–time of flight (MALDI-TOF)/MS. This strategy is suitable for high-throughput analysis and several biomarkers have been reported through glycomic analysis (please see also Chapter 22.1 for defects in *N*-glycosylation).

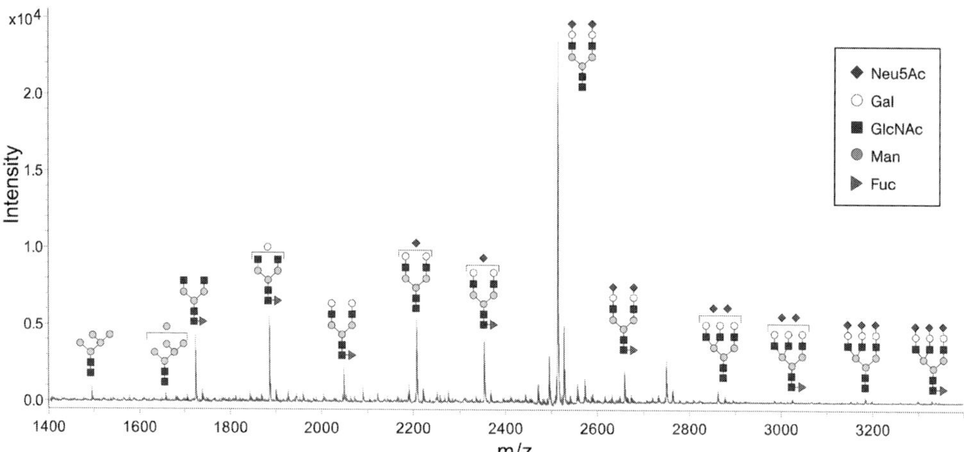

Figure 5.6 Glycomics of human serum. *N*-Glycans were prepared from human serum and analyzed using MALDI-TOF/MS. This strategy is suitable for high-throughput analysis. Sialic acids are methylated to avoid degradation during MS analysis. Data provided by Dr Y. Miura.

5.7 Other Methods of Analysis

To demonstrate the specific characteristics of each analysis method, HPLC and MS are described separately in this chapter. Naturally, valuable information is also obtained by combinations and also additional techniques such as capillary electrochromatography (CE) [16] and electrophoresis [17] with the focus here being given to popular approaches. At present, MS analysis is widely used; however, a number of problems remain (for example, structure-dependent sensitivity and differences in fragmentation according to equipment used). Nevertheless, MS techniques and equipment are steadily improving, and the application of MS to glycan analysis is growing.

Methylation analysis was established by S. Hakomori in 1964 and is still the most basic method of determining linkage positions. Oligosaccharides are hydrolyzed after permethylation and the positions not occupied by a methyl group are assumed to bind to other sugars. Nuclear magnetic resonance (NMR) can determine the detailed structure of whole glycans and remains the method of judgment for confirmation of α or β linkages [18]. These methods require special techniques and a large quantity of the sample glycan, so more convenient methods, such as elution position on HPLC (Figure 22.3), fragmentation pattern on MS, lectin affinity and sensitivity against glycosidases are preferred. However, NMR and methylation analysis offer great potential, and NMR provides insights to the three-dimensional structure of glycans and interaction of proteins (lectins) with glycans in solution, as outlined in Chapter 13.3. This has perspectives for drug design as described in Chapter 28. Lectins, as receptors for distinct glycans, are used as tools, as outlined in detail in Chapters 14.5 and 18.4, with human lectins detecting functional glycans (please see Chapters 19, 25, 27–29). These methods used for analysis as well as diagnostic purposes have been extended to glycopeptides, as shown in the next section.

5.8
Glycopeptide Analysis Using MS

Glycopeptide analysis can reveal the glycosylated amino acid residues and differences in glycosylated pattern between amino acid residues. *N*-Glycans attach to the consensus sequence Asn-X-Ser/Thr (where X is not Pro), whereas *O*-glycan binds to any Ser and Thr residue, as presented in detail in Chapters 6.1 and 7.2. Therefore, glycosylation site analysis is more demanding for *O*-glycans than for *N*-glycans. Analysis of glycopeptides is essential for the resolution of the role of glycans and the identification of biomarkers. MS is the most reliable of the many tools available for the analysis of glycopeptides.

Purification and analysis need to be performed independently. For purification, glycopeptides can be collected using lectins or reversed-phase columns; recently, hydrophilic interaction liquid chromatography (HILIC) columns have been improved and introduced for the same purpose [19, 20]. HILIC columns separate glycopeptides and peptides better than reversed-phase columns, and interact with glycans more generally than epitope-selective lectins (Table 18.1). As the reducing end of glycans is conjugated to peptides, they are no longer available for chemical capture against aldehyde. Therefore, a method by which an aldehyde structure is formed at the nonreducing end of the glycan, allowing capture of the glycopeptide, has been developed [21, 22].

Many forms of the MS analysis of glycopeptides have been reported. One major goal of glycopeptide analysis is to determine the relationships between post-translational modifications (for example competition between phosphorylation and *O*-GlcNAcylation, please see Chapter 7.3.2). Therefore, there is a strong demand for a novel method of analyzing the modification patterns and their sites of conjugation to proteins, as well as their detailed structures in large glycopeptides, preferably in whole glycoproteins. Electron capture dissociation/electron transfer dissociation technologies generally cleave polypeptides evenly without removing post-translational modifications and, therefore, offer a promising approach toward this aim [23].

5.9 Conclusions

Analysis of detailed glycan structure can be accomplished using HPLC alone by comparison with a databank. However, information on glycan structures can be gained directly from MS analysis, although structural work using only MS^n analysis does require purification using HPLC. MS is also suitable for high-throughput glycomic analysis. A large number of analytical methods developed is based on the application of MS technology, with automated annotation [24]. In addition, other powerful methods, using lectins as tools or electrophoresis and NMR, are established. In other words, there is not just one single analytical method covering all aspects. The important task is to select and combine the analytical methods best suited for the purpose of your study, on the basis of the available equipment, sample amount and source. In addition, when using any procedure, it is important that you are properly trained and have undertaken research with appropriate materials as controls, for example with bovine fetuin, bovine ribonuclease B, human IgG or human fibrinogen.

Summary Box

Analytical methods for detailed structural analysis and glycomics are described. In these methods glycans are released from glycoproteins and derivatized to increase the sensitivity of chromatography and MS. Certain purification techniques are used before and after derivatization. Detailed structural analysis is done using HPLC combined with partial degradation and/or MS/MS analysis. These procedures can establish the glycomic profile.

References

1 Altmann F *et al.* Kinetic comparison of peptide: *N*-glycosidase F and A reveals several differences in substrate specificity. *Glycoconj J* 1995;12:84–93.
2 Takasaki S *et al.* Hydrazinolysis of asparagine-linked sugar chains to produce free oligosaccharides. *Methods Enzymol* 1982; 83:263–8.
3 Carlson DM. Structures and immunochemical properties of oligosaccharides isolated from pig submaxillary mucins. *J Biol Chem* 1968;243:616–26.
4 Iwase H *et al.* Analysis of glycoform of *O*-glycan from human myeloma immunoglobulin A1 by gas-phase hydrazinolysis following pyridylamination of oligosaccharides. *Anal Biochem* 1992;206:202–5.
5 Shimizu Y *et al.* Rapid and simple preparation of *N*-linked oligosaccharides by cellulose-column chromatography. *Carbohydr Res* 2001;332:381–8.
6 Niikura K *et al.* Versatile glycoblotting nanoparticles for high-throughput protein glycomics. *Chem Eur J* 2005;11:3825–35.
7 Anumula KR. High-sensitivity and high-resolution methods for glycoprotein analysis. *Anal Biochem* 2000;283:17–26.
8 Viseux N *et al.* Structural analysis of permethylated oligosaccharide by electrospray tandem mass spectrometry. *Anal Chem* 1997;69:3193–8.
9 Tomiya N *et al.* Analysis of *N*-linked oligosaccharides using a two-dimensional mapping technique. *Anal Biochem* 1988;171:73–90.

10 Takahashi N *et al.* Three-dimensional elution mapping of pyridylaminated *N*-linked neutral and sialyl oligosaccharides. *Anal Biochem* 1995;226:139–46.

11 Mechref Y, Novotny MV. Structural investigation of glycoconjugate at high sensitivity. *Chem Rev* 2002;102:321–69.

12 Zaia J. Mass spectrometry of oligosaccharides. *Mass Spectrom Rev* 2004;23:161–227.

13 Harvey DJ. Structural determination of *N*-linked glycans by matrix-assisted laser desorption/ionization and electrospray ionization mass spectrometry. *Proteomics* 2005;5:1774–86.

14 Deguchi K *et al.* Structural assignment of isomeric 2-aminopyridine-derivatized monosialylated biantennary *N*-linked oligosaccharides using negative-ion multistage tandem mass spectral matching. *Rapid Commun Mass Spectrom* 2006;20: 412–8.

15 Miura Y *et al.* BlotGlycoABC™: an integrated glycoblotting technique for rapid and large-scale clinical glycomics. *Mol Cell Proteomics* 2008;7:370–7.

16 Kakehi K *et al.* Analysis of glycoproteins and the oligosaccharide thereof by high-performance capillary electrophoresis–significance in regulatory studies on biopharmaceutical products. *Biomed Chromatogr* 2002;16:103–15.

17 Jackson P. The use of polyacrylamide-gel electrophoresis for high-resolution separation of reducing saccharides labeled with the fluorophore 8-aminonaphthalene-1,3,6-trisulphonic acid. *Biochem J* 1990;270:705–13.

18 Vliegenthart JFG *et al.* High-resolution, ^{1}H-nuclear magnetic resonance spectroscopy as a tool in the structural analysis of carbohydrates related to glycoproteins. *Adv Carbohydr Chem Biochem* 1983;41:209–374.

19 Wada Y *et al.* Hydrophilic affinity isolation and MALDI multiple stage tandem mass spectrometry of glycopeptides for glycoproteomics. *Anal Chem* 2004;76:6560–5.

20 Takegawa Y *et al.* Simple separation of isomeric sialylated *N*-glycopeptides by a zwitterionic type of hydrophilic interaction chromatography. *J Sep Sci* 2006;29:2533–40.

21 Zhang H *et al.* Identification and quantification of *N*-linked glycoproteins using hydrazide chemistry, stable isotope labeling and mass spectrometry. *Nat Biotechnol* 2003;21: 660–6.

22 Kurogochi M *et al.* Reverse glycoblotting allows rapid-enrichment glycoproteomics of biopharmaceuticals and disease-related biomarkers. *Angew Chem Int Ed Engl* 2007;46: 1–7.

23 Hakansson K *et al.* Electron capture dissociation and infrared multiphoton dissociation MS/MS of an *N*-glycosylated tryptic peptide to yield complementary sequence information. *Anal Chem* 2001;73:4530–6.

24 Tissot B *et al.* Glycoproteomics: past, present and future. *FEBS Lett* 2009;583:1728–35.

Part Two
Natural Glycosylation – Glycoproteins

The Sugar Code. Edited by Hans-Joachim Gabius

ISBN: 978-3-527-32089-9

6
N-Glycosylation

Christian Zuber and Jürgen Roth

Previous chapters have described the structural and chemical aspects of glycans. In this chapter, I will be your guide for a journey along the secretory pathway to give you some ideas on how *N*-glycans are synthesized and influence my destiny and function. So let me introduce myself. Some people consider me as a developmentally cell type- and growth state-dependent regulated DNA sequence that gives rise to differently spliced mRNAs. Others view me as a developmentally regulated protein with a versatile set of outfits. You can call me just neural cell adhesion molecule 1 (NCAM1). The ride we will take stands as a model for the route that (glyco)proteins take along the secretory pathway. Although the events occurring during this journey are based on solid experimental evidence, their exact role in the biosynthesis of my *N*-glycans has not been fully elucidated. All human enzymes involved in my maturation process will be named according to the Swiss-Prot database (www.expasy.org; www.uniprot.org) with their entry names (primary protein accession number).

6.1
NCAM1

NCAM1 [NCA1 (P13591; P13592)], one of the most thoroughly studied molecules in the nervous system, is coded by a single copy gene that can give rise up to 30 distinct isoforms (Figure 6.1; for further details on NCAM1, please see Chapter 30.7). As a member of the immunoglobulin superfamily of adhesion molecules, I have five immunoglobulin-like C2-type domains, which are formed by intramolecular disulfide bonds.

All NCAM1 isoforms, as proteins destined for the extracellular space, are targeted by a N-terminal hydrophobic signal peptide so that they can be translated via the sec61 pore across the rough endoplasmic reticulum (ER) membrane into the secretory pathway [1]. The ER lumen, the first luminal cellular compartment I encounter, provides an ideal physicochemical environment for proper folding into a native protein conformation (see below). Nascent NCAM1 polypeptide

The Sugar Code. Edited by Hans-Joachim Gabius

ISBN: 978-3-527-32089-9

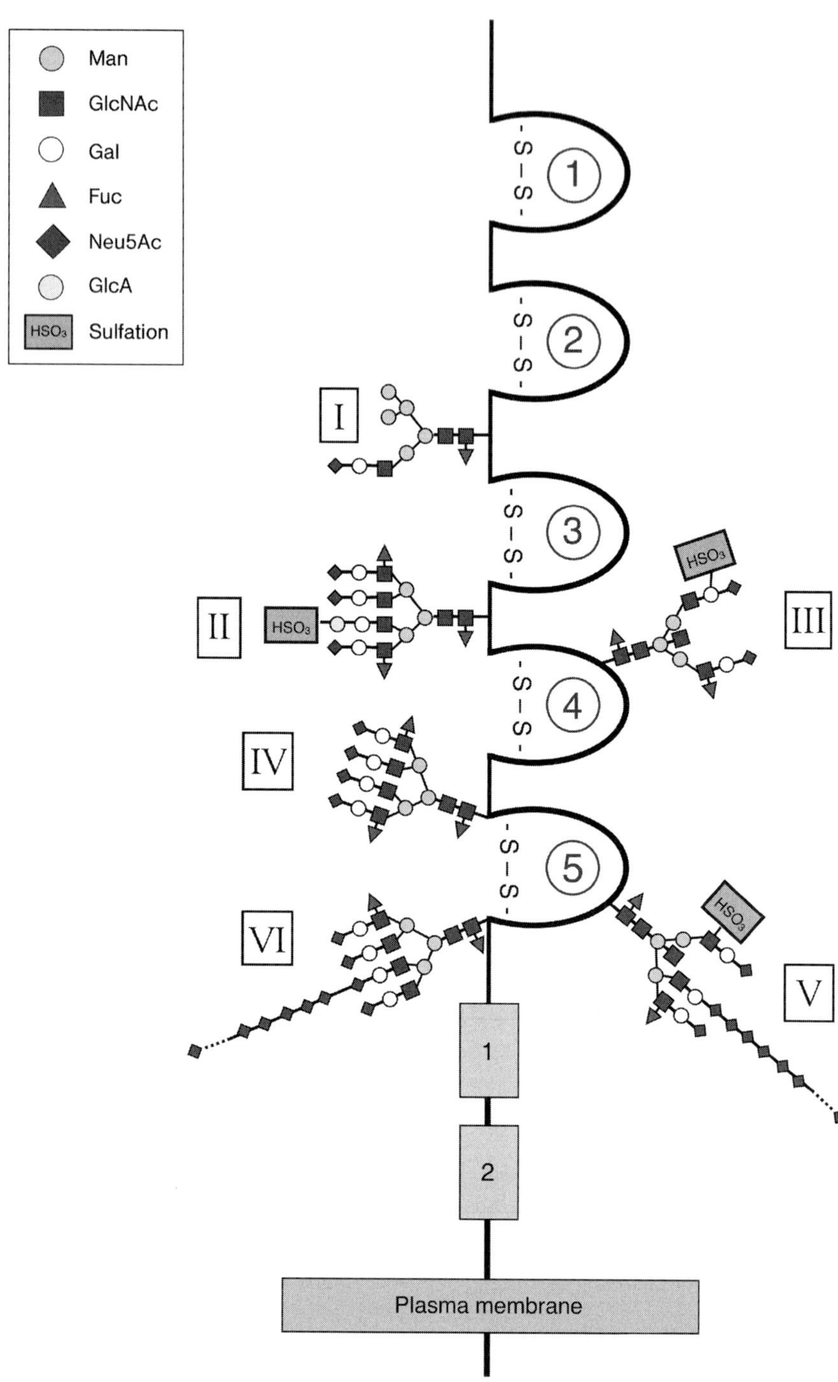

Figure 6.1 Structural organization of NCAM1. The five loop-shaped, immunoglobulin-like C2-type domains (1–5 in circles) that are formed by intramolecular disulfide bonds (S–S) are indicated. The two gray boxes represent the fibronectin type-III-like domains 1 and 2. The NCAM1 carries six *N*-glycans (I–VI in squares), which exhibit an enormous structural diversity. A single polyNeu5Ac-decorated NCAM1 molecule carries all three types of mature *N*-glycans: complex, hybrid and high-mannose types (not indicated). The complex-type *N*-glycans can be bi-, tri- and tetraantennary. Two *N*-glycans of the immunoglobulin-like domain 5 carry the characteristic homopolymers of α2,8-linked Neu5Ac. Sulfation (HSO_3) is a common post-translational modification of the complex-type *N*-glycans and forms the HNK-1 epitope when occurring on glucuronic acid (please see Figure 1.7c for structure).

chains are *N*-glycosylated by oligosaccharyltransferase (OST), a multisubunit enzyme complex, at the β-amino group of six asparagine residues in the consensus sequences Asn-$Xaa_{(\neq Pro)}$-Thr/Ser-$Xaa_{(\neq Pro)}$ (Asn sequon). Asn sequons are highly underrepresented in glycoproteins, although they are often well preserved during evolution. This reflects their local influences during protein folding and assembly [2]. Particular oligosaccharide structures modulate functional properties of membrane proteins, such as those forming membrane channels, and adhesion molecules, including NCAM1 (please see Chapter 30.6–8 for further details). In addition, the *N*-glycans physically stabilize the conformation of mature proteins.

6.2 Initial Steps in Asparagine-Linked Glycosylation

N-Linked glycan (*N*-glycan) biosynthesis starts with the transfer of preassembled lipid-linked oligosaccharide (LLO) to the polypeptide (Figure 6.2). The LLO is synthesized on dolicholpyrophosphate (Dol-PP) by the asparagine-linked glycosylation (ALG) gene products at both sides of the ER membrane. Each of the consecutively acting glycosyltransferases uses as acceptor a specific LLO intermediate and the respective donor substrate to guarantee accurate assembly of the LLOs. All steps of this biosynthetic pathway up to Dol-PP-$GlcNAc_2Man_5$ occur at the cytosolic side of the ER membrane and use activated nucleotide sugars as donor substrates. Synthesis of the oligosaccharide is completed at the luminal side of the ER, but here cytosolically synthesized dolichol-linked membrane anchored monosaccharides serve as donor substrates (Figure 6.2, inset and Info Box 1 on evolutionary aspects; please see also Figure 22.2 for pathway of LLO assembly) [3].

My six potential Asn sequons become glycosylated as soon as they are translocated 40 Å (around 10 amino acids) into the ER lumen. The OST complex forms together with other protein complexes a functional unit working on nascent polypeptide chains. Such tight positioning in close proximity to the entry pore of the nascent polypeptide chain guarantees that OST acts on unfolded polypeptides with their bendable Asn sequons. In most glycoproteins, this bending brings the chemically rather inert asparagine into close proximity with the OH groups of Thr/Ser, or, in some proteins like von Willebrand factor and serum peptide C, with the SH group of Cys. Note that proline inside or just after the Asn sequon prevents this bending and prohibits glycosylation. Glycosylation efficacy may decrease drastically along the last 70 C-terminal amino acids because of the accelerated translocation speed of the ribosome-released polypeptide [2].

The mammalian OST complex is composed of a subset of different subunits that perform the diverse enzymatic functions. Its STT3 subunits A or B recognize Asn sequons and provide catalytic activity in concert with polypeptide-specific subcomponents, such as ribophorin I. OST containing the ubiquitously expressed mammalian STT3B (Q8TCJ2) exerts the basal glycosylation activity. STT3A (P46977) is expressed in professional secretory cells such as hepatocytes in which mature LLOs consumption is extremely high. SST3A containing OST strongly

Info Box 1

Similarities in the process of *N*-glycosylation in all three primary lineages of life strongly suggest common founder events in which the Asn sequon was defined and general principles were arranged. *N*-Glycosylated proteins in prokaryotic cells are rare nowadays, but they still exist in the plasma membrane of Gram-negative bacteria and some Archaea (see Chapter 8.1). In prokaryotes, diverse preassembled oligosaccharide precursors are externalized across the cell membrane, further modified and then used for polypeptide chain glycosylation. In the eukaryotes, however, glycan diversity is created on the glycoproteins themselves, liberating epigenetic pressure on the LLO precursors. Therefore, intracellular *N*-glycans could be structurally conserved, and they have become the basis for the evolution of additional functions.

The bipartite pathway for synthesizing LLOs was already established in cells in which synthesis of the cytosolic intermediates could use soluble donor substrates. For the extracellular extension of the pentamannosyl LLO, non-diffusible membrane-anchored donor substrates (Dol-PP-Man/Glc) had to be utilized to prevent their loss by diffusion into the periplasmic space. Thus, $Glc_1Man_{(7-9)}GlcNAc_2$-PP-Dol has to be considered as the archetypical LLO, already established before the luminal organization of the rough ER evolved from the periplasmic space. The sole use of a sugar nucleotide inside the ER (UDP-Glc) by UGGG1 is an exception that proves the rule, insofar as this process evolved long after compartmentalization had been introduced. The evolutionary origin of the Golgi apparatus is still a mystery, but the use of activated small compounds as donor substrates points to a late appearance of this organelle. Its structural and functional relation with microtubules or actin filaments suggests a time point long after cell polarity was established.

Figure 6.2 Illustration of the biosynthetic and catabolic ER pathway of the *N*-glycan precursor $Glc_3Man_9GlcNAc_2$. (a) Enzymes are named according to their primary entry name in Swiss-Prot (www.expasy.org). Transferases are shown in light-gray boxes and glycosidases in darker gray boxes. Monospecific enzymes (middle gray) act on defined positions. Polyspecific enzymes (darker gray) hydrolyze specific glycosidic linkages at different positions along the *N*-glycan. Enzymatic reactions (arrows) are named according to the enzyme commission (www.brenda.org). Preformed *N*-glycan precursors serve as acceptor and activated sugar nucleotides as donor substrates for glycosyltransferase reactions. (b) Detailed view of the $GCS2_{AB}$–UGGG1 de- and reglycosylation cycle (dashed box B in C). Glycoprotein α1–2-glucosidase cleaves glucose, whereas UDP-glucose: glycoprotein glucosyltransferase 1 catalyzes the reverse reaction. Liberated UDP from the transferase reaction is hydrolyzed immediately to UMP by ENTP5 and exported from the lumen of the ER (see Figure 6.3b). Inset in (b) shows the different human ectonucleoside diphosphatases in the lumen of ER and Golgi apparatus. (c) All glycosyltransferases (light-gray boxes) and glycosidases (middle-dark gray boxes) engaged in the *N*-glycan biosynthesis and trimming reactions are mentioned. Note that for the biosynthesis, the scheme has to be read from right to left. It starts with the transfer of GlcNAc to Dol-PP by ALG7-DPAGT1. The left inset indicates the enzymes for Dol-PP monosaccharide synthesis, the donor substrate for ER-luminal transferases. After transfer to polypeptide, the precursor oligosaccharide is trimmed by mono- (middle-gray boxes) and poly-specific (dark-gray boxes) glycosidases. All glycosidases are exoglycosidases with the exception of MANEA and NGLY1. Gray arrows point to sites of hydrolysis and their thickness is a relative indicator for A, B or C branch preference of the respective glycosidase.

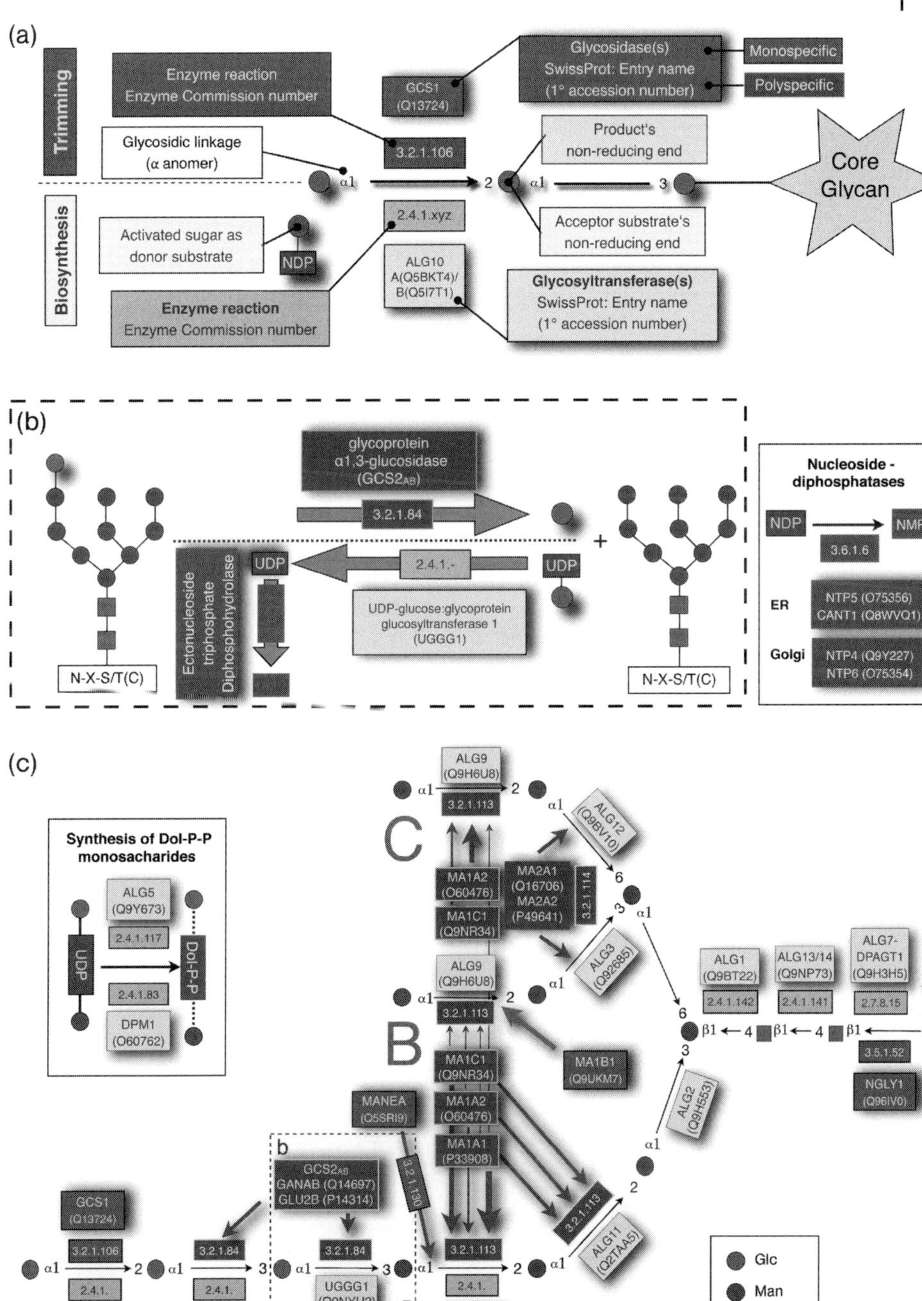

(a)
Trimming
Biosynthesis
Enzyme reaction
Enzyme Commission number
Glycosidic linkage
(α anomer)
GCS1
(Q13724)
Glycosidase(s)
SwissProt: Entry name
(1° accession number)
Monospecific
Polyspecific
3.2.1.106
Product's
non-reducing end
Core
Glycan
Activated sugar as
donor substrate
NDP
2.4.1.xyz
ALG10
A(Q5BKT4)/
B(Q5I7T1)
Acceptor substrate's
non-reducing end
Glycosyltransferase(s)
SwissProt: Entry name
(1° accession number)
Enzyme reaction
Enzyme Commission number
(b)
glycoprotein
α1,3-glucosidase
(GCS2AB)
3.2.1.84
2.4.1.-
UDP
Ectonucleoside
triphosphate
Diphosphohydrolase
UDP-glucose:glycoprotein
glucosyltransferase 1
(UGGG1)
N-X-S/T(C)
Nucleoside -
diphosphatases
NDP
NMP
3.6.1.6
ER
NTP5 (O75356)
CANT1 (Q8WVQ1)
Golgi
NTP4 (Q9Y227)
NTP6 (O75354)
(c)
Synthesis of Dol-P-P
monosacharides
ALG5
(Q9Y673)
2.4.1.117
UDP
Dol-P-P
2.4.1.83
DPM1
(O60762)
ALG9
(Q9H6U8)
3.2.1.113
ALG12
(Q9BV10)
MA1A2
(O60476)
MA2A1
(Q16706)
MA2A2
(P49641)
3.2.1.114
MA1C1
(Q9NR34)
ALG3
(Q92685)
ALG1
(Q9BT22)
2.4.1.142
ALG13/14
(Q9NP73)
2.4.1.141
ALG7-
DPAGT1
(Q9H3H5)
2.7.8.15
3.5.1.52
NGLY1
(Q96IV0)
MA1B1
(Q9UKM7)
MANEA
(Q5SRI9)
3.2.1.130
MA1A1
(P33908)
ALG2
(Q9H553)
GCS2AB
GANAB (Q14697)
GLU2B (P14314)
GCS1
(Q13724)
3.2.1.106
3.2.1.84
2.4.1.
ALG10
A(Q5BKT4)/
B(Q5I7T1)
ALG8
(Q9BVK2)
UGGG1
(Q9NYU2)
ALG6
(Q9Y672)
ALG11?
(Q2TAA5)
ALG11
(Q2TAA5)
Glc
Man
GlcNAc

depends on fully assembled LLOs – a specificity that guarantees accurate *N*-glycosylation under high LLO throughput conditions that are predominant under elevated secretory circumstances. Ribophorin I was necessary for my glycosylation, although it is dispensable for the glycosylation of soluble secretory and polytope membrane proteins [4]. OST activity is crucial in vertebrate development, but its individual subunits may be not. Mutations in OST components such as TUSC3 (Q13454) or MAGT1 (Q9H0U3) cause congenital disorders of glycosylation (CDG) (CDG-Io or CDG-Ip; please see Chapter 22.1 for overview on CDGs).

6.3 Trimming Reactions by α-Glucosidases and Interactions with ER Lectins

Although I get decorated with several $Glc_3Man_9GlcNAc_2$ oligosaccharides inside the ER, I no longer have this glycan structure when I reach the plasma membrane (Figure 6.1). Extensive trimming and elongation reactions occur along the secretory pathway that remodel my *N*-glycans (Figures 6.2 and 6.3; Info Box 2).

Oligosaccharide trimming starts immediately ($t_{1/2} < 2\,\text{min}$) after transfer from glycolipid to my nascent polypeptide chain. Mannosyl-oligosaccharide α-glucosidase-I [GCS1 (Q13724), α-glucosidase-I] hydrolyzes the outermost α1,2-linked glucose to drive and stabilize the oligosaccharide transfer reaction. GCS1 is retained in the ER as a newly synthesized homotetramer, due to its cytoplasmic

Info Box 2

Removal of most sugars, occurring just after transfer of the oligosaccharide from lipid to protein, was the most confusing part of the *N*-glycosylation pathway in vertebrates. These trimming reactions resulted in the removal of not only all three glucoses, as in *Saccharomyces cerevisiae*, but also most of the mannoses before condensing reactions rebuild glycan structures with high species-specific diversity. Why did premetazoan cells conserve such a complicated and energy-consuming synthesis–degradation pathway for *N*-glycans along their evolution? High-mannose glycans evolved when extensive mannosylation at the cell surface was essential. Inside the cells they provided the basis for new pathways introduced during or after the prokaryotic-to-eukaryotic transition. In particular, they participated in the disulfide bond formation processes, and became active compounds for quality control of protein folding, assembly, and degradation. These diverse interrelated processes in the early secretory pathway made conservation of LLO composition vital. Why, however, are almost all mannoses removed in vertebrates after glycoproteins have achieved native conformations? In mammals, oligomannosyl structures play important roles during both cell invasion and induction of innate immunity, suggesting that epigenetic pressure forced early metazoans to remove mannoses to prevent oligomannosyl-dependent cell invasion.

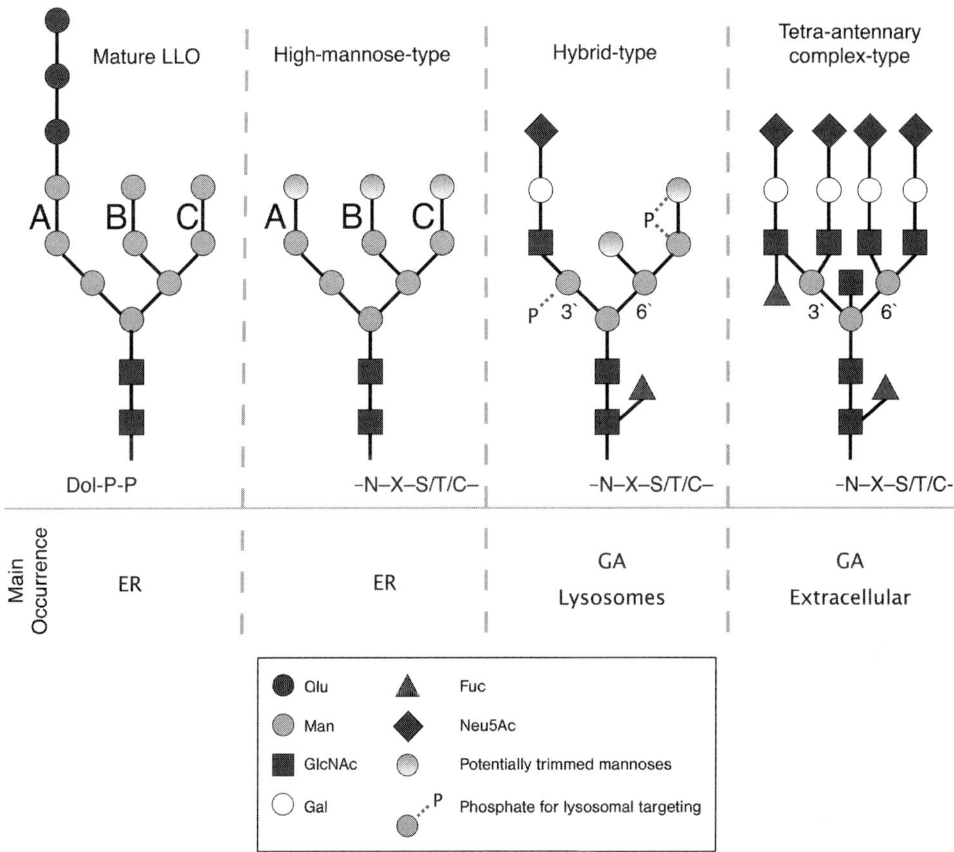

Figure 6.3 Illustration of principal oligosaccharide structures occurring during the biosynthesis of *N*-glycans. The A, B and C branches are indicated in the LLO precursor and high mannose-type *N*-glycan; 3′ and 6′ indicate the respective antenna of the *N*-glycan.

triple arginine motive, and as active compound of the OST–sec61 pore complex. OST transfers preassembled oligosaccharides from a highly mobile lipid to my immobile polypeptide chain anchored in the sec61 pore. This process is entropically highly unfavorable and driven by the hydrolysis of a single diphosphate bond only [5]. Product removal or modifications are methods to increase the output of enthalpically unfavorable reactions. The tight spatial and temporal coupling of glycan transfer by the Glc_3-dependent OST, with Glc_3 to Glc_2 conversion by GCS1, prevents further hydrolytic removal of the protein-bound glycan. Indeed, the absence or inhibition of GCS1 activity results not only in under-glycosylated proteins but also in increased amounts of free oligosaccharides in the ER lumen. This indicates GCS1's importance for proper *N*-glycosylation [6], although it is dispensable for viability of cells in culture, as shown by the $GCS1^{-/-}$ Lec23 Chinese hamster ovary (CHO) (see Info Box 3) cell glycosylation mutant. In

Info Box 3

In 1958, J.H. Tjio and T.T. Puck in the Department of Biophysics at the University of Colorado established cell cultures of CHO cells. The phenotype of CHO cells are highly vulnerable to mutagens since they are monoallelic for diverse gene loci. In 1975, P. Stanley and coworkers at Albert Einstein College of Medicine in New York used chemical mutagenesis and the cytotoxic effect of different lectins as a selection method to establish a whole set of lectin-resistant cells defective in their glycosylation machinery (for further information on lectins, please see Chapters 13–20). These mutants are summarized nowadays as Lec (number) cells. They are still valuable for investigating aspects of *N*-glycosylation, and are used as tools for CDG diagnosis in humans (see Chapter 22.1).

humans, however, absence of GCS1 results in a lethal type of CDG (CDG-IIb, please see also Table 22.1 for listing of CDGs).

The diglucosylated *N*-glycans created on my nascent chains are quickly bound by the ER-resident glucan α1,3-glucosidase complex $GCS2_{AB}$ (α-glucosidase-II). $GCS2_{AB}$ acts on two different glycosidic linkages (Figure 6.2) with different kinetics (Glcα1,3Glc around 5 min; Glcα1,3Man around 20 min), pH optima and inhibitor sensitivities. $GCS2_{AB}$ is a heterotetramer composed of two subunits, both essential in vertebrates. The hydrolytic activity of $GCS2_{AB}$ is related solely to the α-subunit [GANAB (Q14697)], whereas the β-subunit [GLU2B (P14314)] regulates substrate specificity *in vivo*. In addition, GLU2B achieves ER residency either by Ca^{2+}-dependent interactions to the protein matrix or by its C-terminal HDEL retrieval signal. Like GCS1, $GCS2_{AB}$ activity is not essential for viability of cells in culture ($GCS2A^{-/-}$ mouse lymphoma cell line PHAR-2.3), but mutations in one GLU2B allele are apparently related to late-onset polycystic liver disease in humans (PCLD).

$GCS2_{AB}$ binds my first-appearing diglucosylated *N*-glycan co-translationally (Figure 6.4a). However, the α-subunit does not cleave until the mannose-binding mannose-6-phosphate receptor homology (MRH) domain of the β-subunit binds the next *N*-glycan presented (Figure 6.4b; for MRH domains, please see Chapter 19.2). This *trans*-glycan interaction initiates hydrolysis and generates a monoglucosylated *N*-glycan. MRH binding requires an intact 6′-pentamannosyl structure, which explains the observed 90% decrease of $GCS2_{AB}$ activity towards 6′-tetramannosyl-bearing substrates. The second glucose is protected from hydrolysis since the respective linkage (Glcα1,3Man) stays topologically inaccessible for the α-subunit due to the persisting β-subunit association with the distal *N*-glycan. The resulting hopping from one glycan to the next creates monoglucosylated *N*-glycans that are recognized by the ER-resident lectins calnexin [CALX (P27824)] or calreticulin [CALR (P27797)/CALR3 (Q96L12)] (Figure 6.4c; please see Chapter 15.3 for the definition of lectins and Table 19.2 for further details on CALR/CALX). After glycoprotein's release from the sec61 pore, both glucose residues are consecutively hydrolyzed independent of further *N*-glycans. During this process, MRH *cis*-inter-

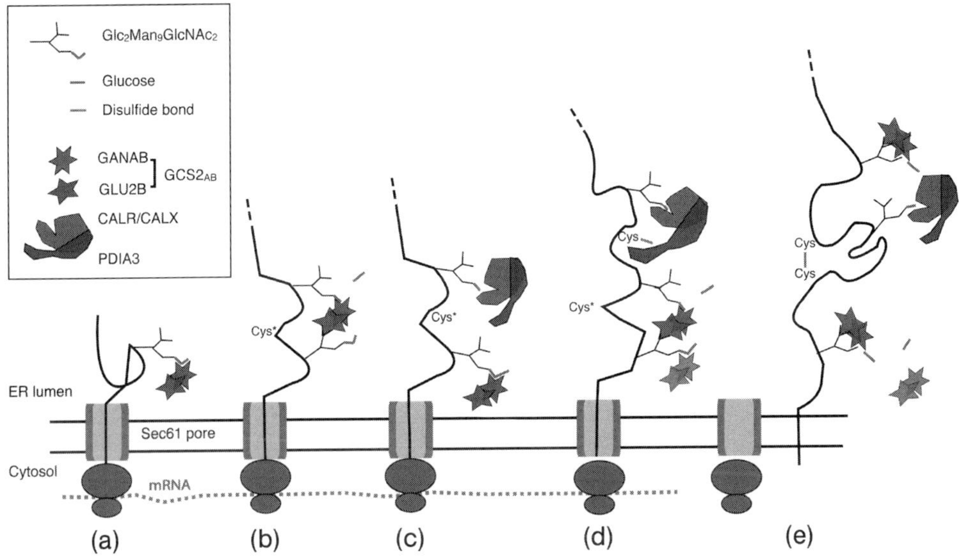

Figure 6.4 Illustration of co- and post-translational glucose-related modifications during initial protein folding. $GCS2_{AB}$ binds diglucosylated *N*-glycans on nascent polypeptide chains (a). Glucose trimming occurs only after *trans*-interaction of $GCS2_{AB}$ with a second *N*-glycan (b). This is followed by interaction of CALR/CALX–PDIA3 with the generated monoglucosylated *N*-glycan (c). The PDIA3 reacts with a single cysteine of the polypeptide chain (d). Disulfide bridge formation and dissociation of CALR/CALX–PDIA3 occur in parallel, which is followed by completion of glucose trimming (e). As indicated in (e), post-translational glucose trimming does not require a *trans*-glycan interaction of $GCS2_{AB}$.

actions with the *N*-glycan to be processed bring the active center of the associated α-subunit into the correct cleavage position (Figure 6.4e). Therefore, only proteins with multiple glycans will bind directly to CALR/CALX [5].

Both CALR and CALX form a complex with protein disulfide isomerase (PDI) A3 [Erp57 (P30101)], their obligate binding partner that exhibits thioredoxin activity similar to that found in other PDIs. Swift CALR–PDIA3 binding to my *N*-glycan at position 222 may be needed to shield the 13-amino acid-distant Cys235 until the correct intrachain binding partner (Cys287) becomes accessible to form immunoglobulin-like domain 3 correctly. Note that proteins with mutations in either a disulfide-forming cysteine or an Asn sequon are often severely misfolded due to aberrant intra- and interchain disulfide bonds.

6.4 Quality Control of Protein Folding and Assembly: Machinery and Principal Mechanism

The mechanisms for the release of the lectins are not yet well defined, but have been shown to relate to disulfide bond formation. However, after their release from the lectins, all of my five monoglucosylated *N*-glycans are fully deglucosylated by GCS_{AB} (around 20 min). At this point, only a minority of glycoproteins is correctly folded and none can be fully assembled into heteromeric structures. This may also be true for my fibronectin-like domain 1, which up to now could not achieve correct folding. This is a typical situation in which another player in the game of protein folding enters the arena, UDP-glucose:glycoprotein-glucosyltransferase [UGGG1 (Q9NYU2)]. This enzyme is a soluble ubiquitous protein containing a C-terminal KDEL ER retrieval sequence. It recognizes fully translated but still immature or not completely assembled glycoproteins. UGGG1 reglucosylates the A-branch on some of their *N*-glycans to produce the ER-retention tag $Glc_1Man_{9-7}GlcNAc_2$. The presence of a single tag is sufficient for ER retention, probably in my case by CALR–PDIA3. Most glycoproteins undergo at least one round of reglucosylation to induce further exposure to ER-mediated protein folding and assembly or degradation events.

UGGG1 is both a folding sensor and a glucosyltransferase, and it is considered to be a decision maker for ER exit of glycoproteins (Figure 6.2b). What are my structural features that define me as a substrate for UGGG1? (i) Glycan VI must have an intact A-branch to be recognized as an acceptor substrate. (ii) Its core GlcNAc has to be accessible for UGGG1 recognition. This is the case for *N*-glycan VI in close proximity to the non-native structures, but not for the others, since in native conformations the core GlcNAc-Asn is buried in a protein grove. (iii) Exposed hydrophobic amino acid patches, as found in the molten-globule stage of nearly native protein conformations, must exist in its vicinity. The obligate interaction with molten-globule-stage domains restricts UGGG1 interaction to near-native glycoproteins, and rules out its action on terminally misfolded or aggregated candidates with accessible core GlcNAc structures. The ubiquitously expressed thioredoxin reductase SEP15 (O60613) is strongly associated with either UGGG1 or its nonenzymatic homolog UGGG2 (Q9NYU1) ($K_d = 20\,nM$). SEP15 may provide the reductase activity that cleaves the disulfide bond in my immunoglobulin-like domain 5 to facilitate refolding. UGGG1 activity is not essential for protein synthesis in cultured cells, but it increases its efficacy drastically and is needed in professional secretory cells. Knockout of UGGG1 in mice (and probably in humans) is embryonic lethal [7, 8].

6.5
ER Exit – Facing a Crucial Decision and What Mannose Has to Do

After a certain time inside the ER, glycoproteins can be either approved for export or designated as waste for degradation (see Info Box 4). *N*-Glycans can directly grant export towards the Golgi apparatus, as in case of the lectin-mediated transport of coagulation factors V and VIII (please see Table 19.2) [9]. However, my export, as in most other cases, is not dependent on *N*-glycan-mediated interactions. Transport is triggered either by default or through mature protein conformation-dependent mechanisms [10]. In general, *N*-glycan structures are not well defined for proteins approved for secretion, but monoglucosylated *N*-glycans are

Info Box 4

A major challenge during the synthesis of secretory proteins is to achieve the correct structures while inside the cell. These structures must be able to resist the harsh physicochemical conditions existing at their final extracellular destinations. In prokaryotes, secretory proteins are synthesized and assembled at their final working location, in the periplasmic space. Further polycistronic genome organization guarantees stoichiometric synthesis of subunits forming heteromeric complexes. During evolution to eukaryotes, and even more pronounced in metazoans, the site of synthesis and final working destination separated. In addition, monocistronic gene organization was only possible after mechanisms were in place that could guarantee stoichiometry in hetero-oligomerization processes outside the cytoplasm. Both requisites were driving forces for the development of quality control of protein folding and assembly along the secretory pathway that decides whether a protein is useful or has to be degraded by ER-associated protein degradation (ERAD).

Different pathways help proteins to fold correctly inside the ER. The more recently evolved, but nowadays major pathway is determined by hydrophobic interactions with ER chaperones (BiP) together with PDIs. They help in folding of nonglycosylated C- and N-terminal regions of NCAM1. This pathway is complemented by the ancient *N*-glycan-dependent pathway, which is engaged in folding processes of the glycosylated middle domains. The archetypical $Glc_1Man_{7-9}GlcNAc_2$ oligosaccharide plays a key role here, and in the subsequent quality control of protein folding and also during protein assembly. Subunits for the heteromeric coagulation factor fibrinogen (α, β, γ) are expressed in nonstoichiometric amounts in hepatocytes. CALX–PDIA3 association with the β-subunit or precomplex $\alpha\gamma$ is stable until the correct isomerization forms heterotrimers. Therefore, CALX/CALR–PDIA3 release is not dependent on the folding stage, but relies on the subsequent formation of stabilizing disulfide bonds, which can occur long after translation has been finished.

Two main classes of degradation-prone glycoproteins exist under normal conditions: surplus components of heteromerization processes and dead-end folding states. Once recognized as waste, the non-native glycoprotein monomers in dead-end folding states are quickly cleared from the ER lumen. They are deglycosylated by a cytosolic *N*-glycosidase [NGLY1 (Q96IV0)], polyubiquitinated and degraded by the proteasomes ($t_{1/2} \sim$ 30–60 min). Native surplus components for heteromers, however, reside much longer inside the ER ($t_{1/2} \sim$ 3–4 h). Fibrinogen monomers (α, γ) are degraded over ERAD, whereas preformed $\alpha\gamma$ subcomplexes follow a different degradation route to lysosomes. In metazoans, how degradation-prone proteins pass the ER membrane to reach the cytosolic glycosidase–proteasome system is still not known. A vesicular pathway defined by the presence of endogenous ERAD factor EDEM1 and transgenic ERAD substrate may be a possible clearing mechanism. Direct ER export through special ER pores may be an alternative method.

Info Box 4 – ctd.

N-Glycans are required for successful glycoprotein synthesis, especially in high-throughput situations, to prevent protein aggregation, cell death and tissue damage. Their synthesis needs a huge energy-consuming machinery to accomplish full functionality. Therefore, protein synthesis in the ER of vertebrate cells is highly vulnerable to metabolic changes, such as amino acid starvation (around 5–15 min) or glucose deprivation, which lowers the *N*-glycosylation capacity due to LLO deficit (around 5 min). Cells grown in such suboptimal conditions induce a so-called unfolded protein response (UPR), not only to save materials, but also to prevent an accumulation of incorrectly folded aggregation-prone underglycosylated proteins. During UPR, cells immediately, and with high efficiency, suppress eIF4-dependent protein synthesis, induce mRNA transcription of all quality control components and guarantee their translation into the ER by use of eIF4-independent ribosomal initiation processes. Molecular components of the ERAD pathway have been characterized biochemically and genetically in yeast. Differences in *N*-glycan processing between yeast and vertebrates, as well as cellular reactions under experimental conditions as mentioned above, complicate research on ERAD in higher eukaryotes.

ERAD is a pathogenetic factor in protein-folding diseases (caused by destabilizing point mutations in glycoproteins) in two ways – either quality control is not able to recognize and remove aggregation-prone mutated glycoproteins (pathological gain of function) or it prevents nearly native, active enzymes from reaching their final destination (loss of function). An example for the former is the apoptosis-inducing myocilin mutations causing open angle glaucoma, and for the latter, Fabry disease-causing mutations in the lysosomal α_1-galactosidase A. In both cases, disease-causing manifestation in cultured cells could be reversed by treatment with appropriate synthetic chaperones.

prohibited. Constitutive glucose cleavage in *N*-glycans released from CALX/CALR–PDIA3, combined with an efficient UGGG1, monitors the folding stage of glycoproteins. When I pass UGGG1's quality control of protein folding, I will be packed into COPII-coated tubulovesicular transport clusters that contrive my transport towards the Golgi apparatus. Cholesterol-enriched transporters are used as carriers for my glycosylphosphatidylinositol-anchored 120-kDa cousins. Alternatively, the acceptor substrate for UGGG1 can be destroyed by the commonly found α1,2 mannosidase MA1A1 (P33908) – the quality control terminator enzyme. MA1A1 preferentially cleaves Man_9-glycans at their A-branch (95%) and produces the ER-export tag $Man_{8-6}GlcNAc_2$-A isomers without, however, determining the fate of the glycoprotein.

Mammalian cells, in contrast to yeast, express several neutral α1,2-mannosidases in the secretory pathway (MA1; glycoside hydrolase family 47) in a tissue-specific manner. They generate glycan structures free of α1,2-linked mannoses – a prerequisite for further trimming and elongation reactions. All processing α1,2 mannosidases are type II transmembrane proteins with long (35–85 amino acids) cytoplasmic tails, typical for ER proteins, and their activity depends on ER Ca^{2+} concentrations. Definite subcellular distributions of α1,2 mannosidases have not been determined yet, but beside ER-resident MA1B1 (Q9UKM7), MA1C1 (Q9NR34) and splice variants of MA1A1 are also localized to the ER. MA1A1, MA1A2 and MA1C1 are capable of hydrolyzing all terminal α1,2-linked mannoses on any of three branches (Figure 6.2). However, all three release B-branch mannose very inefficiently and very slowly, indicating that the $Man_9GlcNAc_2$-dependent B-branch-specific MA1B1 produces the $Man_8GlcNAc_2$-B isomer in most cases. In mammalian cells, MA1B1 action completely destroys the A-branch specificity of the quality control terminator enzyme MA1A1. As a consequence, preceding MA1B1 action increases intact A-branches in MA1A1 products about tenfold. This makes MA1B1 a glycoprotein-saving enzyme in mammalian cells.

Three other members of the glycoside hydrolase family 47 are the ERAD-enhancing α-mannosidase-like proteins EDEM1 (Q92611), EDEM2 (Q9BV94) and EDEM3 (Q9BZQ6), which are cell-type specifically expressed. All three have been identified as ERAD-triggering factors in transgenic cell systems. EDEM3 expression increases partial demannosylation (Man_8 to Man_{7-5}) of *N*-glycans on ERAD substrates inside the ER. This indicates that α1,2-linked mannoses are removed before proteins are dislocated for cytosolic deglycosylation and degradation [11]. Therefore, some Man_{7-5} isomers have to be considered as ERAD-tags for degradation-prone proteins.

Endomannosidase MANEA (Q5SRI9), a post-ER localized endoglycosidase, cleaves A-branch Manα1,2Man linkages in $Glc_3Man_5GlcNAc_2$ oligosaccharides with highest affinitiy [12]. Such glycan structures are rarely seen on glycoproteins, but are highly abundant as free oligosaccharides secreted from Lec23 CHO cells that lack both GCS1 and MANEA [13]. Free oligosaccharides are products of aberrant OST reactions due to slow or missing Glc_3 to Glc_2 conversion (see above). Glucose removal is essential for delivery of free oligosaccharides to lysosomes to prevent their toxic secretion. MANEA may also deglucosylate incompletely trimmed *N*-glycans on proteins. It has been suggested that MANEA functions as an A-branch demannosylation enzyme. However, normal complex glycosylation patterns were found in MANEA-negative vertebrate cells, such as CHO and endothelial cells. On the other hand, defective glycosylation was observed in MANEA-expressing, but $GCS2_{AB}$-defective, PHAR-2.3 cells. The *N*-glycan phenotype observed in these cells contradicts the general MANEA function on glycoproteins and substantiates its function in the clearance of free oligosaccharides.

6.6 How to Become a Mature *N*-Glycan?

Most proteins entering the Golgi apparatus expose identical *N*-glycan structures with α1,2-linked mannoses trimmed. However, when I leave this organelle my glycans exhibit high diversity and differ from those on other proteins traveling along the same route. Four main classes of Golgi apparatus residential proteins are responsible for this diversification process (Figure 6.5): (i) trimming α-mannosidases, (ii) nucleotide sugar transporters, (iii) glycosyltransferases, and (iv) sulfotransferases and *O*-acetylases.

6.6.1 Golgi Mannose Trimming as the Start for *N*-Glycan Elongation

The two Golgi apparatus-located neutral mannosyl-oligosaccharide α1,3/6-mannosidases MA2A1 (Q16706) and MA2A2 (P49641) remove B- and C-branch mannoses to enable the synthesis of complex-type *N*-glycans. Both mannosidases II are able to remove α1,3- and α1,6-linked mannoses from 6′-tetramannosyl structures. This trimming initiates the formation of complex-type *N*-glycans. It strongly depends on prior addition of the A-branch GlcNAc by *N*-acetylglucosaminlytransferase-I (see below). Both α-mannosidases are equally and constitutively expressed in human tissues. Deficiency of MA2A1 activity as observed in HEMPAS (hereditary erythroblastic multinuclearity with positive acidified serum lysis test) patients (congenital dyserythropoietic anemia type II) can be compensated by MA2A2, but only with a substantial loss of *N*-glycan diversity (please see also Table 23.1).

6.6.2 Nucleotide Sugar Transporters Import the Fuel for Oligosaccharide Elongation

Nucleotide sugar transporters (NSTs) and sulfate (3′-phosphate adenosine 5′-phosphosulfate, PAPS) transporters are solute carriers that import activated donor substrates for glycan elongation reactions into the secretory pathway. They are polytope membrane proteins that are targeted by short cytosolic amino acid sequences to their accurate location. For instance, ER-resident UDP-Glc transporter (S35B1) is retrieved by its cytosolic COPI-binding motif (KKTSH), whereas

Figure 6.5 Illustration of the biosynthetic pathway leading to the synthesis of complex-type *N*-glycans. (a) Enzymes are named according to their primary entry name in Swiss-Prot (www.expasy.org). Enzymatic reactions (arrows) are named according to the enzyme commission (www.brenda.org). Activated sugar nucleotides serve as donor and the preformed *N*-glycan precursors as acceptor substrates for glycosyltransferase reactions. (b) Detailed view of all reactions needed for accurate addition of *N*-acetylglucosamine by MGAT5 is shown (dashed box B in C). Catalysis of the reverse reaction (dotted arrow) that is entropically favored is inhibited by the efficient removal of liberated UDP. UDP–UMP conversion by the different human ectonucleoside diphosphatases is coupled to the immediate export of UMP out of the Golgi apparatus lumen by the solute carriers mentioned in the right box in B. (c) Transferases are named in framed boxes adjacent to the enzyme reaction catalyzed. The gray scale of the frame was specifically chosen and reflects sugars transferred. Arrows in gray point to additional distant linkages. The thickness of the arrows for sialyltransferases represents relative specificity towards different positions. Numbers in bold reflect preferred linkage types. Transferases with dotted frames depend on non-linear acceptor substrates (dotted gray box). Different human sulfotransferases are listed with some features in the right upper corner. In case of sulfation, only a few possible linkage types with know function have been illustrated.

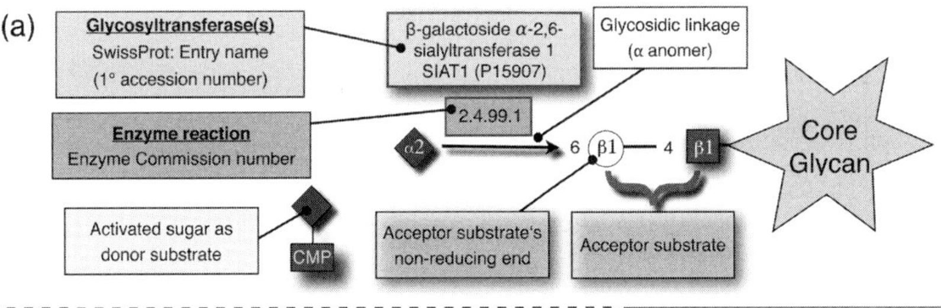
(a)
Glycosyltransferase(s)
SwissProt: Entry name
(1° accession number)
β-galactoside α-2,6-sialyltransferase 1
SIAT1 (P15907)
Glycosidic linkage
(α anomer)
Enzyme reaction
Enzyme Commission number
2.4.99.1
Core Glycan
Activated sugar as donor substrate
CMP
Acceptor substrate's non-reducing end
Acceptor substrate

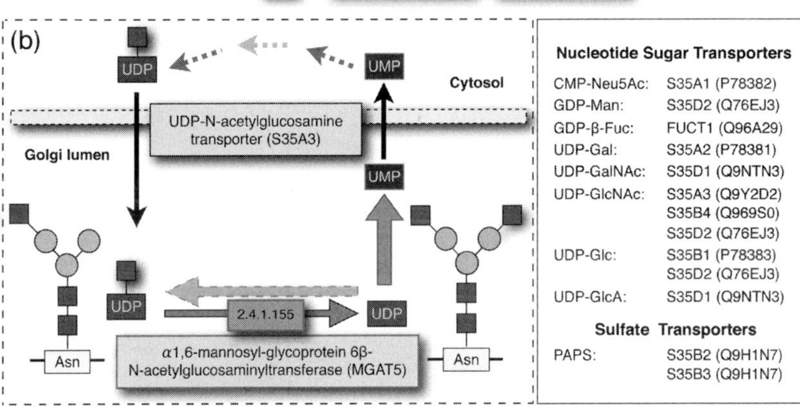
(b)
UDP
UMP
Cytosol
UDP-N-acetylglucosamine transporter (S35A3)
Golgi lumen
2.4.1.155
α1,6-mannosyl-glycoprotein 6β-N-acetylglucosaminyltransferase (MGAT5)
Asn
Nucleotide Sugar Transporters
CMP-Neu5Ac: S35A1 (P78382)
GDP-Man: S35D2 (Q76EJ3)
GDP-β-Fuc: FUCT1 (Q96A29)
UDP-Gal: S35A2 (P78381)
UDP-GalNAc: S35D1 (Q9NTN3)
UDP-GlcNAc: S35A3 (Q9Y2D2) S35B4 (Q969S0) S35D2 (Q76EJ3)
UDP-Glc: S35B1 (P78383) S35D2 (Q76EJ3)
UDP-GlcA: S35D1 (Q9NTN3)
Sulfate Transporters
PAPS: S35B2 (Q9H1N7) S35B3 (Q9H1N7)

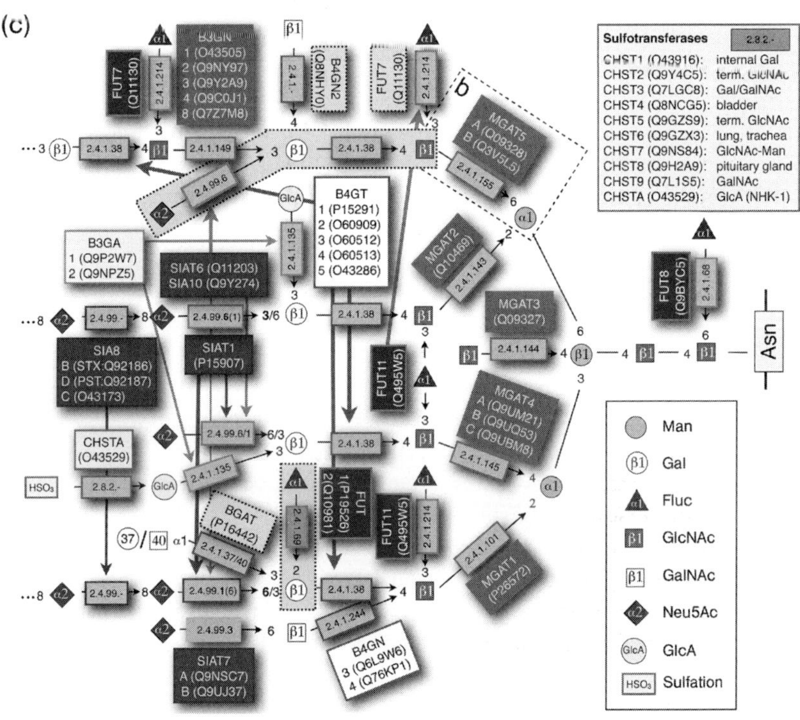
(c)
Sulfotransferases 2.8.2.-
CHST1 (O43916): internal Gal
CHST2 (Q9Y4C5): term. GlcNAc
CHST3 (Q7LGC8): Gal/GalNAc
CHST4 (Q8NCG5): bladder
CHST5 (Q9GZS9): term. GlcNAc
CHST6 (Q9GZX3): lung, trachea
CHST7 (Q9NS84): GlcNAc-Man
CHST8 (Q9H2A9): pituitary gland
CHST9 (Q7L1S5): GalNAc
CHSTA (O43529): GlcA (NHK-1)
FUT7 (Q11130)
B3GN 1 (O43505) 2 (Q9NY97) 3 (Q9Y2A9) 4 (Q9C0J1) 8 (Q7Z7M8)
B4GN2 (Q8NHY0)
MGAT5 A (Q09328) B (Q3V5L5)
B4GT 1 (P15291) 2 (O60909) 3 (O60512) 4 (O60513) 5 (O43286)
B3GA 1 (Q9P2W7) 2 (Q9NPZ5)
SIAT6 (Q11203) SIA10 (Q9Y274)
MGAT2 (Q10469)
MGAT3 (Q09327)
FUT8 (Q9BYC5)
SIA8 B (STX:Q92186) D (PST:Q92187) C (O43173)
SIAT1 (P15907)
FUT11 (Q495W5)
MGAT4 A (Q9UM21) B (Q9UQ53) C (Q9UBM8)
CHSTA (O43529)
BGAT (P16442)
FUT 1 (P19526) 2 (Q10981)
MGAT1 (P26572)
B4GN 3 (Q6L9W6) 4 (Q76KP1)
SIAT7 A (Q9NSC7) B (Q9UJ37)
Asn
Man
Gal
Fuc
GlcNAc
GalNAc
Neu5Ac
GlcA
Sulfation

the CMP-Neu5Ac transporter is exported to the Golgi apparatus by its C-terminal amino acids (IIGV). The NSTs are antiporters since import of activated nucleotide sugars is obligatory, coupled with an equimolar exchange of the corresponding free nucleoside monophosphate. These originated from efficient hydrolysis of nucleotide pyrophosphates that are produced during the sugar elongation reactions (Figures 6.2b and 6.5b). Mutations in NSTs result in drastic changes of glycosylation patterns and may cause severe diseases (CDG-IIf, please see Chapter 22.1). Differences in NSTs between parasites, like *Candida albicans*, and humans have been characterized and defined as targets for therapeutic approaches.

6.6.3
Glycosyltransferases: The Orderly Maturation Reactions

The many Golgi glycosyltransferases that are responsible for glycan's diversity (Figure 6.5) are type II transmembrane glycoproteins with short (5–12 amino acids) N-terminal cytoplasmic tails, an α-helical transmembrane (17–24 amino acids) and luminal stem region, and a C-terminal luminal domain providing catalytic activity. Tail and membrane-spanning regions are responsible for Golgi apparatus localization, and the stem region for the ketosidic linkage specificity of the transferases. Some glycosylation events may be protein specific (see below); however, in general, cell type-specific glycosyltransferases identify discrete mono- or disaccharide structures as acceptor substrates (Figure 6.5a). Therefore, most elongation reactions are independent of the core structure of the *N*-glycan and some glycosyltransferases modify any acceptor saccharide on both *N*- and *O*-glycans, and even process glycolipids. Note that the various glycosyltransferases may compete for the same donor and/or acceptor substrate. Another general principle is that glycosyltransferases depend on each other since one produces the acceptor substrates required by others (Figure 6.5c, dotted border light gray boxes for nonlinear ones). To achieve such orderly trimming and elongation reactions, components of the glycosylation machinery have to be arranged properly. It is still a mystery how the machinery for protein synthesis and glycosylation in the rough ER is organized into subcompartments. In the Golgi apparatus, however, glycoproteins travel along hierarchically arranged enzymes that modify glycans step by step. Two models have been proposed to explain this dynamic process along a rather static enzyme arrangement–vesicular transport or cisternal maturation. In the first, the secretory cargo moves anterogradely along glycosylation machinery stably tied up in Golgi cisternae. In the second, glycosylation machinery moves retrogradely along maturing cisternae filled with secretory cargo. Neither model seems to be correct for animal professional secretory cells, but their combination explains most morphological and biochemical features observed. In the Golgi apparatus running at full capacity, cargo is transported along tubular connections between functionally different Golgi compartments. This transport is complemented with a permanent flow of retrograde-oriented vesicles [14]. Proper localization of the glycosyltransferases in the Golgi apparatus depends on the conserved oligomeric Golgi (COG) complexes. Mutations in COG components, as observed in CDG type Ie and If (see Chapter 22.1), result in wrong localization of at least galactosyltransferases [15].

6.7 Structure Building by *N*-Acetylglucosaminyltransferase-I and Fucosyltransferase-VIII

In the Golgi apparatus my *N*-glycans have lost all α1,2-linked mannoses. Therefore, the exposed non-reducing mannosyl residues are acceptor substrates for subsequent elongation with GlcNAc residues by *N*-acetylglucosaminyltransferases. The α1,3-mannosyl-glycoprotein 2-β-*N*-acetylglucosaminyltransferase [MGAT1 (P26572)] initiates the formation of complex- or hybrid-type *N*-glycans (Figures 6.3 and 6.5c; please see also Figure 23.5). It forms a protein complex with mannosidases-II [MA2A1(2)] by stem region-dependent interactions. Complex formation is highly favorable for preventing formation of hybrid-type *N*-glycans, which indeed are rarely found in human glycoproteins. However, some of my *N*-glycans are protected from being cleaved. Therefore, I carry functional hybrid-type *N*-glycans (Figure 6.1, *N*-glycan I) under certain circumstances (see Chapter 30.7 for functional details). The $MGAT1^{-/-}$ genotype in Lec1 CHO cells resulted in the absence of complex- or hybrid-type *N*-glycans, but did not affect growth of mutant cells in culture. $MGAT1^{-/-}$ knockout in mice is lethal, emphasizing the essential function of this transferase during development (please see also Table 23.1).

The widely expressed core α1,6-fucosyltransferase [FUT8 (Q9BYC5)] modifies exclusively the core GlcNAc of most $GlcNAc_1Man_{(5-3)}GlcNAc_2$ glycans and is dependent on the previous trimming of all α1,2-linked mannoses on the 6′-antennae (see Chapters 2.7 and 8 for structural and Chapter 23.2.4 for functional aspects of core fucosylation). Core fucosylation of *N*-glycans is a common modification on membrane proteins in humans; all my *N*-glycans carry core fucose. The addition of core fucose reduces flexibility of the 6′-antennae in bi- and triantennary complex-type *N*-glycans. The reduced flexibility increases the lifespan of some serum proteins since asialoglycoprotein receptor binding is diminished (please see Chapter 2.6, Chapter 15.4 and Figure 19.1 on this lectin). Conversely, core fucose is missing in short-lived proteins such as the elastase inhibitor α_1-antitrypsin to assure accurate regulation of protein levels.

6.8 Branching and Elongation Reactions

6.8.1 Mannosyl β-*N*-Acetylglucosaminyltransferases

The $GlcNAc_1Man_3(Fuc_{\pm})GlcNAc_2$ glycan structure is the starting point for further elongation and branching reactions. These reactions give rise to a spectrum of *N*-glycans exhibiting an extreme diversity of structure and function (Figures 6.1,

6.3 and 6.5c). The next obligate step in this process is the addition of a GlcNAc residue to the 6′-antenna by MGAT2 (α1,6-mannosyl-glycoprotein 2-β-N-acetylglucosaminyltransferase II; Q10469). The resulting biantennary *N*-glycans may be acceptor substrates for medial Golgi-located MGAT5A (Q09328) or MGAT5B (Q3V5L5), which produce triantennary β1,6-branched *N*-glycans. Alternatively, bisecting MGAT3 (Q09327) may modify biantennary *N*-glycans at the core mannose, which precludes further branching by MGAT5, but not by MGAT4 enzymes. Therefore, spatial and temporal positioning of both MGAT3 and MGAT5 activity regulates biosynthesis of glycans, and influences function at a central point. Effects of β1,6 branches in glycoproteins on cell growth, motility, cell adhesion and differentiation are well documented. They range from regulating metabolism, growth and immune reactions [16], to participation in tissue regeneration [17], as well as a strong correlation between detectable β1,6 branches in primary tumor and survival of colon carcinoma patients [18]. Further branching by three MGAT4 isoenzymes may modify the α1,3-mannosyl branch to form either tri- or tetraantennary *N*-glycans. These isoenzymes are broadly expressed with tissue-specific variations and exhibit distinctive protein-substrate specificities (for phenotypes of KO mice, please see Table 23.1).

6.8.2
N-Acetylglucosaminyl-β-Galactosyltransferases

The GlcNAc added by MGAT3 is the only terminal, nonreducing GlcNAc residue found in mature *N*-linked oligosaccharides in human. All others are elongated by galactose through various β1,3- or β1,4-galactosyltransferases (B3GT and B4GT) to form either type I or II glycans [Galβ1,(3)4GlcNAc: LacNAc] that are susceptible to further condensation reactions. Whether *N*-glycans of type I or type II are synthesized is not encoded in the glycoprotein itself, but depends on the availability of the respective galactosyltransferases. NCAM1 obtained from human brain carries type II *N*-glycans, whereas NCAM1 synthesized in calf brain exhibits both types. The five different human *N*-glycan-processing B4GT isoenzymes act rather nonspecifically. They add galactose to any terminal GlcNAc residue, independent of its linkage and irrespective of its presence on *N*- and *O*-glycans or a lipid. B4GTs are widely expressed with tissue-specific variations and located in *trans* cisternae of the Golgi apparatus in a COG-dependent manner. The ubiquitous B4GT1 (P15291) acts as a cell-surface lectin on germ cells and exists as a soluble enzyme in various body fluids (milk, semen, etc.). In addition to its major role in *N*-glycan processing, B4GT1 is the lactose-synthesizing enzyme in mammary glands. Deficiency of B4GT1 activity cannot be complemented by its homologs in every tissue and therefore causes CDG type IId in humans (see Table 22.1).

6.8.3
Capping Sugars Provide Functions

Terminal reactions on *N*-glycans through which sialic acid, fucose, glucuronic acid and sulfate are added, represent modifications functionally involved in many biological and pathological processes. Sialic acids [Neu5Ac, Neu5Gc, 3-deoxy-D-*glycero*-D-*galacto*-2-nonulosonic acid (KDN)] are negatively charged sugars normally positioned at the nonreducing terminus of glycans. The structural diversity of sialic acids has evolved to greatest extent in the at least 140-nm thick glycocalix that forms a first line of defense that pathogens have to overcome for cell entry. Therefore, it is not surprising that species specificity in sialic acids and their ketosidic linkages are common causes for the interspecies barrier of pathogens (such as viruses, bacteria and others such as malaria-causing *Plasmodium falciparum*) (please see Chapters 17 and 28 for therapeutic aspects). Bird influenza viruses use α2,3-linked sialic acids to dock on target cells in the upper airways (please see Chapter 17.2.1 for detailed information). However, in humans only α2,6-linked Neu5Ac is found in the upper airway epithelia. A virus host expressing both α2,3- and α2,6-linked sialic acids in its upper airways (pig) was of critical importance for a switch in docking specificity of the Spanish influenza virus [19].

6.8.4 Sialyltransferases

Only a minority out of the 20 members of the human sialyltransferase family acts on *N*-glycans. In general, a single α2,6- and two α2,3-sialyltransferases compete for galactose-terminated antennae on my *N*-glycans to fulfill a termination reaction. However, on a limited set of glycoproteins three additional sialyltransferases may elongate α2,3- and α2,6-linked Neu5Ac by oligo/polymers of α2,8 linked Neu5Ac.

All eukaryotic sialyltransferases are expressed species, tissue and cell-type specifically. They are located in the *trans* region of the Golgi apparatus. They share three conserved peptide motifs (sialylmotifs) in their catalytic domains. The 'L-sialylmotif' (Large: 8 invariant residues out of 48–49 amino acids), present in the middle of the luminal catalytic domain, participates in binding of donor substrate CMP-Neu5Ac (KDN/Neu5Gc), while the 'S-sialylmotif' (Small: 3 out of 23 amino acids), closer to the C-terminal end of the enzyme, mediates binding of both donor and acceptor substrates. A distal 'VS-sialylmotif' (Very Small: $HX_{(4)}E$) together with two conserved cysteines in the L- and S-motifs provides catalytic activity. The not-fully conserved amino acids inside the motifs and those between the S- and the VS-motifs may mediate acceptor substrate specificity.

Of note, CMP-*N*-acetylneuraminate-β-galactosamide α2,6-sialyltransferase-I [SIAT1 (P15907)] is the sole ubiquitously expressed mammalian enzyme that catalyzes the addition of α2,6-linked sialic acid to *N*-linked Galβ1,4GlcNAc-R. In tri- and tetraantennary glycans, it elongates preferentially, but not exclusively, the 3′ mannose antennae. Protein-specific CMP-*N*-acetylneuraminate-β1,4-galactoside α2,3-sialyltransferases [SIAT6 (Q11203), SIA10 (Q9Y274)] sialylate *N*-glycans independently of the underlying structures. This explains their broad specificity towards glycoproteins and glycolipids. Which type of ketosidic linkage (α2,3 *versus* α2,6) is formed strongly depends on the species-specific expression patterns of the sialyltransferases. As a case in point, NCAM1 isolated from human or mouse brain carries almost exclusively α2,3-linked Neu5Ac, whereas NCAM1 from bovine brain carries almost exclusively α2,6-linked Neu5Ac. This reflects differential expression patterns of the transferases, revealing the potential of sialic acids in modulating pathogen–host interactions.

Each of three tissue-specific polyNeu5Ac transferases (SIA8B–D) can synthesize homopolymers of α2,8-linked Neu5Ac (polyNeu5Ac) on NCAM1 *in vitro*. All three polysialyltransferases possess a polybasic polysialylmotif (p*I* ~ 12) in front of their S-sialylmotif. Why my *N*-glycans V and VI are 1000 times better acceptor substrates for both SIA8B (Q92186) and SIA8D (Q92187) than others is still incompletely understood, but sequences located inside my immunoglobulin-like domain 5 together with fibronectin type-III-like domain 1 are essential for proper polysialylation. PolyNeu5Ac on NCAM1 modulates cell–cell and cell–substratum interactions, and was essential for brain, muscle, lung and kidney development (please see Chapter 30 for functional details on NCAM1). Activity of the transferases strongly depends on their autopolysialylation capacity, indicating successive *en bloc* transfer of preassembled oligo/polyNeu5Ac stretches to acceptor glycans. Polymorphism of the SIA8B promoter region was linked genetically to schizophrenia in both Japanese and Chinese populations. This relationship was further substantiated by a reduced immunostaining for polyNeu5Ac in brain regions from schizophrenic patients. PolyNeu5Ac is also a diagnostic tumor marker for a variety of highly malignant human tumors.

6.8.5
Fucosyltransferases

In humans, seven different fucosyltransferases have been identified so far [20]. They can attach fucose either to terminal or internal Gal/GalNAc or to internal GlcNAc in any type of glycan found in glycoproteins and glycolipids in α1,2, α1,3, or α1,4-linkage. In functional terms, fucosylated clusters of *N*-glycans are almost equivalent to fucosylated *O*-glycans and similar to those in glycolipids (please see Chapters 7, 10 and 27). In humans, α1,3(4)-fucosylation of *N*-glycans is rarely found, except in peripheral leukocytes (FUT7, FUT9) and kidney (FUT4). Fucosylated glycans trigger cell adhesion events and are essential in distinct biological processes such as L-selectin-mediated leukocyte–endothelial adhesion (for functional aspects, please see Chapter 27).

6.8.6
Glucuronyltransferases

Glucuronyltransferases became famous because of their drug-metabolizing capacities. Golgi apparatus-located glucuronyltransferases are, however, functionally different. They modify glycolipids and glycoproteins, and play a role during chondroitin synthesis (please see Chapter 11). The β1,3-glucoronyltransferases B3GA1 (Q9P2W7) and BG3A2 (Q9NPZ5) act on specific *N*- and *O*-glycans on a restricted set of glycoproteins to initiate human natural killer cell epitope 1 (HNK-1) synthesis (for L2/HNK-1 structure, see Figure 1.7c). Neuronal B3GA1 competes with sialylation on a subset of glycoproteins with known cell–cell contact-mediating properties. In my case, B3GA1 is able to modify *N*-glycan II to initiate HNK-1 synthesis, but has lost the competition with α2,6-sialyltransferase at *N*-glycan V and VI (Figure 6.1). Due to the widely expressed less-specific B3GA2, the HNK-1 is not restricted to cells of neuronal origin.

6.8.7
Sulfotransferases

Sulfation of glycans of secretory proteins occurs in all vertebrates and may be an early evolutionary invention to cap glycans with negative charges. Long known as modifications in extracellular matrix components, sulfates on *N*-glycans were undervalued for a long time [21]. Their function in lymphocyte homing, neural outgrowth and hormone clearance is becoming increasingly clear. Currently, ten different Golgi-associated carbohydrate sulfotransferases (CHSTs, see Figure 6.5) are known, all of which use PAPS to sulfate *O*- and *N*-glycans at terminal (CHST2) or internal GlcNAc (CHST4–6), terminal GalNAc (CHST3, 7 and 8), internal Gal (CHST1 and 3) or GlcA (CHSTA10) residues. Although defined sulfated glycans are bioactive sugars on particular glycoproteins, their relative exclusiveness does not relate to the sulfotransferases themselves. Rather this depends on the availability of acceptor substrates provided by the proximal glycosyltransferases. Clusters of sulfated sialyl Lewis^x (sLe^x) on polyLacNAc structures in both *O*- and *N*-linked glycans act as ligands for L-selectin (for functional aspects, please see Chapter 27.4).

HNK-1 was the first sulfated glycan described, and has been identified on lipids and on several neural cell recognition proteins such as NCAM1, L1CAM or myelin-associated glycoprotein (please see Chapter 30.7 for further details). Its synthesis in neurons is regulated by the controlled expression of the neuron-specific glucuronosyltransferase 1 (B3GA1), but not by the widely found HNK-1-specific sulfotransferase 10 (CHSTA). Although polyNeu5Ac and HNK-1 synthesis could occur at identical positions (*N*-glycan V and VI), only one or the other takes place. However, the respective transferases do not compete with each other. The decision is made by the proximal acceptor substrate-synthesizing transferases (B3GA1 and SIA) competing for terminal galactose residues (Figures 6.1 and 6.5). Knockout mice lacking either HNK-1 or polyNeu5Ac exhibited similar cognitive defects, reflecting the importance of the two structures in similar processes during neuronal outgrowth (for detailed functional aspects on polyNeu5Ac, please see Chapter 30.7).

6.9
Diversity of *N*-Glycans: Structural and Functional Implications

Analogous to the Queen who wears a particular hat on a given day for an unknown purpose, a specific glycoprotein leaving the Golgi apparatus may carry different types of *N*-glycans depending on the situation and the different purposes to fulfill. The glycan diversity on proteins is one of the most interesting, as well as puzzling, issues that glycobiologists investigate. It is based on a particular evolutionary history and intermingled with epigenetic pressure(s). Gene duplication events generated a set of homologous glycosyltransferases that provided our predecessors with redundant enzyme pools. Mutational changes switched either donor (blood group AB) or acceptor substrate specificities of glycosyltransferases or, when affecting gene regulation regions, their cell- and tissue-specific expression patterns. A single point mutation in the respective glycosyltransferase [BGAT (P16442)] induced blood group A–B switch. Such mutational events equipped prevertebrates with a splendid glycosylation machinery serving all upcoming needs for successful development. Note that biological functionalism for *N*-glycan structural versatility often evolved later on. As a case in point, the immune system and brain profited tremendously from this preexisting glycosylation machinery to synthesize functional (poly-)sialylated, fucosylated, sulfated and/or *O*-acetylated (not covered in this chapter) *N*-glycans to realize their tasks according to their functional requirements.

Differences in glycan structures can be explained primarily through cell type-specific expression patterns of glycosylation enzymes and of glycoproteins themselves. Human beings, in contrast to all other mammals including chimpanzees, are not able to synthesize *N*-glycolylneuraminic acid and therefore *N*-acetylneuraminic acids (Neu5Ac) are the only sialic acids synthesized in our tissues. Glycans on human serum α_1-antitrypsin, secreted from human liver, are all terminated by α2,6-linked sialic acids and, therefore, lack α2,3-linked Neu5Ac in sLe^x antigens. However, in α_1-antitrypsin synthesized by leukocytes, one out of three glycans possesses α2,3 Neu5Ac in sLe^x antigens. To refer to NCAM1, I am carrying poly-Neu5Ac-modified *N*-glycans in outgrowing neurons, but mostly sulfated ones in quiescent cells. It is also known that particular motifs on a subset of proteins render them a target for particular glycosylation steps. Hence, only lysosomal proteins, but not for instance NCAM1, are modified with a GlcNAc-P on defined mannoses to guarantee proper targeting to lysosomes. On the other hand, I am the main carrier for polyNeu5Ac, which is not found on lysosomal proteins at all. Last, but not least, *N*-glycan structure can be controlled by earlier glycosyltransferase reactions. The bisecting GlcNAc transferase-III (MGAT3) will prevent GlcNAc transferase-V (MGAT5) from acting. In conclusion, glycosylation patterns are cell-type defined, may be dependent on developmental and growth stage, as well as glycoprotein specific. As a consequence, transgenic glycoproteins may carry abnormal, pathology-causing glycans [22].

Databank entries about expressed sequence tags provide us with mRNA expression information for most enzymes in human tissues (unigene database at www.ncbi.nlm.nih.gov). Data about glycan structures on individual proteins synthesized in particular tissues are still rare, but if available, document a remarkable complexity. For instance, 48 different complex-type *N*-glycans were characterized on a subset of NCAM1 obtained from calf brain (Figure 6.1 for some details). An even greater diversity was found in isolated human neutrophils. Here, 78 different complex *N*-glycan structures have been characterized among others (www.functionalglycomics.org). Such diversity results from glycosyltransferases competing for the same acceptor substrate, which in turn results in a mixture of different *N*-glycans at identical Asn sequons of a given glycoprotein. Diversity cannot be explained solely by purposeful strategies of cells. It may also be the consequence of stochastic events of competing enzyme reactions on suboptimal substrates and of the premature stopping of synthesis due to accelerated intra-Golgi transport.

6.10
Conclusions

NCAM1 was your travel guide along the secretory pathway. Equipped with this experience and knowledge, you know now that *N*-glycosylation is one of the more important post-translational protein modifications. Without the spectrum of *N*-glycans, metazoan and vertebrate development would have resulted in outcomes differing from those we see nowadays. Disulfide bond formation, lysosomal targeting, assembly of heteromeric protein complexes in the secretory pathway and neural plasticity are only some examples of *N*-glycan functions, indispensable for the development of organisms and the maintenance of their function and architecture. Secretion is strongly dependent on different phylum-dependent *N*-glycan tags. Monoglucosylated *N*-glycans provide an ER-retention tag, whereas the Man_8-A isomer confers ER export in vertebrates. Furthermore, the structural diversity of *N*-glycans provided the basis for many coevolutionary events occurring during the evolution to higher eukaryotes and vertebrates. In humans, *N*-glycosylation is directly involved in the pathogenesis of several diseases such as protein-folding diseases, congenital diseases of glycosylation and cancer. Even seemingly subtle modifications can play significant roles such as the core substitutions [23]. Thus, the enormous diversity of *N*-glycans will continue to motivate continued research in this field, since detailed analyses will remain fundamental for future advances to functionally decode the glycome.

Summary Box

The puzzle of *N*-glycosylation is starting to be resolved. The basic molecular and evolutionary aspects of the synthetic and degradative pathways of *N*-glycans are thus defined, but there is still a long way to go to understand the process and implications in their entirety. In the future, system biology approaches may help to fill the empty spaces and complement missing roads on this map.

References

1 Alberts B *et al. Molecular Biology of the Cell*, 4th edn. Garland Science, London, 2002.

2 Helenius A, Aebi M. Roles of *N*-linked glycans in the endoplasmic reticulum. *Ann Rev Biochem* 2004;73:1019–49.

3 Helenius J *et al.* Translocation of lipid-linked oligosaccharides across the ER membrane requires Rft1 protein. *Nature* 2002; 415:447–50.
4 Wilson CM, High S. Ribophorin I acts as a substrate-specific facilitator of *N*-glycosylation. *J Cell Sci* 2007;120:648–57.
5 Deprez P *et al.* More than one glycan is needed for ER glucosidase II to allow entry of glycoproteins into the calnexin/calreticulin cycle. *Mol Cell* 2005;19:183–95.
6 Alonzi DS *et al.* Glucosylated free oligosaccharides are biomarkers of endoplasmic reticulum α-glucosidase inhibition. *Biochem J* 2008;409:571–80.
7 Caramelo JJ, Parodi AJ. How sugars convey information on protein conformation in the endoplasmic reticulum. *Semin Cell Dev Biol* 2007;18:732–42.
8 Roth J *et al.* Protein quality control: the who's who, the where's and therapeutic escapes. *Histochem Cell Biol* 2008;129: 163–77.
9 Appenzeller-Herzog C, Hauri HP. The ER–Golgi intermediate compartment (ERGIC): in search of its identity and function. *J Cell Sci* 2006;119:2173–83.
10 Tang BL *et al.* COPII and exit from the endoplasmic reticulum. *Biochim Biophys Acta* 2005;1744:293–303.
11 Hirao K *et al.* EDEM3, a soluble EDEM homolog, enhances glycoprotein endoplasmic reticulum-associated degradation and mannose trimming. *J Biol Chem* 2006;281:9650–8.
12 Spiro RG. Role of *N*-linked polymannose oligosaccharides in targeting glycoproteins for endoplasmic reticulum-associated degradation. *Cell Mol Life Sci* 2004;61: 1025–41.
13 Durrant C, Moore SE. Perturbation of free oligosaccharide trafficking in endoplasmic reticulum glucosidase I-deficient and castanospermine-treated cells. *Biochem J* 2002; 365:239–47.
14 Marsh BJ *et al.* Direct continuities between cisternae at different levels of the Golgi complex in glucose-stimulated mouse islet beta cells. *Proc Natl Acad Sci USA* 2004; 101:5565–70.
15 Lupashin V, Sztul E. Golgi tethering factors. *Biochim Biophys Acta* 2005;1744:325–39.
16 Mendelsohn R *et al.* Complex *N*-glycan and metabolic control in tumor cells. *Cancer Res* 2007;67:9771–80.
17 Cheung P *et al.* Metabolic homeostasis and tissue renewal are dependent on β1,6GlcNAc-branched *N*-glycans. *Glycobiology* 2007;17:828–37.
18 Seelentag W *et al.* Prognostic value of β1, 6-branched oligosaccharides in human colorectal carcinoma. *Cancer Res* 1998;58: 5559–64.
19 Varki NM, Varki A. Diversity in cell surface sialic acid presentations: implications for biology and disease. *Lab Invest* 2007;87: 851–7.
20 Ma B *et al.* Fucosylation in prokaryotes and eukaryotes. *Glycobiology* 2006;16:R158–84.
21 Monigatti F *et al.* Protein sulfation analysis – a primer. *Biochim Biophys Acta* 2006;1764: 1904–13.
22 Cooper DK *et al.* α1,3-galactosyltransferase gene-knockout pigs for xenotransplantation: where do we go from here? *Transplantation* 2007;84:1–7.
23 André S *et al.* From structural to functional glycomics: core substitutions as molecular switches for shape and lectin affinity of N-glycans. *Biol Chem* 2009;390:557–65.

7
O-Glycosylation: Structural Diversity and Functions

Georgios Patsos and Anthony Corfield

The two preceding chapters have outlined the biosynthesis of *N*-glycans. In addition to the atom in the asparagine amide, the hydroxyl groups of serine/threonine serve as a common target for glycan attachment. This chapter provides a succinct overview of the various forms of *O*-glycosylation, with the *O*-glycan defining the attachment site and these are listed in Table 7.1 (collagens have long been known to constitute a group of proteins found in connective tissue with a unique *O*-glycan disaccharide linked to hydroxyl-lysine, Glcα1,2±Galβ1-*O*-Lys). Table 7.1 shows the structures of the linkages and summarizes general properties of *O*-glycan occurrence. As outlined in Chapter 15.3, mucins were historically the first proteins found to be glycosylated, in the year 1865. We thus start with the mucin-type glycans and then review *N*-acetylglucosamine linked to serine or threonine (β-*O*-GlcNAc; for structures and abbreviations of all letters of the sugar alphabet, please see Chapter 1) on cytosolic and nuclear proteins; *O*-mannosylation detected in mammalian systems and identified in dystroglycan (a highly glycosylated glycoprotein found specifically in muscle and neuronal tissues; please see Chapter 22.4) and brain proteoglycans together with the oligomannosyl glycans found in yeasts (see Chapter 8); and, finally, fucose and glucose *O*-linked to serine or threonine in the epidermal growth factor (EGF) domains in a variety of proteins, notably linked with the Notch signaling pathway. The proteoglycans also belong to this group and are described in Chapter 11. We now consider the structure of the different types of *O*-linked glycans.

7.1
Structure of O-Linked Glycans

O-GalNAc, mucin-type oligosaccharides vary in size from a single *N*-acetyl-D-galactosamine (GalNAc), through disaccharides to chains with 20 or more sugar residues. Three distinct regions can be identified: (i) a core unit linking the oligosaccharide to polypeptide serine and threonine residues (Table 7.2), (ii) a backbone, found in extended glycans, but not always present (Table 7.3), and (iii) a

The Sugar Code. Edited by Hans-Joachim Gabius

ISBN: 978-3-527-32089-9

Table 7.1 Different linkages of *O*-glycans to proteins.

Type of *O*-linked glycan	Structure and peptide linkage	Comments
O-Linked GalNAc (mucin type)	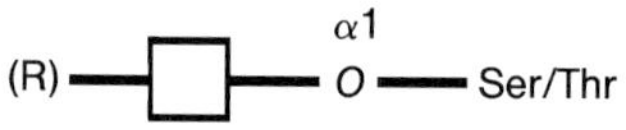 α1; (R)—□—*O*—Ser/Thr; (R)-GalNAcα1-*O*-Ser/Thr	Eight different core structures; Limited peptide consensus sequence for attachment
Glycosaminoglycan	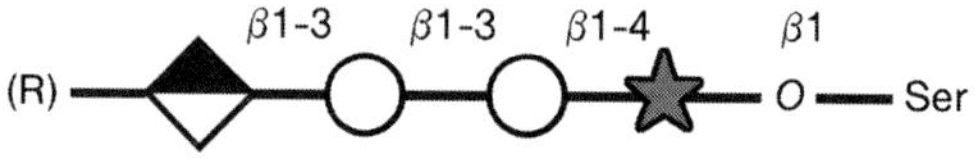 β1-3 β1-3 β1-4 β1; (R)—◆—○—○—★—*O*—Ser; (R)-GlcAβ1,3Galβ1,3Galβ1,4Xylβ1-*O*-Ser	Tetrasaccharide linkage motif to protein
O-Linked GlcNAc	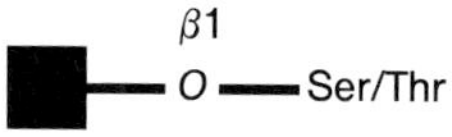 β1; ■—*O*—Ser/Thr; GlcNAcβ1-*O*-Ser/Thr	No evidence for peptide consensus sequence
O-Linked Gal (collagen)	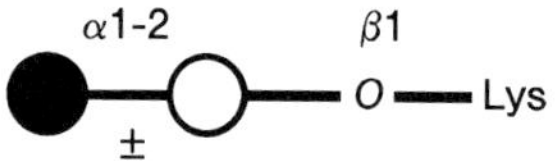 α1-2 β1; ●—○—*O*—Lys; ±; Glcα1,2±Galβ1-*O*-Lys	Hydroxylation of lysines generates glycosylation sites[a]
O-Linked Man	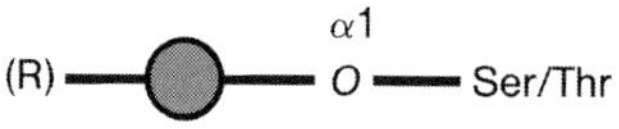 α1; (R)—●—*O*—Ser/Thr; (R)-Manα1-*O*-Ser/Thr	Peptide consensus sequences identified
O-Linked Glc	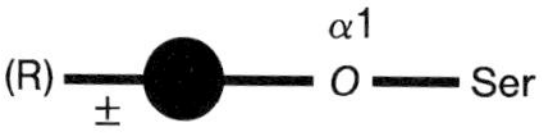 α1; (R)—●—*O*—Ser; ±; (R)±Glcβ1-*O*-Ser	Peptide consensus sequences identified
O-Linked Fuc	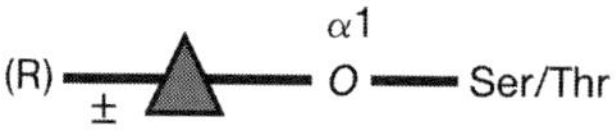 α1; (R)—▲—*O*—Ser/Thr; ±; (R)±Fucα1-*O*-Ser/Thr	Peptide consensus sequences identified
Properties for all *O*-glycans	Differences exist between all groups of *O*-glycan with regard to: • The range of proteins carrying particular *O*-glycan types • The number of chains per protein • The extent of sharing with other glycan types • The size and range of glycan chains. This variation is apparent at the species level, reflecting the process of evolution	

[a] Collagens have long been known to constitute a group of proteins found in connective tissue with a unique *O*-glycan disaccharide linked to hydroxyl-lysine, Glcα1,2±Galβ1-O-Lys.

Table 7.2 Core structures found in *O*-glycans.

Core type	Structure
1	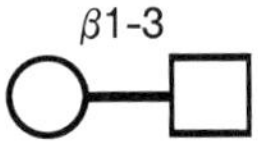 β1-3 Galβ1,3GalNAc
2	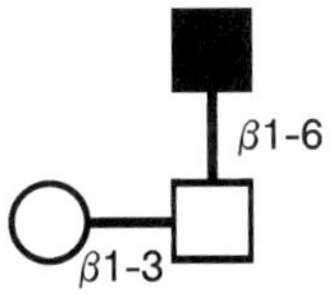 β1-6, β1-3 Galβ1,3(GlcNAcβ1,6)GalNAc
3	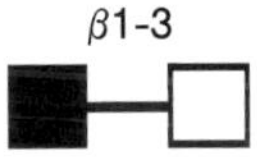 β1-3 GlcNAcβ1,3GalNAc
4	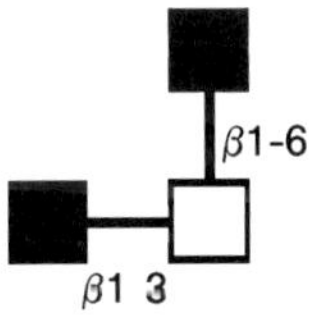 β1-6, β1 3 GlcNAcβ1,3(GlcNAcβ1,6)GalNAc
5	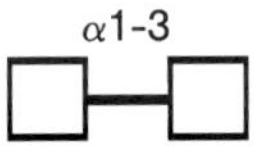 α1-3 GalNAcα1,3GalNAc
6	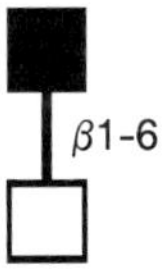 β1-6 GlcNAcβ1,6GalNAc
7	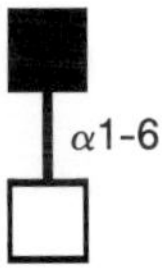 α1-6 GlcNAcα1,6GalNAc
8	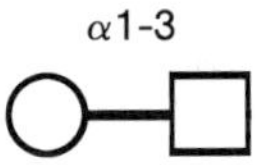 α1-3 Galα1,3GalNAc

Table 7.3 *O*-Glycan chain extension as backbone structures.

Type	Structure	Name
Type 1	β1-3 Galβ1,3GlcNAc	
Type 2	β1-4 Galβ1,4GlcNAc	*N*-acetyllactosamine
Poly *N*-acetyllactosamine type 2	β1-4 β1-3 $($Galβ1,4GlcNAcβ1,3–$)_n$	i-antigen
Branched *N*-acetyllactosamine type 2	β1-4, β1-6, β1-4, β1-3 Galβ1,4GlcNAcβ1,6 \ Galβ1 / Galβ1,4GlcNAcβ1,3	I-antigen

terminal, branch-end domain (Table 7.4). Tables 7.2–7.4 show the sequence structure of these domains and a selection from the many branch-end motifs is collected in Table 7.4.

Eight core structures have been identified and are shown in Table 7.2. Core 1 and core 2 are the most common, with lower occurrence of cores 3 and 4, while the remaining structures are relatively rare. The different cores show a cell- and tissue-specific pattern which relates to their biological function [1–3]. Each core determines the final structure of the extended and completed glycans formed through the common biosynthetic pathways outlined in Figure 7.1. Branched chains are derived from core 2 and core 4, and are extended through the biosynthetic pathways. Larger glycans are extended by Galβ1,3/4GlcNAc repeat units and are shown in Table 7.3. The enormous variety of *O*-glycan oligosaccharide structures derives from the decoration of the core and backbone units. These modifications include fucosylation, sialylation, sulfation, acetylation and methylation. The structures of a selected number of branch-end motifs are given in Table 7.4. In addition to the sequence structural variation, the sialic acids exist as a family of

Table 7.4 Branch-end structures found on *O*-glycans [see also Figure 1.5 for ABH structures and Table 27.2 for Lewis (Le) epitopes].

Type of glycan	Structure
Blood group H type 1 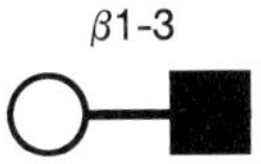 β1-3 (Galβ1,3GlcNAc)	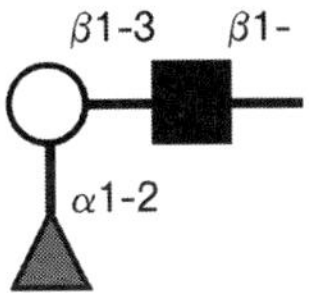 β1-3 β1- α1-2 Fucα1,2Galβ1,3GlcNAcβ1–
Blood group A type 2 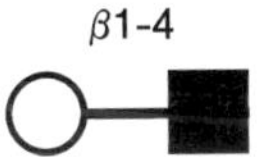 β1-4 (Galβ1,4GlcNAc)	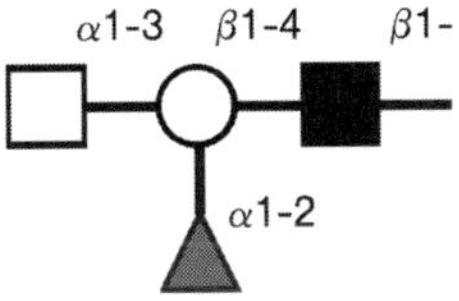 α1-3 β1-4 β1- α1-2 GalNAcα1,3Galβ1,4GlcNAcβ1– | α1,2 Fuc
Blood group B type 1 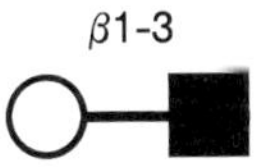 β1-3 (Galβ1,3GlcNAc)	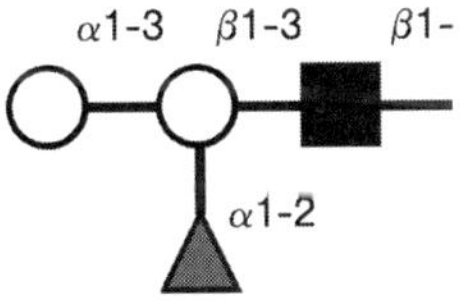 α1-3 β1-3 β1- α1-2 Galα1,3Galβ1,3GlcNAcβ1– | α1,2 Fuc
Le^a and Le^x	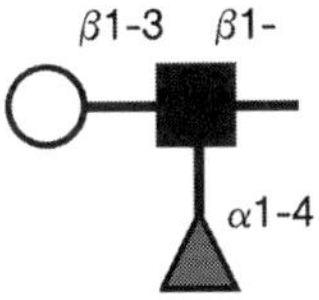 β1-3 β1- α1-4 Galβ1,3GlcNAcβ1– | α1,4 Fuc 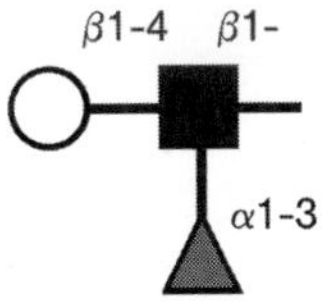 β1-4 β1- α1-3 Galβ1,4GlcNAcβ1– | α1,3 Fuc

continued

Table 7.4 *Continued*

Type of glycan	Structure
Le^{b}	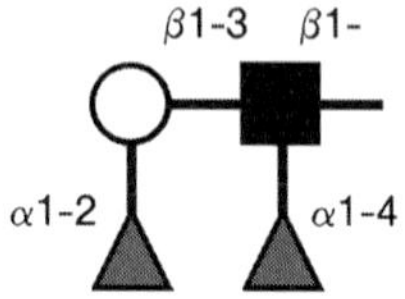 β1-3 β1- α1-2 α1-4 Galβ1,3GlcNAcβ1– | α1,2 | α1,4 Fuc Fuc
Le^{y}	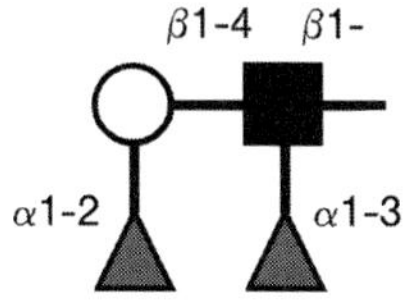 β1-4 β1- α1-2 α1-3 Galβ1,4GlcNAcβ1– | α1,2 | α1,3 Fuc Fuc
Sialyl-Le^{a} and sialyl-Le^{x}	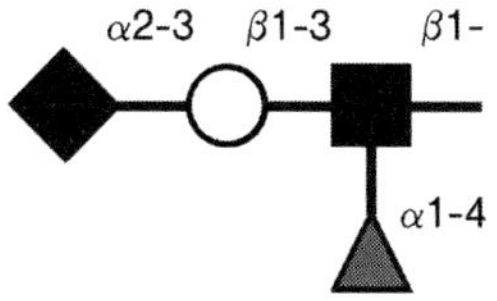 α2-3 β1-3 β1- α1-4 Neu5Acα2,3Galβ1,3GlcNAcβ1– | α1,4 Fuc 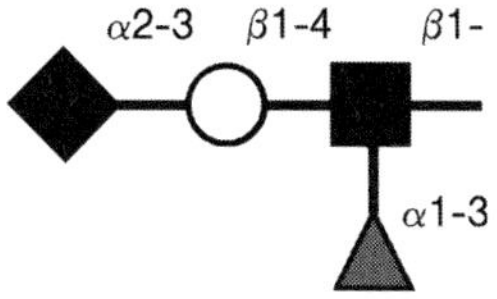 α2-3 β1-4 β1- α1-3 Neu5Acα2,3Galβ1,4GlcNAcβ1– | α1,3 Fuc
6′-Sulfo-Le^{x}	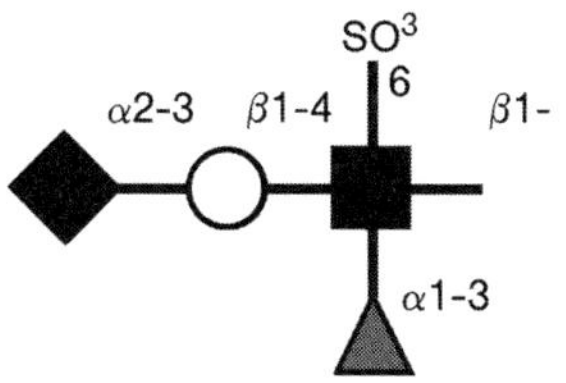 SO^{3} 6 α2-3 β1-4 β1- α1-3 SO^{3-} |6 Neu5Acα2,3Galβ1,4GlcNAcβ1– | α1,3 Fuc

continued

Table 7.4 *Continued*

Type of glycan	Structure
Sialyl-Tn	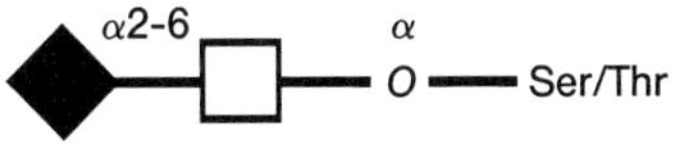 Neu5Acα2,6GalNAc-α-*O*-Ser/Thr
Sialyl-T antigen	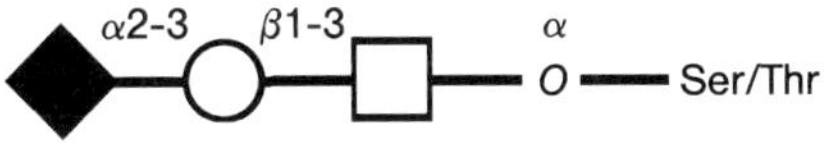 Neu5Acα2,3Galβ1,3GalNAc-α-*O*-Ser/Thr 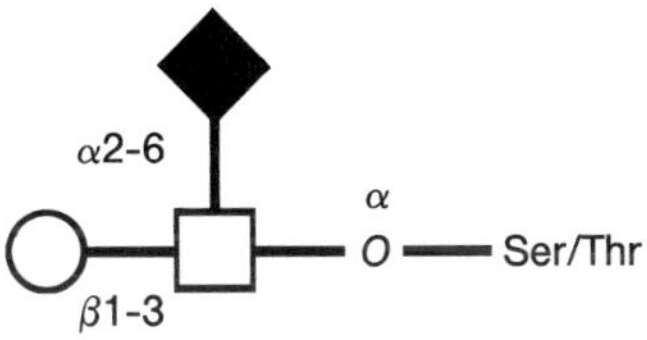 Neu5Ac | α2,6 Galβ1,3GalNAc-α-*O*-Ser/Thr
Monosialylated core 3	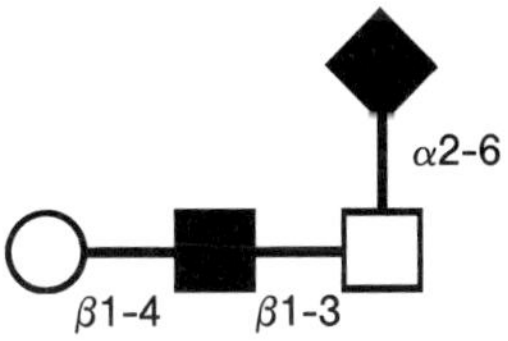 Neu5Ac | α2,6 Galβ1,4GlcNAcβ1,3GalNAc
Sd^{a} antigen (type 1 and 2 chains)	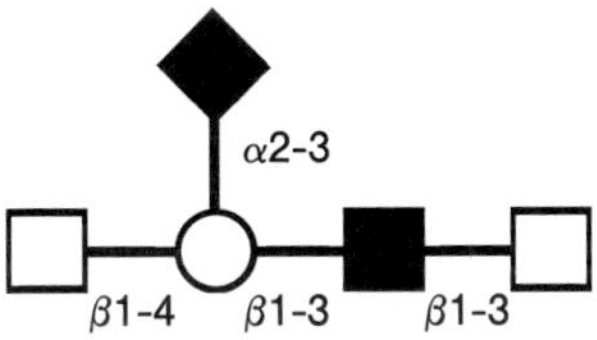 Neu5Ac | α2,3 GalNAcβ1,4Galβ1,3GlcNAcβ1,3GalNAc 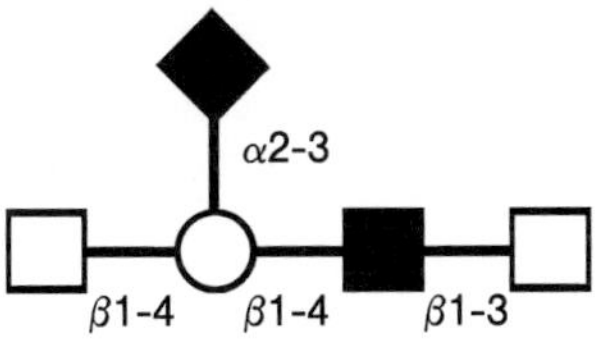 Neu5Ac | α2,3 GalNAcβ1,4Galβ1,4GlcNAcβ1,3GalNAc

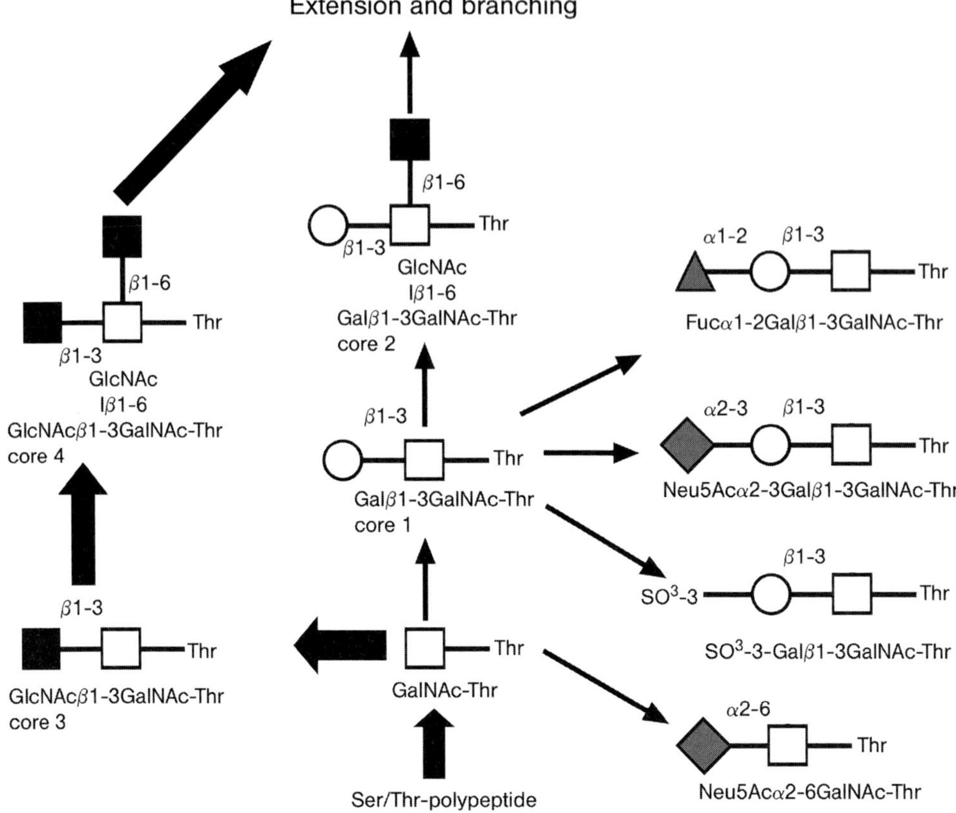

Figure 7.1 *O*-Glycan biosynthetic pathways. The diagram shows the formation of core 1 and core 3 from the addition of the first GalNAc serine/threonine acceptors on the polypeptide chain. The red arrows indicate the favored pathway in gastrointestinal tissues for core 3 biosynthesis, relative to the alternative core 1 pathway in blue. Some examples of branch-end additions are shown (see also Figure 27.4 for the regulation between core 1 extension and core 2 generation).

monosaccharides, further adding to *O*-glycan structural variety and identity (see Info Box 1).

The addition of *O*-glycan chains to polypeptide serine and threonine residues does not follow the tripeptide sequon model for *N*-linked oligosaccharides. Potentially any serine or threonine may be a site and, in contrast, to *N*-glycan sites often appear as clusters. No general consensus sequence has been identified; however, a series of general rules have been identified and an algorithm has been developed to predict *O*-glycosylation sites [4]. The family of UDP-*N*-acetylgalactosamine: polypeptide *N*-acetylgalactosaminyltransferases (ppGaNTases), which add the first GalNAc residues to serine and threonine acceptors, show selection which is regulated at the enzymatic level (see Info Box 2). Common features which determine

Info Box 1

The sialic acids are unique among the monosaccharide components found commonly in *O*-glycans as they are found in Nature as a family of over 50 forms. They were originally discovered in 1930s by Klenk and Blix, and named on the basis of the sources used for isolation. Klenk used brain and termed the products neuraminic acids (amino sugars from neuronal tissue), while Blix used salivary glands and used the Greek root *sialos* ('*saliva*') for sialic acids. Both terminologies remain in use.

The sialic acids are nine-carbon α-keto acids formed through the condensation of pyruvate with *N*-acetyl-D-mannosamine, having a six-membered pyranose ring and adopt a $^2C_5$ chair conformation (please see Figure 1.6 for structure). Natural glycoconjugates are found in the α-configuration except for the glycosyl donor, that is CMP-β-*O*-sialic acids. The family is built due to the substitutions found within the monosaccharide. Substitution of the C5 amino group may be with acetyl or glycolyl groups, and each of these acts as a parent molecule carrying acetyl esters at C4, C7, C8 and C9 as mono, di and tri forms. C9 can also be substituted with phosphate or lactyl esters. Sulfate esters and methyl ethers are found at position 8. 1,7-Lactone and 2,7-anhydro forms exist, and 2-dehydro analogs with a double bond between C2 and C3 are found as free monosaccharides as they have no α-keto group. Substitution of the C5 amino group by a hydroxyl generates 3-deoxy-D-*glycero*-D-*galacto*-2-nonulosonic acid.

The various substitutions have biological importance especially in interactions mediated by the siglecs and selectins (see Chapter 27) where sialic acids form recognition epitopes on receptor proteins. Acetyl esters at positions 4 and 7–9 donate resistance to most sialidases and provide a mechanism for biological protection of these sialic acid forms.

the peptide location of *O*-glycans include the following: (i) A primary peptide sequence preference which varies for serine and threonine and correlates well with higher levels of threonine substitution known. Proline is located adjacent to *O*-glycosylation sites. Specificity windows predict positions for Ser/Thr/Pro and exclude certain amino acids, including Cys, Trp, Met and Asp [4]. (ii) Only those serine and threonine residues exposed on the surface of the molecule will be glycosylated. Thus, a conformational role of the protein must be considered. The recognition of *O*-glycans on β-turns, extended structures where proline residues play a role, the absence of aromatic, bulky amino acids near to *O*-glycosylation sites and lack of *O*-glycans in hydrophobic peptide domains are all significant. (iii) *O*-Glycosylation patterns correlate with tissue location. This is governed by the ppGaNTases and subsequent action of other glycosyltransferases. This has been well demonstrated for the secreted mucin genes (for instance MUC5AC in stomach, respiratory tract and conjunctiva), where the tandem repeat Ser/Thr/Pro domains contain the same amino acid sequence, but where variable glycosylation is found on a tissue-specific basis.

O-β-*N*-Acetylglucosamine (β-*O*-GlcNAc) glycosylation with a single GlcNAc residue attached to serine and threonine hydroxyl residues in a β-linkage (β-*O*-GlcNAc see Table 7.1) has led to the identification of a large family of proteins closely integrated into signaling networks.

Over 500 proteins have been found to carry this modification, including a large number of nuclear and cytoplasmic proteins with a wide range of metabolic functions [5]. The activity of these proteins has been found to be regulated by the presence of this monosaccharide at target protein serine and threonine sites. The presence of β-*O*-GlcNAc on these proteins exists as a dynamic equilibrium with phosphorylation of the same serine and threonine residues in a cyclic manner [5]. The detection of this type of glycosylation reveals novel properties. In contrast to most other glycoproteins, β-*O*-GlcNAc-modified proteins are located in the nucleus and cytoplasm, and not at the cell surface, the extracellular matrix or as luminal components of the secretory pathway [6]. The rapid cycling nature of the glycosylation is also novel and diverges from other glycoproteins where the glycans are more inert. Although few (around 50) sites of β-*O*-GlcNAc addition to proteins have been located there is only limited evidence to suggest any consensus sequence. Some sites have a Pro-Val-Ser motif and many have high Pro-Glu-Ser-Thr (PEST) scores, associated with rapid proteolytic degradation, hinting at a role in the limitation of proteolysis.

O-Mannose modification of proteins has been know for some time and was believed to be limited to yeast mannan structures. However, progress in glycoanalysis has identified linkage domains that are rich in serine, threonine and proline, in common with *O*-GalNAc sites with attachment to the hydroxyl amino acids. The principal structures that have been identified are based on Galβ1,4GlcNAcβ1,2Manα-*O*-Ser/Thr. Extended glycans have also been reported where this core is modified by additional sialic acid, fucose, GalNAc, GlcNAc, mannose and ester sulfate. Some species variation has been reported and human sources contain only the Neu5Ac α2,3Galβ1,4GlcNAcβ1,2Man-α-*O*-Ser/Thr glycans and the non-reducing terminal trisaccharide is the same as that carried by the Fuc-α-*O*-Ser/Thr glycans in human Notch, epidermal growth factor (EGF)-like domains [7, 8]. *O*-Man glycans represent a smaller group of glycoprotein modifications, and are present in only a few examples limited to brain, neural tissue and skeletal muscle where they play an important role in development. Work has mostly been done with α-dystroglycan – a skeletal muscle extracellular protein (Figure 22.4).

No consensus sites have been identified for proteins carrying *O*-Man, but examination of the mucin-like domain in α-dystroglycan revealed two similar peptide sequences which were strong acceptors for *O*-mannosylation by the protein *O*-mannosyltransferase (POMT). Closer examination showed that Thr414 was the principal acceptor site and that the addition of *O*-Man residues was a sequential rather than random process and a preferred peptide acceptor site was proposed. Tandem repeat peptide sequences from MUC1, 2, 3, 4, 7, 11 and 20 were also tested but showed very low acceptor activity, implicating a specific recognition sequence for α-dystroglycan. It is not yet clear whether this peptide recognition sequence is relevant for other *O*-Man-containing proteins.

O-Fuc and *O*-Glc glycosylation, with linkage of fucose or glucose to serine or threonine, was detected through studies on the EGF domains of proteins involved in the regulation of blood clotting. Most work has been concerned with the Notch receptor proteins and their ligands Delta and Serrate (called Jagged in mammalian systems) that contain multiple EGF domains, some of which carry *O*-Fuc and/or *O*-Glc in addition to *N*-glycans [8]. The Notch receptors are large single-pass transmembrane proteins and are coded by one gene in *Drosophila*, but four in mammals. The novel glycans range from single fucose or glucose residues to tetra-, that is Neu5Acα2,3Galβ1,4GlcNAcβ1,3Fuc-α-*O*-Ser/Thr, and trisaccharides Xylα1, 3Xylα1,3Glc-β-*O*-Ser/Thr (Xyl = D-xylose), respectively. The importance of the size of the glycans is related to the signaling pathways mediated by Notch and its ligands, detailed below in Section 7.3. In contrast to other *O*-glycans there are clear peptide consensus sequences associated with both Fuc-α-*O*-Ser/Thr and Glc-β-*O*-Ser/Thr transfer. EGF domains contain six conserved cysteine residues, C^1 to C^6, which contribute to the domain topography through formation of three disulfide bridges. Fuc-α-*O*-Ser/Thr glycans are found attached to the consensus $C^2X_{4\text{-}5}$(S/T)C^3 and Glc-β-*O*-Ser/Thr to the sequence C^1XSXPC^2, where C^1 to C^3 are the respective cysteine residues, S/T are serine and threonine, and X is any other amino acid in the EGF-like domain. A number of proteins have been found which carry these sequences, including the Notch family of receptors, their ligands Delta and Jagged, the Cripto protein, and several serum proteins (factor VII, factor IX, protein Z and thrombospondin).

7.2
Biosynthetic Routes for *O*-Glycans

O-GalNAc, mucin-type biosynthesis of glycan sequences follows a series of pathways located in regions of the ER and Golgi apparatus [1]. The primary structure of glycans is regulated by the substrate specificity of the glycosyltransferases. The first step is the addition of GalNAc to serine and threonine residues in acceptor proteins and is catalyzed by a family of up to 20 ppGaNTases (see Info Box 2). The importance of this family is illustrated by its evolutionary conservation and the selective expression of isoenzymes in a tissue-specific manner. The transferases show substrate specificities detecting acceptor serine or threonine residues in the peptide and whether they are already substituted by a GalNAc residue. The enzymes act in a hierarchical manner dictating the need for coordinated action to achieve full *O*-glycosylation of potential sites. Expression on a tissue-specific manner leads to differential glycosylation patterns. This initial step is seen as a critical, guiding event in *O*-glycan positioning, level of occupancy and oligosaccharide extension in defined proteins.

The synthesis of core, backbone and peripheral regions to generate the immense library of *O*-glycan structures known follows on from the initial action of the ppGaNTases. There is a considerable literature on the families of glycosyltransferases responsible for formation of *O*-glycans [2]. Each possesses a substrate specificity which generates the *O*-glycan structures. An example for some of the *O*-GalNAc glycan pathways is shown in Figure 7.1. In addition, further modification of some sugars occurs generating new epitopes. The most important of these events are sulfation of Gal, GlcNAc and GalNAc and *O*-acetylation of sialic acids (see Info Box 1).

O-β-*N*-Acetylglucosamine monosaccharide metabolism on proteins is governed by GlcNAc modifying enzymes. A specific β-*O*-GlcNAc transferase (OGT) catalyses the transfer of GlcNAc to the serine and threonine residues on protein acceptors. The transferase has been isolated from liver and characterized in detail. It is a heterotrimer of two 110-kDa and one 78-kDa subunits. Some tissue variation of subunit utilization has been found with the 110-kDa unit present in all cases and differential expression of the 78-kDa subunit. The larger 110-kDa unit has two domains linked by a nuclear localization peptide. The C-terminal domain is the catalytic center of the enzyme while the N-terminus comprises multiple tetratricopeptide (TPR) repeats. The TPR domains determine the recognition of protein substrates. The enzyme is itself a carrier of β-*O*-GlcNAc, although the individual sites are not yet known. It shows several K_m values for the donor substrate UDP-GlcNAc. On the one hand, the enzyme could not be saturated by UDP-GlcNAc at maximal physiological concentrations while a low apparent K_m in the range 500nM favors OGT utilization of this substrate over the UDP-GlcNAc transporters in the ER and Golgi which deplete cytosolic levels of the nucleotide sugar. Accordingly, the sensitivity to UDP-GlcNAc levels implicates a sensor role for OGT and a direct link with regulation of the many regulation pathways through the action of the β-*O*-GlcNAc- proteins associated with this control. Interaction of OGT with the huge number of protein substrates is mediated through specific proteins which bind to the TPR domains and target OGT to defined complexes. OGT knockout mice have been generated and exhibit embryonic lethality, underlining the need for OGT in normal cell growth and response to extracellular signals [5].

The removal of β-*O*-GlcNAc from proteins is due to the action of a β-hexosaminidase. This enzyme was show to be specific for β-*O*-GlcNAc. Protein structure analysis revealed two catalytic domains, one for the β-hexosaminidase activity and a second for acetyltransferase, which was also confirmed enzymatically. A caspase-3 cleavage site (caspases are a family of proteinases which play an important part in apoptosis, see Figure 27.2) was reported between these two catalytic domains, but an explanation for the interaction of these two activities remains speculative.

O-Man transfer to serine and threonine acceptor sites on proteins is catalyzed by a family of POMTs. These enzymes are found in the ER and transfer mannose from a dolicholphosphate mannose donor to the protein acceptor. The isoforms of POMT vary in their precise substrate specificities and appear to be analogous to the family of ppGaNTases involved in mucin-type glycan biosynthesis. The main transferases identified in human tissues are POMT1 and 2, and most studies have focused on the properties and action of these isoforms.

Extension of the *O*-Man glycan occurs through the action of a specific transferase, protein *O*-mannose *N*-acetylglucosaminyl transferase (POMGnT1), located in the Golgi apparatus, leading to the formation of a Manβ1,2GlcNAc linkage. The

Info Box 2

The transfer of GalNAc to the serine/threonine residues to initiate the *O*-GalNAc, mucin-type glycans was first identified by Saul Roseman in 1967. Examination of substrate specificity showed that there were sequence requirements for the serine or threonine acceptors and this led ultimately to the identification of a family ppGaNTases (EC 2.4.1.41) largely through the work of the Henrik Clausen (Copenhagen, Denmark) and Lawrence Tabac (Bethesda, USA) groups. The ppGaNTases are type II membrane proteins comprising a short *N*-terminal cytoplasmic tail linked to a small transmembrane anchor, which is in turn associated via a stem domain to a large component located in the lumen of the Golgi. The family has more than 19 members, and reflects the need for generation of clustered *O*-glycan substitution in mucin tandem repeat sequences and the stem domains of membrane glycoproteins. The detection of a number of pseudogenes during cloning work on these transferases emphasized the need to demonstrate enzymatic activity for each protein identified. The large number of family members reflects the importance of the initial transfer of GalNAc to initiate mucin-type *O*-glycans and the need to ensure maximum occupancy of tandem repeat domains in molecules such as the mucins where the carbohydrate content plays roles in physical properties such as viscoelasticity in addition to information conferred by the glycan sequence. Although the ppGaNTases are present in the majority of tissues examined there is a selective expression of members of the family and this matches the main glycoprotein substrates found in these tissues. The substrate specificity shows that different isoenzymes show specificity for peptides or glycopeptides and prior glycosylation of serine or threonine with GalNAc at certain positions in the peptide. Further evidence for selective subcellular expression of ppGaNTase isoenzymes between the rough endoplasmic reticulum (ER) and Golgi apparatus further underlines the sophisticated utilization of the family of transferases to ensure optimal *O*-glycosylation. The data available for mucin substrates suggests that the initial transfer of GalNAc takes place in a hierarchical manner. Thus the complete glycosylation of such substrates relies on the coordinated action of multiple ppGaNTase family members.

enzyme is similar but distinct from other *N*-acetylglucosaminyl transferases leading to the formation of the same linkage in *N*-glycans. Considerable interest has followed the identification of these two glycosyltransferases as their deletion has been associated with a number of congenital abnormalities which are detailed in Chapter 22.2.2.

The completion of the *O*-Man glycans depends on the action of galactosyl and sialyltransferases and the binding of fukutin, itself a glycosyltransferase. Fukutin is found to colocalize with POMGnT1 in the Golgi most likely through the formation of a complex and may play a role in modulating the glycosyltransferase activity. Recent developments have shown that GnT-Vb also plays a role in the branching of *O*-mannosyl-linked glycans, has high activity in brain and testis, and may mediate the adhesion and migration of human neuronal cells through integrin- and laminin-dependent interactions. Tumor-related functions are reviewed in Table 25.2.

O-Fuc and *O*-Glc glycosylation of EGF domains was originally identified through genes regulating Notch signaling and which were found to be glycosyltransferases [8]. The specific protein-linked α-*O*-fucosyltransferase has been cloned from *Drosophila* (OFUT1) and human (POFUT1) sources with transfer to the specific consensus sequences identified in EGF-like domains of protein acceptors. The importance of the consensus sequence is emphasized by the finding that not all EGF domains in Notch proteins are *O*-fucosylated [8]. *Drosophila* OFUT1 is located in the ER, possesses chaperone properties required to achieve the correct folding of the Notch EGF-like domain peptide and is also required for the trafficking of wild-type Notch out of the ER. An extracellular interaction between Notch and OFUT1 is essential for the normal endocytic transportation of Notch. The importance of protein *O*-fucosylation is underlined by the embryonic lethality of POFT1 deletion in mice. Chapter 24 deals with relevant aspects in embryogenesis. A second protein-linked *O*-fucosyltransferase gene, POFUT2, showing substrate specificity for thrombospondin type 1 repeats (TSRs), notably in the ADAMTS proteases and not for EGF domains, has also been reported [7].

The extension of the protein *O*-Fuc unit occurs initially through the action of a specific β1,3-N-acetylglucosaminyl transferase. The gene was detected in *Drosophila* and three mammalian homologs have been identified – Manic Fringe, Lunatic Fringe and Radical Fringe (for phenotypes of KO mice, please see Table 23.1). Recognition of *O*-fucosylated EGF repeat amino acids has been shown for all three mammalian Fringes. The Fringes show substrate activity with the Notch proteins and their ligands Delta and Jagged [7].

In mammalian species the GlcNAcβ1,3Fuc-α-*O*-Ser/Thr disaccharide is a substrate for a β1,4-galactosyltransferase. The investigation of Notch signaling in Chinese hamster ovary (CHO) cells deficient in galactose transfer shows that interaction of Notch with its ligands Delta and Jagged does not occur, and that this can be rescued by the transfection of murine β4GalT1 – a β1,4-galactosyltransferase. Further extension of the trisaccharide Galβ1,4GlcNAcβ1,3Fuc-α-*O*-Ser/Thr by a α2,3 sialyltransferase is proposed, but little information is currently available. In place of the extension of Fuc-α-*O*-Ser/Thr by the action of the fringe genes a β1,3-glucosyltranferase has been reported to yield Glcβ1,3Fuc-α-*O*-Ser/Thr. A second β1,3-glucosyltransferase was found with specificity to TSRs. The biological significance of the Glcβ1,3Fuc-α-*O*-Ser/Thr has not been investigated.

The direct addition of glucose to proteins to generate Glc-β-*O*-Ser/Thr has been reported using recombinant factor VII EGF repeat as the acceptor. This study showed that the enzymatic activity recognized the consensus sequence and also the three-dimensional structure of the EGF repeat. Glc-β-*O*-Ser/Thr should serve as a substrate for further elongation by the action of xylosyltransferases; however, little data is available concerning these activities [7, 8]. Very recent evidence has shown that collagen hydroxylysine glycosylation is initiated by specific β(1-O) galactosyltransferases [9]. Two proteins GLT25D1 and GLT25D2, were identified in man, with constitutive expression of GLT25D1, while GLT25D2 is limited to nervous tissue.

7.3 Regulation of *O*-Glycosylation and Glycan Processing

The biosynthesis and processing for the different families of protein-linked glycans is quite different, and reflects the discrete regulation of structure and biological function. *O*-Glycan chains are formed due to the concerted action of specific glycosyltransferases, and the specificity of the transferases determines the final structure of the glycans and explains the glycoform diversity seen in proteins.

Both biosynthesis and processing of *O*-glycans by glycosyltransferases and glycosidases takes place in the ER and Golgi apparatus. The *O*-glycans are formed and processed in stepwise sequential fashion. The different types of *O*-glycosylation show their own characteristic manipulations in this respect. The formation and processing of *O*-glycans is regulated at the level of gene expression, mRNA and enzyme protein activity. Additional control exists through substrate and cofactor concentrations at the subcellular site of synthesis. These processes are integrated to create and regulate the huge variety of glycan structures utilized by different species, tissue and cell types, and observed in different states of development and differentiation. Table 7.5 summarizes the major points concerning the regulation of *O*-glycosylation.

Table 7.5 Regulation of metabolic routes.

Level of regulation	Specific targets
Intermediary metabolism	
Substrates	Acceptor: protein, glycoprotein Donor: nucleotide sugar
Intermediary metabolism	
Substrate availability	Nucleotide and nucleotide sugar transporters
Cellular and subcellular	
Glycan processing	Protein synthesis and expression Proteolytic processing Subcellular localization of proteins Developmental expression Immune cell expression and response

7.3.1 *O*-GalNAc, Mucin Type

In terms of glycan processing, the Ser (Thr)-GalNAc class of *O*-glycans is assembled through the action of a series of linked biosynthetic pathways located in regions of the ER and Golgi apparatus [2]. The pathways lead to predictable structures with core, backbone and peripheral units that are found linked to specific proteins (see Tables 7.2–7.4 and Figure 7.1). The subcellular organization of the glycosyltransferases into functional units acting on individual proteins delivers the specific glycosylation patterns found for each protein [2]. *O*-Glycan processing leading to biologically active glycoproteins occurs via several routes. First, the sequential action of glycosyltransferases delivers completed structures. 'Incomplete' or truncated structures, which also have biological function, result from the action of a reduced number of glycosyltransferases in the biosynthetic units. Alternatively, the action of catabolic enzymes including glycosidases, sulfatases or esterases on the completed *O*-glycan chains may occur. Thus, *O*-glycans can be processed while attached to proteins analogous to the normal catabolic pathways leading to degradation and recycling of glycoproteins. This involves removal of terminal residues such as sialic acids, fucose or sulfate or total removal of *O*-glycan chains through the combined action of exoglycosidases, endoglycosidases and *O*-glycanases responsible for the internal cleavage of the oligosaccharide chains or removal from the peptide backbone. Table 7.5 indicates that glycan processing acts at different levels and these are described below.

Protein synthesis and expression can be regulated during and after biosynthesis through the presence of glycan chains. The synthesis of mature glycophorin A requires *O*-glycans when expressed in glycosylation-deficient CHO cells (please see Info Box 3 in Chapter 6 on origin of mutant CHO cells). The *N*-glycans normally present are not necessary for expression. The expression of glycophorin A without *O*-glycans was dependent on specific *N*-glycans, and indicates that independent expression of *O*- and *N*-glycans can enable glycophorin A expression.

Proteolytic processing of certain proteins demonstrates a requirement for *O*-glycans at specific sites in order to prevent proteolytic cleavage which eliminates biological activity or prevents continued residence/activity of the intact protein at its designated subcellular location. An example is the cell-surface expression of low-density lipoprotein receptors that require *O*-glycans to prevent proteolytic cleavage of their extracellular domains. The transport of the transferrin receptor (TfR) between the cell surface and endosomes also shows such glycosylation dependency. Soluble TfR is formed by proteolytic cleavage in the endosomes. *O*-Glycosylation of Thr104 prevents the action of the relevant protease. The loss of sialic acids from these *O*-glycans abolishes the protection of the *O*-glycans against proteolytic activity.

Subcellular localization of proteins is mediated by *O*-glycosylation. The transport of proteins to the Golgi apparatus from the ER is an example here. Studies using brefeldin A, which induces a microtubule-dependent back-flow of Golgi components to the ER and β-galactosyltransferase, an established trans-Golgi enzyme, demonstrate disruption of the biosynthesis, maturation and intracellular transport. Targeting of the β-galactosyltransferase to the Golgi apparatus depends on its *O*-glycosylation.

Developmental expression during oogenesis, spermatogenesis, fertilization, preimplantation implantation and post-implantation depends on glycosylation, including expression of *O*-glycans (see Chapter 24). From this large literature two examples have been chosen to illustrate this phenomenon.

The CD8αβ coreceptor has a polypeptide stalk connecting the glycoprotein to the thymocyte surface and this domain carries *O*-glycans. Immature $CD4^+CD8^+$ double-positive thymocytes bind major histocompatibility complex (MHC) class I tetramers more avidly than mature CD8 single-positive thymocytes. The differential binding properties are mediated by a developmentally expressed sialyltransferase (ST3Gal-I) acting on the *O*-glycans. The appearance of CD8β linked core 1 sialic acid on mature thymocytes leads to a decrease in CD8αβ–MHCI avidity and this mechanism accounts for the regulation of binding of dimeric CD8 to MHCI. The cluster designation 'CD' nomenclature identifies groups (clusters) of monoclonal antibody binding to lymphocyte cell surface molecules. CD8αβ refers to a subpopulation of CD8 cells with an αβ T cell antigen receptor (TCR), which is a lineage marker for subpopulations of T cells. Two distinct types of TCR exist, defined as αβ and γδ by virtue of their disulfide linked polypeptide chains (please see Info Box in Chapter 27 for information on the CD system).

During the differentiation of crypt cells in the intestinal brush border the membrane-associated proteins pro-sucrase-isomaltase and dipeptidyl peptidase IV are targeted for apical location. These glycoproteins have both *N*- and *O*-glycans. Full *O*-glycosylation is mediated by the *N*-glycans and the subsequent polarized sorting of these proteins to the apical membrane requires intact *O*-glycans. This example demonstrates interrelated but distinct roles for *N*- and *O*-glycans and their processing in brush border enzyme expression during intestinal cell differentiation. Further details on embryogenesis and fertilization can be found in Chapter 24.

Immune cell expression and response is strongly linked with *O*-glycosylation in the immune system and is apparent at several levels (please see Figure 7.1 and also Chapter 27, especially Figure 27.4, which shows competition between core 2 extension and core 1 sialylation). A regulatory function for T lymphocyte core 2 β-1,6-N-acetylglucosaminyltransferase (C2GnT, see Table 7.2 for the structure of core 2) in the peripheral immune system has been identified. CD43 is the major cell-surface glycoprotein and carries core 2 *O*-glycan structures. It is an activation antigen expressed on both CD4 and CD8 single-positive T lymphocytes. Down-regulation of CD43 occurs in thymic positive, selection suggesting that the C2GnT modulated expression of CD43 glycoforms is associated with thymic selection events. Overexpression of C2GnT in transgenic mice led to impaired cell–cell interaction of T lymphocytes isolated from these mice, thus illustrating a reduced immune response.

7.3.2
β-O-GlcNAc

The presence and biological activity of β-*O*-GlcNAc on proteins is mediated through the action of the β-*O*-GlcNAc transferase and β-*O*-GlcNAc hexosaminidase acting in a cyclic manner and interacting with kinases and phosphatases which phosphorylate the same or adjacent serine and threonine residues in acceptor proteins. The nature of this interplay is complex and includes both same-site and adjacent-site occupancy for GlcNAc and phosphate [5]. Demonstration of same-site occupancy has been shown for c-Myc (c-Myc is a member of the family of *myc* protooncogenes derived from a chicken retrovirus and codes for a transcription factor oncoprotein) and estrogen receptor-β, while adjacent occupancy has been reported in the tumor suppressor p53. The cycling of GlcNAc and phosphate on the same protein, both posttranslational modifications, each with independent regulation, represents a rapid response mechanism with a considerable molecular range for control of protein interaction and function [5].

7.3.3
O-Man

Regulation of *O*-Man glycan synthesis is largely due to the enzyme specificity of the POMT transferases with regard to the few known protein substrates. Most data are available for α-dystroglycan. Extension through the action of POMGnT1 also shows tight substrate specificity and offers an additional site of regulation. Further extension of the GlcNAcβ1,2Manα-*O*-Ser/Thr disaccharide through fukutin and a variety of other glycosyltransferases is probably regulated at species-, tissue- and cell-specific levels, and depends on the complement, enzyme concentration, specificity and activity, leading to competition for substrates.

The subcellular separation of initial *O*-mannosylation by the POMTs in the ER and the extension of the glycans by POMGnT1 and other glycosyltransferases located in the Golgi apparatus is a further and more sophisticated point of regulation depending on complex formation with other transport proteins.

7.3.4
O-Fuc and *O*-Glc

The regulation of *O*-Fuc and *O*-Glc expression is complex and only a brief overview is given here. Both of these protein-linked sugars yield glycan chains that have significant biological activities and much of the data available suggesting regulatory activity for their formation has been tested in biological systems where glycans of different structure have been assessed. Initially the transfer of the sugar to the protein acceptor depends on the existence of consensus sequences in limited protein domains, as detailed earlier. The occurrence of the correct acceptor sequence is evidence of regulation and this is apparent for both the *O*-fucosylation and *O*-glucosylation of EGF and TSR domains of characterized substrates [7, 8].

The extension of the glycans beyond the initial monosaccharide may be regulated through substrate availability at the relevant subcellular location. This will include (glyco)protein acceptor and nucleotide donor, the combination of relevant glycosyltransferases and chaperones or other regulatory proteins or cofactors. For example, the alteration in *O*-Fuc saccharide structure caused by Fringe modulates the response of Notch to its ligands [8]. Thus, glycosylation serves an important role in regulating Notch activity. Finally the regulation of transport of products from their subcellular sites of synthesis is important, as has been demonstrated with POFUT1.

7.4
Functions of *O*-Linked Glycosylation

The functions of the different groups *O*-glycans are expected be wide ranging and varied in view of the large number of *O*-glycosylated proteins known. Table 7.6 identifies the main targets associated with glycan function.

7.4.1
O-GalNAc, Mucin Type

7.4.1.1 Protein Structure and Stability

Protein structure and stability is changed by the addition of a glycan to a peptide or protein backbone and this is evident in the conformation of the completely folded molecule. Many of these properties are shared within one molecule.

7.4.1.2 Protein Conformation and Tertiary Structure

The mucins and molecules with Ser/Thr/Pro-rich domains have served as models to probe changes in protein tertiary structure resulting from *O*-glycosylation. Addition of the first GalNAc residue to the peptide and the elongation of the oligosaccharides leads to an extended conformation often termed the 'bottle brush' structure. Study of the characteristic tandem repeat sequences found in the MUC genes, in particular MUC1, have shown modifications to peptide conformation as a result of glycosylation. This data also correlates with the addition of the initial

Table 7.6 Functions of protein linked *O*-glycans.

Functional target	Specific area of action
Protein structure and stability	Protein conformation, tertiary structure Protein quaternary structure and molecular association Protein stability protease and heat resistance
Recognition phenomena	Cell growth and proliferation Cell fate determination, differentiation, survival and migration Glycoprotein clearance Glycoprotein trafficking Immunological recognition Signaling pathways Regulation of proteolysis Regulation of transcription Stress survival and the cell cycle Disease

GalNAc residues and specificity of the family of ppGalNAc transferases responsible for the transfer. Work with porcine submaxillary mucin has shown that the hydroxyamino acid spacing may contribute to the regulation of glycan length, thereby providing a mechanism for maintaining an optimally expanded mucin conformation.

Nonmucin proteins with 'mucin-like' or *O*-glycosylated 'stalk' domains have a three-dimensional structure with a direct relation to molecular organization and serve to direct functional/binding domains away from the cell surface. A number of examples exist, including nerve growth factor receptor, CD2, LFA-3, CD8αβ coreceptor and calcitonin.

7.4.1.3 Protein Quaternary Structure and Molecular Association

The extended conformations arising from *O*-glycosylation lead to molecular interactions at a quaternary level generating molecular networks and gel structures. Epithelial secreted mucins, which have physiological function in adherent gels at mucosal surfaces, show these properties. In contrast, regulation of molecular function by the prevention of molecular association or aggregation through *O*-glycosylation has also been reported for lactase phlorizin, secretory IgA and CD45 dimerization.

7.4.1.4 Protein Stability: Protease and Heat Resistance

Protective roles against proteolytic degradation and thermal disruption are achieved through *O*-glycosylation. Removal of the glycans results in the loss of these protective properties. Decay-accelerating factor (CD55) is protected in part against proteolysis by *O*-glycosylation and this regulates the half-life for its biological activity [1]. The highly *O*-glycosylated tandem repeat domains of mucus glycoproteins render these glycoproteins resistant to proteolytic degradation. Preparation of mucin glycopeptides relies on this property. Site-specific *O*-glycosylation can also provide protection of protease-sensitive domains. Highly glycosylated proteins such as the mucins and glucoamylase show thermal stability. Improved thermostability could be introduced into glucoamlyase using mutants designed to increase *O*-glycosylation in the already highly *O*-glycosylated belt region.

7.4.1.5 Recognition Phenomena

O-GalNAc glycans play a variety of roles in recognition processes. Such interactions are often complex and may also involve *N*-glycans present in the same molecules (see Chapter 6 and chapters in the final section of this book). Some of the many functions relating to *O*-glycan recognition are listed in Table 7.7.

7.4.2 β-*O*-GlcNAc

Due to the wide range of proteins carrying β-*O*-GlcNAc that play a role in cellular growth and extracellular stimuli there are wide implications for normal and disease-related phenomena.

Table 7.7 Recognition phenomena associated with O-GalNAc, mucin-type glycosylation.

Recognition phenomenon	Examples
Cell growth and proliferation	• Muscle development control by polysialic acid on O-glycans located on the neural cell adhesion molecule • Proliferation of breast cancer cells through O-glycosylation of human sex hormone-binding globulin to achieve binding to its membrane located receptor • Induction of apoptosis by ER stress leading to O-glycosylation of β-catenin and the E-cadherin cytoplasmic domain • Inhibition of O-glycosylation by specific inhibitors leads to inhibition of cell growth through induction of apoptosis and downregulation of proliferation genes
Glycoprotein clearance	Glycoproteins in the circulation are cleared through several different receptors in hepatic, renal, endothelial and lymphatic tissues[a] • The hepatic asialoglycoprotein receptor binds serum glycoproteins with GalNAc-terminated O-glycans • Receptors for the elimination of the glycoprotein hormones and insulin-like growth factor-binding proteins
Glycoprotein trafficking	In polarized cells specific sorting mechanisms are found in the trans-Golgi network • Apical targeting occurs via apical lipid raft carriers and the targeted glycoproteins have glycans mediating their association with the raft and ensuring transport to the correct location • Cellular targeting to the plasma membrane through O-GalNAc glycan recognition has been shown for ceramidase, sucrase/isomaltase, CD44, aminopeptidase N, dipeptidyl peptidase IV, human neurotrophin receptor and C1qRP/CD93
Immunological recognition	Glycosylation has a major role in immune recognition of foreign, nonself antigens functioning at the antibody and T cell levels to generate a protective response • ABH blood groups • IgA1 and IgD hinge region carry short oligosaccharides based on sialylated core 1, Galβ1,3GalNAc, glycans which protect against proteolysis • O-GalNAc glycans play a role in the stabilization of the intracellular MHCII–Ii complex • O-Glycosylated glycoproteins play a role in mechanisms responsible for autoimmunity
Signaling pathways[b]	• O-Glycosylation regulates galectin binding and transduction of cell signaling influencing cell viability in T cells • Inhibition of O-glycosylation leads to an increase of integrin adhesion to fibronectin • Loss of O-glycan sialylation results in increased adhesive capacity of the membrane mucin MUC1 • EGF downregulates the activity of O-glycan core 2 enzymes such as C2GnT and C4GnT due to a Ras–extracellular signal-regulated kinase-dependent mechanism

[a] Information on the relevant lectins, asialoglycoprotein receptor and galectins, is given in Chapter 19.
[b] For information on the effects of N-glycosylation on signaling, see Table 25.2.

7.4.2.1 Protein Structure and Stability

The presence of single or multiple substitution of proteins by β-*O*-GlcNAc units does not play a role in the general stability. However, they may influence conformation and certainly have a role in protein turnover (see below).

7.4.2.2 Recognition Phenomena and Disease

A discrete pattern of recognition events is associated with GlcNAcylation and these are summarized in Table 7.8.

7.4.3 *O*-Man

7.4.3.1 Protein Structure and Stability

Most information concerning *O*-Man glycans has come from studies on α-dystroglycan. This protein contains a characteristic mucin-like domain carrying most of the glycan complement. The functions and disease-related dependency on the *O*-mannosylation of α-dystroglycan is dealt with later. No reports with regard to the stability of the protein itself have been made.

7.4.3.2 Recognition Phenomena

The functions of proteins carrying *O*-Man glycans are largely related to development. The major inferences for these functions have come from the identification of a series of congenital diseases where POMT1, POMGnT1 and fukutin genes are modified or deleted.

The pathology of *O*-Man glycobiology has been linked with a series of congenital abnormalities for the POMT1, 2, POMGnT1 and fukutin (FKRP) genes with reduced enzymatic activity. This heterogenous group of diseases, termed congenital muscular dystrophy, indicate a role for the developmental functions of *O*-Man-linked proteins. Resolution of the disease mechanisms has contributed to the explanation of normal roles for these proteins. It has been proposed that a potential therapeutic approach to these diseases could involve increased and selective glycosylation of the hypoglycosylated α-dystroglycan seen in these patients. Further review of glycosylation deficiency in disease is highlighted in Chapter 22.2.2.

7.4.4 *O*-Fuc and *O*-Glc

There is a wide literature on the functions of proteins carrying *O*-Fuc- and *O*-Glc-linked glycans. This reflects the considerable interest in Notch, its related ligands Delta and Jagged, and their direct involvement in signal pathways mediating evolutionarily conserved developmental pathways in cell differentiation, proliferation, survival and migration [7, 8]. The Notch pathway is one of the relatively few conserved signaling mechanisms associated with the coordination of metazoan

Table 7.8 Recognition phenomena associated with *O*-GlcNAc glycosylation.

Recognition phenomenon	**Examples**
Glycoprotein trafficking	Epithelial cell β-catenin and E-cadherin trafficking is dependent on *O*-GlcNAcylation
Immunological recognition	Peptides substituted with *O*-GlcNAc play a role in the binding to the MHC complex and in the recognition of MHC complex-bound epitopes by the TCR
Signaling pathways	• *O*-GlcNAc–*O*-phosphate regulates signaling molecules • *O*-GlcNAc–*O*-phosphate plays a role in the cell cycle • An inhibitor of *O*-GlcNAcase prevents G_2/M transition in *Xenopus laevis* oocytes • Heat shock protein modulation of signaling pathways related to glucose dependent protection
Regulation of proteolysis	β-*O*-GlcNAc plays a role in the management of protein turnover • β-*O*-GlcNAc linkage sites have a high PEST score, which is coupled with regulation through proteolytic degradation • Proteolytic degradation of proteins is diminished after *O*-GlcNAcylation
Regulation of transcription	RNA polymerase II and many transcription factors are *O*-GlcNAcylated • The β-*O*-GlcNAc units on these proteins have a direct action on their activity and are integrated into overall transcriptional regulation
Stress survival and the cell cycle	Cellular stress is associated with a transient, general increase in *O*-GlcNAcylation • This process may involve *O*-GlcNAcylated heat shock proteins through protein structure stabilization, limitation of protein aggregate formation or direct action on the heat shock protein activity and localization
Disease	Diabetes • *O*-GlcNAc plays a part in insulin signaling and regulation of glucose toxicity • Insulin signaling can be blocked in adipocytes or muscle with increasing *O*-GlcNAcylation • High levels of UDP-GlcNAc promote *O*-GlcNAcylation and glucose toxicity is linked with the hexosamine pathway and the regulation of UDP-GlcNAc levels Neurodegenerative disease Chromosomal loci associated with neurodegenerative disease include the genes for OGT and *O*-GlcNAcase • The formation of neurofibrillary tangles in Alzheimer's disease is associated with abnormal phosphorylation of the tau protein. The protein is *O*-GlcNAcylated and this form prevents the formation of tangles • Brain tissue from Alzheimer's patients shows a global reduction in *O*-GlcNAcylation • Alzheimer's brain tissue shows formation of neuronal amyloid plaques containing the *O*-GlcNAcylated protein amyloid-β precursor protein • Other abnormalities of the *O*-GlcNAcylated proteins have been detected in Alzheimer's disease relating to neurofilament hypo-*O*-GlcNAcylation or reduced *O*-GlcNAcylation of clathrin assembly proteins linked to synaptic vesicle recycling

morphology. Part of this function depends on an integrated hyper-network with highly developed cross-talk to achieve biological significance.

7.4.4.1 Protein Structure and Stability

There is no direct evidence for conformational change in the *O*-fucosylated forms of Notch, Delta or Jagged compared with the nonglycosylated analogs. Analysis of single EGF domains from human blood coagulation factor VII showed little evidence for conformational change on fucosylation. However, as the single EGF domain had limited biological activity, the coordinated action of all EGF domains in the full length factor VII protein may be important for biological recognition. This has been proposed as a possible mechanism in the case of the Notch proteins, where the addition of the *O*-Fuc glycans may mediate conformational change that is apparent at the whole-molecule level [8]. This concept mirrors other glycosylated repeat domain proteins such as the mucins, where the relatively weak individual protein–carbohydrate interactions are enhanced in the form of glycoarray repeat domains. No data is available for O-Glc glycans.

7.4.4.2 Recognition Phenomena

Interactions involving the *O*-glycans on proteins involved in the Notch pathway are summarized in Table 7.9.

7.5 Mucins: A Major Group of *O*-Glycosylated Proteins

The mucin family appears early in metazoan evolution and is a major example of *O*-GalNAc glycosylated proteins. A mucin was the first protein described to be glycosylated in 1865 (please see Chapter 15.3 for details). In humans at least 19 mucin genes have been identified and encode for mucin monomers that are heavily *O*-glycosylated. These belong to a family of large glycoproteins that are either secreted onto mucosal surfaces (for example MUC5AC, MUC5B, MUC2) or are membrane bound (for example MUC1, MUC4). The abundance of serine and threonine in the variable number tandem repeat domains (VNTR), found in central mucin monomer regions and not at the *N*- and C-termini, ensures high *O*-glycosylation of these domains, which can carry hundreds of *O*-glycans. Moreover, in secreted mucins the von Willebrand factor C and D domains, at both N- and C-termini, play an essential role in mucin oligomerization and mucous gel formation [10]. Membrane-bound mucins contain the VNTR domains, but not the von Willebrand factor domains (except for MUC4), and have transmembrane and cytoplasmic domains [11].

Mucins are essentially part of the mucosal protective barrier, an area rich in proliferating cells and continuous challenges by the external environment. It is no wonder that the functions of mucins are diverse, and are apparent in both health and disease [12]. Apart from the involvement of MUC1 and MUC4 in cell signaling the secreted mucins are very important for the protection of epithelial surfaces

Table 7.9 Recognition phenomena associated with O-Fuc and O-Glc glycosylation.

Recognition phenomenon	Examples
Cell fate determination, differentiation, proliferation, survival and migration	The Notch signaling pathway plays a crucial role in specifying cellular fates in metazoan development by regulating communication between adjacent cells • Cell fate determination, where defined populations of cells are constantly being produced from stem cells • Close integration with differentiation, proliferation, survival and migration processes • Adaptation according to gastrointestinal, central nervous system, vascular tissue and cell types • Notch activation is capable of amplifying the intestinal progenitor pool while inhibiting cell differentiation and is required for the maintenance of proliferating crypt cells in the intestinal epithelium
Glycoprotein trafficking	Subcellular trafficking of Notch, Delta and Jagged glycoproteins • Trafficking of the Notch intracellular domain occurs via caveolin-mediated mechanisms in lipid rafts • There is, however, no demonstration that the glycans themselves play any role in these processes
Signaling pathways	The role for O-Fuc glycans in the Notch signal pathway takes place at the cell surface and is upstream of the proteolytic cleavage generating the activated form of Notch • Shown for POFUT1 and Fringe • Nonenzymatic actions of Fringe suggest a recognition process involving the glycans and other modulatory proteins • Cross-talk with wnt, hedgehog and other signal pathways
Disease	Abberations in the glycoproteins associated with the Notch pathway have been found associated with a series of human diseases • T cell leukemia and other cancers • Multiple sclerosis • CADASIL (autosomal vascular disorder with mutations in Notch 1 and 3 genes) • Alagille syndrome (defects in development of eye, heart, kidney and skeleton, abnormal bile duct formation with liver pathology, due to mutations in *jagged1* gene) • Spondylocostal dysostosis

not shielded by keratinocytes (gastrointestinal, airway, genital tracts and ocular surface). Secreted mucins form a negatively charged gel with viscoelastic capacity, which interacts with an apical membrane-anchored glycocalyx. This creates a physical barrier against bacteria and other insoluble material. Additionally, charge repulsion between the sialic acids that terminate the mucin-type O-glycan chains confers a high negative charge that retains water and therefore prevents barrier dehydration. In the gastrointestinal tract this gel lubricates dietary particles, provides mechanical stability during peristalsis, regulates food movement through the tract, and yields protection against gastric acid, proteolytic digestion and bile salts [12].

Due to their high density of glycosylation the mucins have proved to be valuable affinity ligands for the purification of lectins (see Chapters 15.3 and 18.3 for examples) and other carbohydrate-binding proteins.

Finally, a number of proteins are known to interact with secreted mucins. However, the precise nature of *O*-GalNAc glycosylation and glycan recognition on these interactions has not yet been elucidated. A member of the trefoil factor family of peptides can bind to human MUC5AC so that both molecules are secreted together into the stomach, forming a gel that has enhanced mucous viscosity and protective capacity. Also, it is known that a number of interleukins have high affinity for glycolipids and glycopeptides, and they can bind to mucins. It is suggested that the formation of cytokine reservoirs at mucosal surfaces could enable their rapid release affecting local immune reactions.

7.6 Conclusions

This chapter reviews glycans attached to proteins through *O*-glycosidic linkages. Five different types of *O*-glycan links are considered, and the sequence of glycans found in the five groups relates to their distribution in different proteins and their expression at cellular and tissue levels. Elucidating the different glycan structures and their modes of synthesis and function is important to evaluate their biological significance. The range of structures found among the five groups varies considerably. The *O*-GalNAc (mucin-type) constitutes a considerable database of structures, while the *O*-Man, *O*-Fuc/*O*-Glc and, particularly, the β-*O*-GlcNAc, as a single monosaccharide, establish far fewer structures. These variations correlate well with the biological roles of the proteins they are linked to and also reflect the wider functions ascribed to the *O*-GalNAc, mucin-type group.

The continuing accumulation of *O*-glycans in established databases has served to allow a better comparison of occurrence and assessment of possible functional roles.

Summary Box

Glycans attached to proteins through *O*-glycosidic linkages cover a number of discrete types and account for a wide variety of structures that play many roles in the function of the individual glycoproteins in Nature. The metabolic pathways leading to the structures show tight regulation and the potential to process the glycan sequence in relation to biological function.

References

1 Van den Steen P *et al.* Concepts and principles of *O*-linked glycosylation. *Crit Rev Biochem Mol Biol* 1998;33:151–208.

2 Brockhausen I. Glycosyltransferases specific for the synthesis of mucin-type *O*-glycans. In: *Glycobiology* (Eds.: Sansom C, Markman O), pp. 217–34. Scion Publishing, Bloxham, 2007.

3 Corfield AP. Glycobiology of mucins in the human gastrointestinal tract. In: *Glycobiology* (Eds.: Sansom C, Markman O), pp. 248–60. Scion Publishing, Bloxham, 2007.

4 Thanka Christlet TH, Veluraja K. Database analysis of *O*-glycosylation sites in proteins. *Biophys J* 2001;80:952–60.

5 Hart GW *et al.* Cycling of *O*-linked β-*N*-acetylglucosamine on nucleocytoplasmic proteins. *Nature* 2007;446:1017–22.

6 Hart GW *et al.* Glycosylation in the nucleus and cytoplasm. *Annu Rev Biochem* 1989; 58:841–74.

7 Fiuza UM, Arias AM. Cell and molecular biology of Notch. *J Endocrinol* 2007;194: 459–74.

8 Luther KD, Haltiwanger RS. Role of unusual *O*-glycans in intercellular signaling. *Int J Biochem Cell Biol* 2009;41:1011–24.

9 Schegg B *et al.* Core glycosylation of collagen is initiated by two β(1-*O*) galactosyltransferases. *Mol Cell Biol* 2009;29:943–52.

10 Thornton DJ *et al.* Structure and function of the polymeric mucins in airways mucus. *Annu Rev Physiol* 2008;70:459–86.

11 Hattrup CL, Gendler SJ. Structure and function of the cell surface (tethered) mucins. *Annu Rev Physiol* 2008;70:431–57.

12 Hollingsworth MA, Swanson BJ. Mucins in cancer: protection and control of the cell surface. *Nat Rev Cancer* 2004;4:45–60.

8
Glycosylation of Model and 'Lower' Organisms

Iain B. H. Wilson, Katharina Paschinger, and Dubravko Rendić

In the previous chapters of this book, the focus has been on the *N*- and *O*-glycosylation of mammals (see Chapters 6 and 7); when it comes to plants, lower eukaryotes and bacteria, our knowledge about glycosylation is not so extensive, but major advances have been achieved in recent years. Particularly, the wide use of mass spectrometry in glycan analysis (see also Chapter 5) and the ability to clone and knock-out/down glycosylation-relevant genes (see also Chapter 23) has revolutionised our understanding about which glycans are present and which enzymes play a role in their biosynthesis. In this chapter, the reader will be introduced briefly to the basics of glycosylation in bacteria, yeasts, plants, insects, nematodes, protozoans and fish, with the focus being on 'model organisms' such as *Arabidopsis thaliana, Drosophila melanogaster, Caenorhabditis elegans* and *Danio rerio*, as well as on the relationship of non-mammalian glycosylation to biotechnology and host–pathogen interactions. We hope the reader will, thereby, appreciate the structural diversity possible as well as the advantages of using such organisms to understand the biological role of glycosylation, while not ignoring the potential problems if these systems are used.

8.1
Bacterial Glycosylation

First, due to their phylogenetic position, we briefly consider glycosylation in bacteria. Glycoconjugates such as lipopolysaccharides and peptidoglycans are major components of bacterial cell walls, but perhaps the most surprising glycobiological finding in recent years has been the demonstration that bacteria can also glycosylate their proteins [1]. A number of S-layer (surface-layer) glycoproteins have been described from a number of species and show that different species have adopted different means of attaching glycans to their proteins. From a clinical perspective, it is important to note that a number of pathogenic bacteria, which colonise mammals, also glycosylate some proteins which are not S-layer glycoproteins; a well-studied example is the *N*-glycosylation system in *Campylobacter jejuni*. In this

The Sugar Code. Edited by Hans-Joachim Gabius

ISBN: 978-3-527-32089-9

species, a glycan containing bacillosamine (2,4-diacetamido-2,4,6-trideoxyglucose) modified with further GalNAc residues (for abbreviations and structures of monosaccharides, see Chapter 1) is transferred to asparagine; the enzyme (PglB) required to transfer this oligosaccharide displays homology to the catalytic subunit of the eukaryotic oligosaccharyltransferase (STT3; see Figure 8.1) and can be expressed in other bacterial species in order to synthesise novel glycoforms (i.e., new glycosylated forms of proteins). Other examples of bacterial glycosylation include the *N*-glycan of the archaeal species *Methanococcus voltae,* which is a trisaccharide of the form βManNAcA6Thrβ1,4GlcNAc3NAcAβ1,3GlcNAc, and various *O*-glycans, including addition of heptoses to specific proteins in wild-type strains of *Escherichia coli,* of FucNAc to the pilin of *Pseudomonas aeruginosa* (an opportunistic

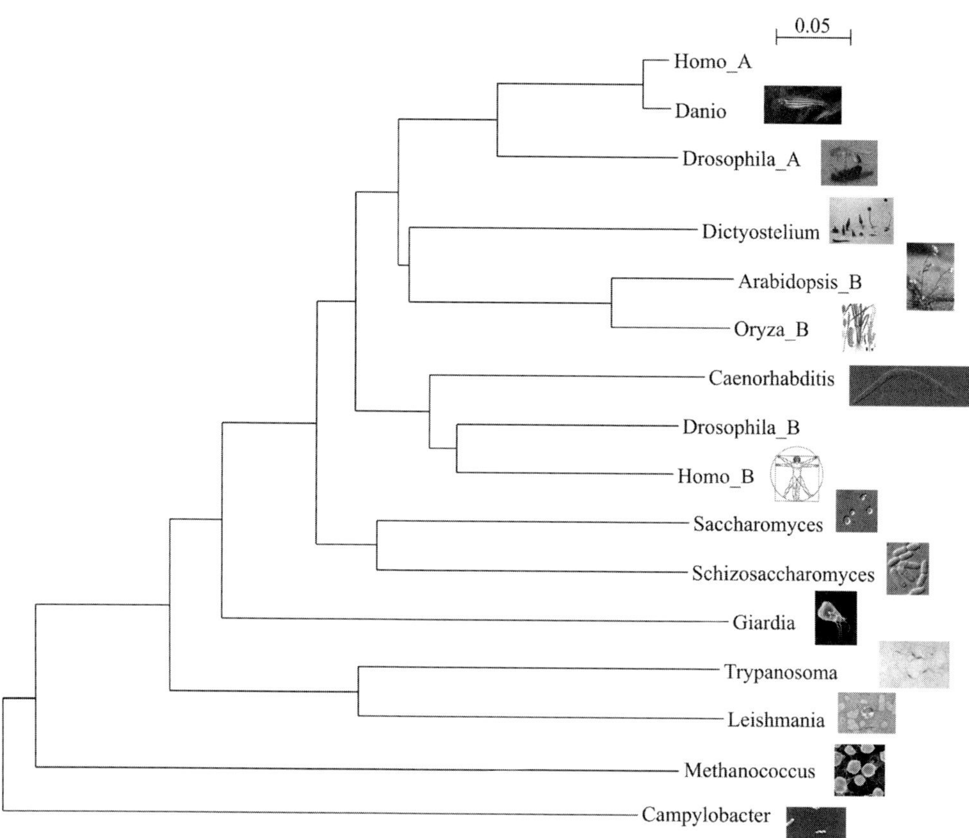

Figure 8.1 Family tree of the STT3 oligosaccharyltransferase catalytic subunit. Oligosaccharyltransferase in eukaryotes, required for the transfer of a dolichol-linked precursor to asparagine residues, is a multi-subunit enzyme; one of these subunits, called STT3 (there are indeed two forms, 'A' and 'B', in many species), has homologues also in those bacteria which also *N*-glycosylate some of their proteins. This computer-derived 'family tree' is based on aligning the STT3 sequences of various prokaryotic and eukaryotic species and features many of the organisms mentioned in this chapter.

pathogen), of glucose to tyrosine in the S-layer glycoprotein of *Thermoanaerobacterium thermosaccharolyticum* (an anaerobic thermophile), of *O*-linked GlcNAc to flagellin in *Listeria monocytogenes* (a food-borne pathogen) and of *O*-linked mannose to proteins of mycobacteria (including the bacterium responsible for tuberculosis).

8.2
Yeast Glycosylation

The simplest eukaryotic systems are obviously unicellular organisms and, amongst these, the use of the baker's yeast *Saccharomyces (S.) cerevisiae* as a glycobiological model has a relatively long tradition. The pioneering work of Phillips Robbins in the 1980s in the use of yeast to isolate *alg* (asparagine-linked glycosylation) mutants with defects in the biosynthesis of the dolichol-linked intermediates for *N*-glycans was important in laying the foundations for our present knowledge about the genetic basis for this pathway in eukaryotes [2]. Today, we know which genes are necessary for each enzymatic step leading to the generation of the dolichol-linked form of $Glc_3Man_9GlcNAc_2$, which serves as the oligosaccharyltransferase substrate in most eukaryotic species (see Chapter 6 on *N*-glycosylation and Chapter 22 on aberrations in the biosynthetic pathway). The ability to generate yeast mutants and complement them with isolated yeast or human genes has been key for this work. Furthermore, these studies have aided uncovering the genetic basis for a number of the human congenital disorders of glycosylation listed in Table 22.1.

It would, though, be a mistake to think of yeast as just meaning *S. cerevisiae*; this is particularly clear when one considers the modifications of the *N*-glycans once they pass through the Golgi apparatus. *S. cerevisiae* has the tendency to hypermannosylate its *N*-glycans; unlike mammals, this yeast has, in addition to those conserved mannosyltransferases in the endoplasmic reticulum (ER) required for the biosynthesis of the dolichol-linked precursor, many Golgi-localised mannosyltransferases, which utilise GDP-Man as a donor substrate [3]. Tens of mannose units (some of which are phosphorylated) may be added, resulting in very large *N*-glycans on some glycoproteins (see Figure 8.2), particularly those of the cell wall. This has the repercussion that recombinant glycoproteins expressed in *S. cerevisiae* also tend to be hypermannosylated, which may negatively affect their physicochemical, biological or enzymatic properties. However, due to their ability to produce high amounts of recombinant proteins, yeasts are biotechnologically interesting; the result is that a number of other yeast species have been examined as potential 'cell factories'. For instance, methylotrophic yeasts such as *Pichia pastoris* or *Ogataea minuta* are used as expression systems. They tend to add fewer mannose residues to their *N*-glycans (structures as large as $Man_{15}GlcNAc_2$ are typical, instead of the sometimes 100 mannose units in some *S. cerevisiae* glycans); these are still not akin to the glycans found in mammals. Thus, the ability to also easily knock-out genes in yeasts has been exploited to re-engineer glycosylation in these species. Particularly, the *och1* (outer chain 1) genes, which encode

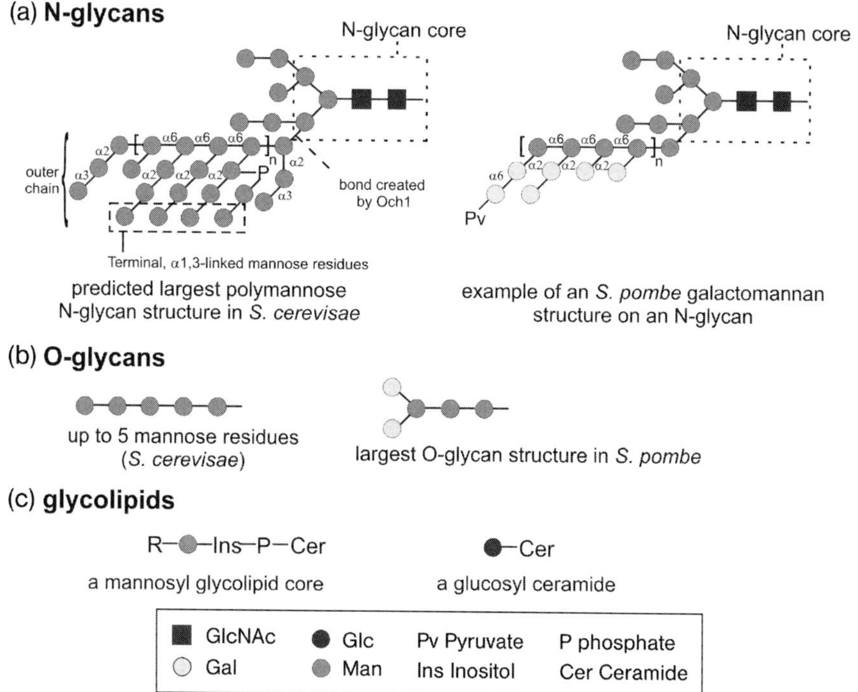

Figure 8.2 The glycosylation potential of yeast. (a) Yeast species add more mannose residues to their *N*-glycans than is the case in other species, but some yeasts also transfer galactose to their *N*-glycans. (b) The tendency to have mannose-rich glycans in yeast is also seen with their *O*-glycans. (c) The glycolipids of yeast also tend to be simple.

the key mannosyltransferases required for outer chain formation in yeast, have been targeted in order to reduce *N*-glycan size. Further efforts to 'humanise' yeast, in order that their glycosylation is closer to that of our own and so facilitate the use of recombinant proteins as therapeutic agents, have included introducing a number of genes required for complex *N*-glycan modifications; this has been accomplished in *Pichia pastoris* [4]. One company, GlycoFi (now a subsidiary of Merck), has been 'built' on this technology. The large size of naturally occurring yeast *N*-glycans is not the only problem; they are also antigenic, which is of distinct disadvantage when one wishes to produce a recombinant protein for pharmaceutical use. Furthermore, some yeast *N*-glycans carry not just poly-α-mannose; the fission yeast *Schizosaccharomyces pombe* produces *N*-glycans modified with galactose residues, which can be pyruvate-substituted (see Figure 8.2), whereas the pathogenic yeast *Candida albicans* has peripheral β-linked mannose on its *N*-glycans [3].

A further complication when considering yeast glycosylation is that their *O*-glycans, unlike those of mammals, tend to contain, dependent on the species, primarily or solely mannose residues [3]. The first mannose residue of such *O*-glycans is transferred by the action of ER-located mannosyltransferases (encoded by essential *PMT* genes), which utilize Dol-P-Man as the donor substrate. These genes are related to human *POMT* genes, defective in some forms of congenital muscular dystrophy, required for the normal post-translational modification of α-dystroglycan (see Chapters 7 and 22 for further discussion, especially Figure 22.4). Interestingly, some fungal species (for example parasitic microsporidia) produce *O*-glycans, but no *N*-glycans. Finally, when considering yeast, one should forget neither the glycolipids (see Figure 8.2), such as hexosylceramides in many fungal species (other than *Saccharomyces*) and mannosylinositolphosphoceramides [5], nor the glycosylphosphatidylinositol (GPI; see Chapter 9) anchors of many fungal proteins. Indeed, just as with the *ALG* genes involved in *N*-glycan biosynthesis, *Saccharomyces* has been valuable in elucidating the function of many of the so-called *GPI* genes involved in GPI anchor biosynthesis [6].

8.3 Plant Glycosylation

When we discuss plant glycosylation, we generally consider only their *N*-linked oligosaccharides. This is due to the relative simplicity of enzymatically releasing this type of glycan using peptide:*N*-glycosidases (especially the *N*-glycosidase from almonds). However, the methods commonly used to release mammalian *O*-glycans (β-elimination) are generally not applicable to those in plants, due to the different alkaline stability of the peptide–glycan bond involved; as a result, our understanding of plant *O*-glycosylation is rather poor. On the other hand, the *N*-glycans of many plant glycoproteins (including many lectins; see Chapter 18) as well as of whole plant organs have been examined over the years. What is remarkable, as compared to animals, is the highly conserved nature of *N*-glycosylation within the plant kingdom – a moss produces the same types of structures as an orange or a rice plant. The range of structures is indeed relatively limited – the main features being, other than the oligomannose-type *N*-glycans found in most eukaryotes, the presence of core α1,3-fucose and β1,2-xylose as well as non-reducing terminal Lewis^a groups (see Figure 8.3), although the Lewis^a epitope is somewhat underrepresented in the 'model' plant *Arabidopsis thaliana*.

The biosynthesis of *N*-glycans in plants shows many similarities to mammals; the biosynthesis of the dolichol-linked intermediates (resulting in $Glc_3Man_9GlcNAc_2$) is probably identical to the process in mammals and yeast. Similar to mammals is the processing by glucosidases and mannosidases to yield $Man_5GlcNAc_2$, if the glycan on the folded protein is accessible to these enzymes. This is then a substrate for *N*-acetylglucosaminyltransferase (GnT)-I, also as in mammals; this modification to yield a hybrid-type structure (sometimes called 'Man5Gn' or 'Man5Gn3') is an important step in the further processing of the glycan. Generally, Golgi α-mannosidase II will remove two mannose residues, GnT-II will transfer another GlcNAc residue, and modification by β1,2-xylosyl-

8.4 Insect Glycosylation

As with plants, the *N*-glycans of insects display immunogenic features which potentially complicate the use of these species as 'glycopharmaceutical factories'. Indeed, with regard to *N*-glycosylation, insects are, in terms of core fucosylation, 'half plant/half mammal', since they display both core α1,3-fucosylation (present in plants) and core α1,6-fucosylation (a feature of many mammalian glycoproteins; see Figure 8.4). The analysis of various bee venom glycoproteins (interesting because they are allergens) also showed the presence of a third form of fucosylation–the fucosylation of antennal 'LacdiNAc' (GalNAcβ1,4GlcNAc-R), a feature also seen in the trematode parasite *Schistosoma mansoni* and akin to the mammalian Lewis^x epitope (for the structure of a sulphated derivative of LacdiNAc, see Figure 1.7b). Generally, though, the recombinant glycoproteins produced in insect cell lines lack the core α1,3-fucose and the fucosylation of LacdiNAc; however, the natural range of *N*-glycans in High Five (*Trichoplusia ni*) cells, one of the commonly used baculovirus hosts, includes a significant proportion of core α1,3-fucosylated structures [12].

The presence of core α1,3-fucose is just one reason that re-engineering of insect cells is also, as with plants and yeasts, necessary to 'humanise' their glycosylation [13]. Indeed, insect cells generally lack the potential to produce complex, sialylated *N*-glycans and also have a hexosaminidase in the secretory pathway, which removes the non-reducing terminal GlcNAc residue transferred by GnT-I and, thus, reduces their ability to produce larger glycans. Therefore, augmentation of the glycosylation pattern of insect cells requires their transformation with a number of mammalian genes as well as identification of the genes encoding the core α1,3-fucosyltransferase and the secretory pathway hexosaminidase. Major steps in this direction have been made using *Drosophila melanogaster* as a tool.

It is interesting that, despite the long tradition of using *Drosophila* in genetics, progress regarding its glycobiology was slow. With the sequencing of its genome, it was possible to begin identifying homologues of mammalian and plant glycosyltransferases of known function. The staining of invertebrate neural tissue with anti-HRP (see Info Box 1) was often used, but without anyone understanding the molecular basis for this cross-reaction. However, in 2001, work in our lab demonstrated the presence of difucosylation of the core GlcNAc (as on bee venom glycoproteins) in flies for the first time, as well as the activity of a recombinant *Drosophila* core α1,3-fucosyltransferase. Recently, mutations in a fly hexosaminidase (encoded by the *fused lobes* gene) and the fly GnT-I have been shown to affect *N*-glycosylation as well as neural anatomy (see also Table 8.1).

It is, though, in terms of *O*-glycosylation that the glycobiological value of *Drosophila* has become most obvious–whether it be through the use of mutants of glycosaminoglycan biosynthesis, of Notch signalling, of *O*-mannosylation or of 'traditional' mucin-type *O*-glycosylation (see Figure 8.4). In all these cases developmental effects or lethality were associated with mutations in these pathways, for which the first steps are conserved as compared to mammals (see also Chapter 7).

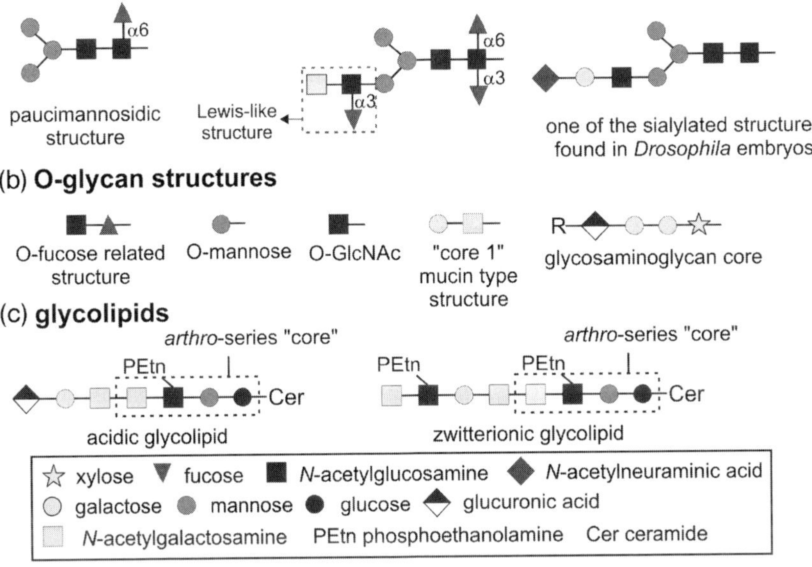

Figure 8.4 The glycosylation potential of insects, especially *D. melanogaster*. (a) Typical features of insect *N*-glycans are the paucimannosidic structures, containing between two and four mannose residues and sometimes fucose; although the presence of two fucose residues on the core is a feature of many insect species, the Lewis-like modification is a feature of bee venom glycoproteins, which has not been found in the fruit fly. (b) Insect *O*-glycans tend to be simpler versions of those types found in mammals. (c) The glycolipids of insects are different from those in mammals and contain, generally, the so-called *arthro*-series core.

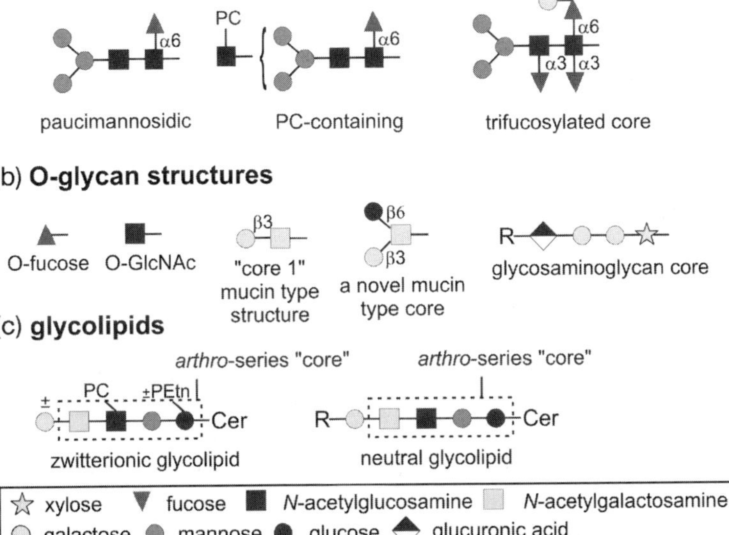

Figure 8.5 The glycosylation potential of nematodes, especially *C. elegans*. (a) As with insects, typical features of worm *N*-glycans are the paucimannosidic structures, containing between two and four mannose residues and sometimes fucose; the core regions decorated with up to three fucose residues and the presence of phosphorylcholine on *N*-glycans is a nematode 'speciality'; (b) Nematode *O*-glycans tend to be simpler versions of those types found in mammals. (c) The glycolipids of nematodes are different from those in mammals and contain, similarly to insects, the so-called *arthro*-series core.

Info Box 2

One reason to use model organisms is to perform otherwise difficult experiments and then draw conclusions as to the human situation (see also Chapter 23). However, ablation of genes that encode proteins with the same biochemical activity can have widely different effects. Taking the example of GnT-I (the first GnT necessary for the generation of complex *N*-glycans; see Table 8.1), a knock-out worm in the lab is still alive, whereas a deletion in this gene in mice is embryonic lethal. Indeed, during evolution, it would appear that the ability of organisms to modify their *N*-glycans by GnT-I has become associated with more and more important processes; on the other hand, even if they display no obvious developmental defects, a plant or a worm lacking GnT-I may not really have a chance in the 'wild' either.

serine or threonine) are much more complex than those in the fruit fly and carry fucose, glucose and galactose residues; the presence of cytosolic *O*-GlcNAc is involved in various signalling events, including a insulin-like pathway, but (unlike in mammals) is not essential under laboratory conditions. The correct formation of glycosaminoglycans, though, is essential for normal worm development as exemplified by the series of eight *sqv* (squashed vulva) mutants, each of which possesses a different defect in chondroitin biosynthesis [23]; mutations in *O*-fucosylation also affect development. Finally, the glycosphingolipids of *C. elegans* are of interest since some of these, as in other nematodes, carry phosphorylcholine; other worm glycosphingolipids act as receptors for one of the *Bacillus thuringiensis* crystal toxins (some of which are used as biological pesticides) and a series of defects in worm glycolipid biosynthesis result in the so-called *bre* (*Bacillus* resistance) mutants [24].

8.6 Protozoan Glycosylation

As with 'worms', many protozoans (unicellular organisms) are also parasites; examples include the trypanosomatids (for example *Trypanosoma brucei*, which causes sleeping sickness), the malaria parasite *Plasmodium falciparum*, the dysentery amoeba *Entamoeba histolytica* and the opportunistic pathogen *Acanthamoeba castellanii* (which causes a corneal keratitis in some contact lens wearers and an encephalitis in some immunodeficient patients). Protozoans express a lot of proteins anchored via GPI anchors and indeed this type of modification was first intensely studied in *T. brucei* (particularly in the context of the anchor of the variant surface glycoproteins of this species; see also Chapter 9). Some species also express forms of GPI anchors that are not linked to proteins, such as the glycoinositol-phospholipids in *T. cruzi*, which are immune activators, or the lipophosphoglycans of *Leishmania* species [25]. With regard to *N*-glycosylation, some protozoans (even though they are eukaryotes) do not synthesise the usual $Glc_3Man_9GlcNAc_2$ dolichol-linked precursor. In the case of *Giardia*, the *N*-glycans and the dolichol-linked precursor contain apparently just two GlcNAc residues and lack all mannoses, whereas trypanosomes do not add glucose to the precursor; this is due to a lack of various *ALG* genes which encode the relevant glycosyltransferases [26].

One protozoan is a very interesting model organism – the non-parasite *Dictyostelium discoideum*, which is sometimes known either as a 'slime mold' or as a 'social amoeba'. In the unicellular state, it is an amoeba; however, upon starvation, the cells aggregate together and form a fruiting body; the resulting spores can survive until food levels are again sufficient. A number of studies have been performed on its *N*-glycosylation; recent studies in our laboratory, using modern techniques, suggest that the most dominant glycan in laboratory axenic cultures consists of $Fuc_1GlcNAc_4Man_8$. The fucose is $\alpha1,3$-linked to the core and constitutes an epitope for anti-HRP. The slime mold also expresses novel *O*-glycans, including a tetrasaccharide present on a cytosolic protein, and, like some trypanosomatids, expresses mucins containing GlcNAc (rather than GalNAc) linked to serine/threonine, as well as protein-linked phosphosugars [27].

8.7
Fish Glycosylation

Once we reach fish in the evolutionary tree (and so the first vertebrates), then we cease to see many of the 'immunogenic' features of glycans from 'lower' organisms. Indeed, it appears that zebrafish (*Danio rerio*) and medaka (*Oryzias latipes*) have a similar *N*-glycomic potential as birds and mammals, with the presence of multiantennary complex, sialylated *N*-glycans having been proven [28]; the *O*-glycans and glycosphingolipids of zebrafish can also be sialylated [28]. Thus, presumably the 'jump' to vertebrates was accompanied by a shift in the glycome. One interesting marker of this shift is not in the realm of *N*-glycans, but is related to glycosaminoglycan biosynthesis: fish (like mammals, but unlike invertebrates) have two peptide *O*-xylosyltransferase genes and also express keratan sulphate, in addition to chondroitin and heparan sulphates. It is also in relation to glycosaminoglycans that perhaps the most 'use' of zebrafish as a glycobiological model has been made. A number of mutants have been characterised, including *jekyll* which has a defect in cardiac valve formation due to a disruption in the gene encoding UDP-Glc dehydrogenase – an enzyme required for production of glycosaminoglycans [29]. In another study, use of morpholino-modified oligonucleotides demonstrated a role for heparan-modifying sulphotransferases in vascular development [30]. Thus, just as in flies and worms, experiments in fish show that heparan and/or chondroitin sulphates are important during morphogenesis.

8.8 Conclusions

There is a wide range of forms of glycosylation across non-mammalian species. Even bacteria glycosylate their proteins, but add sugars not even seen in eukaryotes. However, modern genetic analyses show that *N*-glycosylation has an ancient history. When considering simple eukaryotes, yeast species tend to add many mannose residues to their *N*-glycans; in plants, on the other hand, the repertoire includes structures containing immunogenic xylose and core α1,3-linked fucose. The latter is also seen to some extent in insects and nematodes (Table 8.2), which also contributes to the immunogenicity of these glycans too. By the time, when climbing the evolutionary tree, one reaches the fish, then the forms of *N*-glycosylation are very close to those of mammals. In terms of *O*-glycosylation and glycolipids, there is a wide variety of structures; in particular, there are clear differences in the abilities of plants, animals and yeast to modify the hydroxyl residues of proteins (proteoglycans, for example, are an 'animal' feature also found in insects and worms, whereas arabinogalactans and a wide range of polysaccharides are 'plant specific'). The differences are significant due to the problems of immunogenicity of 'foreign' glycans and must be addressed if we wish to exploit 'lower' eukaryotes as expression systems for the biotechnological production of glycoprotein pharmaceuticals. On the other hand, those similarities which exist

Table 8.2 Comparison of the types of *N*-glycans and the glycogenomic potential of different species.

	C. elegans	*D. melanogaster*	*D. discoideum*	Plants	Mammals
Glycan type					
Paucimannose	Yes	Yes	Yes	Yes	No
Core α1,3-fucose	Yes	Yes	Yes	Yes	No
Core α1,6-fucose	Yes	Yes	No	No	Yes
Sialic acid	No	Low	No	No	Yes
Phosphorylcholine on GlcNAc	Yes	No	No	No	No
Bisected	No	No	Yes	No	Yes
Triantennary	Yes	Yes	No	No	Yes
Tetraantennary	No	No	No	No	Yes
Enzymes					
GnT-I	Yes	Yes	No	Yes	Yes
Mannosidase II	Yes	Yes	No	Yes	Yes
GnT-II	Yes	Yes	No	Yes	Yes
Golgi hexase	Yes	Yes	No	Vacuolar	No

The table offers a summary of the *N*-glycosylation features of *C. elegans*, *D. melanogaster*, *D. discoideum*, plants and mammals. The summary is by no means exhaustive, but considers the presence of paucimannosidic *N*-glycans (see Figures 8.3–8.5), core fucosylation type, sialylation, presence of phosphorylcholine and the branching of the *N*-glycans, as well as the key enzymes (required for formation of complex *N*-glycans) GnT-I and -II, Golgi α1,3/6-mannosidase and processing Golgi/vacuolar β-*N*-acetylhexosaminidases (please see Chapter 6 for a discussion of mammalian *N*-glycosylation, including the structures of 'bisected', triantennary and tetraantennary *N*-glycans).

between 'model organisms' and mammals can aid our understanding of the role of glycosylation in biological processes.

Summary Box

Across species, there is a wide range of variation in the types and structures of glycoconjugates; certainly, the glycans of prokaryotes are fundamentally different from those of eukaryotes, but even within eukaryotes, plants and animals share only two types of glycosylation (*N*-glycans and GPIs). Other than the *O*-linked GlcNAc modification of some cytosolic and nuclear proteins, the forms of *O*-glycans and polysaccharides in plants and animals are quite different; also, when considering non-mammals, the variations in *N*-glycans are significant, and have repercussions with regard to their immunogenicity and functionality as glycopharmaceuticals. Finally, it appears that knocking out glycogenes can have far more serious effects in mammals than in worms, flies or plants.

References

1 Messner P, Schäffer C. Prokaryotic glycoproteins. In: *Progress in the Chemistry of Organic Natural Products* (Eds.: Herz W, Falk H, Kirby GW), pp. 51–124. Springer, Vienna, 2003.

2 Huffaker TC, Robbins PW. Temperature-sensitive yeast mutants deficient in asparagine-linked glycosylation. *J Biol Chem* 1982;257:3203–10.

3 Gemmill TR, Trimble RB. Overview of *N*- and *O*-linked oligosaccharide structures found in various yeast species. *Biochim Biophys Acta* 1999;1426:227–37.

4 Hamilton SR *et al.* Humanization of yeast to produce complex terminally sialylated glycoproteins. *Science* 2006;313:1441–3.

5 Barreto-Bergter E *et al.* Structure and biological functions of fungal cerebrosides. *An Acad Bras Cienc*, 2004;76:67–84.

6 Orlean P, Menon AK. Lipid posttranslational modifications. GPI anchoring of protein in yeast and mammalian cells, or: how we learned to stop worrying and love glycophospholipids. *J Lipid Res* 2007;48: 993–1011.

7 Saint-Jore-Dupas C *et al.* From planta to pharma with glycosylation in the toolbox. *Trends Biotechnol* 2007;25:317–23.

8 von Schaewen A *et al.* Isolation of a mutant *Arabidopsis* plant that lacks *N*-acetyl glucosaminyl transferase-I and is unable to synthesize Golgi-modified complex *N*-linked glycans. *Plant Physiol* 1993;102:1109–18.

9 Gillmor CS *et al.* α-Glucosidase I is required for cellulose biosynthesis and morphogenesis in *Arabidopsis*. *J Cell Biol* 2002;156: 1003–13.

10 Léonard R *et al.* Two novel types of *O*-glycans on the mugwort pollen allergen Art v 1 and their role in antibody binding. *J Biol Chem* 2005;280:7932–40.

11 Benning C, Ohta H. Three enzyme systems for galactoglycerolipid biosynthesis are coordinately regulated in plants. *J Biol Chem* 2005;280:2397–400.

12 Rendić D *et al.* Adaptation of the 'in-gel release method' to *N*-glycome analysis of low-milligram amounts of material. *Electrophoresis* 2007;28:4484–92.

13 Jarvis DL. Developing baculovirus–insect cell expression systems for humanized recombinant glycoprotein production. *Virology* 2003;310:1–7.

14 Okajima T *et al.* Modulation of notch-ligand binding by protein *O*-fucosyltransferase 1 and fringe. *J Biol Chem* 2003;278:42340–5.

15 Lyalin D *et al.* The *twisted* gene encodes *Drosophila* protein *O*-mannosyltransferase 2 and genetically interacts with the *rotated abdomen* gene encoding *Drosophila* protein *O*-mannosyltransferase 1. *Genetics* 2006;172:343–53.

16 Ten Hagen KG, Tran DT. A UDP-GalNAc: polypeptide *N*-acetylgalactosaminyltransferase is essential for viability in *Drosophila melanogaster*. *J Biol Chem* 2002;277: 22616–22.

17 Wandall HH *et al.* Egghead and brainiac are essential for glycosphingolipid biosynthesis *in vivo*. *J Biol Chem* 2005;280: 4858–63.

18 Koles K *et al.* Identification of *N*-glycosylated proteins from the central nervous system of *Drosophila melanogaster*. *Glycobiology* 2007;17:1388–403.

19 Harnett W, Harnett MM. Phosphorylcholine: friend or foe of the immune system? *Immunol Today* 1999;20:125–9.

20 Summers RW *et al. Trichuris suis* therapy in Crohn's disease. *Gut* 2005;54:87–90.

21 Paschinger K *et al.* The *N*-glycosylation pattern of *Caenorhabditis elegans*. *Carbohydr Res* 2008;343:2041–9.

22 Shi H *et al. N*-Glycans are involved in the response of *Caenorhabditis elegans* to bacterial pathogens. *Methods Enzymol* 2006; 417:359–89.

23 Hwang H-Y *et al. Caenorhabditis elegans* early embryogenesis and vulval morphogenesis require chondroitin biosynthesis. *Nature* 2003;423:439–43.

24 Barrows BD *et al.* Resistance to *Bacillus thuringiensis* toxin in *Caenorhabditis elegans* from loss of fucose. *J Biol Chem* 2007;282: 3302–11.

25 Ferguson MAJ. The surface glycoconjugates of trypanosomatid parasites. *Philos Trans Roy Soc Lond B* 1997;232:1295–302.

26 Samuelson J *et al.* The diversity of dolichol-linked precursors to Asn-linked glycans likely results from secondary loss of sets of glycosyltransferases. *Proc Natl Acad Sci USA* 2005;102:1548–53.

27 West CM *et al.* Glycosyltransferase genomics in *Dictyostelium discoideum*. In: *Dictyostelium Genomics* (Eds.: Loomis WF, Kuspa A), pp. 235–64. Horizon Scientific Press, Norwich, 2005.

28 Guérardel Y *et al.* Glycomic survey mapping of zebrafish identifies unique sialylation pattern. *Glycobiology* 2006;16:244–57.

29 Walsh EC, Stainier DYR. UDP-glucose dehydrogenase required for cardiac valve formation in zebrafish. *Science* 2001;293: 1670–3.

30 Chen E *et al.* A unique role for 6-*O* sulfation modification in zebrafish vascular development. *Dev Biol* 2005;284:364–76.

9
Glycosylphosphatidylinositol Anchors: Structure, Biosynthesis and Functions

Hosam Shams-Eldin, Françoise Debierre-Grockiego, and Ralph T. Schwarz

Previous chapters (and also Chapter 11) have illustrated several key points about how protein backbones can be glycosylated, defining both *N*- and *O*-glycosylation as common features in evolution. As will be seen, especially in Chapters 10 and 30, lipids establish the second class of backbone for glycan conjugations. This chapter deals with a hybrid form, in which a glycan structure forms a bridge between a protein and a lipid anchor, which attaches the protein to a membrane. Evidently, this structure substitutes for a membrane-integrating part of the protein. These structures are called glycosylphosphatidylinositol (GPI) membrane anchors. This chapter first defines their structure, then presents methods used for their detection or isolation, outlines their biosynthesis and defects in their biosynthesis, and finally presents a survey of general as well as specialized functions.

9.1
Structure of GPI Anchors

The basic structure of a GPI anchor is given in Figure 9.1 (for abbreviations of sugar names, please see Figure 1.6). It consists of ethanolamine, mannose, nonacetylated glucosamine, inositol (the most prominent, naturally occurring form is *myo*-inositol, *cis*-1,2,3,5-*trans*-4,6-cyclohexanehexol) and a lipid moiety; in part linked by diphosphate bridges. It is commonly abbreviated as ethanolamine6Manα1–2Manα1–6Manα1–4GlcN1–6inositolphosphate lipid (Figure 9.1, upper part). This structure thus connects a protein with the lipid – the integral membrane part. GPIs are widespread among eukaryotes, and the expression of GPI-anchored proteins and free GPIs is particularly abundant among the parasitic protozoa. To date, several hundred GPI membrane proteins are known from protozoa, yeasts, fungi, plants and mammals (Table 9.1). Initially, the discovery of GPIs was not achieved through a single observation, but resulted from a series of serendipitous observations followed by an elegant set of experiments over a number of years by leading laboratories in their respective fields. Highlights are that the sugar part of the GPI is attached to the C-terminal end of a protein via

The Sugar Code. Edited by Hans-Joachim Gabius

ISBN: 978-3-527-32089-9

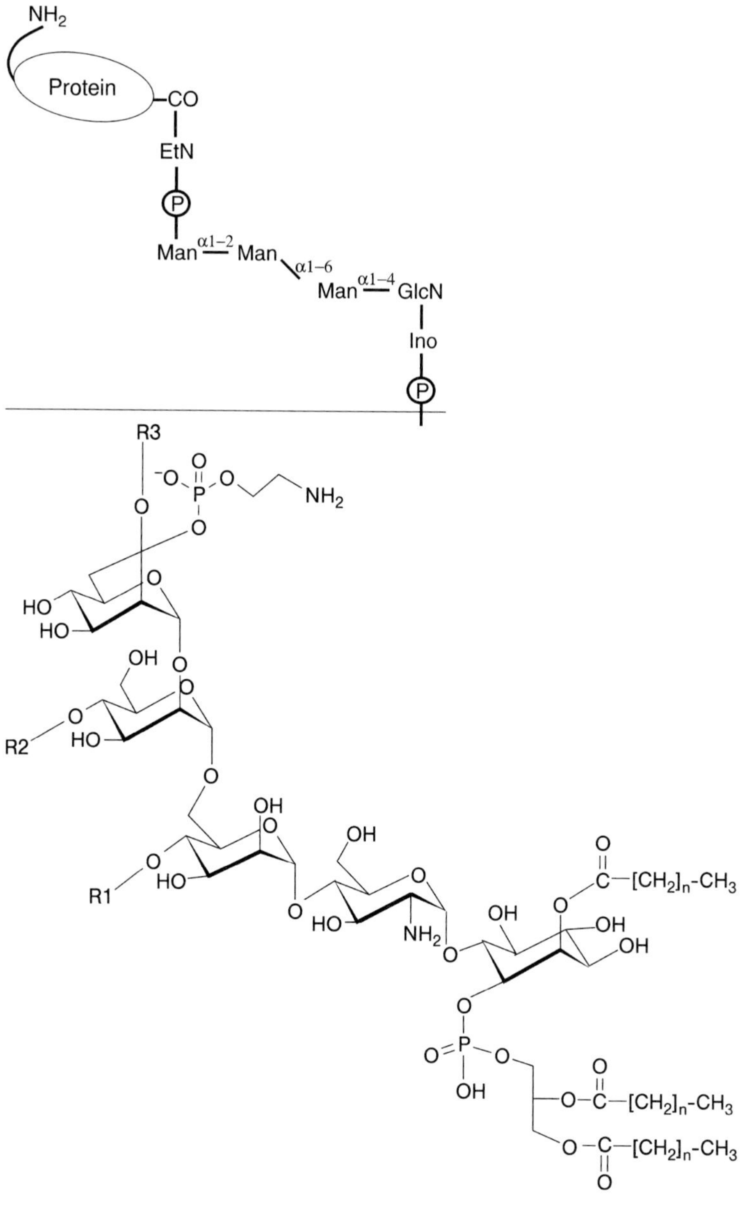

Figure 9.1 Basic structure of a GPI anchor, given as scheme (upper portion) and with formulae (lower portion).

ethanolamine and that their isolation in the so-called 'membrane form' requires special precautions to prevent *in situ* cleavage by phosphatidylinositol (PI)-specific phospholipase C (PLC) – an enzyme that cleaves the phospholipid between the phosphate bound to the inositol ring and the diacylglycerol (DAG) moiety (please see Info Box).

Table 9.1 Examples of GPI-anchored proteins.

Category	Examples
Enzymes	5′-Nucleotidase Alkaline phosphodiesterase I ADP-ribosyltransferase Alkaline phosphatase acetylcholinesterase Aminopeptidase P Carbonic anhydrase IV Carboxypeptidase M *Chlorella* nitrate reductase *Leishmania* surface protease PSP (gp63) NAD^+ glycohydrolase Renal dipeptidase (MDP) Silkworm aminopeptidase N trehalase Yeast aspartyl protease
Receptors	CD14[a] CD16 CD48 Ciliary neurotrophic factor (CNTF) Folate-binding protein *Plasmodium* transferrin receptor Urokinase receptor
Protozoal antigens	*Giardia* GP49 *Paramecium* surface antigens *Toxoplasma* surface antigens (P22, P30 and P43) *Trypanosoma* VSG and PARP (procyclin)
Mammalian antigens	Carcinoembryonic antigen (CEA) CD55 (DAF) Ly6 family (CD59, Ly6A/E), Qa-2, CD24 Thy-1
Miscellaneous	CD58 (LFA-3) Chick F11 Chicken axonin-1 *Dictyostelium* Contact site A Grasshopper REGA-1 Mouse F3 NCAM-120 (the shortest CD56) *Polysphondylium* GP64 Prions (PrP^C, PrP^{Sc}) Squid Sgp-1 and Sgp-2
Adhesion molecules	Heparan sulfate proteoglycan[b] Neural cell adhesion molecule (NCAM)[c]

[a] For explanation of CD nomenclature, please see Info Box in Chapter 27.
[b] The glypican family of heparan sulfate proteoglycans are anchored to the cell surface via a covalent linkage to GPI. For further information on glypicanes, please see Chapters 11 and 23.
[c] For more information on neural cell adhesion molecule, please see Chapters 6 and 30.7.

Info Box

The discovery of GPIs and their functions, their structural elucidation and the unraveling of their complex biosynthesis is a fascinating chapter of glycobiology. As early as 1963, the existence of protein-lipid anchors was suggested. However, the efforts of many leading laboratories and scientists were needed to obtain a clear picture. It was found that bacterial PI-PLC releases alkaline phosphatase from mammalian cells. This was a quite surprising observation. In detail, inositol-containing phospholipid protein anchors were assumed independently by Hiro Ikezawa (Nagoya, Japan) and by Martin Low (New York, USA). Later, Alan Williams (Oxford, UK) found that a cell-surface antigen (Thy-1) showed both the properties of a glycolipid and the behavior of a glycoprotein. The C-terminus of the Thy-1 glycoprotein was then found to contain both fatty acids and ethanolamine. In 1981, Tony Holder and George Cross (Beckenham, UK) showed that the soluble form of the variant surface glycoprotein (VSG) of the African trypanosome *Trypanosoma brucei* contains carbohydrate attached to its C-terminus via an amide linkage involving ethanolamine (since the carbohydrate moiety is recognized by sera against different VSGs it is referred to as immune cross-reactive determinant). Melvin Turner (Cambridge, UK) then found that a membrane-bound form of VSG also exists. In 1985, the groups of Hart and Englund at Johns Hopkins University (Baltimore, Maryland, USA) showed that the lipid anchor on VSG is added within 1 min of the polypeptide's synthesis in the endoplasmic reticulum (ER). They postulated that a preassembled membrane anchor is attached *en bloc* before transfer to the nascent polypeptide chain (in a process now called a transamidation reaction). In 1985 Michael Ferguson and colleagues (Oxford, UK) published a seminal paper on the structural analysis of the glycolipid attached to the membrane-bound form VSG of trypanosomes. These studies defined the term 'GPI'.

9.1.1 Detection and Isolation of GPI-Anchored Proteins

Protocols used for the identification and characterization of GPIs depend mainly upon sufficient amounts of material or metabolic labeling techniques using radioactive GPI precursor molecules, organic solvent extraction procedures, the presence of nonacetylated glucosamine (GlcN; otherwise found in glycosaminoglycan chains,please see Chapter 11), the use of GPI-specific phospholipases and the use of thin-layer chromatography (TLC) analysis. One of the first and the most powerful methods to identify a GPI anchor on a given protein remains the use of the bacterial PI-PLC or GPI-PLC (see Info Box). After treatment with these enzymes, the protein should loose its amphipathic character, which is reflected in a change of solubility. However, GPI anchors carrying a modified inositol ring (for example in the malaria parasite *Plasmodium falciparum* and erythrocytes) are insensitive to cleavage by (G)PI-PLC. Alternative methods to identify a GPI anchor and/or to confirm results obtained by (G)PI-PLC treatment are the use of a mammalian GPI-specific phospholipase D or deamination by nitrous acid (HNO_2 can be prepared *in situ* in a test tube by adding any weak acid to sodium nitrite) (Figure 9.2). Both treatments cleave GPI anchors from solubilized proteins, but marginally from intact cells. The sensitivity of proteins to these GPI-specific treatments can be determined by a shift in mobility of the protein in sodium dodecylsulfate–polyacrylamide gels. The deamination of GPIs (extracted mainly by organic solvent) using nitrous acid (HNO_2) leads to the cleavage of the linkage between the inositol moiety and the non-acetylated glucosamine present in the GPI structure (Figure 9.2). Usually in standard protocols, the anhydromannose generated at the reducing terminus of the GPI glycan is consequently converted to anhydromannitol using sodium borohydride ($NaBH_4$), which resists unspecific destruction as it does not contain a reducing-end sugar moiety. After purification, the hydrophilic fragments can be analyzed using size-exclusion chromatography. Neutral core glycans are prepared by dephosphorylation, deamination and reduction. The released core glycans are completely desalted prior analysis. They are analyzed by high pH anion-exchange chromatography along with an internal standard of β-glucan oligomers. Exoglycosidase (for example mannosidases) treatments are used to confirm the predicted structures of the GPI glycans. In addition to this strategy, alternative methods are also available. For instance, GPI-anchored proteins can also be identified by temperature-induced phase separation in 1–2% Triton X-114. This detergent is normally used to solubilize membranes and whole cells, and the soluble material is submitted to phase separation. This technique distinguishes between GPI-anchored proteins and integral membrane proteins versus cytosolic proteins. It is based on the ability of the nonionic detergent Triton X-114 to partition into two distinct phases at 30 °C – a detergent-rich phase, with the major part of the membrane proteins, and an aqueous phase, containing predominantly nonmembrane proteins. Having thus explained structural aspects including characterization, we now turn our attention to their biosynthesis.

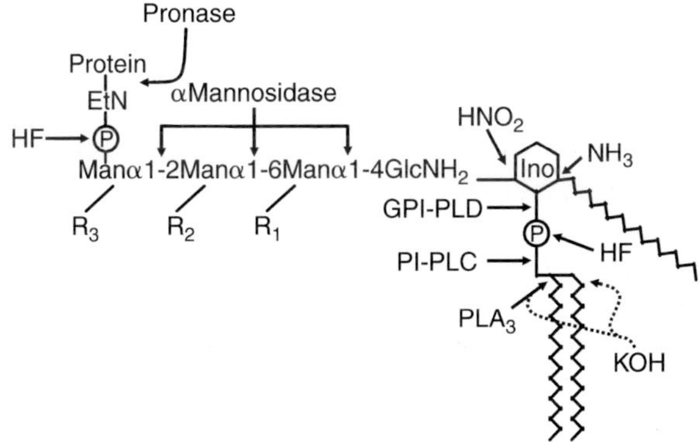

Figure 9.2 Procedures for the identification and characterization of GPIs.

9.1.2
Biosynthesis of GPI Anchors

The basic feature of the biosynthesis of GPI anchors, in contrast to the elucidation of their nature and structures (which took nearly two decades), was elucidated in a few years. The rapid progress was due to the knowledge and experience gained during studying *N*- and *O*-glycan biosynthesis (see Chapters 6–8). In general, co- and posttranslational modifications of proteins by various glycans are initiated in the ER and completed in the Golgi apparatus. Glycosyltransferases in the ER are smaller in number than those in the Golgi, but they are relatively well conserved evolutionarily in a wide variety of eukaryotic cells (for details, please see Chapter 6). The biosynthesis of GPI as well as *N*- and *O*-glycans (in yeast) is characterized alike by sequential addition of mannose residues donated in part by dolichyl mannosyl phosphate synthase (dolichol plays a central role in *N*-glycan biosynthesis; please see Chapter 6). In contrast to the lipid-linked high-mannose oligosaccharide precursor Dol-PP-$(GlcNAc)_2Man_9Glc_3$ (for *N*-glycan synthesis), preassembled GPIs are not activated compounds (such as GDP-Man), as the energy required for the transfer to proteins is generated during the transamidase step (see below). The cloning of genes encoding the enzymes implicated in the biosynthesis of the GPI anchors has been done mostly by using complementation of yeast or mammalian cell mutants lacking GPI proteins at their surface (please see below for further details; Chapter 22.5 provides details on relation to disease). The creation or selection for these cell lines allowed the isolation of PIG and gpi mutants in mammalian cell lines and yeast, respectively. Table 9.2 presents an overview of most of these genes and encoded enzymes. In many cases heterologous expression experiments have shown that genes derived from parasites are also functional in mammalian and yeast cells.

The synthesis of GPI anchors requires at least 12 steps and 23 genes. The numbering of steps in Table 9.2 indicates the most likely sequence of enzymatic reactions based on the structure of the yeast and mammalian GPI intermediates accumulating in gpi or PIG mutants (for further details, please see [1–5]). This numbering correlates with the graphic scheme of biosynthesis presented in Figure 9.3, which summarizes the biosynthetic route step by step. The key roles of GPIs in growth and virulence make this posttranslational modification of protein a potential antimicrobial or antiparasitic target and random screens have already yielded inhibitors of GPI assembly. Indeed, the GPI pathway is an excellent target for the development of new drugs against eukaryotic microbes, although many of the enzymes involved in these steps are conserved. Nonetheless, differences do occur that are critical for protozoa or for yeast, but which are absent from, or of diminished importance in mammals. In this sense, a new set of therapeutics can emerge, underlining the potential of sugars as pharmaceuticals (please see Chapter 28). In what follows, the biosynthetic route will be explained in a stepwise manner (please follow each step in Figure 9.3 and the text) and, if you understand the following steps, you will grasp the working of GPI biosynthesis.

- *Step 1: generation of GlcNAc-PI from UDP-GlcNAc and PI (GlcNAc-PI-transferase).* The first step of GPI anchor biosynthesis occurs at the cytosolic surface of the

ER and is mediated by the GPI-*N*-acetylglucosaminyltransferase (GnT), which is a complex glycosyltransferase consisting of at least six protein subunits (see Table 9.2). The PIG-A (GPI3 in yeast) is the catalytic subunit in the complex, while the human GPI1 (PIG-Q) stabilizes other components of the complex. Deletion of GPI1 does not totally abolish GPI anchoring of proteins in human or yeast cells. This enzymatic step is inhibited irreversibly by reagents blocking SH groups such as *N*-ethylmaleimide (this alkylating compound is a chemical derivative of maleic acid imide or maleic acid and commonly used to modify

Table 9.2 Comparison of yeast and mammalian genes involved in GPI biosynthesis[a].

Steps	Enzymes	Yeast	Mammals	Degree of similarity
1	GlcNAc-PI-transferase	GPI3	PIG-A	46%
		GPI2	PIG-C	21%
		GPI15	PIG-H	24%
		GPI1	PIG-Q	22%
		GPI19	PIG-P	24%
		ERI1	PIG-Y	22%
2	GlcNAc-PI de-N-acetylase	GPI12	PIG-L	32%
3	GlcN-PI-acyltransferase	GWT1	PIG-W	32%
4	GPI-α1,4-mannosyltransferase I (MT-I)	GPI14	PIG-M	38%
		PBN1	PIG-X	16%
5	Ethanolamine-P-transferase	MCD4	PIG-N	30%
6	GPI-α1,6-mannosyltransferase II (MT-II)	GPI18	PIG-V	31%
7	GPI-α1,2-mannosyltransferase III (MT-III)	GPI10	PIG-B	28%
8	GPI-α1,2-mannosyltransferase IV (MT-IV)	SMP3	hSMP3	30%
9	Ethanolamine-P-transferase (GPI-EtN-P-T3)	GPI13	PIG-O	29%
10	Ethanolamine-P-transferase (GPI-EtN-P-T2)	GPI7	hGPI7	25%
		GPI11 (EtN-P-T2)	PIG-F (EtN-P-T3)	30%
11	Transamidase	GPI8	PIG-K	56%
		GAA1	GPAA1	28%
		GPI17	PIG-S	23%
		GPI16	PIG-T	30%
		GAB1	PIG-U	30%
12	GPI-inositoldeacylase	BST1	PGAP1	31%

[a] For schematic illustration of biosynthesis, please see also Figure 9.3.

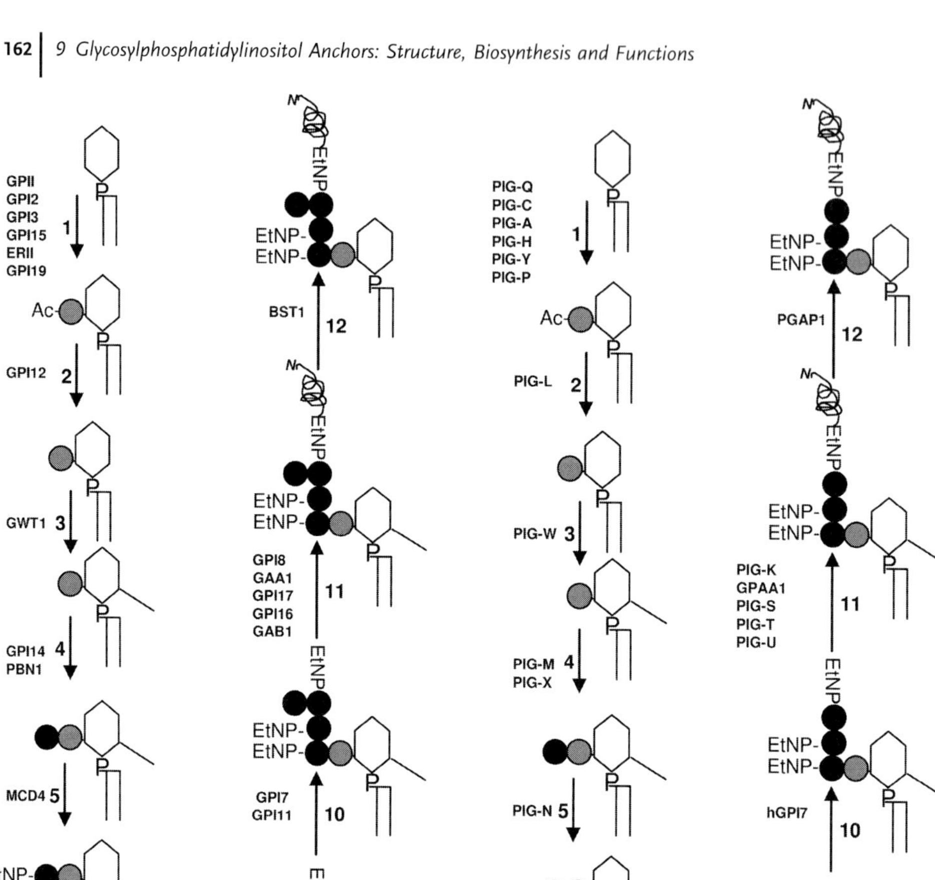

Figure 9.3 The main pathway of GPI biosynthesis in yeast as well as mammals. Biosynthetic steps are numbered like the description in Table 9.2.

cysteine residues in proteins) in all organisms tested thus far, probably by blocking an active site cysteine of GPI-GnT.

- *Step 2: generation of GlcN-PI from GlcNAc-PI.* The second step is mediated by the *N*-acetylglucosaminyl PI (GlcNAc-PI) de-*N*-acetylase (PIG-L, GPI12). PIG-L may interact with PIG-M in the protozoal parasites *Trypanosoma brucei* and *Leishmania major*, but not in mammalian cells. A variety of substrate analogs were used to determine the specificity of the GlcNAc-PI de-*N*-acetylase *in vitro* in the protozoal parasites *L. major*, *T. brucei*, *P. falciparum* and human cells. Substrate analogs are similar in nature to substrates (in this case GlcNR-PI instead of GlcNAc-PI) [6]. However, differences between these molecules and the substrate analogs result in differences in their binding to the active site. Some of these analogs include variations or modification of the D-*myo*-inositol (that is the diastereoisomer with L-*myo*-inositol) substitution of the 2-hydroxyl group of the D-*myo*-inositol with a methyl or octyl group, or GlcNAc-β-PI. All of these are not de-*N*-acetylated by the human enzyme, but were weakly processed by *T. brucei*, *Leishmania* and *Plasmodium* enzymes. Two other GlcNR-PI substrate analogs, $GlcNMe_2$-PI and $GlcNCONH_2$-PI, were found to inhibit both the parasite and human enzymes. Exploiting the differences in substrate specificities of the parasite and human GlcNAc de-*N*-acetylase, two synthetic GlcNAc-PI analogs [$GlcNCONH_2$-β-PI and $GlcNCONH_2$-(2-*O*-octyl)-PI] were shown to be potent *T. brucei*-specific suicide inhibitors, while $GalNCONH_2$-PI was found to be specific for the *Plasmodium* GlcNAc-PI de-*N*-acetylase, which could be used preferentially as a target for developing antiparasitic agent, because this step is mediated only by one gene (for further examples of sugars as pharmaceuticals, please see Chapter 28).
- *Step 3: inositol acylation.* In most organisms the acylation reaction occurs before (except in *T. brucei*, taking place after) the addition of the first mannose residue. The gene encoding the acyltransferase that adds fatty acids to the 2-position of the inositol ring is called PIG-W (GWT1). PIG-W is not essential, where deficient cells are still able to make GPI lipids containing mannose and ethanolamine phosphate (EtN-P) on Man_1. The donor substrate for inositol acylation in the baker's yeast *Saccharomyces cerevisiae* is acyl-CoA, while in mammalian cells an acyl-CoA-dependent as well as an acyl-CoA-independent pathway may coexist. Inositol acylation in *T. brucei* can be inhibited *in vivo* as well as *in vitro* using a serine esterase inhibitor and leads to accumulation of the Man_3-GlcN-PI intermediate in *T. brucei*. This inhibition of trypanosomal inositol acylation was also observed *in vitro* with either GlcN-(2-*O*-methyl)-PI or GlcN-(2-*O*-octyl)-PI. The acylation in mammalian cells, unlike that of *T. brucei*, is inhibited by sulfhydryl alkylating agents but not by the serine protease inhibitor phenylmethylsulfonylfluoride (PMSF) or GlcN-(2-*O*-alkyl)-PI analogs. All of these observations suggest that the mechanism of inositol acylation in the *T. brucei* parasite system is at variance with the host organism, which–again to note–could lead to the development of a novel repertoire of therapeutic molecules targeted against parasitic diseases.
- *Step 4: addition of Man_1.* The first mannose is transferred from dolichol phosphomannose (Dol-P-Man) to GlcN-acyl-PI by PIG-M/Gpi14. Another

subunit, PIG-X, forms a complex with PIG-M, thereby stabilizing it. PIG-X shows 16% identity with the C-terminal part of yeast PBN1. Dol-P-Man is synthesized from dolichol phosphate (Dol-P) and GDP-Man by an enzyme called Dol-P mannose synthase. The antibiotic amphomycin produced by *Streptomyces canis* forms a complex with Dol-P in the presence of Ca^{2+}, which blocks the interaction between the Dol-P-Man synthase and Dol-P, thereby inhibiting GPI biosynthesis *in vitro* (for the role of dolichol in *N*-glycan biosynthesis, please see Chapter 6, especially Info Box 1). Therefore, Dol-P-Man biosynthesis could be an important target for the development of specific inhibitors. The specificity of the α1,4-mannosyltransferase was investigated in *T. brucei*, *P. falciparum* and mammalian (HeLa) cells using various GlcN-PI analogs, which may provide a potential target for drug design, particularly against the malaria parasite *P. falciparum*. Here, in this organism, posttranslational modifications like *N*- and *O*-glycosylation are either absent or present at extremely low levels and GPIs represent its sole carbohydrate modifications.

- *Step 5: modification of Man-GlcN-acyl-PI with a phosphoethanolamine side-chain.* The first mannose is subsequently modified by EtN-P and the reaction is mediated by PIG-N in mammalian cells (MCD4 in yeast). Phosphatidylethanolamine serves as a donor of the EtN-P group. MCD4 is essential in yeast.
- *Step 6: addition of the second mannose.* The second mannose is transferred from Dol-P-Man to position 6 of the first mannose in EtN-P-Man-GlcN-acyl-PI. The reaction is mediated by PIG-V in mammalian cells (GPI18 in yeast). The GPI18 is an essential gene in *S. cerevisiae* and gpi18Δ mutants are rescued also by the expression of mammalian PIG-V.
- *Step 7: addition of the third mannose.* The third mannose is transferred from Dol-P-Man to position 2 of the second mannose by PIG-B (Gpi10 in yeast). Gpi10p requires intermediates containing an EtN-P residue on Man_1, whereas this is not important for Gpi10p homologs of most protozoa, since GPI anchors of these organisms do not have an EtN-P on Man_1. Deletion of PIG-N does not completely prevent the addition of GPIs to proteins, explaining also why an addition of YW3548 does not significantly affect GPI protein expression in mammals or *T. brucei*, whereas it blocks the growth of yeast. YW3548 is a natural terpenoid lactone isolated from *Codinea simplex* and causes an accumulation of the intermediate Man_2-GlcN-(acyl)-PI in yeast, pathogenic fungus *Candida albicans* and mammalian (lymphoma) cells, but not in parasitic protozoa. In yeast, YW3548 blocks also the addition of EtN-P to the first mannose residue, which is a substrate requirement for the addition of the third but not second mannose, therefore, the accumulation of Man_2-GlcN-(acyl)-PI was observed [7]. Similar effects of YW3548 were observed in yeast and mammalian cells using the metalloprotease inhibitor phenanthroline. An accumulation of GPI intermediates that are substrates for EtN-P transferases were identified in cells treated with phenanthroline. Therefore, both phenanthroline and YW3548 are likely to inhibit GPI-phosphoethanolamine transferases in mammalian and yeast cells, but not in protozoa. This selective inhibition is another indication of significant differences between parasite and mammalian/yeast GPI biosynthesis.

- *Step 8: addition of the fourth mannose.* The SMP3 protein, which is encoded by an essential gene, is therefore required for addition of the fourth mannose to yeast GPI precursors. SMP3-related proteins are encoded in the genomes of the fission yeast *Schizosaccharomyces pombe*, *Candida albicans*, fruit fly *Drosophila melanogaster* and humans. However, the human SMP3 homolog (hSMP3) is not expressed in many human cell lines. As Man_4-containing GPI precursors are normally formed in yeast and *P. falciparum*, whereas addition of a fourth mannose during assembly of mammalian GPIs is rare and not required for GPI transfer to protein, SMP3p-dependent addition of a fourth mannose represents a key target for antifungal and antimalarial drugs.
- *Step 9: addition of EtN-P to the third mannose.* As shown in Table 9.2 and Figure 9.3, the EtN-P that links GPI to proteins is transferred from phosphatidylethanolamine to position 6 of the third mannose. Yeast Gpi13 and mammalian PIG-O are required for this step. In yeast, addition of EtN-P to the third mannose occurs only after the addition of the fourth mannose; however, in mammalian cells this is not a perquisite. Furthermore, another subunit, PIG-F, is absolutely required for the addition of EtN-P to Man_3 in mammalian cells, since PIG-O is unstable in its absence. However, the main function of the yeast homolog of PIG-F, an essential gene called GPI11, is unknown since it is not required for such step. In *T. brucei*, the addition of EtN-P is prevented by PMSF, which inhibits the inositol acylation of the intermediate Man_3-GlcN-PI to form Man_3-GlcN-(acyl)-PI – the acceptor of EtN-P.
- *Step 10: addition of EtN-P to the second mannose.* GPI7 and hGPI7 are assumed to encode the transferase that adds EtN-P from phosphatidylethanolamine onto Man_2 of the GPI lipid in yeast and mammals, respectively.
- *Step 11: attachment of the GPI to proteins (GPI transamidase).* Precursor proteins destined to be GPI-anchored posses a C-terminal signal for attachment of GPI anchors, which consists of three portions: a stretch of three amino acids including the amino acid to which GPI attaches (the ω site), a terminal hydrophobic segment of 12–20 amino acids and a hydrophilic spacer segment of usually less than 10 amino acids between them. Well-known characteristics of the signal are the restrictions at the $\omega + 2$ sites that are restricted to six amino acids with small side-chains. The $\omega + 1$ site can tolerate any amino acid except proline and tryptophan. When the precursor proteins are translocated into the ER lumen, a special complex known as the GPI transamidase recognizes the GPI attachment signal sequence, cleaves it and generates an enzyme–substrate protein intermediate linked by a thioester bond. An amino group of the terminal ethanolamine of the GPI backbone attacks the thioester in the intermediate to complete the GPI anchoring. The GPI transamidases of humans and *S. cerevisiae* are complexes of at least five proteins. All these proteins are essential for GPI transamidase as shown by their mutant cells. PIG-U was recently shown as an oncogene in human bladder cancer and showed mRNA overexpression in 36% of primary bladder tumor tissues compared to normal urothelium. Trypanosomatids are not only missing PIG-T and PIG-U, but also possess two unrelated integral membrane proteins of similar membrane orientation, TTA1

and TTA2, instead. The species differences in substrate specificity of the peptide sequence around the ω site and the structure of the GPI anchors, as well as the differences in the protein components of the complex, suggest that there may be exploitable differences between mammalian and parasite GPI transamidase.

- *Step 12: modifications of the GPI anchor after attachment to proteins: removal of the acyl chain from inositol.* In almost all cells except *Plasmodium* and erythrocytes, the acyl group is usually removed from GPI-anchored proteins as result of inositol deacylation that occurs directly after GPI -anchor attachment. Inositol deacylation is mediated by mammalian PGAP1 (post-GPI attachment to proteins 1) and yeast Bst1 (also called dihydrosphingosine phosphate lyase 1). Mutant cells defective in deacylation show a clear delay in the maturation of GPI-anchored proteins in the Golgi and accumulation of GPI-anchored proteins in the ER. Furthermore, BST1 is important for ER-associated degradation of a misfolded, soluble protein, where transport of such proteins to the Golgi may be a prerequisite for its degradation (please see also Chapter 6, especially Info Box 4). The deacylation step can also be inhibited *in vivo* and *in vitro* with diisopropylfluorophosphate (DFP, $C_6H_{14}FO_3P$), a potent serine protease inhibitor such as PMSF mentioned in steps 3 and 9, resulting in only inositol-acylated intermediates. PMSF and DFP can also disrupt the dynamic equilibrium between mature acylated and non-acylated GPI precursors. In *P. falciparum*, once acylated, the GPI intermediate will never be deacylated. Having completed this step, we have reached the final destination, which is the end product of this biosynthetic route.

The preceding explanations covered what is to know about general GPI biosynthesis. In addition, the GPI glycan core undergoes structural variations in various organisms, which arise from different substitutions ('decorations') of the evolutionary conserved core structure. In the following sections we will also see that the lipid moiety of the newly synthesized GPIs may be substantially altered by a process called remodeling. As with all types of natural glycoconjugates, each cellular GPI anchor may be comprised of heterogeneous glycoforms. This heterogeneity can be further enhanced by variations of the lipid moieties. Thus, it is very difficult, if not impossible, as with *N*-glycans from a glycoprotein, to obtain homogeneous GPIs and GPI-linked proteins or glycoproteins from living cells. As a result, their chemical synthesis has attracted significant attention (please see below). Furthermore, the functions of GPIs or how mutant cells helped in gaining insight into their biosynthesis will also be emphasized and discussed.

9.2 Remodeling of Lipid Moieties of GPI Proteins

The lipid moiety of the free GPI lipid or the GPI anchor attached to a protein is modified in a process called remodeling. In yeast cells, the mature GPI-anchored proteins contain mainly ceramide or DAG with saturated long-fatty acids, whereas conventional PI used for GPI biosynthesis contains unsaturated fatty acids. Ceramides are also found in the GPI anchors in *Trypanosoma cruzi* (the causative agent of Chagas disease), the free-living protozoan *Paramecium primaurelia*, the saprotrophic mold fungus *Aspergillus fumigatus* and soil-living amoeba *Dictyostelium discoideum*, sometimes as the sole anchor lipid. The *S. cerevisiae* Cwh43p, whose N-terminal region contains a sequence homologous to mammalian PGAP2, is involved in the remodeling of the lipid moiety of GPI anchors to ceramides (for further details on ceramides, please see Chapters 10 and 30). The remodeling reactions involve the exchange of the fatty acid components of the lipid moiety for different fatty acids or whole-lipid components. *T. brucei* GPI-anchored VSGs contain exclusively myristate as their lipid components. Fatty acid biosynthesis is peculiar in *T. brucei*, since the major product is myristate and is mainly destined to be incorporated into the GPI anchor of VSG protein. This fatty acid pathway could be inhibited with thiolactomycin, a natural product produced by both *Nocardia* and *Streptomyces* spp. that acts as a fatty acid synthesis inhibitor, leading to parasite death. Studies using myristic acid analogs such as 10-(propoxy)decanoic acid or its derivatives showed toxicity towards *T. brucei* (but not to mammalian cells), another target for drug design, here focusing on membrane anchor synthesis, not the glycan part. Next, we turn to the chemical synthesis – a tool to obtain pure material for functional analysis and vaccination.

9.3 Chemical Synthesis of GPIs

The first total synthesis of a GPI was achieved in 1991 by Chikara Murakata and Tomoya Ogawa in Japan, who synthesized the membrane GPI anchor of *T. brucei*. Since then, a number of groups have synthesized several high-profile GPIs (synthetic strategies for glycan synthesis are outlined in Chapter 3). The first automated chemically synthesized glycan of a GPI used for vaccination trials was reported 2002 (please see Chapter 3.9 for further example of synthesis for vaccination) in a study of a GPI-based antimalaria vaccine, in which the malarial GPI toxin was linked to a carrier protein to form a glycoconjugate that has been reported to induce protective immune responses in mice against rodent malaria (for review see [8]). The chemical synthesis is not only helpful for immunological but also for biochemical studies. For instance, a first generation of functional fluorescent GPI probes was used to demonstrate that the ER possesses transporters capable of flipping GPIs in an ATP-independent manner. In the next section we will discuss how mutant cells can help to clone genes involved in GPI biosynthesis (please see above), leading us to present information on natural mutants causing diseases.

9.3.1 Mutant Cells Lead the Way to Identification of Complementation Classes Involved in GPI Biosynthesis

Genes known to be necessary for the biosynthesis of GPI anchors have been cloned in the mammalian system mainly by the group of Taroh Kinoshita in Osaka, Japan, through complementation and/or coimmunprecipitation strategies [9]. The isolation of mutants with complete defects in various reactions in GPI anchor pathway is promising, since cultured cells grow normally and suffer no adverse consequences, even when they are unable to express GPI anchored proteins on their surface. Such a defect in unicellular organisms however is lethal. Ethyl methanesulfonate (EMS) has been extensively used to generate such mutants. EMS generates primarily GC-to-AT transitions, most likely a result of unrepaired O^6-ethyl guanine adducts that mispair with thymine during replication. Subsequently, selection of mutants deficient in GPI-anchored proteins is usually achieved through the use of aerolysin, a secreted bacterial toxin from *Aeromonas hydrophila*, which binds to GPI-anchored protein and kills normal cells by forming pores. Both GPI and *N*-glycan moieties of GPI-anchored proteins are known to be involved in efficient binding of aerolysin. The complementation strategy used in this case depends mainly on the use of mammalian cDNA libraries. Similarly, heterologous complementation of conditional yeast lethal mutants has been successfully used to isolate functional homologs from various species. Based on the high conservation of gene function in cells from different eukaryotic species, many of the known or unknown essential genes from yeast are potential targets for heterologous complementation screens, which depend on the availability of conditional lethal mutants of the gene of interest [10, 11]. So far these mutants have been mainly temperature sensitive mutants, which often revert with certain, sometimes high frequency. The generation of yeast conditional lethal mutants depends mainly on the use of the stringently regulated, glucose-repressed GAL-1 promoter. GAL-1 has been shown to be induced in the presence of galactose and be tightly repressed in the presence of glucose. The expression of the endogenous gene of the yeast mutants can be turned off by shifting the yeast cells from galactose-containing medium, in which the promoter is not repressed, to glucose-containing medium. In the coimmunprecipitation approach, a generic tag like glutathione S-transferase or FLAG (N-DYKDDDDK-C) is usually fused to the protein of interest, stably transfected into the cells and subsequently used to isolate protein–protein complexes based on affinity purification.

9.3.2
Defects in GPI Anchor Biosynthesis

No inherited defects of GPI anchor biosynthesis have been identified up to now. As Chapter 22 on congenital diseases of glycosylation attests, reasons may be the lack of a convenient assay or the labeling or the lethality caused. However, over 100 acquired somatic mutations occur in the X-linked gene PIG-A, which encodes a GlcNAc transferase that initiates this pathway (see Table 9.2). Impaired GPI anchor synthesis causes a defect called paroxysmal nocturnal hemoglobinuria (PNH), a descriptive term for the clinical manifestation of red blood cells breakdown with release of hemoglobin into the urine that is manifested most prominently by dark-colored urine in the morning (please see also Chapter 22.5). Lack of GPI-anchored complement regulatory proteins, such as decay-accelerating factor (CD55) and complement regulatory protein (CD59), results in complement-mediated hemolysis and hemoglobinuria. Paroxysmal nocturnal hemoglobinuria was first observed in 1866 after examination of a man working with chemicals. The disease was initially called 'intermittent hematinuria'. In 1928, the term 'paroxysmal nocturnal hemoglobinuria' was used. Having herewith illustrated defects in biosynthesis and mutant cells, we next turn to the function of GPIs.

9.3.3 Function

A wide variety of proteins are tethered by GPI anchors to the extracellular face of eukaryotic plasma membranes, where they are involved in a number of functions (Table 9.1 presents examples). In general, GPI-anchored proteins represent a large class of functionally diverse proteins. GPI proteins have been found in a wide variety of eukaryotes: mammals (45 in humans), chickens (10), fish, rays, sea urchin, fruit flies (5), silk moth, ticks, grasshopper, protozoa, fungi, slime mold, unicellular green alga, mung bean, even herpes virus (simian surface glycoprotein), but not in bacteria, and oddly nothing reported from nematode (out of 1208 proteins). Even prion proteins from all species sequenced so far (from human to chicken) contain also a C-terminal GPI anchor. Human genetic disorders have been identified in five GPI proteins other than prion. These include fibrillin-1 (Marfan's syndrome, FBN1 gene), Charcot–Leyden crystals (LPPL gene), alkaline phosphatase (infantile hypophosphatasia), lipoprotein lipase (chylomicronemia syndrome), glypican-3 (Simpson–Golabi–Behmel syndrome; for further information on glypicanes, please see Chapters 11.6.2 and 23.2.3). A defect in a GPI biosynthetic enzyme causes paroxysmal nocturnal hemoglobinuria (see above). GPI proteins can be surface antigens, enzymes, adhesion molecules or surface receptors, as seen from Table 9.1. GPI-anchored proteins of various microbial pathogens are reported to be immunogenic and are suggested to be important virulence factors. In addition, GPI-bound proteins can display enzymatic properties, playing an active role in cell wall biosynthesis. A summary of their functions is given in Table 9.3.

Looking at fungi, synthesis of GPI anchors is essential for viability. Effectively, their cell wall mannoproteins require a GPI anchor so that they can be covalently incorporated into the cell wall. In intracellular transport, GPI appears to act as an intracellular signal targeting proteins to the apical surface in polarized cells, since GPI-anchored proteins are sorted into sphingolipid- and cholesterol-rich microdomains, known as lipid rafts, before transport to the membrane surface [12]. Their localization in raft microdomains may explain the involvement of this class of proteins in signal transduction processes. Moreover, substantial evidence suggests that GPI-anchored proteins may interact closely with the bilayer surface so that their functions may be modulated by the biophysical properties of the membrane. The presence of the anchor appears to impose conformational restraints, and its removal may alter the catalytic properties and structure of a GPI-anchored protein. Furthermore, release of GPI-anchored proteins from the cell surface by specific phospholipases (please see Info Box) may play an important role in regulation of their surface expression and functional properties. GPI-anchored proteins play an important role in the biogenesis of the Alzheimer's amyloid β-protein, since one

Table 9.3 Summary of functions of GPI anchors.

Function	Examples
Anchoring of surface proteins	general property
Cell wall component in yeasts	GAS1[a]
Incorporation into glycolipid rafts	general property of in higher eukaryotes
Toxin binding	to α toxin *Clostridium septicum*, aerolysin (a channel-forming bacterial toxin)
Strong apical targeting signal	in polarized epithelial cells
Selective packaging during ER exit	GAS1 in *S. cerevisiae*
Endocytosis by clathrin and caveolin	prion protein, folate receptor
Migration to other membranes	VSG, neural cell adhesion molecule[b]
Involvement in signal transduction	lymphocyte antigen 6 complex (Ly6), Thy-1[c], neural cell adhesion molecule
Parasite toxin	induction of tumor necrosis factor-α and interleukin-1 production by macrophages, increases the inducible nitric oxide synthase gene expression and nitric oxide
Upregulates intercellular adhesion molecules	ICAM-1[d], VCAM-1[e]
Induces low level of apoptosis	induced by *P. falciparum* GPIs in spleen, heart and liver of mice

[a] *gas1* encodes a cell wall-bound β1–3-glucosyltransferase involved in the formation and maintenance of β1–3-glucan, which is the major polysaccharide of the yeast cell wall.

[b] Neural Cell Adhesion Molecule (NCAM, also the cluster of differentiation CD56; for further information on NCAM, please see chapters 6 and 30.7) is a homophilic binding glycoprotein expressed on the surface of neurons, glia, skeletal muscle and natural killer cells. NCAM has been implicated as having a role in cell-cell adhesion, neurite outgrowth, synaptic plasticity, and learning and memory.

[c] Thy-1 or CD90 is a 25–37 kDa heavily *N*-glycosylated, GPI-anchored conserved cell surface protein with a single V-like immunoglobulin domain, originally discovered as a thymocyte antigen. Thy-1 can be used as a marker for a variety of stem cells and for the axonal processes of mature neurons.

[d] ICAM-1 is type of intercellular adhesion molecule continuously present in low concentrations in the membranes of leukocytes and endothelial cells.

[e] VCAM1 (vascular cell adhesion molecule-1).

or more GPI-anchored proteins are essential for the β-secretase activity and Aβ secretion in mammalian cells, and therefore cell-surface GPI-anchored protein(s) involved in Aβ biogenesis may be excellent therapeutic target(s) in Alzheimer's disease. In addition, mammalian and parasite GPI-anchored proteins act as receptors for the bacterial toxins aerolysin or clostridial α toxin. Protozoa tend to express significantly higher densities of cell-surface GPI-anchored proteins than do higher eukaryotes. *Plasmodium* GPIs act as toxins that induce cytokine production by macrophages, increase secretion of nitric oxide and expression of adhesion molecules in host cells, which are implicated in the etiology of the cerebral malaria syndrome. Also, *Plasmodium* GPIs induce apoptosis in spleen, heart and liver of mice [13], indicating that GPI might play a role in myocardial impairment observed in patients suffering from severe malaria (although GPIs from *T. gondii* are unable to trigger apoptosis in human-derived cells). GPIs derived from *Plasmodium* and *Trypanosoma* exert their inflammatory effects by activation of signaling pathway in host cells through Toll-like receptor (TLR)-2. The expression in response to GPIs of host genes implicated in parasite pathogenesis depends on the phosphorylation of cytoplasmic kinases and on the activation of transcription factors (as for example NF-κB) binding to gene promoters after their translocation into the nucleus (Figure 9.4) [14, 15] (for signaling pathway involving transcription factors elicited by lectin–glycan interaction, please see Figure 25.3). On the other hand, GPIs derived from *Toxoplasma* are recognized by TLR-4, while the core glycan and lipid moieties cleaved from these GPIs are recognized by both TLR-4 and TLR-2 [16]. Finally, and as pinpointed in the introduction, GPI anchors are hybrid molecules joining protein and lipids via a glycan hinge for the benefit of protein membrane attachment. Looking at proteins as free headgroups, the next chapter (together with Chapter 30) deals with lipids as backbones for glycan presentations.

9.4 Conclusions

This chapter describes a class of glycophospholipids with a surprising multitude of functions. Such glycophospholipids are called GPI anchors. They are found in yeasts, fungi, parasitic protozoa, other protozoan cells, plants and mammals. Although all have the same evolutionarily conserved backbone, they show a large spectrum of structural modifications in their periphery. The route of their complex biosynthesis has been elucidated; however, many details remain to be worked out – in particular differences between host and parasite – to lay the foundations of an antiparasitic chemotherapy. Most challenging, observations suggest that GPIs ('free' or protein bound) may be involved in different intracellular pathways of signaling and can – in certain circumstances – act as pathogenicity factors in infectious disease. Relatively large amounts of material are necessary to arrive at the structural data through elaborate analytical techniques. Since more individual structures (for example parasitic protozoa, immune cells to name but a few) are required in the future, the elaboration of simplified analytical techniques (and methods of preparations) is essential for further progress.

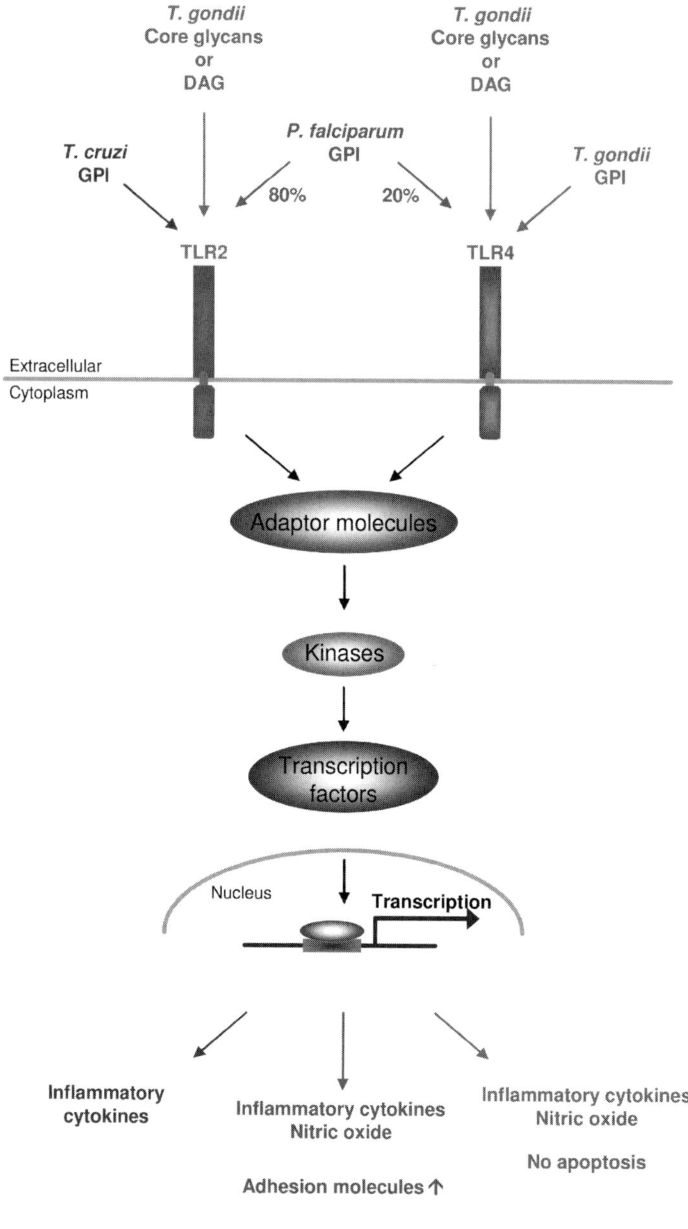

Figure 9.4 Involvement of the Toll/interleukin-1 receptor (TIR) signaling pathway in the activation of the immune response by parasite GPIs. Parasite GPIs lead to the activation of TLR-2 or TLR-4. Activation of the receptors may account for the important role of myeloid differentiation factor 88 in inducing proinflammatory cytokines playing a key role in host resistance to infection with parasites. *Trypanosoma* and *Plasmodium* GPIs trigger phosphorylation of kinases (extracellular regulated kinase-1/-2, mitogen-activated protein kinase kinases and stress-activated protein kinases/p38, please see also Figure 25.3 for kinase pathway) while numerous cytoplasmic proteins are phosphorylated on tyrosine residues in response to *Toxoplasma* GPIs. GPIs of all the three species are able to induce the nuclear translocation of NF-κB leading to the synthesis of proinflammatory cytokines.

Summary Box

The GPI anchor is a glycolipid that can be attached to the C-terminus of a protein during posttranslational modification. It is composed of a hydrophobic PI group linked through a carbohydrate containing linker (glucosamine and mannose glycosidically bound to the inositol residue) to the C-terminal amino acid of a mature protein. The two fatty acids within the hydrophobic PI group anchor the protein to the cell membrane. The GPIs are made by preassembling their components on PI as a scaffold. The synthesis starts on the cytosolic leaflet of the ER and is completed on the lumenal face of the ER. Inhibitors affecting different enzymatic steps in the GPI biosynthesis have a great potential as fungicides and antiparasitic agents.

References

1 Eckert V *et al.* GPI-anchors: structure and functions. In: *Glycosciences: Status and Perspectives* (Eds.: Gabius H-J, Gabius S), pp. 223–43. Chapman & Hall, London, 1997.

2 Delorenzi M *et al.* Genes for glycosylphosphatidylinositol toxin biosynthesis in *Plasmodium falciparum. Infect Immun* 2002; 70:4510–22.

3 Orlean P, Menon AK. Lipid posttranslational modifications. GPI anchoring of protein in yeast and mammalian cells, or: how we learned to stop worrying and love glycophospholipids. *J Lipid Res* 2007;48: 993–1011.

4 Pittet M, Conzelmann A. Biosynthesis and function of GPI proteins in the yeast *Saccharomyces cerevisiae. Biochim Biophys Acta* 2007;1771:405–20.

5 Kinoshita T, Inoue N. Dissecting and manipulating the pathway for glycosylphosphatidyl-inositol-anchor biosynthesis. *Curr Opin Chem Biol* 2000;632–8.

6 Ferguson MA. The structure, biosynthesis and functions of glycosylphosphatidylinositol anchors, and the contributions of trypanosome research. *J Cell Sci* 1999; 112:2799–809.

7 Sütterlin C *et al.* Identification of a species-specific inhibitor of glycosylphosphatidylinositol synthesis. *EMBO J* 1997;16: 6374–83.

8 Seeberger PH. Automated oligosaccharide synthesis. *Chem Soc Rev* 2008;37:19–28.

9 Nagamune K *et al.* GPI transamidase of *Trypanosoma brucei* has two previously uncharacterized (trypanosomatid transamidase 1 and 2) and three common subunits. *Proc Natl Acad Sci USA* 2003;100: 10682–7.

10 Mazhari-Tabrizi R *et al.* Chromosomal promoter replacement in *Saccharomyces cerevisiae*: construction of conditional lethal strains for the cloning of glycosyltransferases from various organisms. *Glycoconj J* 1999;16:673–9.

11 Shams-Eldin H *et al.* The GPI1 homologue from *Plasmodium falciparum* complements a *Saccharomyces cerevisiae* GPI1 anchoring mutant. *Mol Biochem Parasitol* 2002;120: 73–81.

12 Muñiz M, Riezman H. Intracellular transport of GPI-anchored proteins. *EMBO J* 2000;19:10–5.

13 Wichmann D *et al. Plasmodium falciparum* glycosylphosphatidylinositol induces limited apoptosis in liver and spleen mouse tissue. *Apoptosis* 2007;12:1037–41.

14 Tachado SD *et al.* Signal transduction in macrophages by glycosyl-phosphatidylinositols of *Plasmodium, Trypanosoma* and *Leishmania*: activation of protein tyrosine kinases and protein kinase C by inositolglycan and diacylglycerol moieties. *Proc Natl Acad Sci USA* 1997;94:4022–7.

15 Gowda DC. TLR-mediated cell signaling by malaria GPIs. *Trends Parasitol* 2007;23: 596–604.

16 Debierre-Grockiego F *et al.* Activation of TLR2 and TLR4 by glycosphosphatidylinositols derived from *Toxoplasma gondii. J Immunol* 2007;179:1129–37.

Part Three
Natural Glycosylation – Glycolipids, Proteoglycans and Chitin

The Sugar Code. Edited by Hans-Joachim Gabius

ISBN: 978-3-527-32089-9

10
Glycolipids

Jürgen Kopitz

The preceding chapters have provided detailed insights into the glycosylation of proteins. In addition to glycoproteins as a large subgroup of cellular glycoconjugates, glycans can also be a part of conjugates with lipids, as already indicated by the glycosylphosphatidylinositol (GPI) anchors (please see Chapter 9). In this case, the glycan serves as a bridge between the lipid anchor and a protein. Evidently, lipids as membrane constituents present glycans. This chapter–in conjunction with Chapter 30 emphasizing their status as major components in the nervous system–will focus on lipid–glycan conjugates, where the glycans as headgroups are freely accessible for intermolecular interactions, that is the glycolipids. In detail, this chapter will give an overview of glycolipid structures in bacteria, plants and vertebrates, describe the range of their specific cellular functions, and put them into a context to cell biology and medical sciences.

10.1
Classification and General Structures of Glycolipids

The term glycolipid designates any compound containing one or more monosaccharides bound by a glycosidic linkage to a hydrophobic membrane-anchoring compound such as an acylglycerol, a sphingoid, a ceramide (*N*-acylsphingoid) or a prenylphosphate [1]. The primary classification of glycolipids is based on the aglycon part (http://www.lipidlibrary.co.uk/lipids.html). Consequently, the term glycoglycerolipid is used to name glycolipids containing one ore more glycerol residues. Most bacterial and plant glycolipids are such glycerol-containing glycolipids [2] (examples are given in Table 10.1). Glycosphingolipids (GSLs) contain a sphingoid or a ceramide. Sphingoids are long-chain aliphatic amines, containing two or three hydroxyl groups, and often a distinctive *trans*-double bond in position 4. The most abundant of these in animal tissues is sphingosine (2S,3R,4E)-2-amino-4-octadecen-1,3-diol) or 4-sphingenine (Table 10.1). The etymology of the term is depicted in the Info Box. The carbohydrate residue is attached by a glycosidic linkage to O1 of the sphingoid. Fatty acids of varying length are linked via amide

The Sugar Code. Edited by Hans-Joachim Gabius

ISBN: 978-3-527-32089-9

Table 10.1 Principle structures of glycolipids.

	Glycoglycerolipids	**Glycosphingolipids (GSLs)**
Major occurrence	Bacteria, plants	Man, animals
Lipid backbone	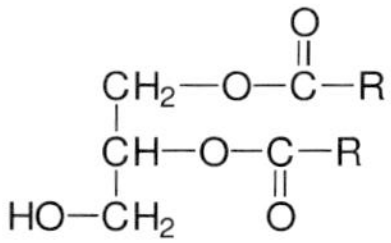 Acylglycerol R = long chain acyl	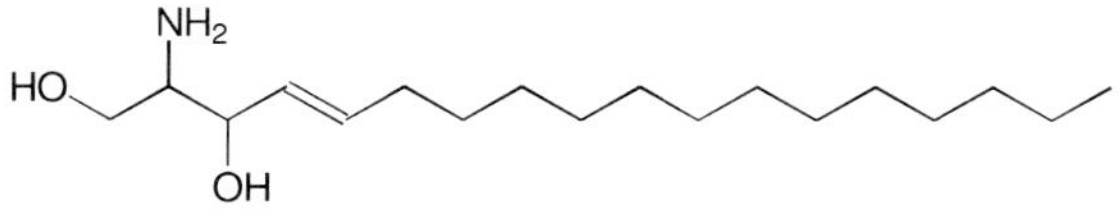 Sphingosine 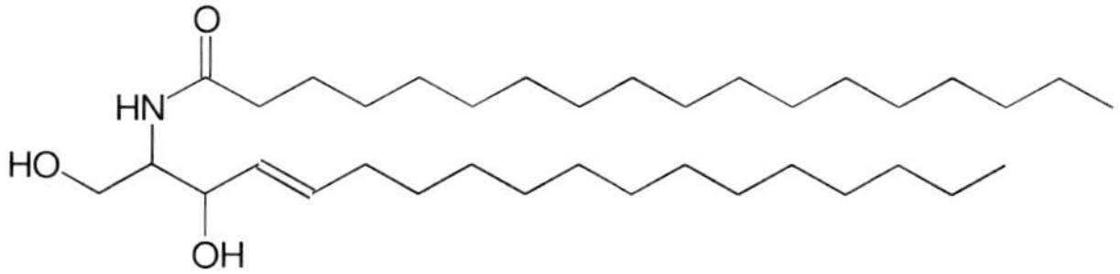Ceramide: fatty acid attached to sphingosine
Typical example	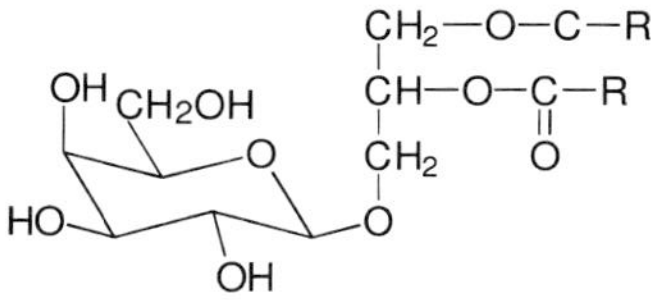 Monogalactosyldiacylglycerol	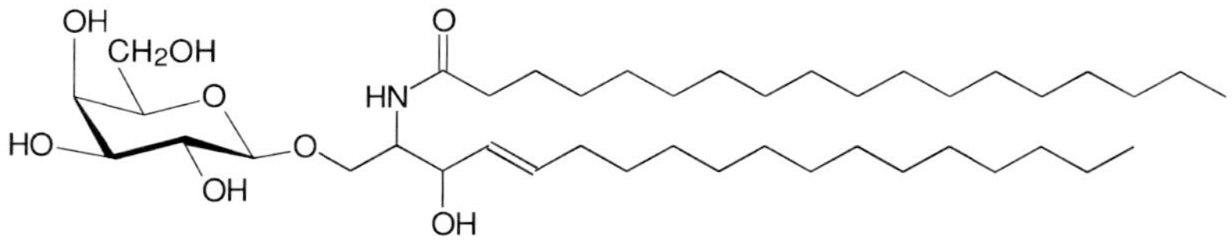 Galactosylceramide

For glycosphingolipids, please see also Figure 30.1.

Info Box

Sphingosine was discovered by Johann Ludwig Thudichum in 1884. He was a pioneer of biochemistry. His visionary understanding of the brain as an organ that could be understood chemically made him famous. In this concept, he postulated an innovative route to treating brain disease by elaborating a chemical profile of the brain. He performed chemical analyses of over 1000 human and animal brain specimens, and isolated, characterized and named numerous brain-derived compounds such as cephalin, galactose or lactic acid. Apparently, Thudichum also spent some of his spare time on studying Greek mythology, intimating why he named the sphingosine backbone of the most enigmatic compounds he discovered, the sphingolipids, after the sphinx: 'In commemoration of the many enigmas which it presents to the inquirer'.

This Greek mythological creature was seated at the gates of Thebes. She posed a riddle to travelers with the condition that only those who could solve it were allowed to pass, but those who failed should be killed. The name 'sphinx' derives from the Greek word '*sphingo*' (to strangle), since the monster had the infamous habit of strangling its victims. No one had yet come up with the right answer when Oedipus arrived at the gate of Thebes and the creature posed her riddle: 'What walks on four legs in the morning, two legs at noon, and three legs in the evening?'. Oedipus solved the riddle, answering that man crawls on all fours in infancy, walks upright on two legs in adulthood and uses a cane as a third leg in old age. The sphinx was so frustrated that Oedipus had answered her riddle correctly that she threw herself from the city walls and died there on the road in front of the city which she had terrorized for so long.

Turning again to today, we now know more than 1000 different sphingolipid structures, the lion's share of structural diversity represented by GSLs. We have also learned over the decades that these are not restricted to the nervous system, rather they are found in virtually all human and animal cells. However, although a wealth of information on molecular behavior and biological functions of this class of biomolecules has been collected since Thudichum's seminal work, it is likely that key functions still appear elusive. Thus, the riddle posed by Thudichum to biological and medical sciences is still an attractive challenge.

bonds to sphingosine's amino group. The formed ceramide constitutes the lipid backbone of the predominant glycolipid structures in man and animals (Table 10.1). These GSLs are further differentiated into neutral charge-free GSLs and acidic GSLs carrying charged saccharides. In gangliosides (sialoGSLs) one or more sialic acids (*N*-acetyl- or *N*-glycolyl-neuraminic acid) are part of the oligosaccharide chain [3]. SulfoGSLs ('sulfatides') carry a sulfate ester group attached to a carbohydrate moiety [4]. PhosphoGSLs contain a phosphodiester bond which either links the saccharide chain to ceramide or esterifies a 2-aminoethyl phosphate to a saccharide residue. The term 'GPI' designates glycolipids which contain saccharides glycosidically linked to the inositol moiety of phosphatidylinositols, including various modifications like *O*-acylation or *O*-alkylation of the inositol residue. The

Bacterial lipopolysaccharides are also of considerable medical interest. When bacteria perish and break up during an infection, lipopolysaccharide is liberated. It then functions as a powerful bacterial toxin that has been termed endotoxin. In particular, the lipid A component is responsible for many of the toxic effects of infections with Gram-negative bacteria. At high concentrations, it induces fever, increases heart rate and, in the worst cases, can lead to septic shock. On the other hand, lipid A is also an active immunomodulator, able to induce nonspecific resistance to bacterial infections. The plasma protein Toll-like receptor 4 was identified as the lipid A signaling receptor of animal and human cells. It triggers the biosynthesis of diverse mediators of inflammation and activates the production of costimulatory molecules required for adaptive immune response (Table 19.3 also lists lectins binding lipopolysaccharides). Therefore, investigations on its downstream signaling pathways are likely to provide new opportunities for therapeutic blocking of inflammation associated with infection [6]. Apart from the lipid A-bound glycan chains, bacterial glycolipids have the potential to induce a defense reaction of the host, which is mainly triggered by glycolipid-induced T-cell activation.

10.4 Bacterial Glycolipids in T-Cell Activation

A well-established mechanism of T-cell activation involves recognition of peptide antigens presented by the major histocompatibility complex proteins (MHC) to T-cell receptors, direct contact between the T-cell receptor with MHC-peptide complexes providing the primary signal for T-cell activation. For a long time, it was assumed that T-cell receptors were solely suited to recognize foreign peptides and largely ignore carbohydrates. However, MHC class I and class II also present glycopeptides that originate from glycoproteins, resulting in glycopeptide-specific activation of T-cells. Recent studies showed that CD1, a MHC class I homolog encoded outside the gene cluster of the MHC, presents bacterial glycolipids to T-cells, thereby initiating T-cell responses. CD1 family members are divided into CD1a–e and all five protein isoforms are expressed in humans (for a primer to the CD nomenclature, please see Info Box in Chapter 27). Glycolipid binding to CD1 is an example for a direct glycan–protein interaction as described in Chapter 13. In particular, natural killer cells display a marked ability to recognize CD1-presented microbe-derived glycolipids. Thus, CD1-presented bacterial glycolipids, including the examples described above, are now considered to play a complementary role to conventional MHC-dependent mechanisms in the initiation and regulation of immune responses to bacterial infections. As a consequence, the inclusion of glycolipids in vaccines to bacteria appears of considerable value: since proteins in bacteria with a short replication time are prone to frequent amino acid alterations by mutations, modifications of antigenic peptides may consistently occur. In contrast, glycolipids are products of complex biosynthetic pathways and therefore are less susceptible to alterations by gene products, thus providing a rather stable repertoire of cellular antigens [7].

Turning back to CD1, its loading with glycolipid takes place in late endosomes and lysosomes with the help of lipid transfer proteins including GM2-activator proteins and saposins, which also act on eukaryotic GSLs. Saposins isolate individual lipids from the membrane surroundings, thus making them more accessible to degradative enzymes (see Section 10.14) or to CD1 binding. Recent studies confirmed that CD1-associated self-glycolipid presentation triggers autoimmune reactions to GSL autoantigens [8]. Moreover, the CD1 presentation of tumor-associated GSLs links these compounds to T-cell-mediated immune responses [9]. These examples illustrate that the profile of GSLs is tied to diverse states, warranting a survey of this class of glycolipids.

10.5 Glycosphingolipids (GSLs)

Glycosphingolipids are found primarily in the plasma membrane of all vertebrate tissues, and they are particularly abundant in the nervous system (for further information on functions in the nervous system, please see Chapter 30.5; for an example of orchestration of synthesis of a distinct ganglioside and its receptor, a lectin, please see Chapter 25.2). The simplest neutral GSL structures are the monoglycosylceramides. The major glycolipid of mammalian brain, for example, is galactosylceramide, constituting about 16% of total lipid in the brain (Table 10.1). In comparison, glucosylceramide is a major constituent of skin lipids, where it is essential for lamellar body formation in the stratum corneum and to maintain the water permeability barrier of the skin. Beyond that, glucosylceramide comprises the core of most glycolipid structures in mammals. It is synthesized on the cytosolic face of the endoplasmic reticulum by a glucosylceramide synthase (UDP-glucose : ceramide glucosyltransferase), which transfers the glucose moiety from UDP-glucose to ceramide. Glucosylceramide is subsequently translocated to the Golgi lumen, where its carbohydrate chain is elongated to more complex GSLs by Golgi-localized glycosyltransferases. Galactosylceramide synthase (UDP-galactose : ceramide galactosyltransferase) initiates the elaboration of the less abundant galactosphingolipid core structures (Figure 10.2). For naming diglycosylceramides it is common to use the designation of the disaccharide, for example, lactosylceramide for β-D-galactosyl-(1,4)-β-D-glucosyl-(1,1)-ceramide. It is the starting point for the biosynthesis of GSLs with extended glycan complexity. The GSLs are divided into two categories which now will be explained.

10.6 Complex Neutral GSLs

The glycan chains of these compounds do not carry a negative charge. It turned out to be rather cumbersome to find systematic names for GSLs with long oligosaccharide chains. Therefore, suitable trivial names were coined for frequently

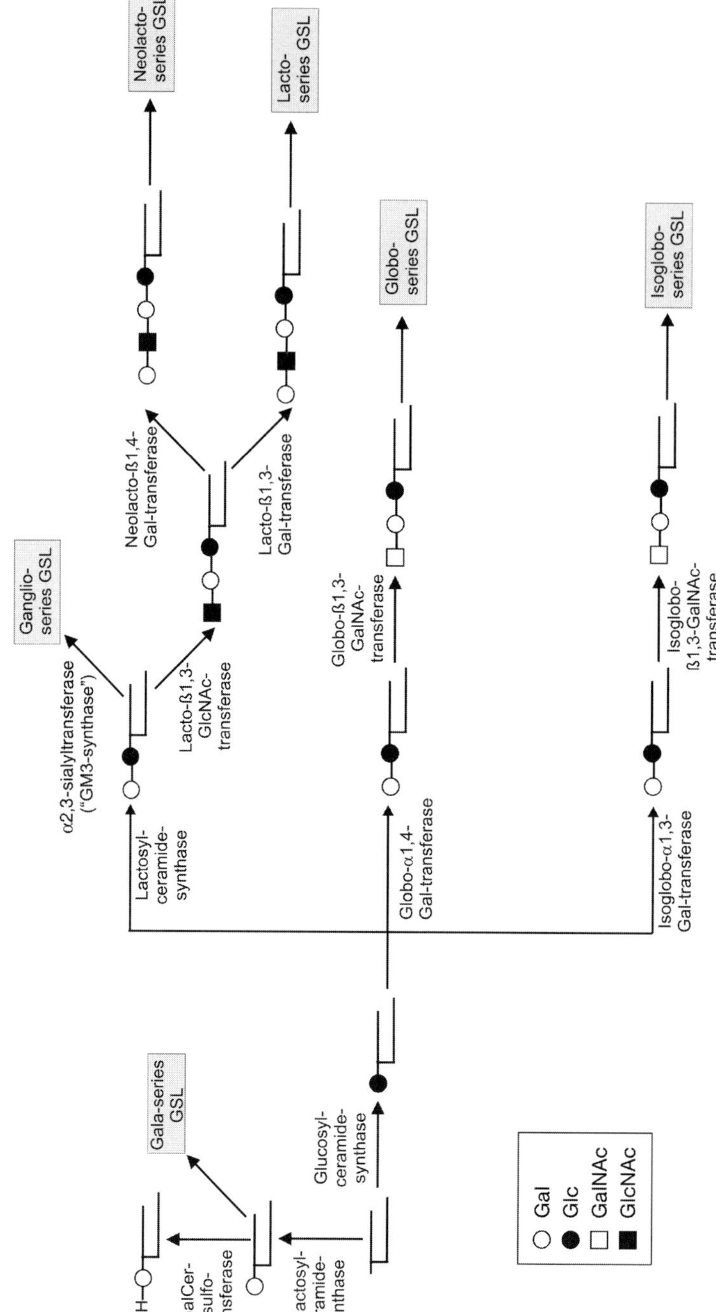

Figure 10.2 Biosynthetic route of the core structures of GSLs (for further information, see Figure 30.3).

Table 10.2 Structures of glycan chains, trivial names and abbreviations of GSLs (for biosynthetic pathways, please see Figure 10.2).

Structure	Trivial name	Symbol
Galα1,4Galβ1,4GlcCer	Globotriaosylceramide	$GbOse_3Cer$
GalNAcβ1,3Galα1,4Galβ1,4GlcCer	Globotetraosylceramide	$GbOse_4Cer$
Galα1,3Galβ1,4GlcCer	Isoglobotriaosylceramide	$iGbOse_3Cer$
GalNAcβ1,3Galα1,3Galβ1,4GlcCer	Isoglobotetraosylceramide	$iGbOse_4Cer$
Galβ1,4Galβ1,4GlcCer	Mucotriaosylceramide	$McOse_3Cer$
Galβ1,3Galβ1,4Galβ1,4GlcCer	Mucotetraosylceramide	$McOse_4Cer$
GlcNAcβ1,3Galβ1,4GlcCer	Lactotriaosylceramide	$LcOse_3Cer$
Galβ1,3GlcNAcβ1,3Galβ1,4GlcCer	Lactotetraosylceramide	$LcOse_4Cer$
Galβ1,4GlcNAcβ1,3Galβ1,4GlcCer	Neolactotetraosylceramide	$nLcOse_4Cer$
GalNAcβ1,4Galβ1,4GlcCer	Gangliotriaosylceramide	$GgOse_3Cer$
Galβ1,3GalNAcβ1,4Galβ1,4GlcCer	Gangliotetraosylceramide	$GgOse_4Cer$
Galα1,4GalCer	Galabiosylceramide	$GaOse_2Cer$
Galα1,4Galα1,4GalCer	Galatriaosylceramide	$GaOse_3Cer$

occurring parent oligosaccharides. General rules for constructing these names are as follows: the number of monosaccharide units is indicated by the suffixes 'biose', 'triaose', 'tetraose' and so on. According to their structure and biogenic relationship, they are grouped in distinct series, for example defined by the prefix globo, muco, lacto, ganglio and gala. Differences in linkage such as 1,4 rather than 1,3 in otherwise identical sequences are highlighted by 'iso' or 'neo' as an additional prefix. The implications of the structural changes for the coding capacity are outlined in Chapter 1. To illustrate this system a summary of common trivial names is presented in Table 10.2. The pathways for the biosynthesis of neutral GSLs are depicted in Figure 10.2. With the introduction of a negative charge to the glycan chains we encounter the second group.

10.7 Complex Acidic (Anionic) GSLs

Sulfation of the saccharide part turns neutral GSLs into charged sulfatides (for examples of sulfation in glycoproteins, please see Figure 1.7 and Chapter 27). The major mammalian sulfatide consists of a monogalactosylceramide in which the 3-OH group of the galactosyl headgroup is sulfated (Figure 10.3a). Most acidic GSLs are gangliosides, which are characterized by the presence of one or more sialic acids (Figure 10.3b). Frequently, the sialic acid is *N*-acetylneuraminic acid, but, for example, *N*-glycolylneuraminic acid is also detected. Additionally, the sialic acid can be modified, for example by *N*-deacetylation or *O*-acetylation. Most

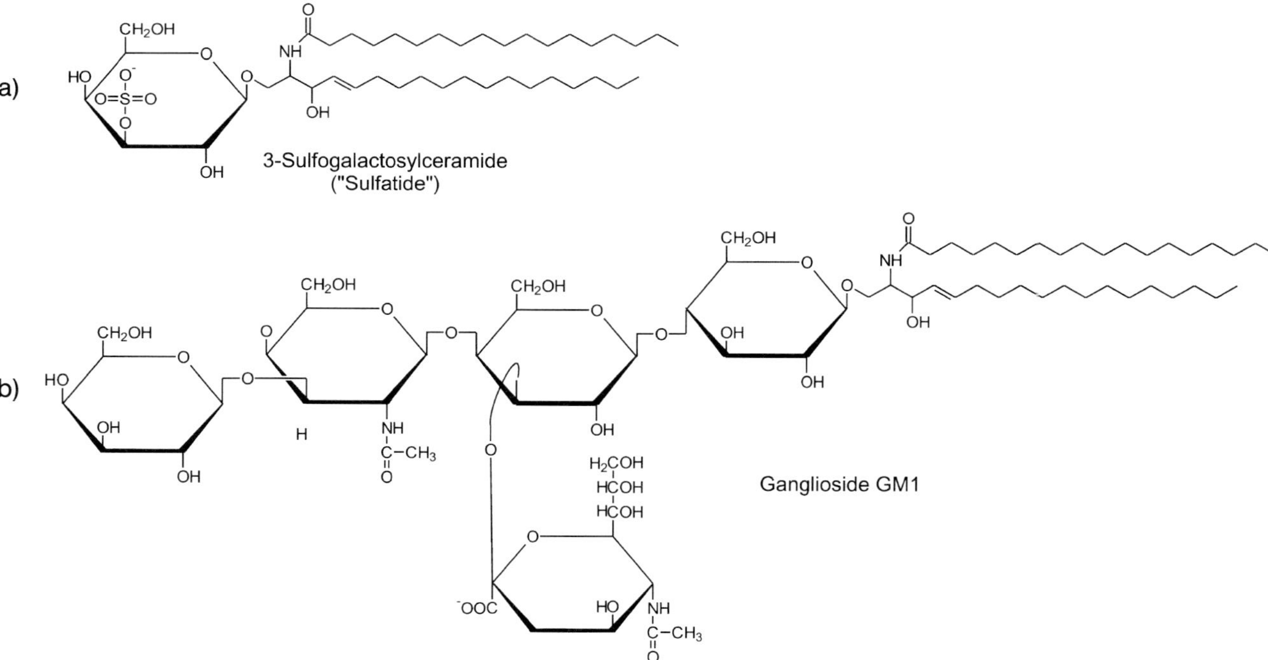

Figure 10.3 Examples of structures of acidic GSLs (for another type of illustration of the pentasaccharide of ganglioside GM1, please see Figure 30.2).

Table 10.3 Examples for the Svennerholm designation of gangliosides (for detailed information on structural and biosynthetic aspects, please see Figure 30.3).

Svennerholm designation	Abbreviated structure
GM3	II^3NeuAc-LacCer
GM2	II^3NeuAc-$GgOse_3$Cer
GM1	II^3NeuAc-$GgOse_4$Cer
GD3	$II^3(NeuAc)_2$-LacCer
GD2	$II^3(NeuAc)_2$-$GgOse_3$Cer
GD1a	IV^3NeuAc,II^3NeuAc-$GgOse_4$Cer
GD1b	$II^3(NeuAc)_2$-$GgOse_4$Cer
GT1a	$IV^3(NeuAc)_2$,II^3NeuAc-$GgOse_4$Cer
GT1b	IV^3NeuAc,$II^3(NeuAc)_2$-$GgOse_4$Cer
GT1c	$II^3(NeuAc)_3$-$GgOse_4$Cer
GQ1b	$IV^3(NeuAc)_2$,$II^3(NeuAc)_2$-$GgOse_4$

gangliosides belong to the gangliotetraose series, but also globo or lacto series members may occur. Principles of naming of GSLs as described above are also applied to gangliosides. Thus, gangliosides are designated as *N*-acetylneuraminyl (or *N*-glycolylneuraminyl) derivatives of the corresponding neutral oligosaccharide structure. The site of the sialosyl residue is indicated by a Roman numeral designating the position of the parent oligosaccharide to which the sialosyl residue is attached. An Arabic numeral superscript specifies the position within the parent chain oligosaccharide to which the sialic acid is attached (Table 10.3). For its simplicity, although being less systematic in structural terms, the Svennerholm designation for gangliosides is often preferred. Therefore, this nomenclature (Table 10.3) is that encountered most often in the literature. In this system, the letter G indicates that we deal with gangliosides. The number of sialic acids is indicated by M = monosialo-, D = disialo-, T = trisialo- and Q = quatrosialo-. The length of the neutral sugar chain is designated by a number following the formula $5 - n$, where n equals the number of neutral sugars in the ganglioside. The position of the sialic acid(s) can be further characterized by a, b or c, indicating the biosynthetic pathway for the ganglioside (for further details, see Figure 30.3). Based on recommendations of the IUPAC using subscripts, like G_{M1}, G_{M2}, G_{D1a} and so on, should no longer be used [1]. Structure and chemical properties of GSL establish the basis for their functions, as will be shown in the next paragraphs.

10.8 Survey of GSL Functions

GSLs are always found in the exoplasmic leaflet of membranes and are positioned, mainly but not exclusively (for functions of GM1 in the nuclear envelope, please in the delivery of proteins to the plasma membrane. In detail, GSL-enriched domains are thought to originate from the Golgi apparatus, where GPI- or acyl-anchored proteins as well as transmembrane proteins associate with rafts. Subsequent targeting of rafts to the plasma membrane serves as a sorting platform for these cell surface proteins. A striking example of such specific sorting functions is the use of GSL domains as a vehicle for translocation of proteins to immune synapses on T-cells [13]. Structurally, GSL/cholesterol composites are thicker than typical phospholipid/cholesterol membranes and transmembrane domains of plasma membrane proteins are longer than those of Golgi proteins. Therefore, proteins for export to the plasma membrane specifically associate with GSL/cholesterol domains in the Golgi. These domains form vesicles that are routed to the plasma membrane, thereby acting as sorting vehicles for the transport of transmembrane proteins to the cell surfaces [14]. Conversely, association of proteins with GSL rafts on the plasma membrane may play a substantial role in their endocytic membrane uptake [15, 16].

In addition to sorting processes, the high degree of spatial organization of glycosyl epitopes in microdomains effects cell adhesion and signal transduction events [17]. Domains performing such functions were termed 'glycosynapse' [18]. This concept provides the key to understanding a variety of functional interactions of GSL with transmembrane receptors or signal transducers involved in cell adhesion and signaling. Three types of glycosynapses are distinguished [18]:

- Type 1 glycosynapse: GSLs organized with cytoplasmic signal transducers and the transmembrane protein tetraspanin with or without growth factor receptors.
- Type 2 glycosynapse: transmembrane mucin-type glycoproteins (for further information on *O*-glycosylation, please see Chapter 7) with clustered *O*-linked glycan for cell adhesion and associated signal transducers, like tyrosine kinases or phosphatases, at the lipid domain.
- Type 3 glycosynapse: *N*-glycosylated transmembrane adhesion receptors complexed with tetraspanin and ganglioside.

In the case of type 1 and type 3 glycosynapses oligosaccharide chains of GSLs are functional, as described below, whereas type 2 applies to glycan chains of mucins (please see Chapter 7). The model for type 1 applies to GSL-domain-associated cytoplasmic tyrosine kinases and to growth factor receptors connected to intrinsic tyrosine kinases (Figure 10.4a and b), for type 3 to integrins associated with members of the tetraspanin family. Oligosaccharide chains of surrounding glycolipids interact with *N*-glycan chains of tetraspanin or the effector molecule itself (Figure 10.4c). For both types of glycolipid functions, two types of interaction with the complementary binding partner are possible: carbohydrate–carbohydrate (please see Chapter 21 for details) and GSL–lectin interaction, with cross-linking by bi- or multivalent lectins (please see Chapter 19). As noted above, the sugar-encoded information of GSLs also facilitates docking as a first step in infections.

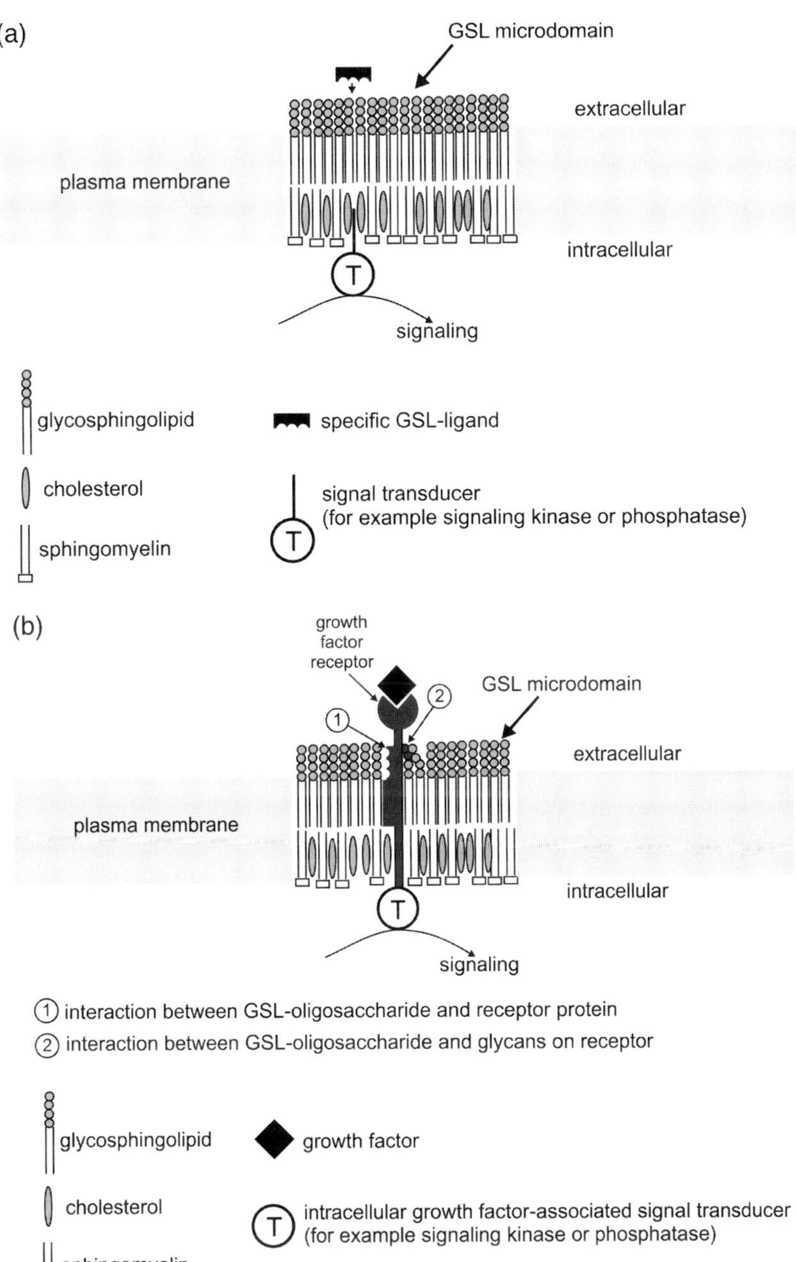

Figure 10.4 Signal transduction in GSL microdomains (adapted from [18]). (a) Nonreceptor signal transduction: ligand-binding to GSL cluster ('activation of microdomain') causes structural changes across the corresponding area of the bilayer, thereby affecting the activity of intracellular membrane-associated signal transducers. (b) Growth factor signaling: growth factor receptor interacts with GSL-oligosaccharides thereby affecting the receptor's affinity or transmembrane signal transduction capabilities. (1) GSL-oligosaccharide directly interacts with the receptor protein; (2) Glycan side-chains of the receptor interact with GSL-oligosaccharides. This interaction may be brought about either by direct oligosaccharide–oligosaccharide interaction (please see Chapter 21 for details) or by cross-linking via bi- or multivalent lectins (please see Chapter 19 for details on human lectins). Such interactions were suggested to be effective in translating altered glycolipid composition of microdomains into modulation of signaling events. (c; please see next page) Integrin signaling: interaction of *N*-glycan chains of integrins with GSLs either mediated by (1) direct oligosaccharide–oligosaccharide contact or by (2) lectin cross-linking modulates integrin signaling (please see Figure 25.3 for an example).

Table 10.5 Typical tumor-associated GSLs.

Name	Tumor expression	Structure
Gb3	Burkitt lymphoma, ovarian cancer	Galα4Galβ4GlcCer 
Globo H	Breast cancer, ovarian cancer	Fucα2Galβ3GalNAcβ3Galα4Galβ4GlcCer 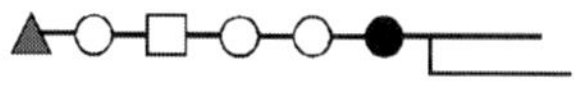
GD3	Melanoma, glioblastoma	NeuAcα8NeuAcα3Galβ3GlcCer 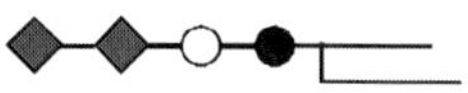
GD2	Neuroblastoma, glioblastoma	NeuAcα8NeuAcα3Galβ4GlcCer 4 GalNAcβ 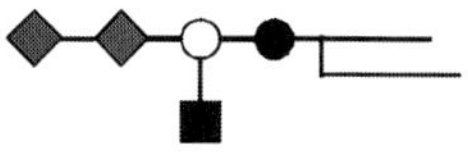
Fucosyl-GM1	Lung small cell carcinoma	Fucα2Galβ3GalNAcβ4Galβ4GlcCer 3 NeuAcα 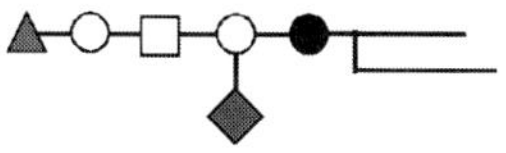
Le^x	Gastrointestinal, colorectal, breast and lung cancers	Galβ4GlcNAcβ3Galβ4GlcCer 3 Fucα 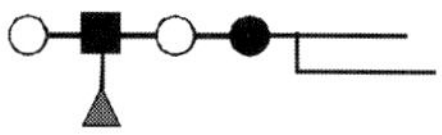
Le^y	Gastrointestinal, colorectal, breast and lung cancers	Galβ4GlcNAcβ3Galβ3GlcCer 2 3 Fucα Fucα 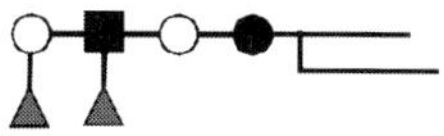
Sialyl-Le^x	Gastrointestinal, colorectal, breast and lung cancers	Galβ4GlcNAcβ3Galβ3GlcCer 3 3 NeuAcα Fucα 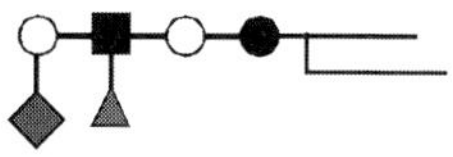

neuroblastoma or GD3 for melanoma) are suitable for both active and passive immunotherapies. With regard to an active immune response to GSLs, it has to be stressed that the glycolipid-specific CD1-dependent T-cell activation mechanism described above is also effective with tumor-associated GSLs. Considering the increasing evidence for GSL functionality in cell adhesion and signaling, a participation of tumor GSLs in tumor cell adhesion and signaling is likewise probable, thus envisioning the development of 'antiadhesion' or 'antisignaling' tumor therapies [9, 22, 23].

10.13 Gangliosides in Neural Tissue

Gangliosides are found in highest concentrations in neural tissue, involved in processes such as synaptic transmission. In diseases, gangliosides have been implicated in various neurological and neurodegenerative disorders. Natural and semisynthetic gangliosides are considered possible therapeutics for neurodegenerative disorders. As an example, considerable interest has developed in the role of autoimmune responses to gangliosides and other GSLs in the pathogenesis of peripheral neuropathies, like Guillain–Barré syndrome, Miller–Fischer syndrome and other autoinflammatory disorders of the peripheral nervous system (for further details on ganglioside occurence and function in the nervous system, please see Chapter 30.2–5). Having so far examined GSL synthesis and function, the final part of this chapter will be devoted to their degradation.

10.14 GSL Degradation and GSL Storage Disorders

In contrast to the situation in glycoprotein biogenesis (please see Chapter 22), except for an infantile-onset symptomatic epilepsy syndrome caused by a homozygous loss-of-function mutation of GM3 synthase, no human diseases resulting from defective GSL biosynthesis are known, emphasizing that GSLs are vital to life. However, a number of inherited human disorders are known to result from defects in the enzymes involved in GSL degradation [24] (for degradation of proteoglycans, please see Chapter 11). Enzymes of GSL catabolism reside in the lysosomal compartment, into which glycolipids are transported by endocytic membrane flow. Figure 10.5 gives an overview of GSL degradation; disorders resulting from deficiency of the corresponding enzyme are also indicated. Soluble lysosomal enzymes involved in GSL degradation cannot degrade membrane-bound lipids directly; instead, the GSL must be solubilized by nonenzymatic 'activator proteins' (saposins). Deficiency of the activator proteins represents an alternative cause of GSL storage disorders [25]. Impairment of a GM2-activator protein, for instance, is an alternative cause of GM2 gangliosidosis. These sphingolipid activator proteins not only facilitate glycolipid digestion, but also act as glycolipid transfer proteins facilitating the association of lipid antigens with proteins of the CD1 family, as given above. In general, most GSL storage disorders cause varying degrees of central nervous damage; neuropathology is often very severe and frequently lethal in early infancy. Of note, the biochemical mechanisms leading from intracellular GSL accumulation to pathogenesis are not yet fully clarified. Detailed study of cellular dysfunction in GSL storage disorders will not only help to suggest new treatment strategies of these diseases, it will also significantly contribute to further understand the roles of GSLs in cell physiology (please see also Chapter 22.4, Table 23.1 and Table 30.1).

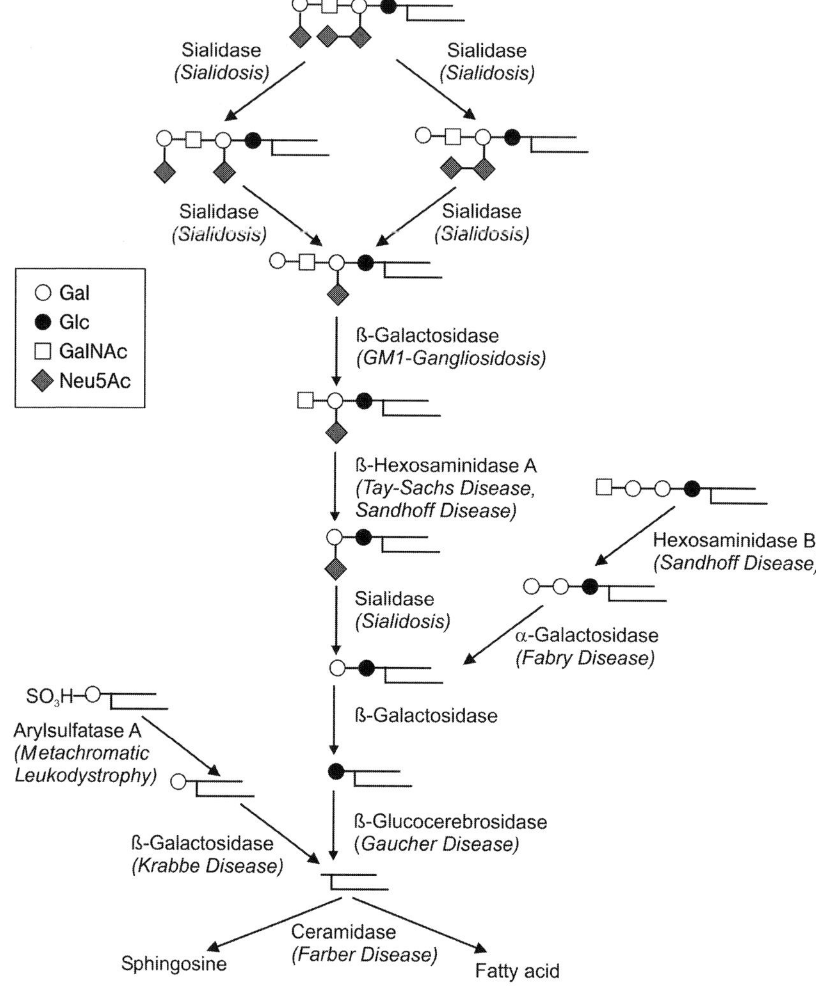

Figure 10.5 Major pathways for GSL degradation. In brackets: disorders resulting from deficiency of the corresponding hydrolase. Disorders are characterized by vast accumulation of the defective enzyme's substrate(s).

10.15 Conclusions

Beyond the fact that glycolipids are indispensable for proper amphiphilic organization of cellular membranes, they are much more than just simple structural elements. Rather, they fulfill, support or modulate membrane-associated cellular functions like intracellular trafficking, signal transduction and cellular interactions. In recent years the microdomain ('lipid raft') model has been shaped, providing a molecular basis to give glycolipid function a further biological dimension. Elucidation of detailed structures and functions of these domains are a major topic of cell biology. In medical terms, the development of new compounds to block interaction of bacterial or viral pathogens with their glycolipid receptors on host cells will help to fight infectious diseases. Next, analysis of the structure and function of tumor-associated glycolipid antigens will advance tumor biology, with the perspective to devise innovative strategies for tumor therapy, as is also the case for autoimmune regulation [26]. Since GSL storage disorders are natural models of GSL-mediated neurodegeneration, their detailed examination on a molecular level is likely to contribute to a general understanding of neurodegenerative processes.

Summary Box

Most bacterial and plant glycolipids are glycoglycerolipids, whereas animal and human glycolipids belong to the class of glycosphingolipids (GSLs). Glycoglycerolipids have been shown to be functional in the photosynthetic process of thylakoid membranes of bacteria and plants. In Gram-positive bacteria glycoglycerolipids are important structural elements of the cell membrane. Lipid A-linked complex oligosaccharide chains are the predominant cell-surface structure in Gram-negative bacteria. During infections bacterial glycolipids act as endotoxins or antigens. GSLs of vertebrate tissues are mainly found in the plasma membrane, clustered in microdomains or 'lipid rafts'. This structure is considered to provide a biochemical platform for interaction with cellular proteins, thereby influencing intracellular trafficking, signal transduction and cellular interactions. Specific changes of GSL patterns during pathological processes, like oncogenic transformation or neurodegeneration, suggest that certain specific GSL structures are disease markers and potential targets for therapeutic intervention.

References

1 Chester MA. IUPAC–IUB Joint Commission on Biochemical Nomenclature (JCBN). Nomenclature of glycolipids–recommendations 1997. *Eur J Biochem* 1998;257:293–8.

2 Holzl G, Dormann P. Structure and function of glycoglycerolipids in plants and bacteria. *Prog Lipid Res* 2007;46:225–43.

3 Hakomori SI. Chemistry of glycosphingolipids. In: *Sphingolipid Biochemistry* (Eds.: Kafner JN, Hakomori SI), pp. 1–165. Plenum Press, New York, 1983.

4 Ishizuka I. Chemistry and functional distribution of sulfoglycolipids. *Prog Lipid Res* 1997;36:245–319.

5 Jones MR. Lipids in photosynthetic reaction centres: structural roles and functional holes. *Prog Lipid Res* 2007;46:56–87.

6 Raetz CR, Whitfield C. Lipopolysaccharide endotoxins. *Annu Rev Biochem* 2002;71:635–700.

7 Bricard G, Porcelli SA. Antigen presentation by CD1 molecules and the generation of lipid-specific T cell immunity. *Cell Mol Life Sci* 2007;64:1824–40.

8 Miyake S, Yamamura T. NKT cells and autoimmune diseases: unraveling the complexity. *Curr Top Microbiol Immunol* 2007;314: 251–67.

9 Swann JB *et al.* CD1-restricted T cells and tumor immunity. *Curr Top Microbiol Immunol* 2007;314:293–323.
10 Merrill AH *et al.* (Glyco)sphingolipidology: an amazing challenge and opportunity for systems biology. *Trends Biochem Sci* 2007;32:457–68.
11 Brown D. Structure and function of membrane rafts. *Int J Med Microbiol* 2002; 291:433–7.
12 Lucero HA *et al.* Lipid rafts–protein association and the regulation of protein activity. *Arch Biochem Biophys* 2004;426:208–24.
13 Jordan S *et al.* T cell glycolipid-enriched membrane domains are constitutively assembled as membrane patches that translocate to immune synapses. *J Immunol* 2003;171:78–87.
14 Bretscher MS *et al.* Cholesterol and the Golgi apparatus. *Science* 1993;261:1280–1.
15 Sillence DJ *et al.* Glycosphingolipids in endocytic membrane transport. *Semin Cell Dev Biol* 2004;15:409–16.
16 Hanzal-Bayer MF *et al.* Lipid rafts and membrane traffic. *FEBS Lett* 2007;581: 2098–104.
17 Prinelti A *et al.* Glycosphingolipid behaviour in complex membranes. *Biochim Biophys Acta* 2009; 1788:184–93.
18 Hakomori SI. The glycosynapse. *Proc Natl Acad Sci USA* 2002;99:225–32.
19 Smith DC *et al.* Glycosphingolipids as toxin receptors. *Semin Cell Dev Biol* 2004;15:397–408.
20 Furukawa K *et al.* Roles of glycolipids in the development and maintenance of nervous tissues. *Methods Enzymol* 2006;417:37–52.
21 Taga S *et al.* Sequential changes in glycolipid expression during human B cell differentiation: enzymatic bases. *Biochim Biophys Acta* 1997;1254:56–65.
22 Hakomori SI. Cancer-associated glycosphingolipid antigens: their structure, organization, and function. *Acta Anat (Basel)* 1998;161:79–90.
23 Slovin SF *et al.* Carbohydrate vaccines as immunotherapy for cancer. *Immunol Cell Biol* 2005;83:418–28.
24 Ginzburg L *et al.* The pathogenesis of glycosphingolipid storage disorders. *Semin Cell Dev Biol* 2004;15:417–31.
25 Kolter T *et al.* Principles of lysosomal membrane digestion: stimulation of sphingolipid degradation by sphingolipid activator proteins and anionic lysosomal lipids. *Annu Rev Cell Dev Biol* 2005;21:81–103.
26 Wang J *et al.* Cross-linking of GM1 ganglioside by galectin-1 mediates regulatory T cell activity involving TRPC5 channel activation: possible role in suppressing experimental autoimmune encephalomyelitis. *J Immunol* 2009;182:4036–45.

11
Proteoglycans

Eckhart Buddecke

Proteoglycans (PGs) make up a family of macromolecules consisting of a protein core to which different high-molecular-weight polyanionic glycosaminoglycan (GAG) chains are linked covalently. GAG chains are made of repeating disaccharide units with sulfate groups linked by ester bonds to certain monosaccharides. The macromolecular structure, the highly charged polyanionic nature of GAG and the variable structure of the protein core make the PGs a widespread and numerous family with essential functions for normal developmental processes in multicellular organisms and in response to injury and disease. PGs play a key role in cellular signaling, as storage depots for growth factors and cytokines, and interact with a broad variety of counterpart molecules. They are involved in maintaining the tensile strength of skin and tendons, the viscoelasticity of blood vessels, and the compressive properties of cartilage by adsorbing large volumes of water.

11.1
Glycosaminoglycans: Components of Proteoglycans (PGs)

11.1.1
Structure

Figure 11.1 gives the major repeating disaccharide units of GAG chains [1] (for structures of monosaccharides, see Figure 1.6). Although four major classes of GAGs are recognized, certain features are common to all. All GAGs form long unbranched chains containing 50–150 disaccharide units. They are attached to the protein core via a carbohydrate–protein linkage region which is either a tetrasaccharide (GlcA-Gal-Gal-Xyl) glycosidically linked to a serine residue (Figure 11.1) or has a glycoprotein-like structure. The xylose residue of the linkage region may be phosphorylated on the 2-position. Chondroitin sulfates (CSs) are the most abundant GAGs in the body. The GlcNAc residues may be sulfated in either the 4- or 6-position. Each CS chain contains between 30 and 50 disaccharide units corresponding to a molecular weight of 15 000–25 000 Da. Dermatan sulfate (DS)

The Sugar Code. Edited by Hans-Joachim Gabius

ISBN: 978-3-527-32089-9

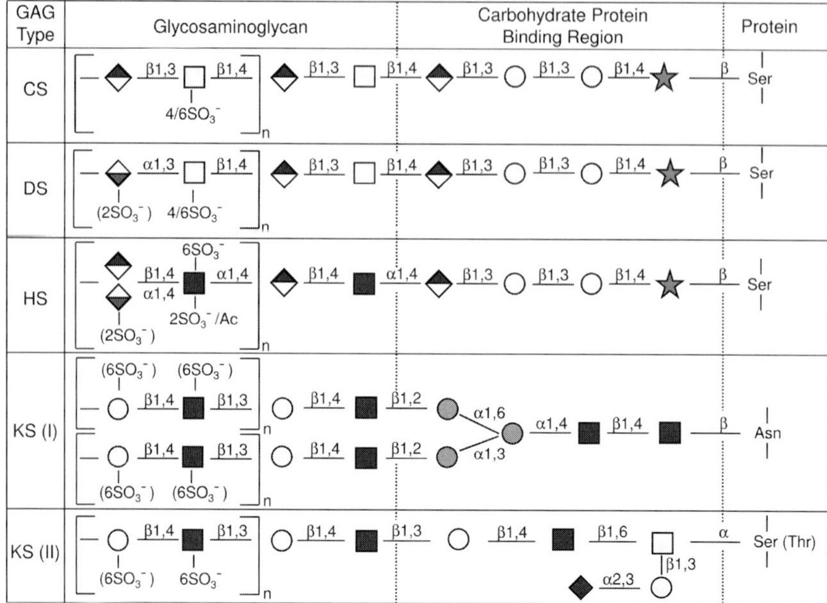

Figure 11.1 Structure of the repeating disaccharide units and the connecting region linking GAGs covalently to their protein core.

differs from chondroitin 4/6-sulfate in that its predominant uronic acid is L-IdoA. In heparan sulfate (HS) the GlcNAc is α-glycosidically linked to D-GlcA or L-IdoA to form characteristic disaccharide repeat units, but HS has a heterogeneous macromolecular structure. HS contains a few sulfate-rich oligosaccharide units (heparin-like clusters) that are separated by longer sulfate-free or sulfate-poor sequences (for the structure of heparin pentasaccharide, see Figure 1.7d). The sulfate-rich clusters of HS contain *N*-sulfate and *O*-sulfate groups on the C6 of glucosamine, but also C2 sulfate groups of the L-IdoA and C3 sulfate groups – a distinctive feature of heparin (for clinical applications, see Chapter 28). Keratan sulfate (KS) is characterized by molecular heterogeneity. The polysaccharide is composed principally of a repeating disaccharide unit of GlcNAc and Gal with no uronic acid in the molecule. The sulfate content is variable with ester sulfate present on C6 of both Gal and GlcNAc. Two types of KS have been distinguished that differ in their overall carbohydrate content and tissue distribution. They contain as additional monosaccharides Man, Fuc, sialic acid and GalNAc. KS I isolated from cornea is linked to protein by a GlcNAc–asparaginyl bond, typical of *N*-glycoproteins (for further details on *N*-glycosylation, please see Chapter 6). KS II isolated from cartilage is attached to protein through GalNAc in *O*-glycosidic linkage to either serine or threonine (for further details on *O*-glycosylation, please see Chapter 7).

11.1.2 Biosynthesis

The formation of the core protein of the PGs is the first step in this process and proceeds according to the general mechanisms of protein biosynthesis. The molecular size of the individual protein core varies over a broad range from 20 up to 400 kDa (Table 11.1). The GAG chain is synthesized by a posttranslational modification of the protein. The carbohydrate–protein linkage region (Figure 11.1) is assembled by the sequential action of UDP-xylosyltransferase, UDP-galactosyltransferase and UDP-glucuronyltransferase. Polymerization of the GAG chain then results from the concerted action of glycosyltransferases (*N*-acetylgalactosaminyltransferase, *N*-acetylglucosaminyltransferase, glucuronyltranferase and β-galactosyltransferase). Different GAG protein-binding regions are synthesized for KS I and II (see Figure 11.1). Substrates of the glycosyltransferases are nucleotide (UDP)-activated monosaccharides imported from the cytoplasm which are transferred to an appropriate acceptor which is either the nonreducing end of a growing GAG chain or an amino acid residue of a polypeptide chain. Epimerization of GlcA to IdoA is coupled with the process of sulfation and occurs along chain elongation. The sulfate donor in this reaction is 3′-phosphoadenosine-5′-phosphosulfate, which is formed from ATP and sulfate in two steps. A single enzyme is responsible for removal of the *N*-acetyl moiety of the glucosamine residue (HS) and its

Table 11.1 Main PGs in multicellular organisms (acronyms and chromosomal location refer to human genes).

PG location	PG type	Acronym (access to gene bank)	MW (kDa) core protein	GAG chain		Chromosomal location
				Type	Number	
Extracellular	Aggrecan	ACAN	210–250	CS/KS	>100/>30	15q26.1
HA-binding	Brevican	BCAN	145	CS	0–3	1q31
	Neurocan	NCAN	136	CS	1	19q12
	Versican	VCAN	260	CS	15–17	5q14.3
Extracellular	Biglycan	BCN	38	DC	2	Xq28
Leucine-rich	Decorin	DCN	36	DC	1	20q21.33
	Fibromodulin	FMOD	42	KS	4	1q32
	Lumican	LUM	38	KS	4	12q21.3
Basement membrane	Agrin	AGRN	215	HS	2	1p36.33
	Collagen XVIII	COL-18A1	2451 aa[a]	HS	2	21q22.3
	Perlecan	HSPG2	450	HS	3	1p36.1–3
Cell surface	Syndecan-1	SDC1	32	HS/CS	3/2	2p24.1
	Syndecan-2	SDC2	22	HS	3	8q22–23
	Syndecan-3	SDC3	46	HS/CS	4/2	1p22.3
	Syndecan-4	SDC4	35	HS/CS	3/1	20q12
	Glypican 1–6	GPC1–6	59–65	HS	2–5	see Text

[a] Number of amino acids.

replacement by *N*-sulfate. The biosynthesis of HS (Figure 11.2) exemplifies the principle of GAG synthesis (for information on animal models with deficiency, please see Table 23.1D).

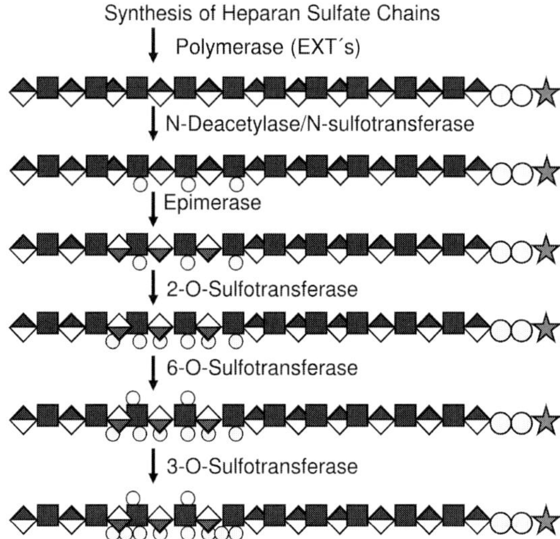

Figure 11.2 HS biosynthesis. The linker region consists of a xylose, two galactose and a glucuronic acid residue. Enzymes of the EXT polymerase family (glycosyltransferases) add alternating UDP-activated *N*-acetylglucosaminyl and glucuronyl residues (UDP-GlcNAc and UDP-GlcA) followed by the action of specific enzymes (epimerase and sulfotransferases) to form selected highly sulfated domains of the final HS chain.

11.1.3
Catabolism

PG degradation is usually initiated in the extracellular environment where proteases [matrix metalloproteinases (MMPs), membrane-type MMPs, ADAMS (metalloproteinases with a disintegrin domain) and cathepsins] cleave the protein core. The MMPs are the major mediators of extracellular PG degradation. In addition, particularly under inflammatory conditions, reactive oxygen species released by mononuclear leukocytes and macrophages can cause protein and polysaccharide cleavage within the extracellular matrix. For further degradation the PG fragments are internalized by tissue cells, and within the lysosomes a complete repertoire of proteases, sulfatases and glycosidases converts both the protein and polysaccharide component into their constituent units. Disassembling of GAG chains starts from the nonreducing end by stepwise cleavage of glycosidic linkages of monosaccharides. Sulfate ester and *N*-sulfate groups have to be removed prior to this process (Figure 11.3). The lysosomal glycosidases and sulfatases have been the subject of extensive studies since mutation in them results in a block of degradation, accumulation of oligosaccharides and lysosomal storage diseases (please see Info Box).

Info Box

Much of the information about the catabolism of PG has been derived from the study of *mucopolysaccharidoses* (Type I–VII, see below). These inborn human genetic disorders are characterized by lysosomal accumulation in tissues and excretion in urine of oligosaccharides derived from incomplete degradation of GAGs due to deficiency of one or more lysosomal hydrolases. Clinical manifestations vary from mild symptoms up to severe cases resulting in early death. The disorders are amenable by prenatal diagnosis; the incidence is 1/30 000 births. Another group of lysosomal storage diseases with similar symptoms, the *sphingolipidoses*, comprises disorders characterized by single sphingolipid accumulation in bone marrow, liver, brain and other cells. Deficiencies of galactosidase, glucosidase and hexosaminidases cause the storage of incompletely degraded glycocerebrosides and gangliosides within the lysosomes (please see also Figure 10.5 and Table 30.1).

Type	Name	Impaired degradation of	Symptoms	Enzyme deficiency
I	Hurler disease (IH) Scheie disease (IS)	DS, HS	Dwarfism, gargoyle face, mental retardation, cornea cloudiness	α-Iduronidase
II	Hunter disease	DS, HS	As in type I but less severe	Iduronide 2-sulfate sulfatase
III	Sanfilippo disease (polydystrophic oligophrenia) Type A, Type B, Type C, Type D	HS	Severe mental retardation, low physical alterations	(A) HS sulfamate sulfatase (B) α-N-Acetyl-glucosaminidase (C) Acetyl-CoA α-glucosamide N-acetyl-transferase (D) HS N-Acetylglucosamine 6-sulfate sulfatase
IV	Moquio-Brailsford disease (osteochondodystrophy)	KS, CS 6-sulfate	Dwarfism, skeletal deformities, cornea cloudiness	(A) N-Acetyl-galactosamine 6-sulfate sulfatase (B) β-Galactosidase
VI	Marotaux-Lamy disease	DS, CS 4-sulfate	Skeletal deformities, cornea cloudiness, normal mental development	N-Acetylgalactosamine 4-sulfate sulfatase
VII	β-Glucuronidase deficiency	CS, DS, HS	As in type I but variable symptoms	β-Glucuronidase

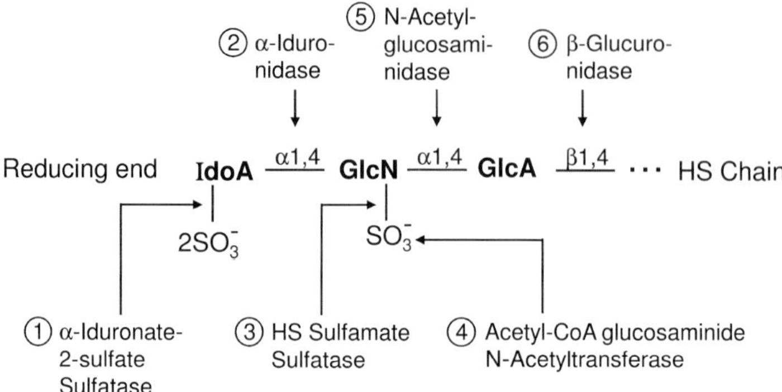

Figure 11.3 Subsequent enzymatic steps during the lysosomal degradation of a HS chain.

GAG-degrading enzymes are also secreted from many prokaryotes (that is *Arthrobacter aurescens, Flavobacterium heparinum*). These enzymes (chondroitinase ABC, several forms of heparitinase and heparinase) are important tools for the structure elucidation of GAG. They cleave glycosidic bonds by removing a water molecule (lyases) forming an unsaturated uronate residue, while the mammalian enzymes are hydrolases cleaving the glycosidic bonds by the addition of water.

11.2 PGs

Basic data about PGs are given in Table 11.1. They are listed according to their location to multicellular organisms. PGs are highly conserved in evolution and are found from lower animals (*Drosophila melanogaster, Caenorhabditis elegans, Xenopus laevis*) up to mammals and humans. The main functions of PGs are summarized in Table 11.2 [2, 3].

11.3 Large Aggregating (Hyaluronan-Binding) PGs

The large aggregating PGs aggrecan (ACAN), versican (VCAN), neurocan (NCAN) and brevican (BCAN) exist as extracellular PG aggregates [4–8]. Each aggregate is composed of a central filament of hyaluronan (HA) with a various number of HA-binding PGs. The interaction of HA and PGs (ACAN, VCAN, NCAN, BCAN) is stabilized by the presence of a link protein (for its function as lectin, please see Chapter 19 and the upper right part of Figure 19.1).

Table 11.2 Main functions of PGs.

PG type	Main functions
Extracellular PGs (ACAN, VCAN, BCAN, NCAN)	High viscosity and elasticity causes adsorption of large volumes of water. Maintenance tensile strength of skin and tendon and compressive properties of cartilage. Mineralized matrix of bones. Involved in cellular attachment, cell proliferation and differentiation by interacting with cell surfaces and extracellular matrix molecules.
SLRPs (DCN, BCN, LUM)	Interact with collagens (types I, III, VI, XII and XIV) that form the framework of the extracellular matrix. Core protein of SLRPs interacts with fibrillar collagen, fibril/fibril interaction of DS chains (DCN and BCN) regulates collagen fibril diameter and stabilizes the collagen network. DNC/BCN protein core binds growth factors and influences their bioactivity.
Basement membrane PGs (HSPG2, AGRN, Collagen XVIII)	Bind to and cross-link themselves and glycan-free matrix components to form flexible thin mats with a wide range of regulatory functions including permeability control. Selective barrier for macromolecules. Control invasion of cancer cells. Adhesive matrix for endo- and epithelial cells.
Cell-surface PGs (SDC1–4, GPC1–6)	Act as coreceptors for growth factors, mediate cell–cell and cell–matrix adhesion, modulate the activity of enzymes and their inhibitors, are low-affinity receptors of enzymes and serve as attachment sites for blood cells and viruses. The ectodomain can be enzymatically released ('shedding') and thereby modulates various biological processes. The cytosolic domain is involved in intracellular signaling, adhesion of cell skeleton structures and facilitating focal adhesions.

11.3.1 Aggrecan

The ACAN molecule (Figure 11.4) contains multiple functional domains. The globular domains G1 and G2 are separated by a short interglobular domain, and the G2 and G3 regions are separated by a long GAG attachment region which consists of adjacent domains rich in KS and CS, but also possesses sites for the attachment of *O*-linked oligosaccharides which with age may become substituted with KS. The G1 region is at the N-terminus of the core protein, and can be further subdivided in three functional domains termed A, B1 and B2. The aggregates are composed of a central filament of hyaluronic acid with up to 100 ACAN molecules radiating from it with each interaction able to be stabilized by the presence of a link protein. ACANase (a member of the ADAMTS family–metalloproteinases

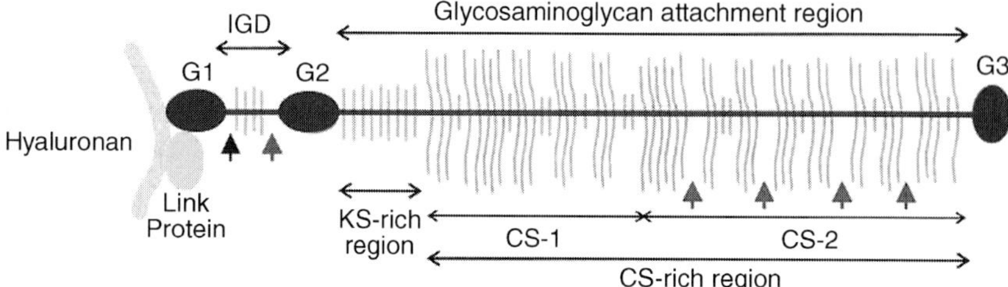

Figure 11.4 ACAN monomer with attached KS and CS chains. The interaction with hyaluronate involves the globular G1 domain and the stabilizing link protein.

with a disintegrin domain and thrombospondin motifs) and MMPs are associated with ACAN proteolysis. Five ACANase cleavage sites have been described in ACAN, with one residing in the interglobuar domain and four in the CS-2 domain. Mutation in the ACAN gene leads to chondrodysplasia described in human, mouse and chicken. In human, a single base pair insertion in exon 12 causes a frame shift and results in a form of spondyloepiphyseal dysplasia. Human chondrodysplasia is also associated with an undersulfation of ACAN due to gene defects in sulfate transporter.

Link proteins involved in the formation of large aggregating PGs possess three domains (A, B1 and B2). The A domain is responsible for interaction with the G1 region of ACAN whereas the B domains interact with hyaluronate. As noted above, the link proteins belong to the family of human lectins (further information in Chapter 19). MMPs are implicated in proteolytic degradation or modifications of link proteins – a process that takes place throughout life. The human cartilage link protein gene (chromosome 5q13–14) possesses five exons. Link protein knockout mice exhibit cartilage defects resulting in dwarfism and craniofacial abnormalities.

HA is a nonsulfated and protein-free GAG that does not belong to the group of PGs, but has been shown to be essential for maintaining the physiological structure of cartilage and other tissues containing large aggregating PGs. HA is characterized by its large molecular weight and its unique mode of synthesis which occurs at the plasma membrane of the cell via a HA synthase. This enzyme incorporates in a stepwise synthesis alternatively GlcA and GlcNAc up to a large length molecule, and is present as a coat surrounding all cells synthesizing HA. The formation of PG aggregate occurs initially in the pericellular compartment. A decrease of HA involves the action of hyaluronidases or free radicals. Intracellular HA is involved in inflammation [9].

11.3.2
Versican

In the family of large aggregating PGs VCAN is the most versatile member with regard to its structure and tissue distribution. Several VCAN core proteins have been identified. The structural diversity originates from alternative splicing processes which generate four splice variants in humans (V0, V1, V2 and V3) with molecular masses of 370, 262, 180 and 72 kDa. The number of CS attachment sites varies from 17–23 (V0) and 12–15 (V1) to 5–8 (V2). Due to the absence of the central GAG carrying domain V3 is devoid of CS side-chains and is therefore not a PG. In the isoforms V0–V2 chondroitin 6-sulfate and chondroitin 4-sulfate are found in a ratio of 2:1. Apart from HA, VCAN interacts with various ligands. Binding is mediated by both the DS chains and the protein. In particular, the C-terminal epidermal growth factor (EGF)-like repeats, the C-type lectin domain and the suchi element are altogether capable of binding D-Man, D-Gal, L-Fuc, D-GlcNAc, HS, fibronectin and collagen I. All these binding activities contribute to the high organizational level of the extracellular matrix of all internal organs including fibrous and elastic cartilage, tendon, skeletal muscle, and the dermis.

11.3.3 Neurocan, Brevican

The nervous tissue PG NCAN has a double-loop protein core for HA binding and a C-terminal EGF-like and C-type lectin-type domain showing 60% homology to those of ACAN and VCAN. The middle domain has no homology to any known protein in adult, it has only one CS chain and a cleaved form of its core protein. NCAN binds to neural cell adhesion molecules, and inhibits neural adhesion and neurite outgrowth. Together with NCAN, BCAN is found in the brain, it is much shorter and shows little homology to the other HA-binding PGs. A significant amount of BCAN is devoid of any GAG chains, indicating that BCAN is a 'part-time' PG. In brain it is present predominantly in primary cerebellar astrocytes, but not in neurons.

11.4 Small Leucine-Rich PGs

The small leucine-rich PGs (SLRPs) form a large family of leucine-rich repeat proteins [10–13]. All SLRPs have 10–12 tandem repeats of 24 amino acids with hydrophobic residues in conserved positions. The main members of the family are decorin (DCN), biglycan (BCN), fibromodulin (FMOD) and lumican (LUM). They can be divided into several subfamilies based on their gene organization, the number of leucine-rich repeats and the type of GAG chain substituent. The four main SLRPs have 10 leucine-rich repeats which are flanked by disulfide-bonded domains. DCN and BCN are classified as DS PGs, whereas FMOD and LUM are KS PGs. All SLRPs possess *N*-linked oligosaccharide chains within their central leucine-rich repeats. The CS/DS attachment sites of DCN and BCN are within the extreme N-terminus of their core protein. In DCN, one attachment site at amino acid residue 4 is present; in BCN, two attachment sites at amino acid residues 5 and 10 are present. GAG-free forms of BCN, FMOD and LUM are the result of proteolysis within the N-terminal region of the core protein, and accumulate with age.

A predominant function of SLRPs depends on their interaction with fibrillar collagen that forms the framework of the extracellular matrix. The core proteins of DCN, BCN, FMOD and LUM interact with the fibrillar collagen to multiple but different binding sites, while the GAG side-chains form interfibrillar bridges and stabilize the collagen network. According to John E. Scott ([14] and references cited therein) these bridges are based on the property of GAG to form 2-fold helices in solution that are completely complementary, allowing duplexes to form spontaneously provided the chains are oriented antiparallel to each other. This type of interaction also limits access of the collagenase to the unique cleavage site on each collagen molecule, in this way helping to protect the fibrils from proteolytic damage.

The interaction of the SLRPs is not confined to fibrillar collagen. DCN, FMOD and LUM have been reported to interact with many other macromolecules, including type VI, XII and XIV of collagen, fibronectin and elastin, and growth factors, such as EGF, transforming growth factor-β and tumor necrosis factor-α. In the case of DCN the interaction with fibronectin occurs in a Zn^{2+}-dependent manner and possibly also other members of the SLRPs act as Zn^{2+}-binding proteins. DCN has been shown to possess antitumorogenic properties. Depletion in SLRP production can influence tissue properties. DCN null mice show most pronounced alteration of skin collagen with a more loose and irregular packed fibrillar network and a reduced tensile strength leading to an abnormal skin fragility phenotype. BCN knockout results in an osteoporosis-like phenotype with a reduced growth rate and a decreased bone mass. Absence of LUM produces skin laxity and corneal opacity with an increased proportion of abnormally thick collagen fibrils and delayed corneal epithelial wound healing. In humans a frameshift mutation of DCN causes dystrophy of the cornea and a deficiency of a galactosyltransferase has been associated with a progeriote form of Ehlers-Danlos syndrome.

FMOD is a most abundant member of the leucine-rich repeat protein family and has the highest sequence homology with LUM. The protein core contains four potential attachment sites for KS all present in the leucine-rich region. The KSs are short in length (on average nine disaccharide units). Tyrosine sulfation sites have been identified in the N-terminal part of the molecule.

FMOD is known to interact with type I and II collagen. FMOD is present on collagen fibers and tendons. An interaction relevant for the inflammation reaction is that of FMOD interacting with C1q. In rheumatoid and osteoarthritic cartilage, degraded fragments of the core proteins of FMOD were reported, but an increased mRNA and total amount of translated protein indicate that repair of damaged cartilage may be attempted. In FMOD deficient mice ($FMOD^{-/-}$) an early osteoarthritis of the knee and an impaired function of the ligaments were reported. In explant cultures of cartilage treated with interleukin-1 FMOD fragments cleaved by MMP13 are found. LUM is glycosylated by four *N*-linked KS chains. It has been isolated from cornea and as undersulfated variant from aorta.

11.5 Basement Membrane PGs

Basement membranes are flexible thin (40–120 nm) mats of a specialized extracellular matrix that underlie all endothelial and epithelial cell sheets, and surround many other cells including Schwann cells (peripheral nerve cell axons). Basement membranes contain three PGs: perlecan (HSPG2), agrin (AGRN) and collagen XVIII [3, 15].

Perlecan is a large multidomain PG (Figure 11.5) that binds to and cross-links many extracellular components and cell surface molecules. It has a wide range of regulatory controls, binding properties and interactions. Perlecan has a complex gene structure that spans five domains. In the N-terminal domain three attachment sites (Ser-Gly-Asp sequences) for large HS chains, each 70–100 kDa, are located in close proximity to one another. In addition, the remaining four domains contain eight Ser-Gly-Asp/Glu and 30 Ser-Gly sites which are potential GAG attachment sites. In the basement membrane specific interactions are formed between perlecan and other components of the basement membrane (type IV collagen, laminin, entactin and fibronectin). Binding is mediated by both the HS side-chain and the protein core domains. Perlecan has also been shown to interact with growth factors (fibroblast growth factors (FGF)-1, -2, -4, -5, -8 and -9), which require HS for signaling, and to promote dimerization of the growth factors. Trimolecular complexes of HS, FGF and FGF receptor are formed. However, perlecan shares this ability with syndecans and glypicans. Other counterparts of

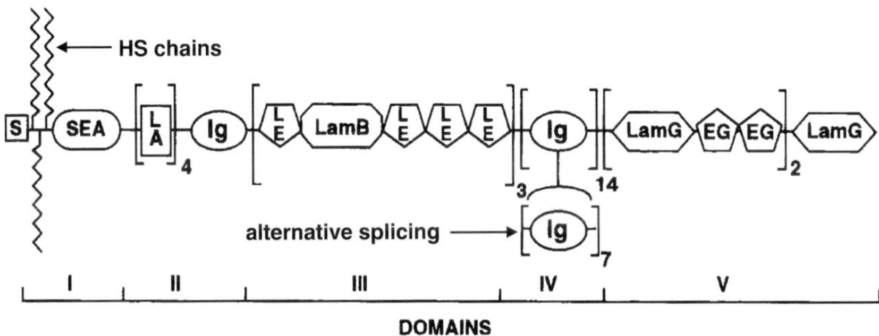

Figure 11.5 Domain structure of the perlecan protein core. S, signal sequence; SEA, SEA module; LA, low-density lipoprotein receptor class A module; LE, laminin EGF-like module; LamB/G, laminin B/G domain-like module; Ig, immunoglobulin-like module; EG, EGF-like module [1].

perlecan are the isoforms of platelet-derived growth factor, interferon-γ, and the integrins β1 and β3.

Perlecan endows basement membranes with fixed negative electrostatic charge and is responsible in part for the charge selective ultrafiltration properties of this extracellular matrix. The occurrence of perlecan is not restricted to basement membranes. It plays a key role in the development of pulmonary, intestinal, cartilagenous, muscular and vascular maturation. Studies on perlecan null mice ($HSPG2^{-/-}$) have shown that perlecan is essential for the development of cartilage and encephalic regions; 40% of knockout mice died at embryonic day 10.5 due to defective encephalic development, while the remaining 60% of knockout mice died shortly after birth due to skeletal dysplasia. Perlecan has been found to be associated with all amyloid plaques in many diseases including Alzheimer's disease [associated with the β-amyloid protein (Aβ) plaques] and infectious prion diseases. Down syndrome patients produce HS immunoreactivity in neurons as early as one day after birth although absence of any Aβ reactivity has been found.

ARGN is a large PG required for development of the neuromuscular junction during embryogenesis. ARGN is named based on its involvement in the aggregation of acetylcholine receptors and other post-synaptic proteins during synaptogenesis. The ARGN core protein (215 kDa) contains three potential HS attachment sites, but only two of these actually carry HS chains when the protein is expressed. During development, the growing end of motor neuron axons produce and secrete ARGN. It binds to several receptors required for formation of neuromuscular junction such as the muscle-specific kinase (MuSK) receptor that induces cellular signaling. In addition, the MuSK–ARGN complex binds muscle surface proteins including dystroglycan and laminin [this interaction by a domain with a jelly-roll motif of the LNS family (LNS = laminin A, neurexins and sex hormone binding globulin) can depend on lectin activity, see Chapter 19 for details, and Chapters 13 and 16 for illustration of this structural motif]. The neuromuscular junction does not form in mice deficient for AGRN ($ARGN^{-/-}$) or MuSK. The AGRN-induced MuSK activation stimulates the clustering of post-synaptic proteins including acetylcholine receptor.

Collagen XVIII (COL-18A1) has structural properties of both a collagen and a proteo-HS. Proteolytic cleavage within its terminal domain releases endostatin – a potential inhibitor of angiogenesis in the tumor growth and of endothelial cell proliferation. Inactivating mutations in the human collagen XVIII/endostatin gene ($COL\text{-}18A1^{-/-}$) lead to Knobloch syndrome characterized by age-dependent vitro retinal degeneration and occipital encephalocele. The collagen XVIII/endostatin has an essential role in ocular development and the maintenance of visual function. Age-dependent loss of vision in mutant mice is associated with pathological accumulation of deposits under the retinal pigment epithelium as seen in early stages of age-related macula degeneration in humans. In addition to AGRN and perlecan, collagen XVIII may be involved in the ionic filtration accomplished by the glomerular basement membrane.

11.6 Cell-Surface (Transmembrane) PGs

11.6.1 Syndecans [16–22]

The syndecan (SDC) family of HSPGs comprises four members which are expressed by different genes on four different chromosomes in mammals (see Table 11.1). Their expression is strongly regulated in a tissue-specific and developmentally-dependent manner. SDC1 is most abundant in endothelial cells and epithelial cells, SDC2 (fibroglycan) is the predominant syndecan in fibroblasts and smooth muscle cells, SDC3 (*N*-syndecan) is highly expressed in the central nervous system, and SDC4 (amphiglycan, ryudocan) is more ubiquitously expressed by multiple cell types. In mammals SDC1 and 3 and SDC2 and 4 share stronger sequence homologies, and therefore represent two subfamilies.

11.6.1.1 Structure

The genes of the syndecans are divided into five exons which code for three different domains (Figure 11.6). (i) The extracellular domain contains GAG attachment sites (between 3 and 8) for HS (N-terminal) and CS chains (in proximity to the plasma membrane). Various specific functions of the syndecans depend on the presence of sugar chains. (ii) The transmembrane domain is highly conserved among the syndecans (24 amino acid residues) and promotes dimerization. (iii) The cytoplasmic domain of syndecans can be divided into three regions – two constant regions (C1 and C2) that are highly conserved among the syndecans separated by a variable (V) region that is specific for each family member. The last four amino acids of the cytoplasmic domain (EFYA) are identical in all syndecans and represent the binding sequence for PDZ (PSD-95/Disk large/ZO-1)-containing proteins. The cytoplasmic part also contains four tyrosine residues which may serve as putative phosphorylation sites: two tyrosines are located in C1 and each one in the V and C2 sections, respectively.

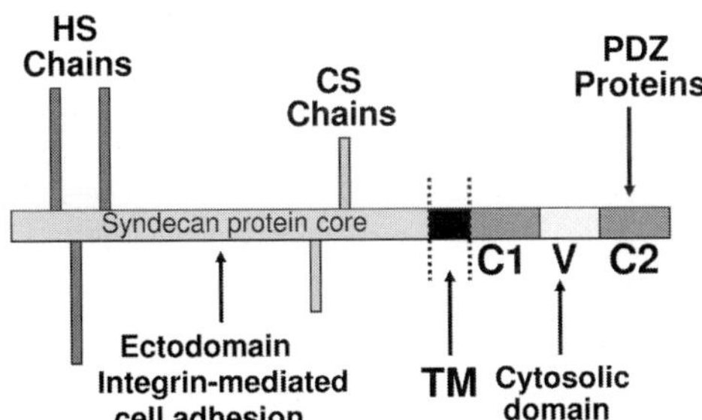

Figure 11.6 Domain structure of syndecan protein core. TM: transmembrane domain; C1 and C2: constant regions of the cytosolic domain; V: variable region of the cytosolic domain.

11.6.1.2 Functions

Many proteins depend for their function on interaction with HS, raising the perspective for respective drug design. A very large number of proteins and pathogens have been shown to interact with HS by their HS-binding region. These proteins are protected from degradation by the presence of bound HS. Examples of the various functions of syndecans are given in Figure 11.7 [23].

- The syndecans can immobilize ligands, present them to a specific receptor, and prevent their degradation and contribute to the dimerization of growth factors and/or modify their interaction with their receptor, but the syndecan action is concentration-dependent concerning the promotion or inhibition of cell proliferation. One example is the interaction between FGF-2 and HS that is required for binding of FGF-2 to its high-affinity tyrosine kinase receptor FGFR-1. A ternary complex of growth factor, HS and receptor seems to maximize the signaling potential of FGF-2, but an excess of free HS chains can counterbalance the mitogenic activity of FGF-2.
- Syndecans are low-affinity receptors for enzymes. They also bind lipoproteins and can mediate uptake, lysosomal delivery and degradation of the lipoproteins independently of specific lipoprotein receptors.
- Syndecans supply attachments sites for viruses (herpes simplex virus, Pseudorabies virus glycoprotein C and HIV-1 are bound).
- Syndecans bind proteases and protease inhibitors. The interaction of the protease thrombin and its inhibitor antithrombin III that controls blood coagulation is regulated by various membrane-bound proteo-HS. Based on these findings their role in wound healing processes was the subject of many studies.

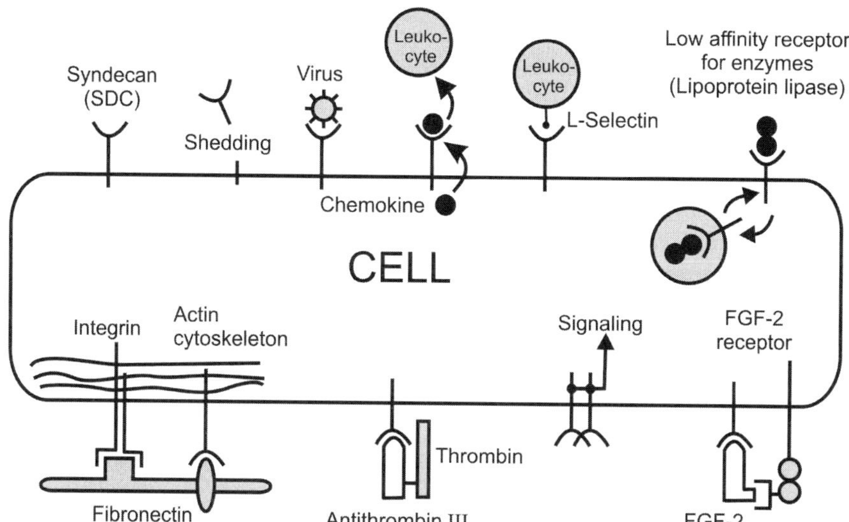

Figure 11.7 Pleiotropic functions of syndecans, adopted and extended from [23].

- The protein core of syndecans also contributes to their function. That is, the syndecans have properties beyond the ability to bind and concentrate HS-binding ligands at the cell surface. Thus, the isolated ectodomain of SDC1 and 4, free of HS side-chains, promotes the adhesion of human fibroblasts and B-lymphoid cell lines. Moreover, the cytoplasmic domain of SDC4 may initiate a signal cascade by formation of a ternary complex which comprises two dimers of syndecan ectodomains, phosphatidylinositol 4,5-bisphophate and protein kinase C. This complex acts as a phosphorylate regulator of Rho family GTPases and actin-associated proteins, and leads to activation of Rho kinases.
- Syndecans are shed from the cell surface by enzymatic cleavage of the core protein. The enzyme responsible (sheddase, secretase, convertase) was identified as a TIMP-3-sensitive MMP. Several pathways (receptor activation, mitogen-activated protein kinase, c-Jun N-terminal kinase and protein tyrosine kinase) are involved in triggering the shedding process. Shedding can be accelerated by exogenous proteases like thrombin and plasmin. The shedding process releases a soluble fragment corresponding to the ectodomain which can act as a dominant-negative form competing for binding partners with the membrane-integrated syndecan. The cleavage sites are located in immediate proximity to the transmembrane domain. Shed syndecans have been found in serum and in body fluids during wound healing and cancer.

Syndecans have the ability to replace each other in functional terms, at least partially. SDC-1 and SDC-4 null mice ($SDC1^{-/-}$ and $SDC4^{-/-}$) exert a normal phenotype, they are viable and fertile, and develop well. However, under strain conditions such as injury or in disease models some loss of function becomes evident – retarded wound healing and reepithelization, defective angiogenesis in granulation tissue, and atypical binding of leukocytes to vascular endothelium.

11.6.2 Glypicans

Glypicans (GPC) [24] are a family of HSPGs which are anchored to cell membranes by a glycosylphosphatidylinositol (GPI) linkage (for further information on GPI anchors, see Chapter 9). So far six members (GPC1–6) are known in vertebrates. The chromosomal location has been established. A schematic structure of GPC1 is given in Figure 11.8. The glypican protein components anchored in this manner are called 'glypiated'. The GPI anchor generally consists of a phosphatidylinositol phosphoglyceride whose inositol group has been modified by acylation and an oligosaccharide. A phosphoethanol group present on the terminal mannose serves as an acceptor site for the protein. Glypican family member are selectively expressed on different cell types with only GPC1 present in vascular cells. The proteo-HS of glypican is mainly targeted to apical surfaces and this process is partially dependent upon the extent of glypiation. In human brain glypican is accumulated in DIG (detergent-insoluble glycophospholipid)-enriched domains of cell membranes and regularly colocalizes with Aβ particles in human brain of

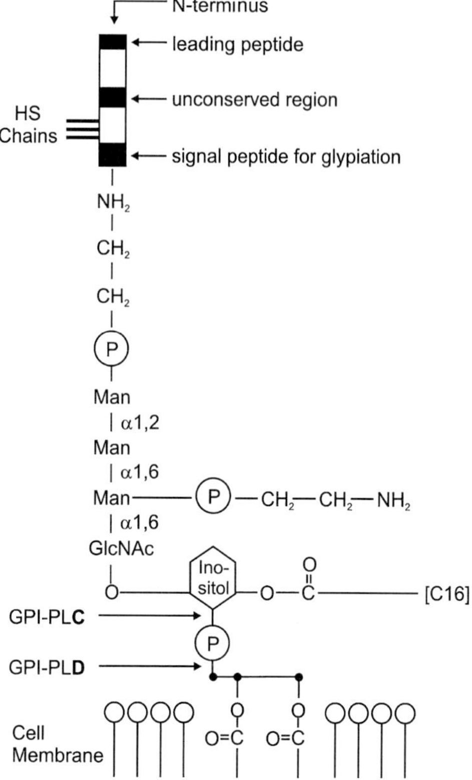

Figure 11.8 Structure and attachment of glypican by its GPI anchor. GPI-PL **C** and GPI-PL **D**, two GPI-specific phospholipases, use cleavage sites for shedding of glypican [1].

Alzheimer's disease patients. Glypican is involved in the suppression/modulation of growth and the activity of growth factors. Removal of glypican renders the cells insensitive to many growth factors but SDC1 does not compensate for glypican loss. Most germinal mutations of glypican are deleterious involving different exons, nonsense as well as splicing site mutations. In somatic mutations the expression of glypican is altered. In hepatocellular carcinoma, Wilm's tumor and metastatic colorectal malignancy, GPC3 is upregulated; in breast and pancreatic cancer, GPC1 is upregulated. Other mutations in glypican genes cause dysmorphy and overgrowth syndromes in humans and mice.

11.7 Conclusions

The elucidation of the primary structure of PGs has been established, the cDNA of the protein core of the main PGs has been cloned, and the principle of biosynthesis of the polymeric GAG chain and the enzymes involved are known, but the field is still in rapid evolution and there remain many challenges to study the function and impact of PGs in development, biology and pathophysiology. The tasks concern the knowledge of the three-dimensional conformation of the protein core, the binding domains and affinity of protein and GAG chain to different counterparts, the function-dependent variation of GAG chain length, the degree and location of sulfation, and the development of experimental animal models to study PG functions and deficiency diseases.

Summary Box

PGs comprise a collection of macromolecules that are composed of a protein to which long unbranched GAG chains made of repeating disaccharide units are covalently linked by posttranslational modification. PGs are widespread in the animal kingdom. PGs perform three major functions: (i) PGs are essential constituents of an extracellular framework needed for structural organization of all organs in multicellular organisms; (ii) PGs play a key role as storage depot for cell membrane associated growth factors and cytokines and their bioactivity; (iii) Transmembrane or GPI-anchored PGs act as coreceptors for growth factors, as attachment sites for blood cells, matrix components and viruses, and as modulators of enzyme activity. The cytosolic domain of transmembrane PGs is involved in intracellular signaling.

References

1 Iozzo RV (Ed.), *Proteoglycans – Structure, Biology, and Molecular Interactions.* Marcel Dekker, New York, 2000.
2 Ayad S *et al.* (Eds.), *The Extracellular Matrix Facts Book.* Academic Press, London, 1994.
3 Iozzo RV. Basement membrane proteoglycans: from cellar to ceiling. *Nat Rev Mol Cell Biol* 2005;6:646–56.
4 Day AJ. The structure and regulation of hyaluronan-binding proteins. *Biochem Soc Trans* 1999;27:115–21.
5 Hascall VC, Heinegard D. Aggregation of cartilage proteoglycans. I. The role of hyaluronic acid. *J Biol Chem* 1974;249:4232–41.
6 Iozzo RV, Murdoch AD. Proteoglycans of the extracellular environment: clues from the gene and protein side offer novel perspectives in molecular diversity and function. *FASEB J* 1996;10:598–614.
7 Sajdera SW, Hascall VC. Proteinpolysaccharide complex from bovine nasal cartilage. A comparison of low and high shear extraction procedures. *J Biol Chem* 1969;244:77–87.
8 Spicer AP *et al.* A hyaluronan binding link protein gene family whose members are physically linked adjacent to chondroitin sulfate proteoglycan core protein genes: the missing links. *J Biol Chem* 2003;278:21083–91.
9 Hascall VC *et al.* Intracellular hyaluronan: a new frontier for inflammation? *Biochim Biophys Acta* 2004;1673:3–12.
10 Bengtsson E *et al.* The primary structure of a basic leucine-rich repeat protein, PRELP, found in connective tissues. *J Biol Chem* 1995;270:25639–44.
11 Iozzo RV. The biology of the small leucine-rich proteoglycans. Functional network of interactive proteins. *J Biol Chem* 1999;274:18843–6.

12 Kinsella MG *et al.* The regulated synthesis of versican, decorin, and biglycan: extracellular matrix proteoglycans that influences cellular phenotype. *Crit Rev Eukaryot Gene Expr* 2004;14:203–34.

13 Matsushima N *et al.* Super-motifs and evolution of tandem leucine-rich repeats within the small proteoglycans – biglycan, decorin, lumican, fibromodulin, PRELP, keratocan, osteoadherin, epiphycan, and osteoglycin. *PROTEINS* 2000;38:210–25.

14 Scott JE. Supramolecular organization of extracellular matrix glycosaminoglycans, *in vitro* and in the tissues, *FASEB J* 1992;6: 2639–45.

15 Hallmann R *et al.* Expression and function of laminins in the embryonic and mature vasculature. *Physiol Rev* 2005;85:979–1000.

16 Alexopoulou AN *et al.* Syndecans in wound healing, inflammation and vascular biology. *Int J Biochem Cell Biol* 2007;39:505–28.

17 Bishop JR *et al.* Heparan sulphate proteoglycans fine-tune mammalian physiology. *Nature* 2007;446:1030–7.

18 Götte M. Syndecans in inflammation. *FASEB J* 2003;17:575–91.

19 Iozzo RV. Heparan sulfate proteoglycans: intricate molecules with intriguing functions. *J Clin Invest* 2001;108:165–7.

20 Isacke CM, Horton MA (Eds.), *The Adhesion Molecule FactsBook*. Academic Press, London, 2000.

21 Schmidt A *et al.* Plasmin- and thrombin-accelerated shedding of syndecan-4 ectodomain generates cleavage sites at Lys^{114}–Arg^{115} and Lys^{129}–Val^{130} bonds. *J Biol Chem* 2005; 280:34441–6.

22 Couchman JR. Syndecans: proteoglycan regulators of cell-surface microdomains? *Nature* 2003;4:926–37.

23 Bernfield M. *et al.* Biology of the syndecans. *Annu Rev Cell Biol* 1992;8:365–93.

24 Fransson LA *et al.* Novel aspects of glypican glycobiology. *Cell Mol Life Sci* 2004;61: 1016–24.

12
Chitin

Hans Merzendorfer

Among Nature's many biopolymers the sugar polymers take on a special position as they possess distinctive biological and physicochemical properties that make them highly attractive for industrial applications. One well-known example is cellulose, which is actually considered the most abundant organic compound on Earth. It is the major constituent of paper and textiles, but is also converted in coatings, laminates, pharmaceuticals and food. However, cellulose is not the only prevalent biopolymer on Earth. With an estimated annual production of at least 10 gigatons (1×10^{13} kg) chitin is presumably the second most abundant organic compound [1]. In contrast to cellulose, which is obtained from terrestrial sources, such as wood or cotton, chitin is regarded a marine biopolymer, as oceanic crustaceans, in particular those of the pelagic zooplankton, produce most of its biomass. Krill has an exceptional position as a chitin-producing organism, because these small crustaceans appear in gigantic swarms in every ocean of the world. Nevertheless, chitin has remained an unused biomass resource for a long time, although it was in fact discovered earlier than cellulose (see Info Box 1). In more recent times, however, attention has increased dramatically, because chitin and its derivatives have unique physicochemical properties allowing a broad spectrum of technical applications. Meanwhile, more than 10,000 tons are extracted every year from shellfish waste accumulating in the seafood processing industry. Most of the worldwide chitin production is used to obtain glucosamine and various oligosaccharides by acidic hydrolysis (see also Chapter 3). Accelerated alkaline hydrolysis leads to the deacetylation of chitin yielding chitosan–a more soluble polymer consisting mainly of glucosamine units (Figure 12.1). Chitosan is a cationic polymer at acidic or neutral pH, and has antimicrobial properties and a relatively low toxicity profile. Therefore, it can be employed in the food industry for preservation and as a dietary supplement, in water treatment plants for flocculation and adsorption of heavy metal ions, and in the medical/pharmaceutical industry as an excipient and for wound dressings. Furthermore, chitosan is used as base material in biopolymer research, because it is comparatively easy to chemically modify the primary amino and the secondary hydroxyl groups. A multitude of chitosan derivatives have therefore been synthesized allowing the design of polysaccharide-based intelligent materials with specific functions.

The Sugar Code. Edited by Hans-Joachim Gabius

ISBN: 978-3-527-32089-9

Info Box 1

Chitin was first isolated from mushrooms in 1811 by Henry Braconnot (1781–1855), a plant chemist and Director of the botanical garden of Nancy (France), and therefore originally termed 'fungine'. While analyzing the alkali-resistant fraction, Braconnot recognized that 'fungine' was different from woody materials, as it contained much more nitrogen than wood. Later on, in 1823, Augustus Odier isolated the same substance from the cuticles of beetles. He coined the term 'chitin' from the Greek word '*chiton*' meaning tunic or envelope. However, Odier originally believed that his chitin preparation did not contain any nitrogen, which finally would turn out to be an error of measurement. Rather, he thought that insect cuticles were made of the same substance that forms the cell walls of the plants (that is cellulose). The controversy on the differences between chitin and cellulose continued for quite a while until 1875, when Georg Ledderhose (1855–1925), a German student, later physician, treated lobster shells with hot concentrated hydrochloric acid (see also Chapter 1). After evaporating the solution characteristic crystals remained, which contained acetate and a novel nitrogen-containing sugar named glucosamine. As acetate and glucosamine were formed in equimolar amounts, it became clear that chitin consists of acetylated glucosamine and hence it was different from cellulose. Already, in 1859, Charles Rouget had tested different conditions to bring chitin into solution and found that unlike chitin the substance obtained after the treatment with potassium hydroxide dissolved in acids. The resulting compound was later named chitosan by Felix Hoppe-Seyler.

12.1 Occurrence

Chitin is widely distributed among living creatures and is detected in a striking variety of taxonomic groups ([2], see also Table 12.1). Next to its well-documented presence in the cell walls of fungi and in the shells and cuticles of arthropods, it is also found in the cyst walls of protists, in the calyces of sponges, in the egg shells from rotifers and nematodes, in oocytes, cocoons and peritrophic matrices of insects, in squid pens and cuttlefish bones, in the shells from brachiopods and mollusks, and in the radulae of mollusks. Less known is its presence in chordates such as tunicates, where it is a component of the test and the peritrophic matrix, and in a few lower vertebrates, as chitin has been detected in the cuticles of pectoral fins from certain bony fishes. However, it seems as if the ability to produce chitin has been lost at the root of the deuterostome lineage. In higher deuterostomes, chitin appears to be replaced either by other sugar polymers, such as hyaluronan and chondroitin, or by proteinous polymers, such as collagen or keratin. From a

Table 12.1 Functions of chitinous structures in various chitin-synthesizing organisms.

Organism	Localization	Function
Bacteria (Rhizobiacea)	Secreted, extracellular space	Nodulation of leguminous plant roots
Protozoa	Cyst	Physical and chemical resistance
Fungi	Cell wall	Compensation of the turgor pressure
	Spore wall	Physical and chemical resistance
	Septa	Stabilization of cell division zones
Nematodes	Pharynx	Mechanical breakdown of food
	Egg shell	Physical and chemical resistance
	Gut peritrophic matrix	Multiple protective functions
Arthropods	Epidermal cuticles, shells	Cuticle differentiation, skeletal functions
	Tracheal cuticle	Tracheal development, skeletal functions
	Gut peritrophic matrix	Multiple protective functions
	Egg shells	Physical and chemical resistance, oogenesis
Bivalves	Shell	Shell and nacre formation
Gastropods	Snail shell, radula	Shell formation, grazing
	Gut peritrophic matrix	Multiple protective functions
Cephalopods	Squid pen, cuttlefish bone	Endoskeleton
	Stomach cuticle	Protective functions
Tunicates	Integument (test)	Reinforcement
	Gut peritrophic matrix	Multiple protective functions
Vertebrates (bony fish)	Fin cuticle	Reinforcement

phylogenetic point of view it is remarkable that some bacteria belonging to the *Rhizobium/Agrobacterium* group possess the biosynthetic machinery to produce chito-oligosaccharides, which are involved in the nodulation of leguminous plant roots and thus have a morphogenic function (for further information on nodulation factors and their sensing, please see Chapter 18). Since the NodC enzyme, which is responsible for chito-oligosaccharide synthesis, is related to chitin-synthesizing enzymes of eukaryotes, it is tempting to speculate that all chitin-synthesizing systems are phylogenetically related and hence may originate from a common ancestral progenitor. However, the sequence similarities are comparably low and therefore may be the result of convergent evolution as well.

12.2 Structure

Chitin is regarded the functional equivalent of cellulose, as both are long, unbranched sugar polymers used to support extracellular structures. Actually, they are pretty similar in their basic structures: the monomeric sugar residues are linked by β1,4-glycosidic bonds and rotated with respect to each other by 180° (Figure 12.1). The proper repetitive units in cellulose and chitin are therefore

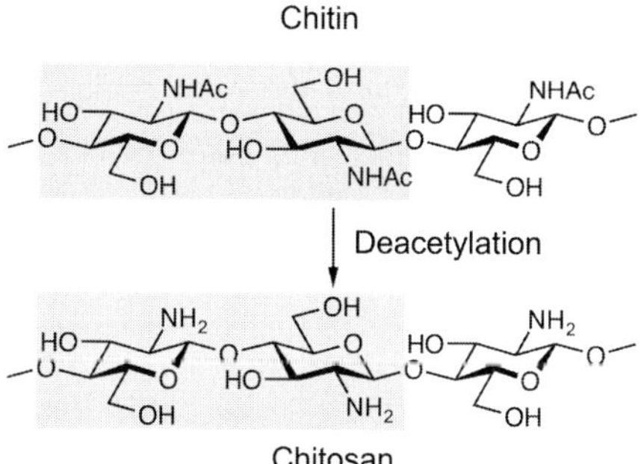

Figure 12.1 Chemical structures of chitin and chitosan. Chitin and chitosan are polymers of GlcNAc and glucosamine (GlcN) units, respectively. The repetitive units of chitin or chitosan are disaccharides named chitobiose (indicated in gray).

disaccharides called cellobiose and chitobiose, respectively (for structural depiction of cellobiose and monosaccharides, please see Figures 1.4/5). However, while cellulose is composed of β-D-glucose, chitin is made of 2-acetamido-2-deoxy-β-D-glucose [*N*-acetylglucosamine (GlcNAc)]. The three-dimensional structure of chitin was investigated extensively in the second half of the last century. It is usually found in the form of crystalline microfibrils of varying diameters and lengths. On the basis of X-ray diffraction patterns, chitin is an anisotropic polymer in the $^4C_1$ conformation that occurs in three allomorphic crystalline forms named α-, β- and γ-chitin [3]. Additionally, non-crystalline, transient chitin states have been reported in fungi [4]. The crystalline forms mainly differ in the degree of hydration, in the size of the unit cell and in the number of chitin chains per unit cell. α-Chitin is the most widely distributed chitin allomorph in Nature, mainly found in the cell walls of fungi and in the cuticles of arthropods. The sugar chains in the crystal exhibit an antiparallel orientation with respect to the reducing and nonreducing ends facilitating tight packaging. α-Chitin is stabilized by two types of intramolecular hydrogen bonds [C(3)=OH···O=C(5) and C(6)=OH···O=C] and two types of intermolecular hydrogen bonds [NH···O=C and C(6)=OH···OH=C(6)] [5]. Accordingly, two types of hydrogen bonds exist that involve the carbonyl groups: (i) exclusively intermolecular hydrogen bonding which account for 60% of all hydrogen bonds, and (ii) a mixture of inter- and intramolecular hydrogen bonding which account for about 40% of all hydrogen bonds as determined by ^{13}C solid-state nuclear magnetic resonance (Figure 12.2) [6]. This rather high number of intermolecular hydrogen bonds, which resemble those of cellulose II, seems to explain why α-chitin is thermodynamically very stable and unable to swell after soaking it in water.

β-Chitin is less widely distributed in Nature than α-chitin, and is found in squid pens, in the skin of (pogonophoran and vestimentiferan) tubeworms, and in the spines of diatoms and some insect cocoons. The sugar chains are arranged in a parallel orientation reducing the packaging tightness and the number of intermo-

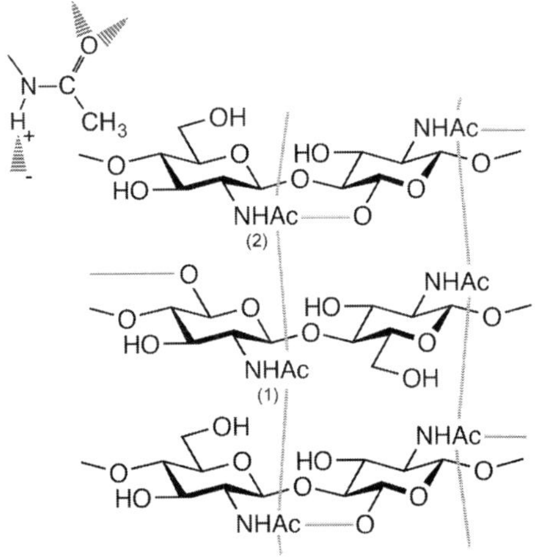

Figure 12.2 Hydrogen bonding in α-chitin. Two types of hydrogen bonds (shown in gray) involve the carbonyl groups (top left), those which form exclusively intermolecular hydrogen bonds (1) and a mixture of inter- and intramolecular hydrogen bonds (2). In β-chitin the hydrogen bonds are frequently formed with intercalating water molecules.

lecular hydrogen bonds [7]. Therefore, β-chitin is less stable than α-chitin. Upon hydration water molecules can intercalate to a greater extent resulting in intracrystalline swelling [8]. While the diffraction patterns of α- and β-chitin are well documented, and the deduced structures are widely accepted, the precise nature of γ-chitin is still unsure because of the unusual nature of the raw material (that is, stomach linings of squids, beetle cocoons and some insect peritrophic matrices). Based on X-ray diffraction patterns obtained for stomach cuticles from the sepia *Loligo*, Rudall suggested that γ-chitin is made of sets of two parallel strands that alternate with single antiparallel strands [9].

12.3 Function

Chitin is used a structural component of many biological composites, which may be divided into three categories depending on whether they contain predominately proteins (arthropod cuticles, peritrophic matrices, cocoons), carbon hydrates (fungal cell walls) or inorganic minerals (squid pens, cuttlebones, crustacean shells, diatom spines). In many cases chitin function is similar to that of a steel lattice in reinforced concrete, which makes the skeletal structures remarkably tough and durable (see also Table 12.1). In the following we will look at several chitinous structures found in Nature to provide closer insights into chitin function.

12.3.1
Fungal Cell Walls

The cell walls of fungi consist mainly of polysaccharides (glucans, mannans, chitin) that account for more than 90% of the cell wall and only minor amounts of glycoproteins [10]. The common skeletal core structure is composed of branched β1,3- and β1,6-glucans to which chitin is linked via β1,4-linkages. In addition, amorphous polysaccharides (β1,6-, α1,3-glucans and mannans) are present, which vary between the different species as much as the chitin content. While in *Saccharomyces cerevisiae* chitin accounts for only 1–2% of the cell wall's dry weight, the cell walls of *Aspergillus fumigatus* or *Neurospora crassa* contain up to 20% chitin. Chitin is predominantly deposited close to the plasma membrane, whereas the β1,3-glucans extend throughout the cell wall. Most of the proteins found in the cell wall are in transit towards the extracellular milieu and only part of them are real cell wall proteins. The latter proteins are frequently linked to the plasma membrane via glycosylphosphatidylinositol anchors (see Chapter 9), but are not considered to be structural components. Rather, they appear to be involved in remodeling or become covalently bound to polysaccharides in order to fulfill their biological functions at the cell wall's surface.

12.3.2
Arthropod Cuticles and Shells

The major components of arthropod cuticles are chitin, which is predominately in the α-form, a variety of different cuticle proteins, which significantly determine the physical properties (soft or hard cuticles), and calcium salts (mainly calcium carbonate present as calcite), which particularly harden crustacean shells. The relative amounts of chitin and protein vary considerably between the different arthropods. The shells from crabs and shrimps, for instance, consist of 30–40% protein, 20–30% chitin, 30–50% calcium carbonate, and less than 1% lipids and astaxanthin, the red pigment of crustaceans (values are given on the basis of dry weight, [11]). Soft insect cuticles with low stiffness (1 kPa to 1 MPa) commonly contain chitin and protein in equal amounts, and 40–75% water. Hard cuticles (up to 60 MPa) contain higher amounts of protein, 15–30% chitin and only about 10% water [12]. Cuticles are initially formed as rather soft structures, but become rigid after sclerotization, a process that leads to the nonenzymatic cross-linking of cuticle proteins by *ortho*-chinones, which are generated from *N*-acylated catecholamines in an enzymatic reaction catalyzed by diphenoloxidases. Chitin is present in the procuticle, but not in the epicuticle, which instead contains sclerotized proteins impregnated with lipids, lipoproteins, cements and waxes. In the procuticle chitin is deposited in about 3-nm thick microfibrils of indefinite length, each containing about 20 single sugar chains. Parallel-aligned microfibrils are embedded in a proteinous matrix forming single layers. The procuticle is built of many of these layers, which are horizontally arranged. Frequently, the layers are slightly rotated against each other resulting in typical helicoidal textures (Figure 12.3a)

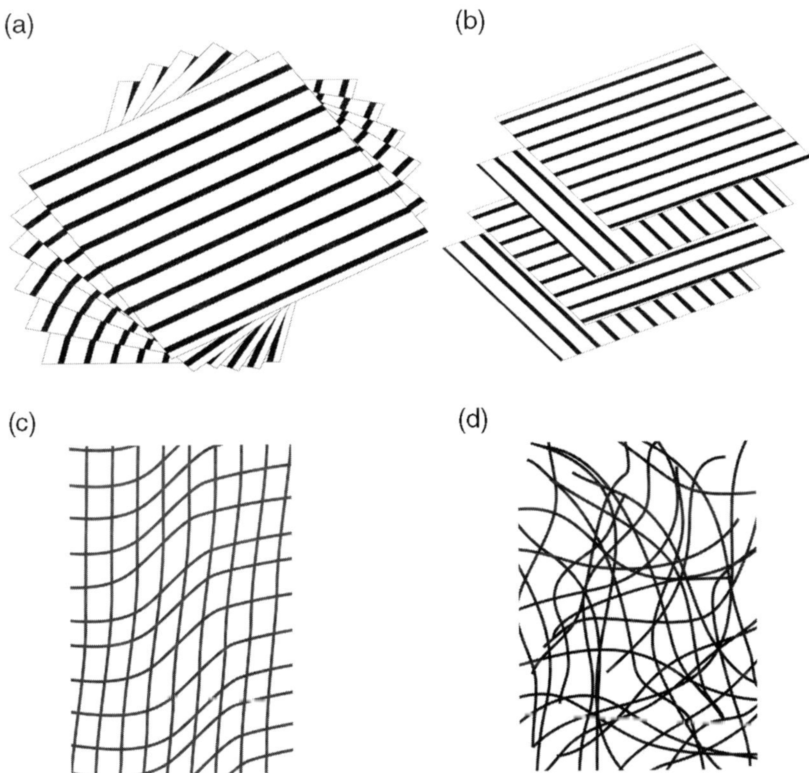

Figure 12.3 Helicoidal (a) and plywood-like (b) arranged layers in arthropod cuticles and shells; ordered (c) and random-felt like (d) microfibril textures of arthropod peritrophic matrices.

[13]. However, depending on cuticle function, other patterns have also been observed, including unidirectional and plywood-like arrangements of the fibers (Figure 12.3b).

12.3.3
Peritrophic Matrices and Cocoons

The peritrophic matrix is a membrane-like secretion product of gut epithelial cells and lines the inner surface of the intestine of many different organisms, including insects, nematodes, annelids and mollusks [14]. Its precise function is still unclear, but it might aid digestion and could protect the gut epithelium from mechanical damage, radical oxygen species and infection by pathogenic microorganisms. Peritrophic matrixes can be formed by either the entire midgut (type I peritrophic matrices) or by specialized cells of the anterior midgut (type II peritrophic matrices). In most cases chitin has been detected as a structural component of the peritrophic matrix. The chitin microfibrils (α-, β- or γ-chitin) are embedded in a gel-like protein/proteoglycan matrix, containing covalently bound proteins to a minor extent, tightly associated proteins (so-called peritrophins) and weakly associated proteins. In contrast to cuticles, peritrophic matrices are frequently organized in random felt-like or less highly-ordered structures (Figure 12.3c and d). Interestingly, some beetles excrete peritrophic matrices to form cocoons. In contrast to cocoons produced by glands such as silk cocoons, these cocoons are chitinous structures [3].

12.3.4 Other Functions

Next to chitin's well-documented structural function, it appears to act as a kind of alarm molecule signaling the invasion by pathogens and triggering immune responses in those organisms that do not produce chitin themselves like plants or mammals. For this purpose chitin-binding proteins are expressed, including lectins and chitinases, which are part of the natural defense against pathogenic fungi or parasites. For instance, wheat germ lectin is toxic to a variety of insect pests and the small chitin-binding protein hevein from the latex of the rubber tree has a significant antifungal activity [15, 16] (for further information on lectins binding chitin and the hevein-like domain, please see Chapters 15, 18 and 19). Moreover, many plants synthesize chitinases predominately to protect themselves from fungal growth. However, plant chitinases serve also physiological functions and are involved in symbiotic interactions [17]. As chitin does not exist in mammals, it had been assumed that chitinolytic enzymes are also restricted to lower life forms. Hence, it came as a big surprise that chitinases and chitinase-like proteins are even expressed in humans, where they appear to contribute to host antiparasite responses and asthmatic T helper type 2 inflammation (please see also Chapter 19).

12.4 Metabolism

Chitin synthesis is a sequence of metabolic reactions that requires different enzyme activities. Certainly, the key enzyme in this process is the chitin synthase (EC 2.4.1.16), a membrane-integral protein, which has been grouped into family II of processive, polymerizing glycosyltransferases – a family that includes closely related enzymes like hyaluronan synthases and cellulose synthases [18]. The chitin synthase utilizes UDP-GlcNAc as the activated sugar donor, which is synthesized in a sequence of enzymatic reactions following a variant of the Leloir pathway (see Info Box 2). The chitin synthase transfers the sugar moiety of UDP-GlcNAc to the nonreducing end of the growing polymer. Due to the lack of structural data on chitin synthases the precise catalytic mechanism is not known to date. In general, catalysis of glycosyltransferases occurs with two possible outcomes – retention and inversion of the anomeric configuration of the donor sugar (Figure 12.4). From a

Info Box 2

The general route for incorporating sugar units into polysaccharides starts with their conversion into activated sugar nucleotides. In the case of chitin the activated sugar nucleotide is UDP-GlcNAc, the sugar portion of which is transferred to the nonreducing end of the growing polymer in the transglycosylation step catalyzed by the enzyme chitin synthase (EC 2.4.1.16). The biosynthetic steps to produce UDP-GlcNAc follow a variant of the Leloir pathway, which was named after Louis Federico Leloir (1906–1987), who received the 1970 Nobel Prize in Chemistry. One of Leloir's major discoveries was the finding that the biosynthesis of basically all glycosylated molecules including glycoproteins, glycolipids and polysaccharides requires an activated sugar nucleotide as donor for the transglycosylation reaction. In eukaryotes UDP-GlcNAc is synthesized from Fru-6-P in a sequence of four enzymatic reactions: (i) conversion of Fru-6-P into GlcN-6-P catalyzed by GlcN-6-P synthase (EC 2.6.1.16), which transfers the ammonia from the cosubstrate L-glutamine to Fru-6-P and isomerizes the resulting fructosimine-6-phosphate to GlcN-6-P; (ii) transfer of an acetyl group from coenzyme A by GlcN-6-P acetyltransferase (EC 2.3.1.4) to obtain GlcNAc-6-P; (iii) isomerization of GlcNAc-6-P to GlcNAc-1-P catalyzed by phoshoacetylglucosamine mutase (EC 5.4.2.3) which is autophosphorylated and dephosphorylated during the reaction cycle in an ping-pong mechanism; (iv) uridylation of GlcNAc-1-P by UDP-GlcNAc pyrophosphorylase (EC 2.7.7.23) resulting in the end product UDP-GlcNAc, which inhibits the first step of the pathway in a negative-feedback loop in eukaryotes.

stereochemical point of view chitin synthesis follows an inverting reaction mechanism, in which the nucleophilic attack by the acceptor hydroxyl group leads to an inversion of the anomeric carbon of the donor substrate [19]. The underlying catalytic mechanism for inverting glycosyltransferases was deduced from crystal structures of bacterial enzymes. The reaction cycle involves an oxocarbenium ion-like transition state and requires a catalytic base, which deprotonates the incoming nucleophilic acceptor facilitating S_N2 displacement of the nucleoside diphosphate (Figure 12.4c). Frequently, a divalent metal ion acts as a Lewis acid catalyst in the reaction cycle by stabilization of the leaving nucleoside diphosphate. Next to their catalytic function the chitin synthase may exhibit also a transport function. As the catalytic site of the chitin synthase faces the cytoplasm, the growing polymer has to be translocated across the membrane to reach the extracellular space where it is deposited. It is likely that the extended transmembrane regions typically found in family II glycosyltransferases are involved in this transport process [22].

Chitin is rapidly degraded in Nature by three different types of chitinolytic enzymes: (i) endochitinases splitting chitin into oligosaccharides of different chain length, (ii) exochitinases splitting oligosaccharides into diacetylchitobiose and (iii) chitobiases splitting diacetylchitobiose into GlcNAc monomers. Chitinases do not

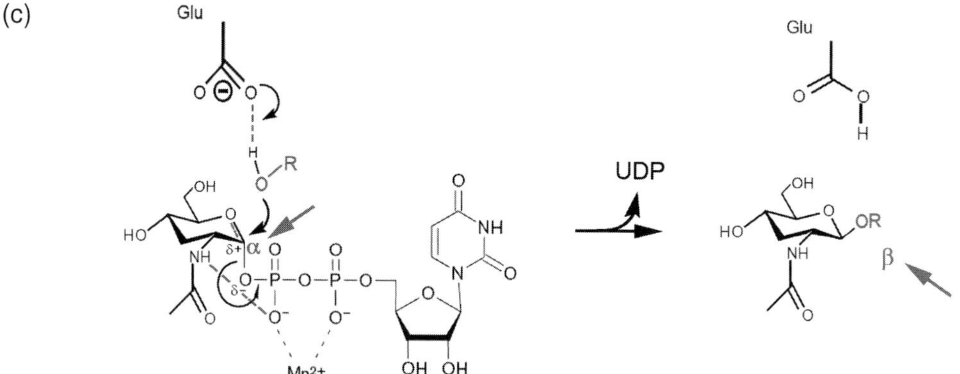

Figure 12.4 Crucial steps in the catalytic reaction cycles of retaining family 18 (a) and inverting family 19 (b) chitinases, and of inverting β-glycosyltransferases (c); modified according to [19–21]. Details of the reaction mechanisms are described in the text. The anomeric configurations of the C1 atoms are marked with arrows.

only play a role in chitin decomposition by chitinolytic microorganisms and in the defense against fungi and parasites, but are also expressed by chitin-synthesizing organisms themselves to allow re-organization of their chitinous structures during growth and development. Endochitinases are grouped into families 18 and 19 of glycosyl hydrolases attacking preferentially the β1,4-glycosidic linkage between the GlcNAc residues [20]. Family 18 are widely distributed in Nature and produced by many organisms including bacteria, fungi, nematodes, insects and vertebrates. They produce chito-oligosaccharides exhibiting GlcNAc at their reducing ends, which discriminate them from family 19 glycosyl hydrolases, which can also contain glucosamine at their reducing ends. The catalytic mechanism of chitinolytic enzymes has been discussed for a long time on the basis of the crystal structure of lysozyme, which also shows some activity on chitin. However, with the availability of crystal structures for family 18 and 19 chitinases it became obvious that the catalytic mechanism differs from that of lysozyme. Moreover, family 18 and 19 chitinases also differ with respect to structure and catalytic mechanisms. Family 18 chitinases are basically $(\alpha/\beta)_8$ barrels (for a comparison with similar structural motifs of lectins, please see Chapter 19.1), while family 19 chitinases are bilobal structures with an catalytic core composed of a three-stranded β-sheet and two α-helices. Family 18 chitinases yield a β-anomeric configuration and hence are retaining enzymes (Figure 12.4a), whereas family 19 chitinase result in the α-anomer and hence are inverting enzymes (Figure 12.4b). The reaction cycle of family 18 chitinases appears to follow a substrate-assisted mechanism, in which the oxocarbonium ion intermediate is stabilized by anchimeric assistance of the sugar *N*-acetyl group after donation of the proton from the catalytic carboxylate (Figure 12.4a) [23]. By contrast, the reaction cycle of family 19 chitinases follows a single-displacement mechanism involving two glutamates of the enzyme acting as the general acid or base, respectively (Figure 12.4b) [24]. While one glutamate functions as a proton donor, the other activates a water molecule, which in turn attacks the C1 atom of the intermediate sugar.

Since the chitin metabolism is crucial for growth and development of pathogenic fungi, parasites, agricultural pests and blood-sucking arthropods transmitting infectious diseases, inhibition or deregulation of its key enzymes is an important objective for the development of fungicides and insecticides, which virtually do not harm amphibians, reptiles, birds or vertebrates. Such inhibitors affect either chitin synthesis, chitin degradation or the control of both processes [22]. Current available inhibitors of chitin metabolism include peptidyl nucleosides (nikkomycins, polyoxins), which effectively inhibit fungal chitin synthases, allosamidin and the peptide CI-4, which mimic the transition state of family 18 chitinases, and acyl ureas (diflubenzuron and its derivatives) that might affect chitin synthesis at a non-catalytic step. Some polyoxins are effective weapons against phytopathogens and acyl ureas are commercial insecticides used in pest control. Allosamidin has been extremely useful in elucidating the catalytic mechanism of family 18 chitinases by crystallographic analysis of the hevamine/inhibitor complex [23].

12.5 Conclusions

Almost 200 years have passed since the discovery of chitin, and enormous progress has been made in understanding its structure, function and metabolism. As chitin is a central component of many biological composites with incredible properties regarding stiffness, toughness and durability, the gained knowledge promotes biomimetic approaches to obtain more powerful synthetic composites. However, not everything is answered yet. The biggest gap in our knowledge may concern the question how fungal and animal cells actually produce chitin – a process that is elusive for other biopolymers such as cellulose as well. The main reason for this lack of information is that the involved family II glycosyltransferases catalyzing their polymerization have stubbornly resisted any attempt at structural analysis. It may require the development of new methods by biochemists and structural biologists to solve these long persisting problems.

Summary Box

Chitin is a polymer of GlcNAc and widely distributed in Nature, where it is found in biological composites such as cell walls, cuticles, shells or peritrophic matrices. Due to its easy availability and unique physicochemical properties, chitin is attracting increasing interest as a renewable resource for the chemical and pharmaceutical industry. Moreover, inhibitors affecting different enzymatic steps in chitin metabolism have great potential as fungicides and insecticides.

References

1 Muzzarelli RA. Native, industrial and fossil chitins. *EXS* 1999;87:1–6.
2 Wagner GP. Evolution and multi-functionality of the chitin system. *EXS* 1994;69: 559–77.
3 Rudall KM, Kenchington W. The chitin system. *Biol Rev* 1973;48:597–636.
4 Vermeulen CA, Wessels JG. Chitin biosynthesis by a fungal membrane preparation. Evidence for a transient noncrystalline state of chitin. *Eur J Biochem* 1986;158:411–5.
5 Minke R, Blackwell J. The structure of α-chitin. *J Mol Biol* 1978;120:167–81.
6 Kameda T *et al.* Hydrogen bonding structure and stability of α-chitin studied by ^{13}C solid-state NMR. *Macromol Biosci* 2005;5: 103–6.
7 Blackwell J. Structure of β-chitin or parallel chain systems of poly-β-(1–4)-*N*-acetyl-D-glucosamine. *Biopolymers* 1969;7:281–98.
8 Saito Y *et al.* Structural data on the intracrystalline swelling of β-chitin. *Int J Biol Macromol* 2000;28:81–8.
9 Roberts GAF. *Chitin Chemisty.* MacMillan Press Ltd., London, 1992.
10 Latge JP. The cell wall: a carbohydrate armour for the fungal cell. *Mol Microbiol* 2007; 66:279–90.
11 Hirano S. Chitin and chitosan (6th ed.). In: *Ullmann's Encyclopedia of Industrial Chemistry* (Ed.: Gerhartz W), pp. 231–2. Wiley-VCH, Weinheim, 2002.
12 Vincent JFV. Design and mechanical properties of insect cuticle. *Arthropod Struct Dev* 2004;33:187–99.

13 Bouligand Y. Twisted fibrous arrangements in biological materials and cholesteric mesophases. *Tissue Cell* 1972;4: 189–217.

14 Lehane MJ. Peritrophic matrix structure and function. *Annu Rev Entomol* 1997;42: 525–50.

15 Chrispeels MJ, Raikhel NV. Lectins, lectin genes, and their role in plant defense. *Plant Cell* 1991;3:1–9.

16 Parijs J *et al.* Hevein: an antifungal protein from rubber-tree (*Hevea brasiliensis*) latex. *Planta* 1991;183:258–64.

17 Kasprzewska A. Plant chitinases – regulation and function. *Cell Mol Biol Lett* 2003;8:809–24.

18 Coutinho PM *et al.* An evolving hierarchical family classification for glycosyltransferases. *J Mol Biol* 2003;328:307–17.

19 Lairson LL, Withers SG. Mechanistic analogies amongst carbohydrate modifying enzymes. *Chem Commun (Camb)* 2004;20: 2243–8.

20 Fukamizo T. Chitinolytic enzymes: catalysis, substrate binding, and their application. *Curr Protein Pept Sci* 2000;1:105–24.

21 Unligil UM, Rini JM. Glycosyltransferase structure and mechanism. *Curr Opin Struct Biol* 2000;10:510–7.

22 Merzendorfer H, Zimoch L. Chitin metabolism in insects: structure, function and regulation of chitin synthases and chitinases. *J Exp Biol* 2003;206:4393–412.

23 Terwisscha van Scheltinga AC *et al.* Stereochemistry of chitin hydrolysis by a plant chitinase/lysozyme and X-ray structure of a complex with allosamidin: evidence for substrate assisted catalysis. *Biochemistry* 1995; 34:15619–23.

24 Brameld KA *et al.* Substrate assistance in the mechanism of family 18 chitinases: theoretical studies of potential intermediates and inhibitors. *J Mol Biol* 1998;280: 913–23.

Part Four
Protein–Carbohydrate Interactions

The Sugar Code. Edited by Hans-Joachim Gabius

ISBN: 978-3-527-32089-9

13
Protein–Carbohydrate Interactions: Basic Concepts and Methods for Analysis

Dolores Solís, Antonio Romero, Margarita Menéndez, and Jesús Jiménez-Barbero

The structural diversity of carbohydrates underlies the potential of this class of biomolecules for storing biological information, as explained in Chapter 1. Turning this potential into an operative sugar code entails the existence of efficient decoding devices. As complex carbohydrates decorate the surface of cells, complementary binding between carbohydrates on neighboring cells is possible, although the role of direct carbohydrate interactions in biological processes has been reported in only a few cases (please see Chapter 21). On the contrary, recognition of carbohydrates by proteins has been shown to be central to a myriad of intra- and extracellular physiological and pathological processes (please see Table 19.2 for functions of animal lectins). Proteins exhibiting carbohydrate-binding ability include sugar-specific antibodies, carbohydrate-active enzymes (which, in addition to the catalytic module, may also contain non-catalytic carbohydrate-binding modules), transport/sensor proteins for free sugars and lectins (for definition of the term 'lectin', please see Chapter 15). The scope and relevance of these protein–glycan recognition systems is extensively illustrated throughout this book. This chapter will focus on the fundamental structural and thermodynamic features of protein–carbohydrate interactions revealed by different methodological approaches. A detailed knowledge of the intricacies of carbohydrate recognition by proteins is fundamental to a rational design of new carbohydrate-based drugs (please see Chapter 28 for details).

13.1
Atomic Features of Protein–Sugar Interactions

Basic atomic features of protein–sugar interactions stem from the chemical properties of carbohydrates (Figure 13.1). First, the presence of freely rotatable hydroxyl groups facilitates the formation of directional hydrogen bonds. The sp^3-hybridized oxygen atom can participate in two hydrogen bonds as acceptor while the proton can act as donor (Figure 13.1a). Thus, taking into account the abundance of hydroxyl groups in carbohydrates, a major role of hydrogen bonding is foreseeable.

The Sugar Code. Edited by Hans-Joachim Gabius

ISBN: 978-3-527-32089-9

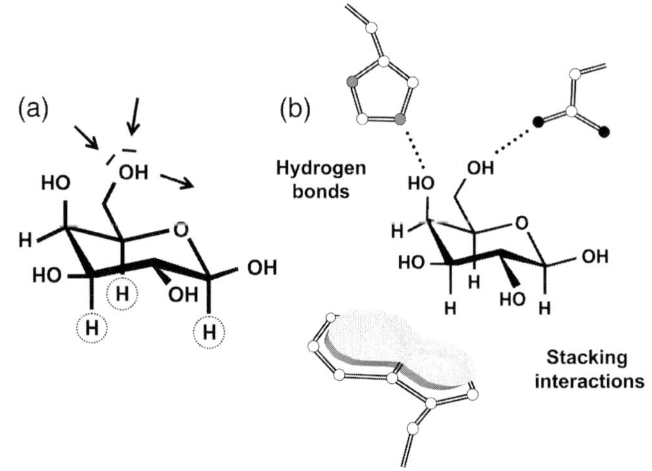

Figure 13.1 Illustration of the chemical character of saccharides underlying basic atomic features of protein–sugar interactions. (a) The arrows indicate the capacity of hydroxyl groups to be engaged in hydrogen bonding and point from donor to acceptor. The formation of aliphatic C—H patches (in this case in β-D-Gal) is highlighted. (b) Sugar-binding sites contain polar residues able to participate in hydrogen bonding and a suitably positioned aromatic residue (here Trp) that offers a delocalized π-electron cloud for stacking interactions with the C—H patches of positively polarized character.

Second, the configuration of aliphatic C—H patches invites hydrophobic interactions with proteins (Figure 13.1b). Furthermore, other non-polar groups of the sugar, such as the methyl moiety of acetamido groups, may be involved in van der Waals contacts. Finally, negatively charged sugars (containing carboxylate, sulfate or phosphate groups) can be engaged in strong electrostatic interactions. Evidence for the occurrence of all these types of interactions has been obtained from the X-ray crystal structures of protein–sugar complexes [1–3] (please see also Chapter 16).

X-ray crystallography is the most powerful tool to obtain information on the overall architecture of the protein (Figure 13.2) and on the topology of the carbohydrate-binding site (Figure 13.3), as exemplarily illustrated for human galectin-1 [4] (for information on the family of galectins, please see Chapters 19, 25 and 27). A brief outline of the steps followed for resolution of the structure is given in Figure 13.4. The overall folding of human galectin-1 establishes a β-sandwich motif (please see Table 19.1) and the design of the binding site follows the general depiction shown in Figure 13.1b. Folds of carbohydrate-binding proteins may, however, differ considerably, as detailed for animal lectins in Chapters 16 and 19, and the shape of the binding site may also be significantly different, ranging from shallow grooves close to the protein surface, as observed for many lectins, to deep pockets, as found for the active site of some exoglycosidases. The size of the sites also differs, reflecting the number of contiguous sugar units involved in oligosaccharide binding. Nevertheless, a number of common features can be discerned.

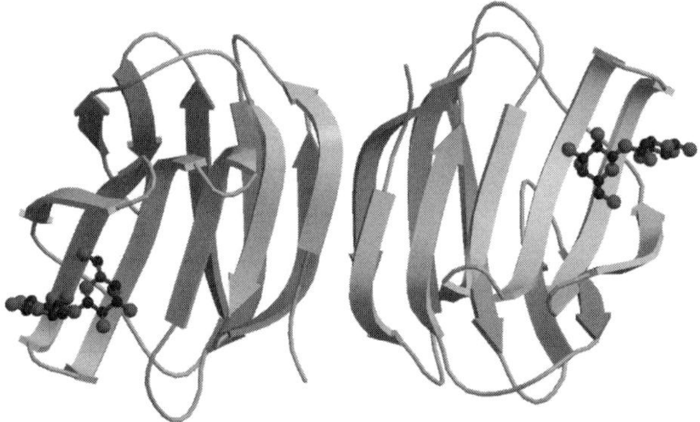

Figure 13.2 Ribbon diagram of the homodimeric human galectin-1. The N- and C-termini of each monomer are positioned at the dimer interface, and the carbohydrate-binding sites are located at the far ends of the same face of the β-sandwich (two lactose molecules are shown in ball-and-stick mode).

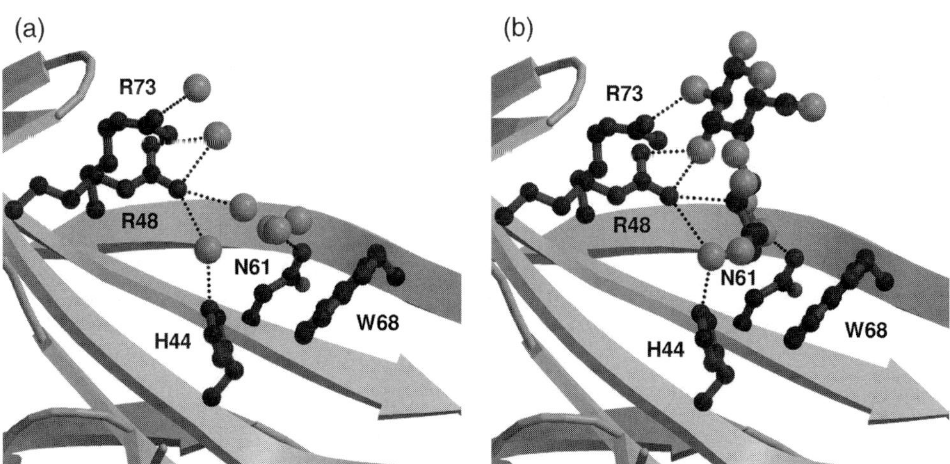

Figure 13.3 The carbohydrate-binding site of the C2S mutant of human galectin-1. (a) The binding site in the ligand-free C2S mutant of human galectin-1, showing bound water molecules as spheres. In the absence of a ligand, this water network helps to stabilize the spatial arrangement of the residues involved in sugar recognition. (b) The binding site of the C2S mutant in complex with lactose. The hydroxyl groups O4, O6 of galactose and O3 of glucose are placed at positions formerly occupied by water molecules.

Sugar-binding sites profusely contain polar residues with planar side-chains able to participate in hydrogen bonding (for example Asn, Asp, Gln, Glu, Arg, His and Lys) (please see Figures 13.1 and 13.3). Complex formation usually involves only a few hydroxyl groups of the sugar directly engaged in highly organized bidentate or cooperative hydrogen bonds with those side-chains, conferring specificity to the

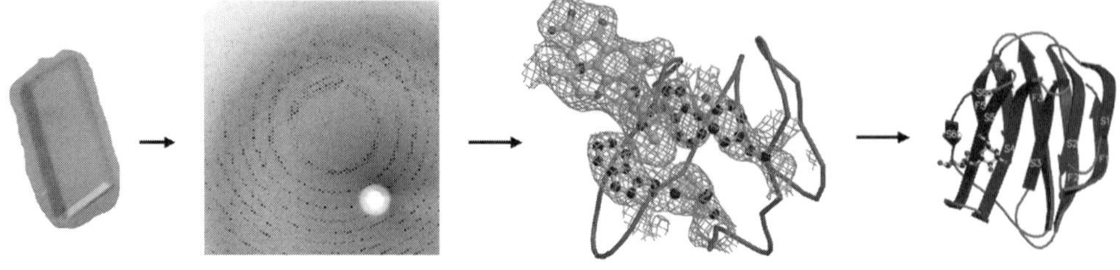

Figure 13.4 Schematic diagram of the necessary steps for solving the structure of a macromolecule by X-ray crystallography. After succeeding in the growth of suitable crystals, diffraction data is collected. Next, the X-ray diffraction pattern has to be converted into a three-dimensional model of the density of electrons within the crystal. The positions of the atomic nuclei in the amino acid sequence of the protein are then deduced from this electron density and after a series of refinements the final model is produced.

interaction. The contribution of water-mediated hydrogen bonds bridging protein and sugar is also frequent, and in some cases coordination with a metal ion is involved (please see Chapter 16). Numerous van der Waals contacts are generated between atoms of the polar residues and the sugar. In addition, the presence in the binding site of an aromatic side-chain of Trp, Tyr and, less commonly, Phe is detected. The aromatic ring of these residues offers a delocalized π-electron cloud for interaction with the C—H patches of positively polarized character (Figure 13.1b), thereby establishing planar or partly twisted stacking interactions with the sugar (please see Figure 13.3). Furthermore, the orientation of the aromatic side-chain is a significant determinant of specificity because it impedes the binding of incorrect epimers. Finally, positively charged Arg and Lys residues are particularly prominent in the binding sites of proteins recognizing negatively charged saccharides, evidencing that charge–charge interactions dominate recognition.

As already mentioned, usually only a few hydroxyl groups of the sugar are involved in key hydrogen bonds with the protein. Other hydroxyls participate in weaker hydrogen bonds while the rest are exposed to the solvent and do not interact with the protein. The contribution of each group to the binding can be estimated through chemical mapping studies in solution [5, 6]. This strategy is based on the specific modification of the sugar at selected positions and evaluation of the binding properties of the derivatives compared with the parent compound. The binding potential of different groups present in synthetic carbohydrate derivatives is compiled in Figure 13.5. In detail, replacement of a particular hydroxyl group by hydrogen abrogates the hydrogen-bonding potential at that position. Thus, if the absent hydroxyl is involved in the hydrogen-bonding network, deoxygenation results in a reduction in binding affinity. Alternatively, the hydroxyl group can be replaced by a fluorine atom, which may serve as hydrogen-bond acceptor. Therefore, the behavior of the fluorodeoxy derivative reveals the involve-

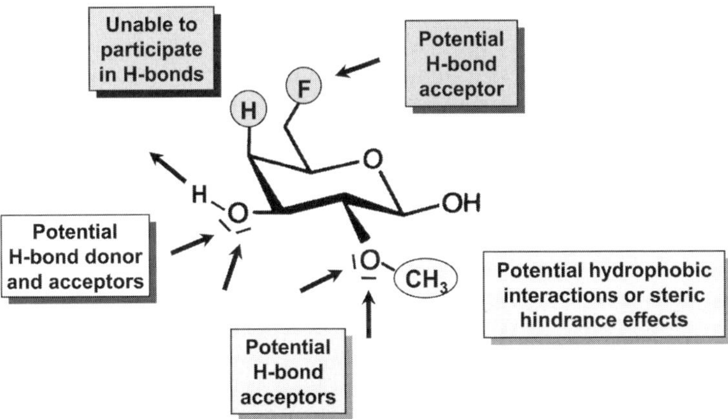

Figure 13.5 Binding potential of different functional groups in synthetic carbohydrate derivatives. The arrows indicate the capacity to be engaged in hydrogen bonding, as specified in the legend to Figure 13.1.

ment of the hydroxyl group as donor and/or acceptor. *O*-Methyl derivatives may also help to probe the hydroxyl group activity and also provide information on the steric requirements of the binding site, as the bulky methyl substituent may engender steric hindrance. On the other hand, the methyl group may generate favorable nonpolar interactions with the protein, thereby increasing the binding affinity. In this way, by systematic chemical mapping studies using an array of synthetic ligand derivatives with different binding abilities, it is possible to dissect the contribution of every sugar hydroxyl group to the binding, and even to delineate the topology and distribution of nonpolar regions of the binding site. This information can be extremely useful for tailoring superior ligands in drug design processes. In addition, it helps to validate the list of interactions observed in the crystal structures of protein–sugar complexes. The need for such a validation originates from the frequent use of far-from-physiological conditions for crystallization and the potential influence of crystal packing on the structure.

13.2 Role of Water in Protein–Sugar Interactions

Water plays an essential role in protein–carbohydrate interactions [7]. In protein–sugar complexes, water molecules are often found bridging sugar hydroxyl groups with side-chains of the binding site, as mentioned above. In their uncomplexed forms, both the protein and the sugar are fully solvated. As observed, for example, in the crystal structure of the ligand-free form of human galectin-1 (here a C2S mutant designed for increased stability against oxidative inactivation) (Figure 13.3a), the carbohydrate-binding site is occupied by water molecules that stabilize the orientation of the protein residues involved in sugar binding through a network of interactions, so that no significant displacement of the side-chains is required for sugar binding. Some of these water molecules strategically occupy the positions that will accommodate the key hydroxyl groups involved in the recognition (Figure 13.3b) [4]. Similarly, the polar groups of the sugar are expected to be extensively hydrogen bonded to water molecules. Water molecules directly bonded to protein and sugar groups show restricted motion compared to those in the bulk and cannot freely adopt the most favorable orientation for hydrogen bonding to neighboring water molecules. As a result, water molecules surrounding polar surfaces are perturbed [7]. On the other hand, organized layers of molecules are formed over the nonpolar patches of both protein and sugar surfaces. It is clear that for complex formation water molecules solvating the contact surfaces of protein and sugar have to be released and returned to the bulk, where they establish new water–water interactions. This reorganization of the solvent, involving a large number of molecules, contributes significantly to the binding thermodynamics (see below). The importance of solvent reorganization reflects the dynamism of protein–carbohydrate recognition systems, far from the static picture provided by crystal structures. Another aspect to be considered in this context is the flexibility of carbohydrates (please see Chapter 2), which engenders a repertoire of conformations in solution. The question arises whether proteins recognize selectively a distinct conformer, that is, perform conformer selection.

13.3 Selection of Carbohydrate Conformers by Proteins

Nuclear magnetic resonance (NMR) spectroscopy is the key technique for unraveling the three-dimensional structure of biomolecules in solution. Most of the basic NMR parameters can be employed to monitor the binding of potential ligands to their putative receptors and, moreover, to provide well-defined information on the mode of binding, including the three-dimensional shape of the bound conformer. In some favorable cases, by combining state-of-the-art protein labeling techniques with modern instrumentation, the complete three-dimensional structure of the protein–sugar complex can also be derived by NMR methods [8].

Information on the bound-state conformation of the carbohydrate ligand can be obtained by transferred (TR) nuclear Overhauser effect (NOE) spectroscopy [9]. The NOE provides information on the distances between different proton pairs of a molecule. The basis of TR-NMR experiments is that carbohydrate-binding proteins and their small carbohydrate ligands have distinct physical and spectroscopic properties. In the complex, the bound ligand effectively adapts to physical properties of the protein and develops strong NOEs. They can be observed when they are transferred to the sharp resonances of the free ligand (Figure 13.6). Therefore, the experiment relies on the exchange of all ligand molecules between free and bound states; consequently, the binding kinetics, particularly the off-rate relative to the experiment time, are important. In addition, there is an optimal carbohydrate ligand to protein ratio for the observation of TR-NOEs, typically using a low

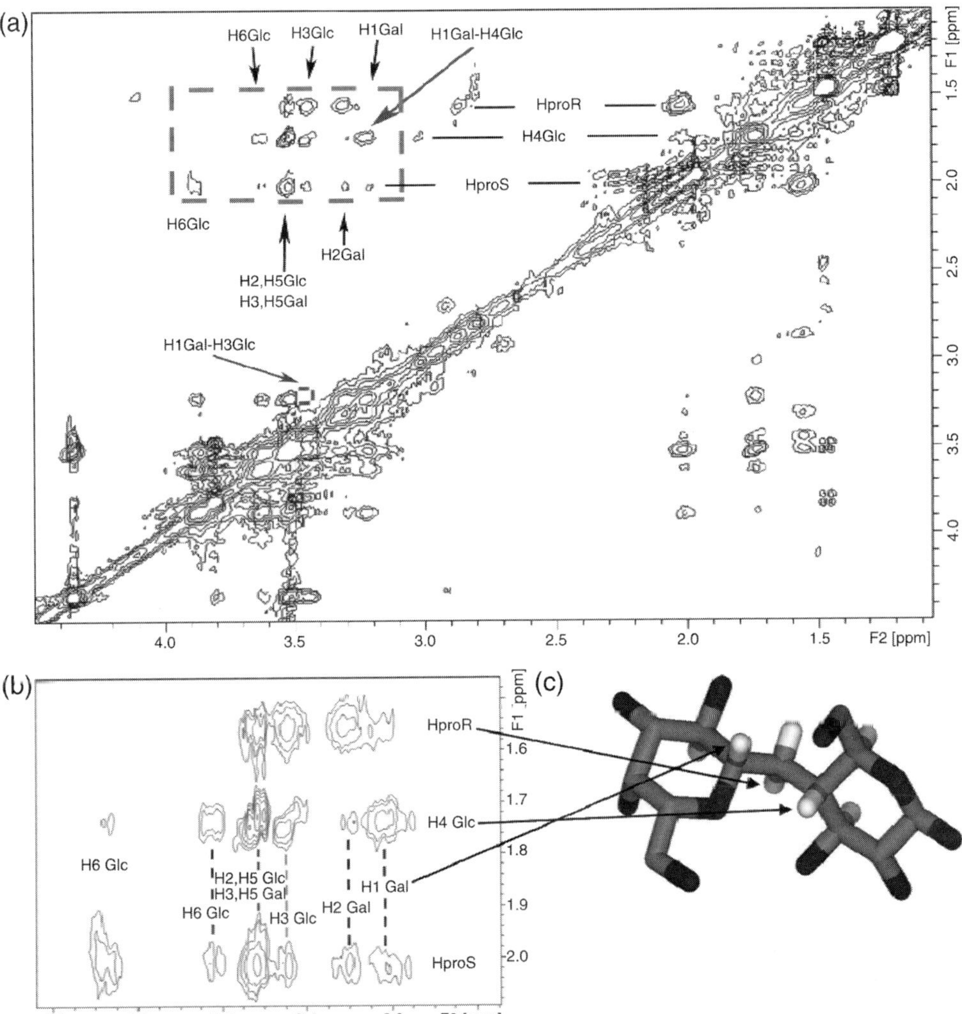

Figure 13.6 TR-NOE spectroscopy spectrum (mixing time 150 ms) for a 25 : 1 molar ratio of the C-glycoside of *N*-acetyllactosamine : human galectin-1 (0.1 mM of lectin). Cross-peaks have the same sign as diagonal peaks (a). No H1Gal—H3GlcNAc contact is observed, indicating that the *anti*-Ψ conformer is not bound by this lectin. (b) Expansion of the key region of the TR-NOE spectroscopy spectrum. Preferential recognition of the *syn*-Ψ conformer (c) is evident due to the presence of a strong H1Gal—H4GlcNAc cross-peak.

concentration of lectin and an excess of ligand. Fortunately, the conditions required for the successful use of this technique are frequently fulfilled by protein–sugar systems and, since 1990 (please see Info Box), TR-NOE experiments have been widely used to investigate the bound-state conformation of the sugar [9, 10].

Frequently, sugar-specific proteins bind the global-minimum-energy conformation of a saccharide and, as with the ricin/lactose complexes (see Info Box), no

Info Box

The first TR-NOE-based NMR study in the carbohydrate field investigated the binding of methyl β-lactoside to the B-chain of the potent plant toxin ricin (please see Chapters 15 and 18 for further information on ricin). Using one-dimensional TR-NOE experiments and a selectively deuterated substrate, Prestegard and coworkers showed that only minor changes in the conformation of free methyl β-lactoside took place upon binding [V.L. Bevilacqua *et al.* Conformation of methyl β-lactoside bound to the ricin B-chain: interpretation of transferred nuclear Overhauser effects facilitated by spin simulation and selective deuteration. *Biochemistry* 1990; **29**, 5529–5537]. Later, Asensio *et al.* showed that the protein caused a slight conformational variation in the glycosidic torsion angles of methyl α-lactoside upon binding [J.L. Asensio *et al.* Studies of the bound conformations of methyl α-lactoside and methyl β-allolactoside to ricin B chain using transferred NOE experiments in the laboratory and rotating frames, assisted by molecular mechanics and dynamics calculations. *Eur J Biochem* 1995; **233**, 618–630], although the recognized conformer was still within the lowest energy region. In the same study, it was found that different conformations around the ϕ, ψ and ω glycosidic bonds of methyl β-allolactoside were recognized by the lectin. In fact, for this complex, only the TR-NOE spectroscopy cross-peaks corresponding to the protons of the galactose residue were negative, as expected for a molecule in the slow motion regime. In contrast, the corresponding cross-peaks for the glucose residue were nearly zero, as expected for a molecule with motional properties practically independent from the overall motion of the protein.

major variations in the conformational behavior of the sugar are observed upon protein binding. Galectin-1, as further example, recognizes the *syn*-conformer (ϕ 50–60°, ψ 0°) of lactose and different β-galactosyl xyloses [11], which represents around 90% of the free sugar population (for angle designation of glycosidic linkage, please see Figure 2.2; for information on the conformational behavior of lactose, please see Figure 2.3; for biological activity of galectin-1, please see Fig. 25.1). Indeed, only this conformation allows the establishment of favorable contacts of the Glc/Xyl unit with the protein (Figure 13.7). Similarly, the *syn*-Φ conformer of *C*-lactose, a non-hydrolyzable lactose analog, is bound by galectin-1 (Figure 13.6). In this case, this is not the predominant conformer in solution, as *C*-lactose exhibits much higher flexibility than the *O*-analog and three different conformational regions (*syn*, *anti* and *gauche-gauche*) are significantly populated. However, only the *syn*-Φ, *syn*-Ψ conformer makes the formation of three hydrogen bonds between the Glc residue and the protein possible. Thus, this is an instructive example of selection of a minor conformer from an equilibrium mixture in solution. In contrast, *C*-lactose is bound to ricin B-chain in the *anti*-ψ conformation, whereas the enzyme *Escherichia coli* β-galactosidase recognizes the *anti*-Φ conformer ([10] and references therein), hereby showing that different carbo-

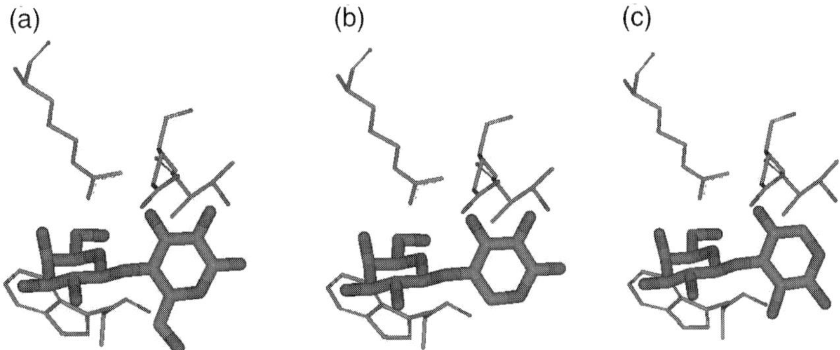

Figure 13.7 The recognition of *syn*-Φ conformers of lactose by galectin-1 (Protein Data Bank file 1GZW). The Gal moiety presents both CH–π and hydrogen bonding interactions, but the Glc residue is the key for the exclusive recognition of *syn*-Φ, *syn*-Ψ geometries. Indeed, only for this geometry, hydrogen bonds involving O3 of the Glc moiety with Glu71 and with Arg48 take place. This is the case for lactose (a) and the analogous Galβ1,4Xyl compound (b). Galectin-1 also recognizes the Galβ1,3Xyl isomer in the same conformation (c), but now the hydrogen bonds involve O4 of the Xyl residue. The mentioned hydrogen bond interactions are not possible for the other conformational families.

hydrate-binding proteins may recognize selectively different conformations of the sugar [10]. In contrast, when protein–sugar contacts are limited to one monosaccharide residue, the binding site may accommodate different conformations around the glycosidic bond, providing that there are no steric impediments, and, consequently, there is no selection of a single conformer of the sugar, as observed for methyl β-allolactoside binding to ricin (see Info Box).

13.4 Thermodynamics of Protein–Carbohydrate Interactions

Considering all the preceding information, it becomes clear that, in addition to the specific protein–sugar contacts, other factors govern protein–carbohydrate interactions. The binding affinity or, in thermodynamic terms, the Gibbs free energy of binding ($\Delta G°$), is the corollary of the enthalpy/entropy gains and penalties derived from the different events associated with the recognition process. In detail, the net enthalpy change comprises the favorable contribution of protein–sugar contacts and all the new water–water hydrogen bonds generated upon desolvation of protein and sugar, together with the unfavorable contribution of the breakage of protein–water and sugar–water interactions. Similarly, the net entropy change includes the entropy gain of water molecules ordered over nonpolar regions or directly bonded to protein and sugar groups when they return to the bulk, along with an entropy decrease of the less-ordered perturbed water molecules that surrounded polar surfaces. In addition, other events imposing entropic penalties are the restrictions in the translation and overall rotation of the molecules, and the loss of conformational freedom of the sugar and the protein upon complex formation. At this point, it is opportune to bring up the observations that, in solution, the overall protein structure of human galectin-1 becomes more compact after accommodating the ligand [12] and that small structural changes in the protein overall fold induced by introduction of single-site mutations at positions far away from the binding site have an impact on the binding parameters [4]. Thus, although the carbohydrate-binding site can be mostly preshaped and may not undergo major conformational changes upon ligand binding, any change in the flexibility and internal mobility of the whole protein can affect the binding thermodynamics. The specific weight of the enumerated contributions is surely particular for each protein–sugar pair. Nevertheless, a simple inventory of gains and penalties suggests that the binding may be enthalpically driven, with a favorable net enthalpy counteracted by an unfavorable entropy term, and this is actually the case for most protein–carbohydrate interactions.

Many biophysical techniques have been used to measure the affinity of these interactions, including NMR spectroscopy. The basic NMR parameter is the chemical shift, which depends on a number of factors, including the environment of the corresponding nucleus. Protein–ligand interactions can subtly change the environment of the NMR-active nuclei (that is ^{1}H, ^{13}C, ^{15}N and so on) of both entities, leading to changes in the chemical shifts. Therefore, titration of a carbohydrate-binding protein with a sugar ligand (or vice versa) following chemical shift changes, by use of single one-dimensional ^{1}H-NMR experiments, is a common mean to determine the binding affinity. The principal drawback is the overlapping of lectin and ligand resonances, particularly at low ligand : lectin ratios. For that reason, different NMR methods using two-dimensional-like experiments have been developed to simplify the task (Table 13.1) [10]. Another NMR method successfully used for measuring the binding affinity and also the vicinity between the

Table 13.1 Different NMR experiments used for the titration of protein–sugar systems.

System	Nucleus observed	Experiment
Unlabeled protein and ligand	^{1}H	^{1}H one-dimensional
^{19}F-labeled ligand, unlabeled protein	^{19}F	^{19}F one-dimensional
^{13}C-labeled ligand, unlabeled protein	^{1}H, $^{1}H + ^{13}C$	$^{1}H(^{13}C)$-edited one-dimensional, $^{1}H,^{13}C$ heteronuclear single-quantum coherence two-dimensional
^{2}H-labeled protein, unlabeled ligand	^{1}H	^{1}H one-dimensional
^{15}N-labeled protein, unlabeled ligand	$^{1}H+^{15}N$	$^{1}H,^{15}N$ heteronuclear single-quantum coherence two-dimensional

protein and the ligand is the saturation transfer difference technique. With galectin-1 and its growth-regulatory interaction with ganglioside GM1 as an example (for details on ganglioside and tumor biology of this interaction, please see Chapters 25 and 30), saturation transfer from the protein to the carbohydrate ligand was picked up for the terminal Gal and GalNAc units [13]. This technique has been adapted to use the fluorine atom (^{19}F) as sensor [14]. Close contact between lectin and ligand underlies this transfer process, allowing us to map spatial parameters. However, the only technique that directly determines binding enthalpies is isothermal titration calorimetry (ITC) (for techniques to measure carbohydrate-binding specificities, please see Chapter 14).

ITC measures the heat evolved after addition of a ligand to a receptor as a function of ligand concentration [15]. A single titration experiment can yield a complete binding isotherm from which the enthalpy of binding ($\Delta H°$), the association constant (K_a) and the stoichiometry (n) can be directly determined. Then, the entropy of the binding ($\Delta S°$) can be calculated from the expression:

$$\Delta G° = \Delta H° - T\Delta S° = -RT \ln K_a,$$

where R is the universal gas constant and T is the absolute temperature. By way of illustration, Figure 13.8 shows the calorimetric titration of human galectin-1

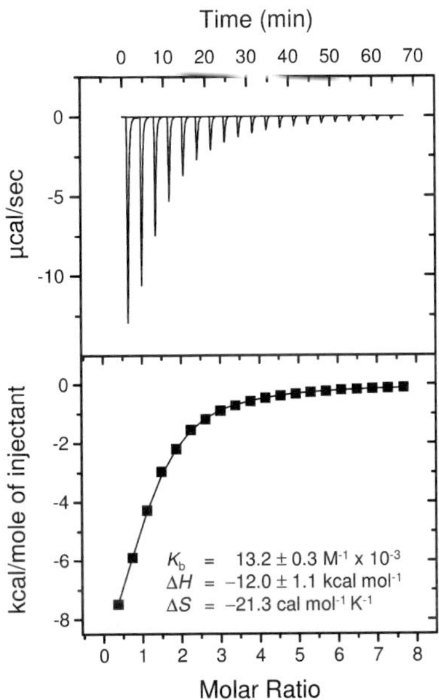

Figure 13.8 Calorimetric titration of human galectin-1 with *N*-acetyllactosamine. The upper panel shows the experimental measurement of the heat released after sequential addition of 5-µl aliquots of 10 mM *N*-acetyllactosamine to a 0.1 mM solution of human galectin-1 at 298.08 K. The lower panel shows the plot of this heat as a function of the sugar : protein molar ratio. The continuous line corresponds to the best fit of the experimental data in a one-set-of-sites model. The thermodynamic parameters derived from the analysis are given.

with *N*-acetyllactosamine [4]. The thermodynamic analysis evidences the existence of enthalpy–entropy compensation, as typically observed for protein–carbohydrate interactions [16]. Thus, although changes in enthalpy and entropy may be considerable, the net free energy of binding or, in other words, the binding affinity is relatively small, in the millimolar range, as commonly observed for the binding of mono- or disaccharides by lectins.

The affinity for oligosaccharides can be increased up to the micromolar range as a result of the binding of several sugar units to extended binding sites in proteins. Accommodation at the secondary binding subsites is also driven by a favorable enthalpy, but the entropy cost may be smaller because the overall motion and conformational freedom of the oligosaccharide has already been restricted, at least in part, upon binding at the primary binding subsite. Multivalency can also result in considerably higher binding affinities in protein–carbohydrate interactions [16–18]. The simultaneous binding of clustered binding sites in carbohydrate-binding proteins to multivalent ligands, as those presented on cell surfaces, can increase the affinity up to nanomolar levels (please see Chapters 17–19 for different ways of clustering carbohydrate recognition domains in lectins, and Chapter 4 for details on synthetic polyvalent ligands). After the first contact is established, ligand binding to the other binding sites may be facilitated by spatial vicinity and the entropy cost of these interactions is progressively smaller. Furthermore, the first contact itself may be affected by a smaller entropic penalty compared to a monovalent interaction if the particular presentation in the multivalent ligand restricts the flexibility and freedom of the carbohydrate.

13.5 Conclusions

The important role played by protein–carbohydrate interactions in many biomedically relevant processes makes them interesting targets for the development of new carbohydrate-based therapeutics. A rational design of ligands with improved binding affinity should be grounded on the understanding of the atomic features and forces governing the binding thermodynamics. A detailed knowledge of the carbohydrate-binding site architecture and the sugar groups involved in the recognition may facilitate the optimization of the binding enthalpy by increasing the number and strength of hydrogen bonds and van der Waals contacts between the protein and the sugar. On the other hand, elucidation of the conformational properties of the sugar in the free and bound states may help to decrease the entropic penalty of the binding by designing ligands with reduced flexibility and maximum complementarity with the protein's binding site. Finally, the entropy gain associated with desolvation of non-polar surfaces could be exploited by tailoring hydrophobic ligands.

Summary Box

The chemical properties of carbohydrates justify why protein–carbohydrate interactions are mainly stabilized by hydrogen bonds together with stacking interactions and other van der Waals contacts. In addition to the basic atomic contacts, the reorganization of the solvent and the restriction of ligand flexibility strongly influence the binding thermodynamics. As a result, protein–carbohydrate interactions are enthalpically driven and exhibit enthalpy–entropy compensation.

References

1 Quiocho FA. Atomic structures of periplasmic binding proteins and the high-affinity active transport system in bacteria. *Phil Trans Royal Soc London Ser B Biol Sci* 1990;326:341–51.

2 Weis WI, Drickamer K. Structural basis of lectin–carbohydrate recognition. *Annu Rev Biochem* 1996;65:441–73.

3 Imberty A *et al.* Structural view of glycosaminoglycan–protein interactions. *Carbohydr Res* 2007;342:430–9.

4 López-Lucendo MF *et al.* Growth-regulatory human galectin-1: crystallographic characterisation of the structural changes induced by single-site mutations and their impact on the thermodynamics of ligand binding. *J Mol Biol* 2004;343:957–70.

5 Solís D, Díaz-Mauriño T. Analysis of protein carbohydrate interaction by engineered ligands. In: *Glycosciences: Status and Perspectives* (Eds.: Gabius HJ, Gabius S), pp. 345–54. Chapman & Hall, London, 1997.

6 Audette GF *et al.* Mapping protein: carbohydrate interactions. *Curr Protein Pept Sci* 2003;4:11–20.

7 Lemieux RU. How water provides the impetus for molecular recognition in aqueous solution. *Acc Chem Res* 1996;29:373–80.

8 Kogelberg H *et al.* New structural insights into carbohydrate–protein interactions from NMR spectroscopy. *Curr Opin Struct Biol* 2003;13:646–53.

9 Siebert HC *et al.* Describing topology of bound ligand by transferred nuclear Overhauser effect spectroscopy and molecular modeling. *Methods Enzymol* 2003;362:417–34.

10 Jiménez-Barbero J, Peters T (Eds.), *NMR Spectroscopy of Glycoconjugates.* Wiley-VCH Verlag GmbH, Weinheim, 2002.

11 Alonso-Plaza JM *et al.* NMR investigations of protein–carbohydrate interactions: insights into the topology of the bound conformation of a lactose isomer and β-galactosyl xyloses to mistletoe lectin and galectin-1. *Biochim Biophys Acta* 2001;1568:225–36.

12 He L *et al.* Detection of ligand- and solvent-induced shape alterations of cell-growth-regulatory human lectin galectin-1 in solution by small angle neutron and X-ray scattering. *Biophys J* 2003;85:511–24.

13 Siebert HC *et al.* Unique conformer selection of human growth-regulatory lectin galectin-1 for ganglioside GM1 versus bacterial toxins. *Biochemistry* 2003;42:14762–73.

14 Diercks T *et al.* Fluorinated carbohydrates as lectin ligands: versatile sensors in ^{19}F-detected saturation transfer difference NMR spectroscopy. *Chem Eur J* 2009;15:5666–8.

15 Christensen T, Toone EJ. Calorimetric evaluation of protein–carbohydrate affinities. *Methods Enzymol* 2003;362:486–504.

16 Dam TK, Brewer F. Thermodynamic studies of lectin–carbohydrate interactions by isothermal titration calorimetry. *Chem Rev* 2002;102:387–429.

17 Lee RT, Lee YC. Affinity enhancement by multivalent lectin–carbohydrate interaction. *Glycoconj J* 2001;17:543–51.

18 Gabius HJ *et al.* The chemical biology of the sugar code. *ChemBioChem* 2004;5:740–64.

14
How to Determine Specificity: From Lectin Profiling to Glycan Mapping and Arrays

Hiroaki Tateno, Atsushi Kuno, and Jun Hirabayashi

The preceding chapter provided insights into the interactions of proteins with glycans and biological methods to analyze this recognition process. Historically, the agglutination of erythrocytes was used in the pioneering studies to detect lectin activity (for details, please see Chapter 15, discussing the history and the definition to lectins, and also Figure 26.4 for an illustration; for details on lectins, please see Chapters 16–26). As this assay is generally assessed by serial twofold dilution and by I_{50} in inhibition, but not by K_i (or K_a), the obtained results are neither quantitative nor absolute values. Therefore, determination of lectin–carbohydrate binding strength in terms of either K_d, K_a or K_i is of fundamental importance from an analytical viewpoint. In this chapter, we explain analytical methods representing both conventional and advanced techniques developed for lectin–carbohydrate interaction analysis – the hemagglutination assay (a conventional semiquantitative method), equilibrium dialysis (a quantitative method) and frontal affinity chromatography (FAC) (a further quantitative analysis). The first two are low-throughput and the third high-throughput methods.

The hemagglutination assay is performed in a serial twofold dilution manner to determine the lectin dilution showing the minimum hemagglutination activity. To function in this assay, lectins must be polyvalent to cross-link glycan structures expressed on erythrocytes (Figure 14.1). Therefore, the results possibly include relatively large experimental errors (50–100%). When lectin specificity is assessed in terms of I_{50} (defined as the saccharide concentration showing 50% inhibition of the hemagglutination), various saccharides (such as lactose, methyl-α-D-mannoside) are added to the hemagglutination assay system, also in a serial twofold manner, while the lectin is maintained at the lowest concentration to show hemagglutination activity. Therefore, I_{50} values may also include 50–100% experimental errors. This method is time consuming and requires a large amount of sugars to determine their inhibitory effects, particularly if they have only weak inhibitory activity. Therefore, from a practical viewpoint, this method is not suitable for expensive oligosaccharides or relatively rare sugar derivatives. The method also cannot be applied to monovalent lectins. Despite these deficiencies, the method is superior over others in its extreme simplicity and economy.

The Sugar Code. Edited by Hans-Joachim Gabius

ISBN: 978-3-527-32089-9

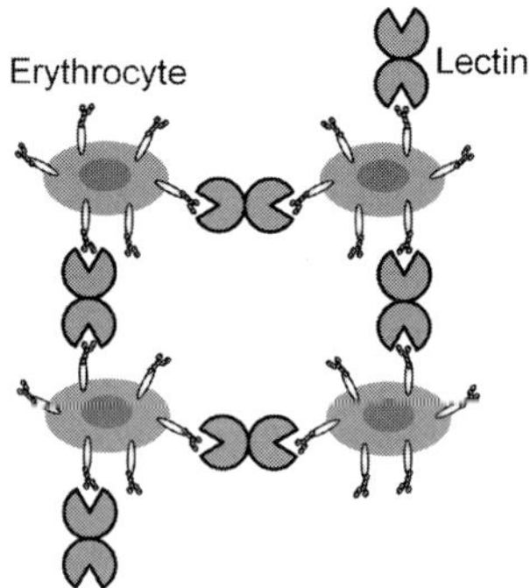

Figure 14.1 Cross-linking of glycans on the surface of different erythrocytes leads to formation of large aggregates. This process is called hemagglutination and has been used to detect sugar-binding (lectin) activity.

14.1 Quantitative Aspects of Lectin Affinity

For a more quantitative determination of lectin–carbohydrate interactions, i.e., in terms of K_a or K_d, several methods are available such as equilibrium dialysis, isothermal calorimetry (please see Chapter 13.4), binding studies with (neo)glycoprotein (see Table 25.1) and surface plasmon resonance (SPR) (please see Chapter 21.4.2.1). In principle, equilibrium between bound and free forms of lectins and saccharides (i.e., affinity constant, K_a) is defined by the following equation (also see Figure 14.2a):

$$K_a = [\text{Lectin–Saccharide (complex)}]/[\text{Lectin (free)}][\text{Saccharide (free)}]. \quad (14.1)$$

And K_d is defined as:

$$K_d = [\text{Lectin (free)}][\text{Saccharide (free)}]/[\text{Lectin–Saccharide (complex)}]. \quad (14.2)$$

Therefore, K_d is the inverse of K_a:

$$K_d = 1/K_a. \quad (14.3)$$

From a kinetic viewpoint, the above equilibrium consists of two processes, association and dissociation, which are defined by association (k_{ass}) and dissociation (k_{dis}) rate constants.

$$\text{Lectin (free)} + \text{Saccharide (free)} \underset{k_{dis}}{\overset{k_{ass}}{\rightleftarrows}} \text{Lectin–Saccharide (complex)}. \quad (14.4)$$

When association and dissociation rates are expressed as v_{ass} and v_{dis}, respectively, they are defined as follows:

(a)

$$\text{Lectin (free)} + \text{Saccharide (free)} \underset{k_{dis}}{\overset{k_{ass}}{\rightleftharpoons}} \text{Lectin-Saccharide (complex)}$$

$$K_d = \frac{\text{[Lectin (free)][Saccharide (free)]}}{\text{[Lectin-Saccharide (complex)]}} = 1/K_a$$

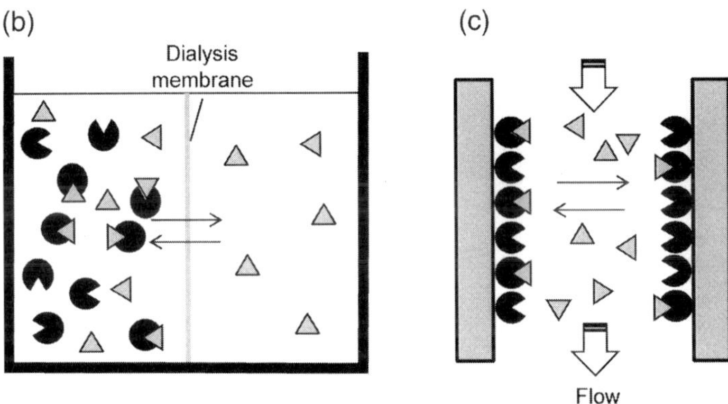

Figure 14.2 (a) Equilibrium between complex and free forms of lectins and saccharides. (b) Illustration of static equilibrium in the equilibrium dialysis analysis. (c) Illustration of dynamic equilibrium in the FAC analysis.

$$v_{ass} = k_{ass}\text{[Lectin (free)][Saccharide (free)]}. \tag{14.5}$$

$$v_{dis} = k_{dis}\text{[Lectin–Saccharide (complex)]}. \tag{14.6}$$

Since v_{ass} equals v_{dis} at equilibrium, the following equation applies:

$$k_{ass}\text{[Lectin (free)][Saccharide (free)]} = k_{dis}\text{[Lectin–Saccharide (complex)]}. \tag{14.7}$$

Hence, affinity constant (K_a) is a simple expression of k_{ass}/k_{dis}:

$$k_{ass}/k_{dis} = \text{[Lectin–Saccharide (complex)]/[Lectin (free)][Saccharide (free)]} = K_a. \tag{14.8}$$

In SPR analysis, both association and dissociation processes are monitored in terms of the 'resonance unit', and their rate constants, k_{ass} and k_{dis}, are determined using various concentrations of analyte molecules (lectins *versus* immobilized saccharides or saccharides *versus* immobilized lectins) along with curve-fit analysis.

Therefore, in SPR analysis, equilibrium constants, K_a and K_d, are determined in an indirect manner.

In an equilibrium dialysis experiment (Figure 14.2b), K_d is determined by measuring a free saccharide concentration, i.e., [Saccharide (free)] of the right chamber of Figure 14.2b under equilibrium. Since the initial amount of the saccharide is known, the concentration of the remaining form of saccharide, i.e., [Lectin–Saccharide (complex)], can be obtained by calculation. Similarly, [Lectin (free)] can be determined by subtracting the concentration [Lectin–Saccharide (complex)] from the initial concentration of the lectin. Hence, the equilibrium constant K_a (or K_d) is obtained from Equation 14.1. The method is simple and straightforward, but requires a relatively long time and large amounts of saccharides for each experiment. To improve the latter point, a modified procedure has been developed using micro-dialysis equipment and pyridylaminated (PA) oligosaccharides [1]. In fact, this improvement enabled systematic determination of K_d of concanavalin A (ConA; for its crystal structure, please see Figure 16.1a) for a series of PA-oligosaccharides. However, the method inherently has limitations in throughput and speed, since it depends on a measurement principle dealing with 'static equilibrium', which takes at least several hours. To overcome this, measurement in 'dynamic equilibrium' is an alternative. To achieve appropriate throughput, speed as well as precision, a chromatographic procedure is considered to be a suitable approach. FAC, originally discovered in 1975, was then turned into a lectin profiling system [2].

14.2 Frontal Affinity Chromatography (FAC) for Sugar–Protein Interactions

Affinity chromatography is not only a convenient method for protein purification, it is also applicable for quantitative analysis of lectin–glycan interactions (please see Chapters 15.3 and 18.3 for lectin purification by affinity chromatography). For this purpose, the lectin is immobilized on an affinity support and a buffer containing a glycan is added to the immobilized lectin. If the glycan binds to the immobilized lectin, elution is delayed. In this manner, the approximate binding specificity of the lectin can be assessed. As a more sophisticated version, FAC was developed for quantitative analysis of sugar–protein interactions, where precise K_d and K_a values can be obtained. An excess volume of diluted glycan solution is applied to a lectin-immobilized column and the elution front is monitored by a detector (Figures 14.2 and 14.3). If glycans bind to the column, the 'frontal volume' of the glycans (V) is delayed compared to control glycans (V_0), since it repeatedly interacts with the immobilized lectins as it passes through the column. Thus, its name, 'frontal' is derived from the 'frontal volume (V)', the factor that is obtained to determine quantitative interaction data.

Two different types of FAC have been developed for analysis of sugar–protein interactions, which involve mass spectrometry (MS) and fluorescence detection (FD). For FAC-MS, a lectin-immobilized micro-tube (PEEK) column is connected to an electrospray ionization mass spectrometer for online detection of terminally

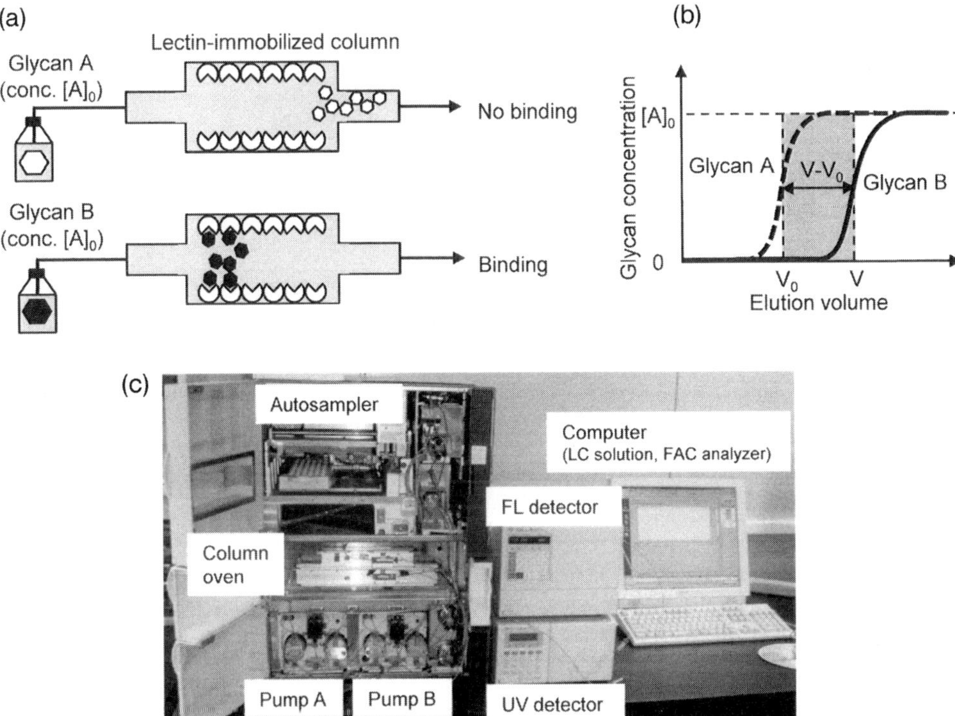

Figure 14.3 (a, b) A diluted labeled glycan solution is continuously applied to a column containing immobilized lectin. If Glycan A has no affinity to the immobilized lectin, its elution front (V_0) is observed immediately. In contrast, if Glycan B has affinity to the immobilized lectin, its elution front (V) is retarded. $[A]_0$ is the initial glycan concentration (M). (c) Photograph of the automated FAC system (FAC-1). FAC-1 consists of two isocratic pumps, an autosampler, a column oven, two miniature columns connected to a fluorescence (FL) or ultraviolet (UV) detector, and a computer with control (LCsolution) and data analysis (FAC analyzer) software.

modified (e.g., alkylated) oligosaccharides, allowing sensitive and simultaneous determination of K_d of even multiple components with different m/z values [3]. FAC-FD consists of conventional high-performance liquid chromatography (HPLC) connected to a fluorescence detector, which allows high-sensitivity as well as high-throughput analysis of sugar–protein interactions [4, 5]. In this method, PA-oligosaccharides are often used for the following reasons: PA-glycans can be detected at low concentrations (e.g., 5 nM), show no nonspecific interactions with gel matrices (e.g., agarose) and are stable compared to other fluorescence-labeled derivatives (please see Figure 5.3 and Chapter 5.3 for pyridylamino labeling).

In FAC, K_a (or K_d) of sugar–protein interactions can be calculated as follows; if the volume of the elution front of Glycan A that has no affinity to the immobilized lectin is indicated by V_0 (negative control) and that of Glycan B is indicated by V, the dissociation constant (K_d) of Glycan B with respect to the immobilized lectin is calculated from the basic equation of FAC:

$$[A]_0(V - V_0) = B_t[A]_0/([A]_0 + K_d), \tag{14.9}$$

where B_t is the effective lectin content expressed in mol, and $[A]_0$ is the initial glycan concentration in M. Equation 14.9 can be simplified to Equation 14.10, when $[A]_0$ is negligibly small compared with K_d:

$$K_d = B_t/(V - V_0), \quad \text{if} \quad K_d \gg [A]_0. \tag{14.10}$$

According to Equation 14.10, the $V - V_0$ value is proportional to the affinity of a glycan for an immobilized lectin, where the K_a is the inverse of K_d ($K_a = 1/K_d$). Since PA-glycans used for FAC-FD are of negligible concentration (5.0×10^{-9} M) relative to the K_d values (10^{-3} to 10^{-7} M) of general sugar–protein interactions, the requirement for Equation 14.10 is fulfilled in most cases.

14.3 Automated FAC-FD System

In 2003, a specialized system for FAC-FD, 'FAC-1', was developed in collaboration with Shimadzu in the course of the so-called Structural Glycomics (SG) project supported by New Energy and Industrial Technology Organization of Japan. A photograph of FAC-1 is shown in Figure 14.3c. The system consists of two isocratic pumps, an autosampling system (up to 210 samples), a column oven, two miniature columns connected to a fluorescence detector, and a computer equipped with control (LCsolution) and data analysis (FAC analyzer) software [5, 6]. The parallel column-switching system reduces analysis time by 50% compared to the previous manual injection system, with no substantial 'time loss'; while one column is in analysis, the other is in washing mode. Use of a specially devised miniature column (2×10 mm; bed volume, 31.4 μl) reduces both injection and analysis times as well as the amounts of lectins required for the analysis. Using the FAC-1, the analysis time for 100 sugar–protein interactions is reduced to less than 10 h. Thus, FAC is a high-throughput, powerful and reliable method capable of providing accurate K_d simultaneously for a number of sugar–protein interactions (application of calorimetry is explained in Chapter 13.4).

Using the FAC-1, we undertook comprehensive analysis of the interactions between more than 100 lectins and more than 100 oligosaccharides, referred to as the '100×100' (hect-by-hect) project, along with the SG project mentioned above. To achieve this goal, we performed FAC analyses based on the following scheme. After preparation of the lectin-immobilized columns, the first screening is carried out to validate the prepared column using an initial set of 49 PA-oligosaccharides (designated '49ers'), which represent the glycan structures of a wide range of glycoconjugates. If sufficient binding (retardation in terms of $V - V_0$) is observed for a particular group of glycans, the second screening with a full panel of standard

PA-oligosaccharides is performed for a more detailed and precise analysis. In parallel with these interaction analyses, concentration-dependence analysis is necessary to determine an 'effective ligand content' (B_t) for each column, which is required for the calculation of dissociation constants (K_d) between immobilized lectins and the series of PA-oligosaccharides. We have now obtained quantitative interaction data (in terms of K_d) for more than 120 lectins toward more than 100 PA-oligosaccharides [7]. The collective interaction data is now open for public in the Lectin Frontier database (http://riodb.ibase.aist.go.jp/rcmg/glycodb/LectinSearch).

The relationship between glycan and lectin is analogous to that between a key and a lock (please see Figure 16.1 for examples). In general, however, lectins tend to show broader sugar-binding specificity compared to antibodies, which often have a narrow specificity to antigens (for a view on specificity of plant lectins, please see Table 18.1). This is one basic reason why lectins are generally more useful than antibodies for the purpose of glycan profiling (for details, see Info Box). This broad specificity of lectins is sometimes confusing when evaluating experimental results obtained from lectin-binding studies. Thus, the mannose-specific ConA is not only specific for high-mannose-type structures, but also shows significant affinity to the trimannoside core of biantennary complex-type *N*-glycans (please see Figure 5.4) [1, 8]. In this context, comprehensive FAC data can contribute to provide a useful map to find a key structural element required for each lectin recognition.

Info Box

Why are lectins better than antibodies for the purpose of glycan profiling? Since glycan structures are extremely diverse and antibody production toward glycans is basically difficult, it seems at this time impossible to prepare a complete set of antiglycan antibodies. As yet, the size of the glycome is likely to exceed 10^4 structures including glycoproteins, glycolipids and glycosaminoglycans of mammalian and non-mammalian origins. If bacterial glycans such as those found in lipopolysaccharides are included, the likely glycome size will rise to about 10^5. It should also be taken into consideration that glycan structures are significantly affected by the structures of backbone moieties (core protein, lipid, etc.) as well as their own densities and association states on the cell surface. On the other hand, such actual differences are considered to be discriminated in a living system most possibly through lectin–glycan interaction machineries. Most currently available lectins are from plants. They have proved useful for profiling purposes in various biological and glycobiological fields (for specificity data, see Table 18.1). If commercial plant lectins prove not to be satisfactory for glycan profiling, endogenous animal lectins or artificially engineered probe lectins will be tested to draw physiological conclusions on bioactivity and biomarker status (see Chapter 25; for recent application of plant and human lectins together with antibodies, please see G. Patsos *et al.* Compensation of loss of protein function in microsatellite-unstable colon cancer cells (HCT 116): gene-dependent effect on cell surface glycan profile. *Glycobiology* 2009, 19, 726–734.

14.4
From 'Lectin Profiling' to 'Glycan Mapping'

In the above sections, we described high-throughput FAC analysis of lectin–glycan interactions. Through accumulation of critical information on 'lectin profiles', which contain extensive fundamental data on how individual lectins interact with extensive glycans, we have demonstrated that lectins can discriminate various forms of structural isomers (i.e., enantiomers, diastereomers, anomers) at the monosaccharide level, as well as linkage and positional isomers at a glycan level (for details on carbohydrate structure, please see Chapter 1). Thus, lectins can be useful for the task of glycan profiling. The use of lectins shows significant advantages over other analytical methods, such as MS (for details on MS in glycan analysis, please see Chapter 5.5), as discrimination between the above isomers is relatively easily achieved on the basis of biological rather than physicochemical principles. Advanced studies on lectin–glycan interactions have thus pointed to a turning point from the previous lectin-profiling to a new 'glycan-mapping' era.

The features of such a 'paradigm shift' can be described using a model example with four lectins (designated here I, II, III, and IV) and four glycans (A, B, C, and D). Binding affinities of each of the four lectins to a set of the four glycans are expressed in a three-dimensional bar graph in terms of relative affinity (see the upper left panel of Figure 14.4). The binding patterns of the four lectins to the four glycans are different and, thus, are clearly distinguishable in terms of 'lectin profiles'. Once the binding patterns of an extensive range of lectins are registered in a database, an appropriate lectin profile can be searched for as necessary. Such a database can provide different 'glycan profiles' in terms of relative affinities towards the four lectins (see the upper right of Figure 14.4). Although this is a model explanation using a few lectins and glycans, it is of practical worth in substantiating lectin-based glycan profiling, since lectins are expected to span the structural diversity of glycans (as described in the Info Box). Such glycan profiling has now been realized in terms of the lectin microarray. This emerging technique utilizes a series of (more than 10) lectins immobilized on an appropriate substrate, e.g., a glass slide or a plastic plate (see bottom of Figure 14.4), enabling simultaneous lectin–glycan interaction analysis in a rapid, sensitive and high-throughput manner never previously achieved [9]. The method is also exceptional in its applicability to many medical fields – demonstrating that the lectin microarray is not only an emerging technique for glycan profiling, but also a new tool for the exploration of other aspects of glycomics.

14.5
Lectin Microarray Enables Multiplexed Lectin–Glycan Interaction Analysis

A combination of multiple lectins with different specificities is essential to achieve the profiling of the diverse structures exhibited by glycans. In principle, as the multiplicity of lectins increases, both accuracy and coverage will increase.

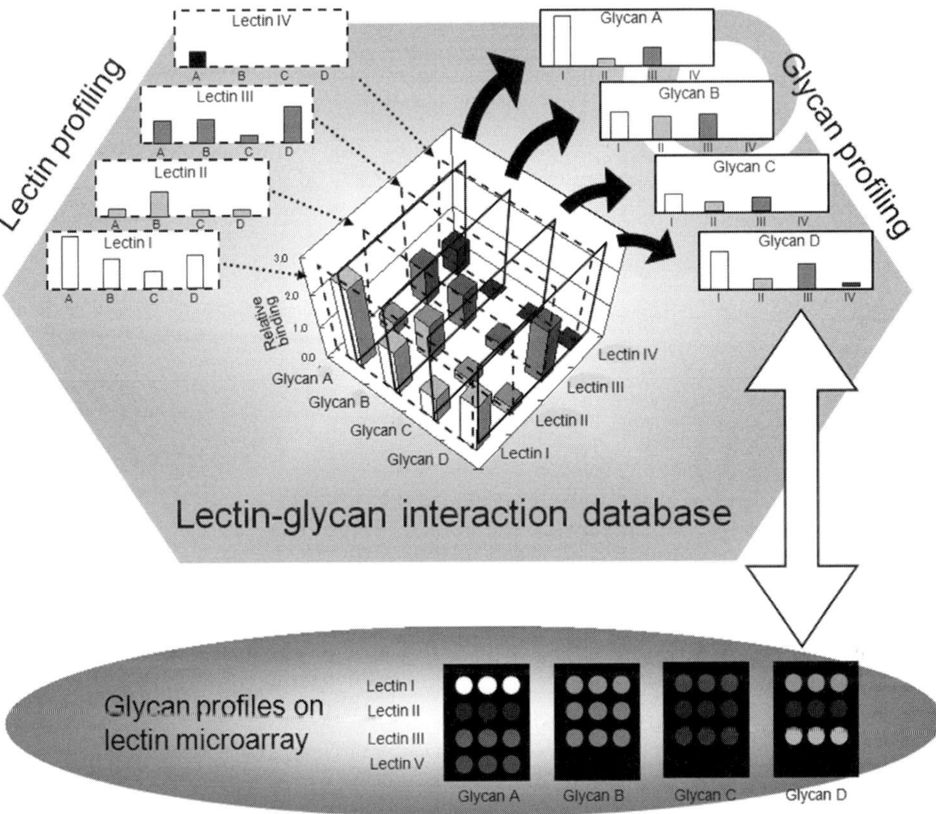

Figure 14.4 Schematic representation of the relationships between lectin profiling and glycan profiling. A model example with four lectins (designated here I, II, III and IV) and four glycans (A, B, C and D) is shown. Binding affinities of each of the four lectins to a set of the four glycans are expressed in a three-dimensional bar graph in terms of relative affinity. The binding pattern of each of the four lectins to the four glycans is the 'lectin profile' (upper left panel), whereas the signal pattern of each of the glycans to the four lectins is the 'glycan profile' (upper right panel).

An approach using the lectin microarray would therefore be expected to succeed in targeting both *O*-glycans and *N*-glycans (for details on structures of *N*- and *O*-glycans, please see Chapters 6–8). The lectin microarray strategy is distinct in several aspects from other conventional methods. Unlike MS and HPLC (including FAC), the lectin microarray does not require prior glycan liberation. This is a great advantage for non-specialized researchers [10, 11]. Moreover, the method is directly applicable not only to purified glyco-materials (such as a glycoprotein), but also to crude samples containing various glycoconjugates, such as cell lysates, body fluids (sera, urine) and bacteria [12–15]. Through differential profiling by lectin microarray, structural differences can be directly detected as the changes in signal patterns on the lectin microarray. The system is useful for the quality control of various glycoprotein products (e.g., antibody drugs). Another promising application is in the differential analysis of clinical samples to investigate glycan-related biomarkers (described later).

Nevertheless, there is a basic problem with conventional lectin microarrays, which, as with other microarray techniques, generally require a prior washing process to remove unbound fluorescence probes (Figure 14.5a). However, as affinities of lectin–glycan interactions are relatively weak (in terms of K_d, 10^{-4} to 10^{-7} M), approximately two orders of magnitude lower than those for antigen–antibody (10^{-6} to 10^{-9} M), lectin–glycan complexes are likely to be easily dissociated by the microarray washing processes usually applied. This results in significant reduction in the signal intensity. If such a washing process could be eliminated, much stronger signals should be obtained. Accordingly, we have recently developed a novel lectin microarray system based on an evanescent-field assisted FD principle [10]. In this system, an evanescent wave is used to directly excite a fluorescent group covalently linked to a glycoconjugate such as a glycoprotein or a glycopeptide. When the fluorescence-labeled glycoconjugates bind to the lectins immobilized on the array, these molecules come close to the array surface. The excitation

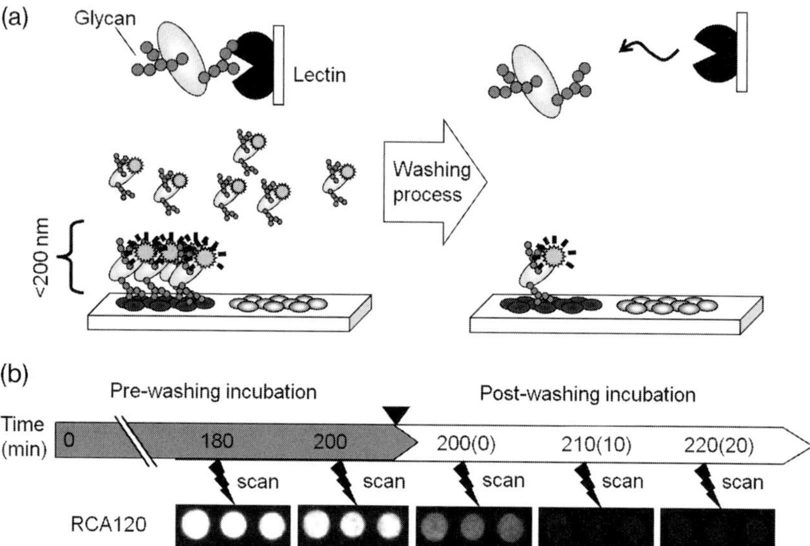

Figure 14.5 (a) Effect of washing steps to remove unbound glycoproteins on a lectin microarray. Since lectin–glycan interactions are relatively weak (in terms of K_d, 10^{-4} to 10^{-7} M), washing steps to remove unbound glycoproteins could significantly reduce binding signals on the lectin microarray. The evanescent-field fluorescence-assisted scanner requires no washing steps before detection and thus even weak signals can be sustained, unlike a confocal-type scanner used for DNA microarray that requires washing steps to remove unbound molecules for detection. An evanescent wave is propagated only within a wavelength distance from the sensor surface (below 200 nm) so that only bound fluorescently labeled glycoproteins can be excited. (b) Binding of TMR-labeled biantennary galactosylated *N*-glycan on the spots of the plant lectin RCA120 is shown before and after washing processes.

light is injected from both sides of the glass slide, so that the generated evanescent wave is emitted from the glass surface to an extremely restricted space, within approximately 200 nm. As a result, only the bound fluorescent molecules are substantially excited by this near-optic-field excitation principle. Good signal-to-noise ratios are obtained, without washing, in a manner preserving binding equilibrium. Assuming that the above principle is valid, it should be possible to detect weak interactions that other methods cannot reveal.

To test this method, tetramethylrhodamine (TMR)-labeled biantennary galactosylated *N*-glycan (generally called NA2) was applied to the lectin microarray (for *N*-glycan structures, please see Chapters 6 and 8) [16]. After incubation, the glass slide was directly measured by the evanescent-type scanner. The fluorescent signal was specifically observed on *Ricinus communis* agglutinin (RCA120) (see the first image on the left in Figure 14.5b), which is known to recognize the Lac/LacNAc unit. The RCA120 intensity was preserved even when the scan was made later (the second image from the left in Figure 14.5b). On the other hand, if the incubation bath (well on the glass slide) was rinsed with buffer, the detectable signals dramatically decreased in intensity (the third image from the left in Figure 14.5b). This observation unambiguously shows that the developed system has the great advantage of being able to detect relatively weak interactions in an equilibrium state.

14.6 Practice in Differential Glycan Profiling: Approaches and Applications

To observe lectin–glycan interactions in terms of fluorescent signals on the lectin microarray, two approaches to fluorescence labeling of glycoconjugates can be used, direct and indirect methods. In the direct-labeling method, appropriate fluorescent reagents (e.g., Cy3-succinimidyl ester) are used to label target glycoconjugates, directly prior to a binding reaction. For this purpose, various chemical cross-linkers (e.g., succimidyl ester and maleimide) are commercially available, which enable profiling of a mixture of glycoproteins from whole-cell lysates. It is a general consensus of glycobiologists that each cell type (origin of species and tissue) and state (differential stage and malignancy) directs different expression of glycoconjugates on cells. In other words, each cell is defined by its own glycome. Based on this concept of differential glycomics, the lectin microarray approach can be used efficiently to discriminate complex and heterogeneous features of glycans expressed on cells [14, 15]. As an application of lectin microarray, a novel procedure was developed for direct cell analysis, in which the fluorophore Cell-Tracker Orange CMRA is incorporated and metabolized to generate a fluorescent compound (Figure 14.6) [17]. This method enables direct profiling of cell-surface glycans. The developed methodology should also contribute to the elucidation of physiological functions of endogenous animal lectins in the context of 'real glycomics' (please see also Table 25.1 and [18] for design of glycan arrays to analyze lectin specificity).

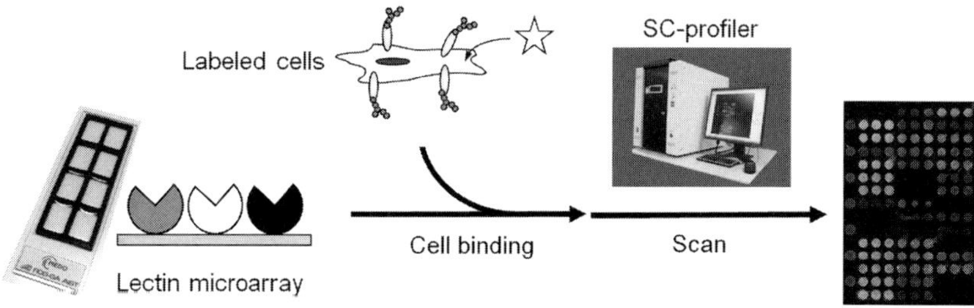

Figure 14.6 A strategy for live cell glycome profiling. Fluorescently labeled live cells can be directly profiled by the scanner in the liquid phase.

14.7 Conclusions

A central issue in the study of lectins has always been elucidation of their sugar-binding specificity, which is closely associated with their physiological functions. In this sense, the history of lectin research can be regarded as that of advancement in technologies for the investigation of lectin–carbohydrate interactions. Beginning with the most commonly used method, the hemagglutination assay, various methods have been adapted to glycotechnology. A robust technique is FAC, which was first introduced by Kasai as a quantitative chromatographic method for biomolecular interactions. This simple technique is refined by strategic combination of a HPLC technique with an FD system using PA-oligosaccharides (synthetic access by strategies outlined in Chapter 3). As a result, lectin–carbohydrate interaction analyses are feasible in a high-throughput manner. Thus, the concept of glycan profiling has become a reality by means of lectins, which had previously been regarded as merely 'useful tools', as a supplement to glycan analysis (please see also Chapter 25.2, especially Figure 25.2, for an example from tumor biology). As tool in lectin microarrays for lectin-based glycomics they play important roles in studies of glycoproteins and cells, and in biomarker discovery, with specificity analysis performed by glycan microarrays.

Summary Box

Conjugation of lectins to a resin does not only enable glycan purification. Frontal affinity chromatography (FAC), together with calorimetry and other binding assays, is a systematic means to investigate lectin specificity and to determine K_d values. FAC thus contributes to the construction of a fundamental lectin database–an essential prerequisite for the second stage (the lectin microarray). The lectin microarray is an emerging technique for glycome analysis with medical perspective. It requires no particular skill on the part of non-specialized researchers.

References

1 Mega T, Hase S. Determination of lectin–sugar binding constants by microequilibrium dialysis coupled with high performance liquid chromatography. *J Biochem* 1991;109:600–3.

2 Kasai K, Ishii S. Quantitative analysis of affinity chromatography of trypsin. A new technique for investigation of protein–ligand interaction. *J Biochem* 1975;77: 261–4.

3 Ng ES *et al.* Frontal affinity chromatography–mass spectrometry. *Nat Protoc* 2007; 2:1907–17.

4 Hirabayashi J *et al.* Frontal affinity chromatography as a tool for elucidation of sugar recognition properties of lectins. *Methods Enzymol* 2003;362:353–68.

5 Tateno H *et al.* Frontal affinity chromatography: sugar–protein interactions. *Nat Protoc* 2007;2:2529–37.

6 Nakamura-Tsuruta S *et al.* Comparative analysis of carbohydrate-binding properties of two tandem repeat-type Jacalin-related lectins, *Castanea crenata* agglutinin and *Cycas revoluta* leaf lectin. *FEBS J* 2005;272:2784–99.

7 Hirabayashi J *et al.* Oligosaccharide specificity of galectins: a search by frontal affinity chromatography. *Biochim Biophys Acta* 2002;1572:232–54.

8 Ohyama Y *et al.* Frontal affinity chromatography of ovalbumin glycoasparagines on a concanavalin A–Sepharose column. A quantitative study of the binding specificity of the lectin. *J Biol Chem* 1985;260: 6882–7.

9 Hirabayashi J. Lectin-based structural glycomics: glycoproteomics and glycan profiling. *Glycoconj J* 2004;21:35–40.

10 Kuno A *et al.* Evanescent-field fluorescence-assisted lectin microarray: a new strategy for glycan profiling. *Nat Methods* 2005;2: 851–6.

11 Rosenfeld R *et al.* A lectin array-based methodology for the analysis of protein glycosylation. *J Biochem Biophys Methods* 2007;70: 415–26.

12 Zheng T *et al.* Lectin arrays for profiling cell surface carbohydrate expression. *J Am Chem Soc* 2005;127:9982–3.

13 Hsu KL *et al.* Analyzing the dynamic bacterial glycome with a lectin microarray approach. *Nat Chem Biol* 2006;2:153–7.

14 Ebe Y *et al.* Application of lectin microarray to crude samples: differential glycan profiling of lec mutants. *J Biochem* 2006;139: 323–7.

15 Pilobello KT *et al.* A ratiometric lectin microarray approach to analysis of the dynamic mammalian glycome. *Proc Natl Acad Sci USA* 2007;104:11534–9.

16 Uchiyama N *et al.* Development of a lectin microarray based on an evanescent-field fluorescence principle. *Methods Enzymol* 2006;415:341–51.

17 Tateno H *et al.* A novel strategy for mammalian cell surface glycome profiling using lectin microarray. *Glycobiology* 2007;17: 1138–46.

18 Song X *et al.* Novel fluorescent glycan microarray strategy reveals ligands for galectins. *Chem Biol* 2009;16:36–47.

15
The History of Lectinology

Harold Rüdiger and Hans-Joachim Gabius

A salient precondition for progress and uncomplicated communication in research is the general agreement on definitions and terms. As the headline attests, the term 'lectin' is commonly accepted. This chapter will explain how this term is rooted in the experimental work of the field's pioneers. The impressive demonstration of the capacity of agglutinins for se**lectin**g distinct blood-group epitopes, like an antibody, was a cornerstone for lectins to broadly enter laboratories as tools. Further discoveries, for example lectin-dependent cell stimulation, paved the way to realizing the enormous scope of lectin–carbohydrate interactions documented in this book. Our historical survey, summarized in Table 15.1, focuses on early developments leading to coining the term. Selected more recent advances which have spurred the publication rate, as a measure of the field's momentum, are also included in Table 15.1. It presents a graphic overview of the chronology of how lectinology developed.

15.1
How Lectinology Started

A seminal report of lasting impact was published in 1860. Therein, the physician S. W. Mitchell gave a detailed account on what happened when performing the following experiment: 'One drop of venom was put on a slide and a drop of blood from a pigeon's wounded wing allowed to fall upon it. They were instantly mixed. Within three minutes the mass had coagulated firmly, and within ten it was of arterial redness' [1]. In the catalog of his works printed in 1894, Mitchell described the situation encountered during the preparation of the cited treatise as follows: 'this quarto with its many drawings was the result of four years of such small leisure as I could spare amidst the cares of constantly increasing practice. The story of the perils and anxieties of this research, embarrassed by want of help and by its great cost, is untold in its pages'.

The Sugar Code. Edited by Hans-Joachim Gabius

ISBN: 978-3-527-32089-9

Table 15.1 Brief historical account of lectinology (from [11], extended, updated and modified).

1860	Observation of blood 'coagulation' by rattlesnake (*Crotalus durissus*) venom (S.W. Mitchell)
1888	Detection of erythrocyte agglutination by a toxic protein fraction from castor beans (termed ricin) and seeds of related plants (H. Stillmark)
1890	Detection of a toxic lectin in the bark of black locust (*Robinia pseudoacacia*) (O. Power, O. Cambier)
1891	Toxic plant agglutinins applied as model antigens (P. Ehrlich)
1898	Introduction of the term 'haemagglutinin' or 'phytohaemagglutinin' for plant proteins that agglutinate red blood cells (M. Elfstrand)
1902	Detection of bacterial agglutinins (R. Kraus, S. Ludwig) and demonstration that blood 'coagulation' by snake venom observed in 1860 (seven to nine decades later shown to depend on the presence of a C-type lectin) was due to cell agglutination but not to blood clotting (S. Flexner, H. Noguchi)
1906	Detection of an agglutinin in bovine serum (later characterized as the C-type lectin conglutinin) by use of activated complement-coated erythrocytes (J. Bordet, F.P. Gay); detection of a hemagglutinin in mushrooms (*Amanita* spp.) (W.W. Ford)
1907/09	Detection of nontoxic agglutinins in plants, of their nature as proteins and of 'deagglutination' of erythrocytes by hog gastric mucin (K. Landsteiner, H. Raubitschek)
1913	Use of intact cells for the purification of the agglutinin ricin (R. Kobert)
1919	Crystallization of a globulin from jack bean, concanavalin A, which was later defined as lectin and used in pioneering studies (please see below) (J.B. Sumner)
1935/36	Concanavalin A identified as jack bean hemagglutinin; precipitation of starch, glycogen and mucins by concanavalin A defines carbohydrate as ligand and points to 'a carbohydrate group in a protein' as binding partner on erythrocytes (J.B. Sumner, S.F. Howell)
1941	Detection of viral agglutinins (G.K. Hirst, L. McClelland, R. Hare)
1944	Description of anti-0(H) hemagglutinating activity in serum of *Anguilla anguilla* (B. Jonsson), following earlier work on a similar activity in *Anguilla japonica* (S. Sugishita; 1935)
1947/48	Detection of plant agglutinins specific for the human histo-blood group A (W.C. Boyd, K.O. Renkonen), 'good keeping qualities' and low 'cost of producing them' are emphasized as advantageous properties
1952	Carbohydrate nature of histo-blood group H(0) determinant proven by eel serum-mediated agglutination of respective erythrocytes and its α-L-fucose-dependent inhibition (W.M. Watkins, W.T.J. Morgan)
1954	Introduction of the term 'lectin' for plant (antibody-like) agglutinins, primarily for those which are specific for a distinct histo-blood group (W.C. Boyd)
1956	Detection of an agglutinin specific for the human blood group B in the seeds of the African shrub *Griffonia (Bandeiraea) simplicifolia* (O. and P. Mäkelä)

continued

Table 15.1 *Continued*

1960	Detection of the mitogenic potency of lectins toward lymphocytes (P.C. Nowell)
1963	Introduction of affinity chromatography for the isolation of lectins, published in 1965 (I.J. Goldstein, B.B.L. Agrawal)
1968–1974	Detection of rapid serum clearance of asialoceruloplasmin in rabbits and isolation of a Gal/GalNAc-specific lectin (asialoglycoprotein receptor) from liver – the first mammalian lectin (G. Ashwell, A.G. Morell *et al.*)
1972	Determination of the amino acid sequence and the three-dimensional structure of a lectin – concanavalin A or ConA (G.M. Edelman, K.O. Hardman, C.F. Ainsworth *et al.*)
1972–1977	Detection of impaired synthesis of a marker for glycoprotein (lysosomal enzymes) routing as cause for a human disease (mucolipidosis II) and its identification as mannose-6-phosphate, the ligand for P-type lectins (E.F. Neufeld *et al.*; W.S. Sly *et al.*)
1978	First conference focusing on lectins and glycoconjugates, termed Interlec (T.C. Bøg-Hansen)
1979	Detection of endogenous ligands for plant lectins (H. Rüdiger)
1981–1988	Further refinements of the definition of the term 'lectin' as carbohydrate-binding protein, separated from antibodies and carbohydrate-processing enzymes/sensor or transport proteins for free sugars (S.H. Barondes, I.J. Goldstein, J. Kocourek *et al.*)
1982	Introduction of serial lectin affinity chromatography as analytical tool for structural analysis of glycans from glycoproteins (R.D. Cummings, S. Kornfeld)
1983	Detection of the insecticidal action of a plant lectin (L. L. Murdock)
1984	Isolation of lectins from tumors (H.-J. Gabius; R. Lotan, A. Raz)
1985	Immobilized glycoproteins as pan-affinity adsorbents for lectins (H. Rüdiger)
1989	Detection of the fungicidal action of a plant lectin (W.J. Peumans)
1992/93	Detection of impaired synthesis of lectin (selectin) ligands by defective fucosylation as cause for leukocyte adhesion deficiency type II, a congenital disorder of glycosylation (CDG IIc) (A. Etzioni *et al.*)
1995	Structural analysis of a lectin–ligand complex in solution by nuclear magnetic resonance spectroscopy (J. Jiménez-Barbero *et al.*)
1996–2003	Detection of differential conformer selection by plant, bacterial and animal lectins (H.-J. Gabius *et al.*; L. Poppe *et al.*)
2001–2005	Development of glycan/lectin microarrays for specificity analysis of lectins/structural analysis of glycans and glycoproteomics (various laboratories worldwide)
2001–2009	Advances in lectinology and glycosciences honored by devoting special issues in *Advanced Drug Delivery Reviews, Biochemical Society Transactions, Biochimica et Biophysica Acta, Biochimie, Biological Chemistry, Cells Tissues Organs, Chemical Reviews, Current Opinion in Structural Biology, Glycoconjugate Journal, Immunology and Cell Biology, Journal of Agricultural and Food Chemistry* ('Liener symposium'), *Journal of Proteome Research, Nature, Proteomics, Science* and *Seminars in Cancer Biology* to these topics

His later experiments using washed erythrocytes and then studies by S. Flexner and H. Noguchi, which S. W. Mitchell himself inspired, revealed that the noted 'coagulation' did not result from procoagulants (clotting factors) in blood. As stated by these authors, 'the value of the use of washed corpuscles comes especially from the fact that the suspension of lytic phenomena is eliminated. Agglutination, therefore, may be studied purely' [2]. The 'venom-agglutination' was especially strong with rabbit erythrocytes, with swine and ox cells being less susceptible, and akin to the reaction with 'intermediary bodies' (today known as immunoglobulins) [2]. Its biochemical nature was defined in 1984 after purifying the agglutinin of *Crotalus durissus* venom, a C-type lectin [3]. This feat underscored the pioneering character of S.W. Mitchell's research for lectinology. Groundbreaking as it was, the spirit and attitude with which he carried out his work, as captured on p. 1 of his report [1], also continue to set a commendable example: 'for the researches which form the novel part of the following essay, I claim only exactness of detail and honesty of statement. Where the results have appeared to me inconclusive, and where further experimental questioning has not resolved the doubt, I have fairly confessed my inability to settle the matter. This course I have adhered to in every such instance, thinking it better to state the known uncertainty thus created than to run the risk of strewing my path with errors in the garb of seeming truths'. The range of study was extended to plants as sources for agglutinins by a medical thesis in 1888 (Table 15.1).

Using the same technique of hemagglutination, extracts of plant seeds, initially from the toxic castor beans (*Ricinus communis*), were also shown to be active. H. Stillmark described a toxin with agglutinating activity, termed ricin, in his MD thesis in 1888. This thesis was prepared under the guidance of R. Kobert at the University of Dorpat (now Tartu) in Estonia, then belonging to the Russian empire [4]. He defined 'ricin' as a protein ('Eiweisskörper, sog. Phytalbumose'), conglomerating (or agglutinating) the red blood corpuscles in defibrinated serum-containing blood ('Zusammenballung der rothen Blutkörperchen') [4]. Such an activity was also found in *Abrus precatorius* seeds ('abrin'), the respective protein fractions from these extracts were produced and tested in R. Kobert's pharmacological institute. Preparations of ricin and abrin, made commercially available by the Merck Company in Darmstadt on Kobert's initiative, found immediate use beyond that of lectinology. They substituted for bacterial toxins in P. Ehrlich's fundamental studies on the immune response [5]. Once that activity was measured and the degree of purity increased, the need for a name became obvious.

15.2 Early Definitions

On the initiative of R. Kobert, who left Dorpat for Germany in January 1897 due to the russification of the Baltics unleashed by an attempted assassination on Czar Alexander III, the issue of determining a name was addressed in 1898. M. Elfstrand introduced the term 'Haemagglutinin (Blutkörperchenagglutinin)' into the literature [6]. He also noted the 'striking similarity' between agglutinating proteins from plants and from human/animal sera [6]. Indeed, exactly this period was characterized by an equally dynamic development in serology. It led to the recognition of the AB0 blood group system based on detecting and monitoring isoagglutination (for historical reviews, please see [7, 8]). The discovery of the agglutination and lysis of erythrocytes by serum compounds is especially linked with three investigators, A. Creite (a medical student in Göttingen in 1869), L. Landois (director of the Physiological Institute at the University of Greifswald in 1875) and K. Landsteiner (a Nobel Laureate in 1930, who in 1900 was working at the Institute for Pathological Anatomy in Vienna) [7, 8]. Their studies and the work on plant agglutinins (K. Landsteiner referred to them as 'Normalantikörper' – normal antibodies) revealed that the two protein classes also shared: (i) activity as precipitin, (ii) selectivity for erythrocytes from different species, and (iii) inhibition of agglutination by haptens (the Info Box 2 in Chapter 1 tells the story how an agglutinin from eel serum was instrumental in defining α-L-fucose as a structural determinant of the H epitope). On these grounds it seemed rather obvious to look at plant (and other) agglutinins as antibody-like proteins. That they are capable of selecting epitopes figures as a central factor in coining a suitable generic term.

15.3 The Current Definition of the Term 'Lectin'

Work on blood-group-specific proteins laid the foundation for the new definition (Table 15.1). It was given by W.C. Boyd in 1954 as follows: 'it would appear to be a matter of semantics as to whether a substance not produced in response to an antigen should be called an antibody even though it is a protein and combines specifically with a certain antigen only. It might be better to have a different word for the substances and the present writer would like to propose the word *lectin* from Latin *lectus*, the past principle of *legere* meaning to pick, choose or select' [9]. The author thus intended – in his own words – 'to call attention to their specificity without begging the question as to their nature' [10]. Specificity for lectins today means binding activity for sugars. Since lectins are not alone in exhibiting this property, nowadays they have to be strictly distinguished from (i) carbohydrate-specific immunoglobulins, (ii) enzymes using carbohydrates as substrates (glycosyltransferases, glycosidases and any enzymes introducing or removing substitutions such as sulfotransferases/sulfatases) and (iii) sensor/carrier proteins for free mono- or oligosaccharides [11].

Historically, selectivity of lectins for glycans was unambiguously delineated first in the case of concanavalin A (ConA) by J.B. Sumner and S.F. Howell in 1936 (for the protocol how to prepare the crystalline lectin, please see Info Box in Chapter 16) [12]. These authors surmised that dissolving concanavalin A in solutions with sucrose may have masked this activity in previous studies. In this context, K. Landsteiner and H. Raubitschek had observed 'deagglutination' of erythrocyte aggregates (formed by ricin, abrin or bean extracts) by hog gastric mucin in 1909 [13]. This observation, viewed in retrospect, can now be interpreted as early evidence for the carbohydrate-binding activity of the tested lectins. In fact, mucins, actually the first protein types detected to be conjugates with sugars in 1865 ('... *daß das Mucin einen gepaarten Stoff darstelle'* [that the mucin may represent a conjugated compound]) [14, p. 206], are potent lectin binders exploitable in one-step purification of lectins (please see Chapter 18.3). Initially, affinity chromatography for lectins used cross-linked dextran (Sephadex®) as a matrix for the isolation of concanavalin A [15]. The enormous potential of this strategy was not immediately realized. A reviewer judged the report after its initial submission to represent 'a modest advance in an obscure area' [16]. In effect, this technical advance was instrumental in markedly increasing the publication activity in this field, yielding a surge in the number of papers on new lectins and applications [17]. The weak reactivity to cross-linked agarose of the Charcot-Leyden crystal protein is of interest in this respect, because it indicates reactivity for sugars of this protein, whose typical hexagonal bipyramidal crystals were first described in the post-mortem spleen of a leukemia patient in 1853 [18] and in sputum of asthmatics in 1872 [19]. Autocrystallization *in situ* of this protein, which constitutes 7–10% of the protein content of a mature blood eosinophil (about 8.5 pg/cell), may thus be considered as a physiological purification step.

The technical breakthrough of establishing affinity chromatography in lectin research also opened the door for the first experimental purification of a mammalian lectin (Table 15.1) [20]. How its presence and functional significance was traced is recounted in the next section.

15.4 Recent Developments

This first purification of a mammalian lectin was part of a line of research intended to elucidate the role of ceruloplasmin in maintaining the copper level and details on the metabolism of this transport protein for copper ions. As a reagent to address these issues, a radioactive form of the glycoprotein was produced by tritiation. This reaction required desialylation of *N*-glycan chains (for structural details on these chains, please see Chapter 6) and oxidation of galactose at its C6 atom, catalyzed by galactose oxidase [21]. Amazingly, the performed engineering of the *N*-glycan chains, which unmasked galactose residues, was not without consequences. It dramatically altered ceruloplasmin's serum clearance: "evidence is presented to show that, in contradistinction to homologous, native ceruloplasmin, which survives for days in the plasma of rabbits, intravenously injected asialoceruloplasmin disappears from the circulation within minutes and accumulates simultaneously in the parenchymal cells of the liver. The rapidity of this transfer of asialoceruloplasmin from plasma to liver has been shown to be dependent upon the integrity of the exposed, terminal galactosyl residues" [22]. Thus, desialylation turned *N*-glycans into ligands and, fortunately, the enzymatic oxidation did not impair the bioactivity for the endocytic hepatic C-type lectin (for further information on C-type lectins, please see Chapters 16, 19, 20 and 27). It became a role model for targeted drug delivery by (neo)glycoproteins [23] (for conjugation chemistry in neoglycoprotein synthesis, please see Table 4.1, for further application of neoglycoproteins and examples of sugars as pharmaceuticals, please see Chapters 25 and 28).

What is presently covered by the umbrella term 'lectin' is outlined throughout this book. As exemplarily emphasized in Table 15.1, intriguing clinical correlations between the status of glycosylation and functional implications via lectins have turned up (please see for example Chapters 25.2, 27 and 29). Moreover, the molecular details of glycan recognition are unraveled inspiring drug design, and new technologies for measuring lectin specificity are being developed (please see Chapters 13 and 14). Combined, these examples for dynamic research lines in lectinology afford efficient driving forces, to make sure that this field maintains its currently acquired prominent status, honored by a series of special feature issues (please see bottom of Table 15.1), and its momentum.

15.5 Conclusions

The ability to agglutinate cells is common to different types of proteins. Respective assays with erythrocytes led to an early convergence of work on antibodies and on proteins from diverse sources that are not produced in response to an antigen. In particular, the crucial role of certain hemagglutinins from plants and eel serum in the process of defining the biochemical basis of the AB0 blood group epitopes attest to their selectivity, rivaling that of antibodies. Close inspection of the literature teaches the lesson that this context entailed coining the term 'lectin'. It embodies the aspect of molecular selectivity. At the same time, it separates the agglutinins terminologically from immunoglobulins. As a consequence of the haptenic inhibition of agglutination by sugars, 'lectin' is now the generic name for carbohydrate-specific proteins, differing from (i) antibodies, (ii) enzymes acting on the ligand and (iii) sugar sensor/transport proteins. The current wide scope of structural and functional studies including promising medical perspectives, described throughout this book, ensures that lectinology will properly address the challenges of deciphering the sugar code.

Summary Box

The discovery of nonantibody proteins with inherent selectivity rested on observing agglutination of erythrocytes. The detection of AB0 blood-group-selective activities different from immunoglobulins led to the introduction of the term 'lectin' by W. C. Boyd in 1954. The current definition distinguishes these carbohydrate-binding proteins from sugar-specific immunoglobulins, enzymes acting on the ligand, and sensor/transport proteins for free mono- and oligosaccharides.

References

1 Mitchell SW. Researches upon the venom of the rattlesnake. *Smithson Contrib Knowl* 1860;XII:89–90.

2 Flexner S, Noguchi H. Snake venom in relation to haemolysis, bacteriolysis and toxicity. *J Exp Med* 1902;6:277–301.

3 Gartner TK, Ogilvie ML. Isolation and characterisation of three Ca^{2+}-dependent β-galactoside-specific lectins from snake venoms. *Biochem J* 1984;224:301–7.

4 Stillmark H. Ueber Ricin, ein giftiges Ferment aus den Samen von *Ricinus comm.* L. und einigen anderen Euphorbiaceen. MD Thesis, Kaiserliche Universität zu Dorpat (now Tartu), Schnakenburg's Buchdruckerei 1888.

5 Ehrlich P. Experimentelle Untersuchungen über Immunität. II. Ueber Abrin. *Dtsch Med Wschr* 1891;17:1218–9.

6 Elfstrand M. Ueber blutkörperchenagglutinierende Eiweisse. In: *Görbersdorfer Veröffentlichungen* (Ed.: Kobert R), pp. 1–159. F. Enke-Verlag, Stuttgart, 1898.

7 Hughes-Jones NC, Gardner B. Red cell agglutination: the first description by Creite (1869) and further observations made by Landois (1875) and Landsteiner (1901). *Br J Haematol* 2002;119:889–93.

8 Schwarz HP, Dorner F. Karl Landsteiner and his major contributions to haematology. *Br J Haematol* 2003;121:556–65.

9 Boyd WC. The proteins of immune reactions. In: *The Proteins* (Eds.: Neurath H, Bailey K), pp. 756–844. Academic Press, New York, 1954.

10 Boyd WC. The lectins: their present status. *Vox Sang* 1963;8:1–32.

11 Gabius H-J *et al.* Chemical biology of the sugar code. *ChemBioChem* 2004;5:740–64.

12 Sumner JB, Howell SF. The identification of a hemagglutinin of the jack bean with concanavalin A. *J Bacteriol* 1936;32:227–37.

13 Landsteiner K, Raubitschek H. Ueber die Adsorption von Immunstoffen. V. Mitteilung. *Biochem Z* 1909;15:33–51.

14 Eichwald E. Beiträge zur Chemie der gewebbildenden Substanzen und ihrer Abkömmlinge. I. Ueber das Mucin, besonders der Weinbergschnecke. *Ann Chem Pharm* 1865; 134:177–211.

15 Agrawal BBL, Goldstein IJ. Specific binding of concanavalin A to cross-linked dextran gel. *Biochem J* 1965;96:23c.

16 Sharon N. Lectins: from obscurity into the limelight. *Protein Sci* 1998;7:2042–8.

17 Rüdiger H *et al.* Medicinal chemistry based on the sugar code: fundamentals of lectinology and experimental strategies with lectins as targets. *Curr Med Chem* 2000;7:389–416.

18 Charcot JM, Robin C. Observation de leucocythemie. *C R Mem Soc Biol* 1853;5:44–50.

19 Leyden E. Zur Kenntnis des Bronchialasthma. *Arch Pathol Anat* 1872;54:324–44.

20 Hudgin RL *et al.* The isolation and properties of a rabbit liver binding protein specific for asialoglycoproteins. *J Biol Chem* 1974;249: 5536–43.

21 Morell AG *et al.* Physical and chemical studies on ceruloplasmin. IV. Preparation of radioactive, sialic acid-free ceruloplasmin labeled with tritium on terminal D-galactose residues. *J Biol Chem* 1966;241:3745–9.

22 Morell AG *et al.* Physical and chemical studies on ceruloplasmin. V. Metabolic studies on sialic acid-free ceruloplasmin *in vivo*. *J Biol Chem* 1968;243:155–9.

23 Yamazaki N *et al.* Endogenous lectins as targets for drug delivery. *Adv Drug Deliv Rev* 2000;43:225–44.

16
Ca^{2+}: Mastermind and Active Player for Lectin Activity (Including a Gallery of Lectin Folds)

Hans-Joachim Gabius

A set of monosaccharides is the biochemical equivalent of an alphabet for code word generation by oligomerization (please see Chapter 1), and the resulting messages are decoded by lectins (please see Chapters 13 and 15 for basic aspects, and Chapters 17–19 and 24–30 for lectin functions). To accomplish this feat functional groups of amino acids and water molecules as well as stacking between aromatic and sugar rings cooperate (please see Chapter 13 for details). What's more, Ca^{2+} can be recruited for assignments in sugar binding. This ion exerts its activity in remarkably different ways. This chapter systematically illustrates crystallographic evidence for the various natural strategies of teaming up Ca^{2+} with amino acid residues and water molecules to let the respective protein gain optimal lectin activity. At the same time, the illustrations provide a view on folding of Ca^{2+}-dependent bacterial, plant and vertebrate lectins. We start with the function of Ca^{2+} to structurally organize the lectin site without direct contact to the sugar, calling on the best-known lectin. The classical protocol for its crystallization is given in the Info Box.

16.1
Ca^{2+}: Organizing the Active Site

The agglutinin concanavalin A (ConA) from *Canavalia ensiformis* has a prominent position in the history of lectinology (please see Chapter 15 for details). Its fold is established by an antiparallel β-sandwich evocative of a jelly-roll (a Danish pastry) (Figure 16.1a). ConA is the structural role model for many leguminous lectins (please see Chapter 18 for functions and applications of plant lectins). It requires the presence of a calcium ion and a further metal ion, naturally Mn^{2+}, at distinct preformed sites for lectin activity [1]. Invariably, four amino acids and two water molecules are engaged for contact, and the driving force for the orchestrated structural rearrangements with loop stabilization and a *trans-/cis*-isomerization of a peptide bond is provided upon ion accommodation [2, 3]. As shown in Figure 16.1a, occupancy of the two sites for metal ions, for example, directs the amide

The Sugar Code. Edited by Hans-Joachim Gabius

ISBN: 978-3-527-32089-9

Info Box

The preparation of crystalline concanavalin A: 'Stir 100 grams of finely powdered jack bean meal with 500 cc. of 32-per-cent acetone and filter in an ice chest for 4 or 5 hours, or overnight. Place the residue in a beaker, stir with 500 cc. of 30-per-cent alcohol and filter at room temperature. Now extract the concanavalin A from the residue with 400 cc. of a solution containing 1% sodium chloride and 0.1-per-cent neutral phosphate and filter. Extract the residue with 250 cc. of 5-per-cent sodium chloride and filter. Combine the two last filtrates, add toluene and dialyze against distilled water in three collodion bags for 24 hours. Centrifuge off the crystals of concanavalin A and wash once with 10-per-cent sodium chloride solution. Dissolve the crystals in a small amount of 0.1 N hydrochloric acid. After standing for about one-half hour add a small quantity of neutral phosphate solution and then exactly enough 0.1 N sodium hydroxide to neutralize the hydrochloric acid previously added. Dialyze the material for several hours and centrifuge off the crystals of concanavalin A' [J.B. Sumner, S.F. Howell. The identification of the hemagglutinin of the jack bean with concanavalin A. *J Bact* 1936; **32**, 227–237]. In a previous publication [J.B. Sumner. The globulins of the jack bean, *Canavalia ensiformis*. *J Biol Chem* 1919; **37**, 137–142] the crystals were characterized as follows: 'Dr. A.C. Gill has been kind enough to examine them, and has provisionally declared them to be bisphenoidal in shape, probably belonging to the rhombic system, optically biaxial, with a large optical angle, and negative in optical character.'

nitrogen of Asn14 to the sugar and lets Asp208 flip into its position for contact via acquisition of the Ala207–Asp208 *cis*-peptide bond. Obviously, Ca^{2+} acts as mastermind of structural rearrangements to optimize ligand contact.

The β-sandwich fold is not only encountered in the leguminous agglutinins. Lectins involved in quality control of folding of nascent glycoproteins (calnexin, calreticulin) and in further sorting and transport of maturing glycoproteins in the endoplasmic reticulum–Golgi intermediate compartment (ERGIC) such as ERGIC-53 share this general structure, together with other proteins to be mentioned below (please see Chapter 6.3 for details on quality control). The conspicuous changes in the positions and/or number of Ca^{2+}-binding sites are signs for independent evolutionary ancestry of this apparently useful compact fold. This assumption is backed by a lack of sequence homology among respective proteins (please see also below). Going into details of the number of bound cations and their positioning, the endoplasmic-reticulum-resident calnexin has only one putative Ca^{2+}-binding site. Ca^{2+} is coordinated to Ser75, Asp118 and Asp437 (Figure 16.1b). It is located rather far away from the site interacting with α-D-glucose, as this ligand accepts hydrogen bonding with Tyr165, Lys167, Tyr186 and Glu217 [4]. The role of the calcium ion may in this case thus be more indirect than for ConA.

As mentioned above, the Ca^{2+}-dependent glycoprotein cargo transporter ERGIC-53 (p58) in animals is another β-sandwich protein. This intracellular lectin has two cation-binding sites, but only one can be set into obvious structural relation to the constellation of the respective Ca^{2+}-binding section present in leguminous lectins (Figure 16.1c). However, the detected structural changes upon complexation are reminiscent of those in the mentioned plant proteins, and the selection of complex mannose structures as ligands by ERGIC-53 (please see Chapter 6 for details) may have actually required the intramolecular shift of one site [5]. Of note, the yeast orthologs of ERGIC-53, that is Emp47p and Emp46p, accomplish their mission as glycan cargo receptors without dependence on Ca^{2+}, as the proteins of the galectin family (please see Figures 13.2/13.3 for folding and ligand binding, Chapter 19, for example Table 19.2, for information on functions and Table 19.3 for listing of ligands) and carbohydrate-binding modules (CBM) of polysaccharide-degrading microbial enzymes (CBM families 4, 6 and 22) constitute further cases for Ca^{2+}-independent β-sandwich lectins [6–8]. Evidently, this type of scaffold has turned out to be a highly flexible platform for lectin generation, and this multipurpose character goes along with broadening the role of Ca^{2+} in ligand binding. By turning to the pentraxins (phylogenetically conserved proteins arranged into a disk-like structure of five non-covalently associated protomers), the next aspect of Ca^{2+} functionality will be profiled.

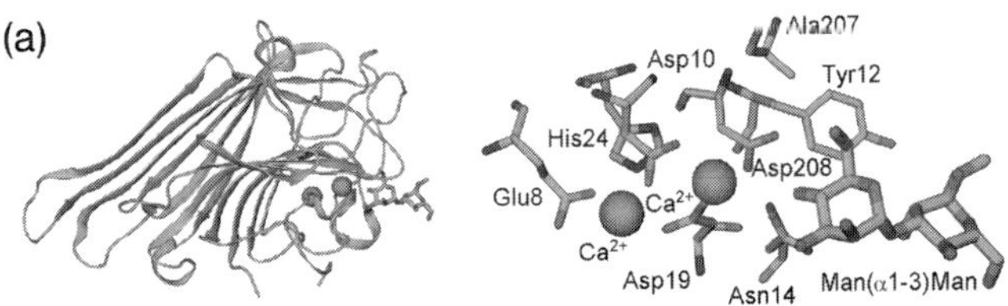

Figure 16.1 Illustration of X-ray crystallographic evidence [with the Protein Data Bank (PDB) codes] for the functional versatility of Ca^{2+} in lectins. In each case, the general folding of the lectin (left) and the anatomy of the Ca^{2+}-binding site(s) or carbohydrate recognition domain (CRD) including the crucial Ca^{2+} with, if available, the sugar ligand (right) are presented: Ca^{2+}-loaded ConA in complex with Manα1-3Man (a; 1DQO), the luminal domain of the canine lectin chaperone calnexin (b; 1JHN), the CRD of the rat cargo-transporting lectin ERGIC-53 (p58/MR60) (c; 1R1Z), one subunit of the human pentraxin serum amyloid P component in complex with MOβDG (d; 1GYK), the fifth laminin G-like module of the mouse laminin α_2-chain (e; 1QU0), human annexin A2 starting at Asn31 in complex with a heparin tetrasaccharide (f; 2HYU; for details of heparin structure and conformations of IdoA, please see Figures 1.6 and 1.7d), the fucose-specific agglutinin from the European eel *(Anguilla anguilla)* (g; 1K12), the construct of CRD/epidermal-growth-factor-like domain of human C-type lectin P-selectin in complex with sialyl Le^x tetrasaccharide (CD62P) (h; 1G1R), the CRD of human C-type lectin dendritic-cell-specific ICAM-3-grabbing non-integrin (DC-SIGN, CD209) in complex with the pentasaccharide lacto-*N*-fucopentaose III (i; 1SL5), soluble extracytoplasmic domain (Asn81/STOP155) of the bovine cation-dependent mannose-6-phosphate receptor in complex with pentamannosyl phosphate (j; 1C39), the *Pseudomonas aeruginosa* lectin I (PA-IL) in complex with galactose (k; 1OKO) and the *Pseudomonas aeruginosa* lectin II in complex with fucose (PA-IIL) (l; 1GZT). Processing of PDB entries kindly provided by H.-C. Siebert.

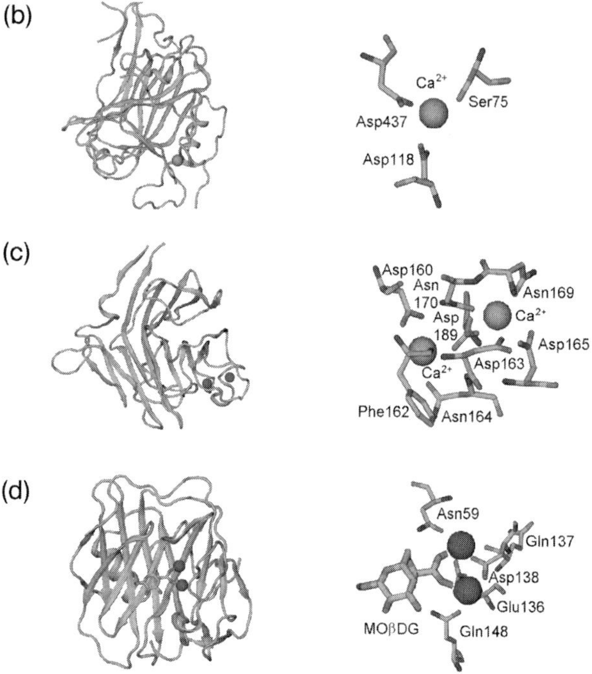

Figure 16.1 *Continued*

16.2 Ca^{2+}: Contacting Charged Ligands

The human pentraxin serum amyloid P component (SAP) contains two Ca^{2+} about 4 Å apart and bridged by amino acid side-chains (Figure 16.1d). As methyl 4,6-*O*-(1-carboxyethylidene)-β-D-galactopyranoside (MOβDG) can displace SAP from amyloid fibers, this substance was used as ligand. Its carboxyl group is in contact to both Ca^{2+} (distances of 2.39 and 2.45 Å, respectively) [9, 10]. Beyond that, Gln148 and Asn59 share assignments in binding Ca^{2+} and galactose (here the O6 atom) [10]. Thus, the calcium ions ensure the correct positioning of side-chain amides and sugar oxygen atoms. We have herewith witnessed an example of a direct contact between Ca^{2+} and a negatively charged group of the ligand. It opens a new aspect for Ca^{2+} functionality beyond its eminent structural role.

Moving from a serum protein to basement membranes, a pertinent protein domain mediating binding to heparin and α-dystroglycan is found in laminin, agrin and perlecan (please see Chapter 11.5). With a root mean square deviation of 2.7 Å relative to SAP, this laminin G-like module is also classified as a β-sandwich protein (Figure 16.1e) [11, 12]. Binding-site positioning has special features: whereas the two Ca^{2+} are at the center of one face of the β-sandwich in SAP (Figure

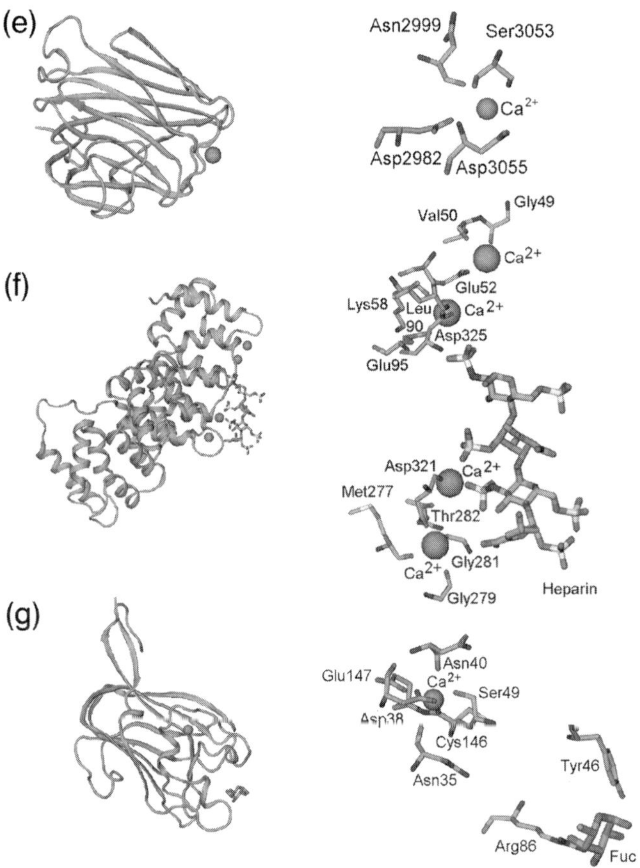

Figure 16.1 *Continued*

16.1d), only one Ca^{2+} is present in this module, and this one at the edge of the β-sandwich (Figure 16.1e). Akin to the situation in leguminous lectins, the Ca^{2+}-binding site appears preformed by four amino acid side-chains, one water molecule and, notably, an inorganic sulfate ion to complete an octahedral coordination [11]. The rather small distance of 2.5 Å between Ca^{2+} and SO_4^{2-} prompts to suspect an association with anionic carbohydrates or α-dystroglycan *in vivo* at this site [11, 12]. The latter interaction is pivotal for the integrity of the dystrophin–glycoprotein complex, because aberrations lead to muscular dystrophy [13] (Figure 22.4; for summary table on plant/animal lectins with β-sandwich fold, please see Table 18.2 and Table 19.1). Heparin is a known ligand for this module (for structure of the anticoagulant pentasaccharide, please see Figure 1.7 (bottom panel); for an overview on proteoglycans, please see Chapter 11 and for therapeutic application, see Chapter 28.5). How this type of glycosaminoglycan and Ca^{2+} act in concert to facilitate tight binding to a protein is illustrated by the case of annexin A2.

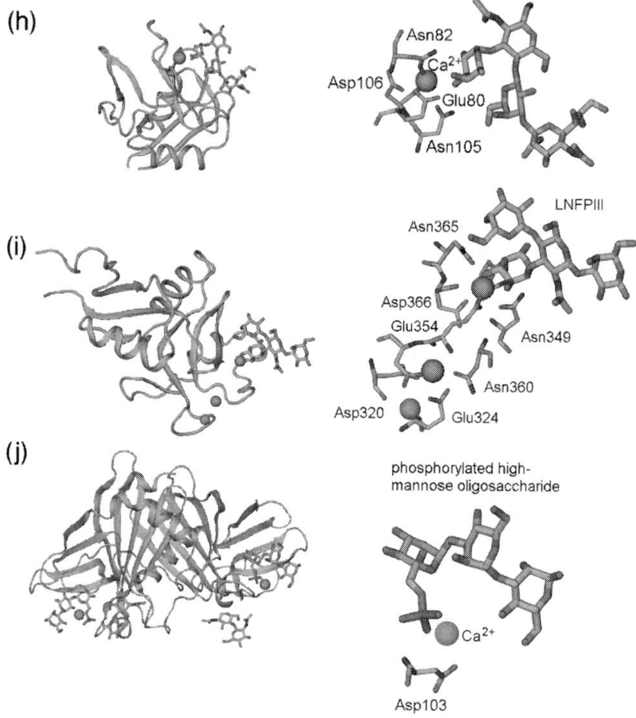

Figure 16.1 *Continued*

A direct ligand contact, as discussed in the previous paragraph, is also essential for the Ca^{2+}-dependent affinity of annexin A2 to heparin oligosaccharide [14]. In this case, two calcium ions, together with main- and side-chain nitrogen atoms, participate to create the equivalent of otherwise common basic clusters for heparin accommodation – one of them hosting two sulfate groups in its coordination shell (Figure 16.1f). Thus, Ca^{2+} in this case functions as a structural organizer and at the same time as a direct binding partner for carbohydrate/sulfate groups. In addition to coming into contact with anionic compounds, the reader will now wonder whether a lectin-presented Ca^{2+} might be engaged to neutralize negative charges of the protein or even to make contact with neutral sugars such as a fucose moiety. Indeed, there is no chemical reason why this should not be the case. As follows, insight is provided that the superfamily of lectins has different examples in store for the role of Ca^{2+} to prevent a clash between anionic charges in ligand and protein and to coordinate oxygen atoms of sugars. The *Anguilla anguilla* (European eel) agglutinin, famous for its key role as a probe to define fucose as building block of the histo-blood group H substance in 1952 (please see Info Box 2 in Chapter 1), appeared to belong to this group. Having completed crystallographic analysis, it yet turned out not to meet this criterion, with at best indirect effects of Ca^{2+} (Figure 16.1g) [15]. Direct contacts, the next topic of this chapter, are detected in other lectin types.

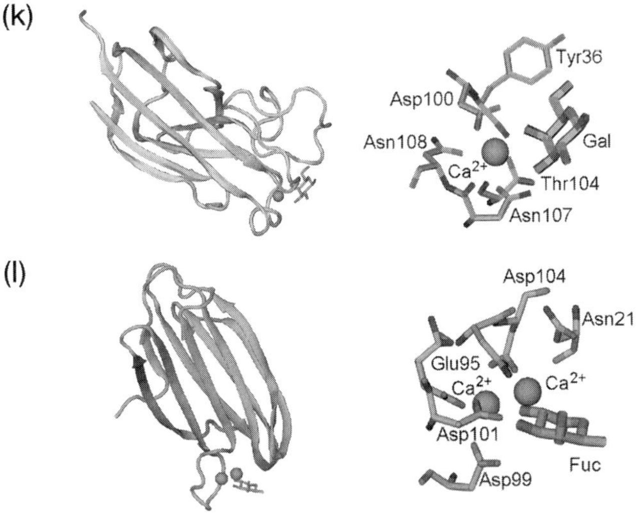

Figure 16.1 *Continued*

16.3
Ca^{2+}: Neutralizing Negative Charges and Contacting Neutral Ligands

The Ca^{2+}-dependent lectin activity has given reason to refer to a large lectin (and lectin-like) family as C-type (please see Chapter 20 for its evolutionary divergence). The common structural trait in each member is a fold with two antiparallel β-strands and two α-helices (Figure 20.1) [16]. A direct contact of a Ca^{2+} to hydroxyl groups underlies this activity. Only the correct topological signature of hydroxyl groups makes the directional coordination bonds possible. Ca^{2+} in P-selectin (for details of C-type lectins in inflammation, please see Chapter 27 and Figure 29.5) accepts the 3- and 4-hydroxyl groups of L-fucose of the sialylated Lewis^x (Le^x) tetrasaccharide and is furthermore an organizer of additional contacts between the ligand and amino acid residues (Figure 16.1h) [17]. This principal Ca^{2+} site fulfills the same role for instance in the related lectin DC-SIGN (Figure 16.1i; for abbreviation, please see legend) [18] and the collectin mannan-binding lectin (Figure 20.3).

A variation of the strategy to engage a divalent cation in ligand binding is realized by the cation-dependent mannose-6-phosphate receptor, a P-type lectin [16, 19]. It routes enzymes with the 'postal code' mannose-6-phosphate to their final destination in lysosomes (please see Figure 1.7a and Chapter 19 for further information on structure and functions). Its extracytoplasmic region was crystallized as a homodimer, visualizing a comparatively deep binding pocket for the ligand with extensive contacts [19]. In this case, Ca^{2+} has a neutralizing effect. The divalent cation comes close to the phosphate group of the ligand and, even more important, shields it from the negatively charged side-chain of Asp103 (Figure 16.1j). It thereby suppresses an electrostatic repulsion within the network of multiple interactions, which explains the quantitative effect of its presence on lectin activity.

Extensive contacts to the carbohydrate ligand are also characteristic of the two Ca^{2+}-dependent lectins of *Pseudomonas aeruginosa* (for further details, please see Chapter 17). The Gram-negative pathogen is infamous for causing morbidity and mortality in cystic fibrosis patients as well as for being the culprit of nosocomial infections in immunocompromised patients. As in C-type lectins, calcium ion(s) are engaged to bind neutral sugars. The two lectins PA-IL (gene *lec*A) and PA-IIL (gene *lec*B) target galactose and L-fucose, respectively (for chair conformations of both sugars, please see Figure 1.6). PA-IL adopts a β-sandwich fold where the contact site for galactose is at the apex [20]. The sole Ca^{2+} has no notable structural role on the protein. Its major function is to participate in the intricate network of contacts which comprises all hydroxyl groups of galactose except for the O1 atom [20]. Presented in the loop constituted by amino acids 100–108, the calcium ion includes the O3/O4 atoms of the sugar ligand into its coordination sphere, the axial hydroxyl group of galactose with its additional hydrogen bonds to the carboxylate of Asp100 serving to distinguish the 'letter' galactose from glucose/mannose residues (Figure 16.1k). The strong involvement of the O6 atom in the bonding network accounts for the separation of this binding mode from that of the C-type lectins mentioned above.

Moving on to the second bacterial lectin PA-IIL, it is special among agglutinins due to its strong binding affinity to monosaccharides in the micromolar range (5.6 to 8.3×10^{-6} M) [21, 22]. Its nine-stranded β-sandwich strategically positions two Ca^{2+}. They are crucial for the high affinity combined with a less pronounced specificity (Figure 16.1l) [20, 22, 23]. Although the jelly-roll motif defined above as a frequently encountered fold is thus shared by both PA lectins, an evolutionary relationship by divergence cannot be traced on the level of sequences. This is reflected, too, when analyzing the mode of sugar binding. In detail, one Ca^{2+} and a network of hydrogen bonds endow PA-IL with specificity to galactose. In contrast, the two calcium ions of PA-IIL figure prominently in ligand contact(s) to fucose. Amazingly, the two cations rather rigidly dock sugar ligands with equatorial/axial hydroxyl groups by virtue of a total of four coordination bonds: O2/O4 atoms of L-fucose to one Ca^{2+} each and the O3 atom to both Ca^{2+}, with O4, O3 and O2 atoms of D-mannose being sterically equivalent to O2, O3 and O4 atom of L-fucose (Figure 16.1l; please see also Figure 1.6) [22, 23]. The conspicuously high affinity stems from an enthalpy-driven process with no entropic penalty [21] (for details on thermodynamics of sugar binding, please see Chapter 13.4). In contrast to the pentraxins no negative charge on the ligand is required; in contrast to PA-IL and to C-type lectins, whose affinity for monosaccharides is in the millimolar range, coordination bonds are established with both Ca^{2+} in a topologically precise manner. They substitute for the otherwise operative van der Waals interactions [24] (please see also Chapter 13).

Table 16.1 The strategic roles of Ca^{2+} in lectin activity.

Function	Type of lectin
Structural role in organizing lectin site (no direct contact to ligand)	Leguminous lectins homologous to ConA, possibly lectin chaperones in quality control (calnexin), animal lectin-type cargo receptor (ERGIC-53) but not yeast orthologs, *Anguilla anguilla* agglutinin
Structural role and direct contact to anionic group(s) of ligand or neutralization of anionic charges	Pentraxins, laminin G-like module (?), annexin A2, cation-dependent mannose-6-phosphate receptor
Direct contact to neutral group(s) of ligand with/without structural role	C-type lectins, *Pseudomonas aeruginosa* lectin I (two coordination bonds), *Pseudomonas aeruginosa* lectin II (four coordination bonds)

16.4 Conclusions

This chapter documents how calcium ions fulfill their mission in lectins. The different strategies are summarized in Table 16.1. This survey inspires the notion that lectin activity might be regulatable by local/transient changes in Ca^{2+} concentration or the lectin's affinity to Ca^{2+}. In this respect, the detection of a pH-sensitive switch for Ca^{2+} and sugar binding in the endocytic asialoglycoprotein receptor of rat liver (please see Chapters 15 and 19 for its function) when exposed to the endosomal pH of 5.4 teaches us a salient lesson on regulation of binding activity by the microenvironment [25]. Beyond lectins, Ca^{2+} is operative in other sugar-binding proteins, such as in enzymes degrading glycosaminoglycans and pectate or glycosyltransferases. Therefore, when talking about sugar-binding activities in lectins and also sugar-processing enzymes, the remarkable talents of Ca^{2+} deserve to be appreciated.

Summary Box

Lectins single out distinct sugar epitopes from the glycomic complexity for binding. This process is often solely determined by functional groups of amino acids and water molecules. In a series of cases, Ca^{2+} is involved. It is a versatile means toward the following ends: organizing the binding site without contact to the ligand, neutralizing the negative charge in mannose-6-phosphate as well as interacting with charged and even neutral ligands.

References

1 Loris R *et al.* Legume lectin structure. *Biochim Biophys Acta* 1998;1383:9–36.
2 Naismith JH, Field RA. Structural basis of trimannoside recognition by concanavalin A. *J Biol Chem* 1996;271:972–6.
3 Bouckaert J *et al.* The structural features of concanavalin A governing non-proline peptide isomerization. *J Biol Chem* 2000; 275:19778–87.
4 Schrag JD *et al.* Lectin control of protein folding and sorting in the secretory pathway. *Trends Biochem Sci* 2003;28:49–57.
5 Velloso LM *et al.* The crystal structure of the carbohydrate-recognition domain of the glycoprotein sorting receptor p58/ERGIC-53 reveals an unpredicted metal-binding site and conformational changes associated with calcium ion binding. *J Mol Biol* 2003;334:845–51.
6 Czjzek M *et al.* The location of the ligand-binding site of carbohydrate-binding modules that have evolved from a common sequence is not conserved. *J Biol Chem* 2001;276:48580–7.
7 López-Lucendo MF *et al.* Growth-regulatory human galectin-1: crystallographic characterisation of the structural changes induced by single-site mutations and their impact on the thermodynamics of ligand binding. *J Mol Biol* 2004;343:957–70.
8 Satoh T *et al.* Structures of the carbohydrate recognition domain of Ca^{2+}-independent cargo receptors Emp46p and Emp47p. *J Biol Chem* 2006;281:10410–9.
9 Emsley J *et al.* Structure of pentameric human serum amyloid P component. *Nature* 1994;367:338–45.
10 Thompson D *et al.* The structures of crystalline complexes of human serum amyloid P component with its carbohydrate ligand, the cyclic pyruvate acetal of galactose. *J Mol Biol* 2002;320:1081–6.
11 Hohenester E *et al.* The crystal structure of a laminin G-like module reveals the molecular basis of α-dystroglycan binding to laminins, perlecan, and agrin. *Mol Cell* 1999;4:783–92.
12 Tisi D *et al.* Structure of the C-terminal laminin G-like domain pair of the laminin α2 chain harbouring binding sites for α-dystroglycan and heparin. *EMBO J* 2000; 19:1432–40.
13 Michele DE, Campbell KP. Dystrophin–glycoprotein complex: post-translational processing and dystroglycan function. *J Biol Chem* 2003;278:15457–60.
14 Shao C *et al.* Crystallographic analysis of calcium-dependent heparin binding to annexin A2. *J Biol Chem* 2006;281:31689–95.
15 Bianchet MA *et al.* A novel fucose recognition fold involved in innate immunity. *Nat Struct Biol* 2002;9:628–34.
16 Gabius H-J. Animal lectins. *Eur J Biochem* 1997;243:543–76.
17 Somers WS *et al.* Insights into the molecular basis of leukocyte tethering and rolling revealed by structures of P- and E-selectin bound to sLe^x and PSGL-1. *Cell* 2000;103: 467–79.
18 Feinberg H *et al.* Structural basis for selective recognition of oligosaccharides by DC-SIGN and DC-SIGNR. *Science* 2001;294: 2163–6.
19 Dahms NM, Hancock MK. P-type lectins. *Biochim Biophys Acta* 2002;1572:317–40.
20 Cioci G *et al.* Structural basis of calcium and galactose recognition by the lectin PA-IL of *Pseudomonas aeruginosa*. *FEBS Lett* 2003; 555:297–301.
21 Mitchell EP *et al.* High affinity fucose binding of *Pseudomonas aeruginosa* lectin PA-IIL: 1.0 Å resolution crystal structure of the complex combined with thermodynamics and computational chemistry approaches. *PROTEINS* 2005;58:735–46.
22 Mitchell E *et al.* Structural basis for oligosaccharide-mediated adhesion of *Pseudomonas aeruginosa* in the lungs of cystic fibrosis patients. *Nat Struct Biol* 2002;9:918–21.
23 Mishra NK *et al.* Molecular dynamics study of *Pseudomonas aeruginosa* lectin-II complexed with monosaccharides. *PROTEINS* 2008;72:382–92.
24 Gabius H-J. The how and why of protein–carbohydrate interaction: a primer to the theoretical concept and a guide to application in drug design. *Pharm Res* 1998;15:23–30.
25 Wragg S, Drickamer K. Identification of amino acid residues that determine pH dependence of ligand binding to the asialoglycoprotein receptor during endocytosis. *J Biol Chem* 1999;274:35400–6.

17
Bacterial and Viral Lectins

Jan Holgersson, Anki Gustafsson, and Stefan Gaunitz

The preceding chapters illustrated that cell-surface glycans are ideal to store biological information, and have described their structural complexity and biosynthesis. Then, it was illustrated how well glycans interact with receptors (for definition of the term 'lectin', see Chapter 15.3, with principles explained in Chapter 13 and crystal structures shown in Chapter 16). What is timely now is to apply this knowledge to the first step in infection and provide information on bacterial/viral lectins.

In this chapter, lectins of bacteria and viruses are described with regard to their structural organization, carbohydrate specificity and distribution between species. Infectious diseases caused by bacteria and viruses involving particular lectin are briefly mentioned. In addition, carbohydrate-binding bacterial toxins are discussed to illustrate versatility of lectin functions. In the last part of this chapter, carbohydrate-based anti-infectives on the market as well as under development are described. The importance of multivalency for high-avidity binding of carbohydrate-based inhibitors (please see Chapter 4 for synthetic aspects of glycodendrimers and also applications), and how it can be accomplished, is discussed.

17.1
Bacterial Lectins

Bacterial lectins exist in different forms. They may be presented at the tip and/or along the shaft of a fimbriae or pili, they can be expressed directly on the bacterial surface and they can be secreted as toxins. Their common denominator, however, seems to be to facilitate host cell entry and to promote infection. The existence of bacterial lectins has been known since 1902 (please see Table 15.1), but their full complexity with regard to carbohydrate specificity (for details on how to determine lectin specificity, please see Chapters 13 and 14) and its influence on host cell infection and pathogenicity are being unraveled. Table 17.1 lists bacterial lectins and their carbohydrate specificity together with diseases caused by bacterial infection.

The Sugar Code. Edited by Hans-Joachim Gabius

ISBN: 978-3-527-32089-9

Table 17.1 Carbohydrate specificity of bacterial fimbriae/pili lectins, surface adhesins and toxins, bacterial cell and tissue tropism, and infections they cause.

Lectin group	Lectin	Species	Carbohydrate specificity[a]	Tissue tropism	Infectious diseases
Type 1 fimbriae	FimH	*Escherichia coli*	α-Mannose, trimannose, high-mannose	Urothelial umbrella cells of the urethra, bladder and ureters	Urinary tract infections, cystitis
		Klebsiella pneumoniae	α-Mannose	Gastrointestinal tract, eyes, respiratory tract, genitourinary tract	Transfusion-mediated sepsis, pneumonia, urinary tract infections, surgical site infections, conjunctivitis
		Salmonella typhimurium	Oligomannose	Gut epithelium, enterocytes, macrophages	Diarrhea, vomiting, abdominal cramps, fever
Type P fimbriae	PapG	*Escherichia coli*	Globotriaosylceramide	Urinary tract	Upper urinary tract infections, cystitis
	FsoG, FstG	*Escherichia coli*	Globotetraosylceramide	Uroepithelial cells	Upper urinary tract infections
	Prs	*Escherichia coli*	Forssman antigen, GloboA	Uroepithelial cells	Cystitis
Type S fimbriae	SfaS-I	*Escherichia coli*	Siaα2,3Gal	Urinary tract	Urinary tract infections
	SfaS-II	*Escherichia coli*	Siaα2,3Gal (on fibronectin)	Epithelium and endothelium of the choroid plexus	Meningitis
Type IVa pili	PilA	*Pseudomonas aeruginosa*	asialo-GM1, asialo-GM2	Epithelial cells	Infects patients with cystic fibrosis, cancer or injured corneas
		Neisseria gonorrhoeae	Unknown	Epithelial cells	Gonorrhoeae

Type IVb pili	PilA	*Vibrio cholera*	Mannose	Intestine	Antibiotic-associated diarrhea
		Enteropathogenic *Escherichia coli*	*N*-Acetyllactosamine	Epithelial cells in the intestine	Diarrhea
Surface lectins	BabA	*Helicobacter pylori*	Le^{b}, H type 1, ALe^{b}	Stomach epithelial cells	Peptic ulcer disease, MALT lymphoma or gastric cancer
	SabA	*Helicobacter pylori*	SLe^{x}	Stomach epithelial cells and progenitor cells	Peptic ulcer disease, MALT lymphoma or gastric cancer
	LecA (PA-IL)	*Pseudomonas aeruginosa*	Galα1,4Gal (in P1, p^{k}), Galα1,3Gal (in blood group B antigen)	Epithelial cells	Infects patients with cystic fibrosis, pneumonia, cancer or injured corneas
	LecB (PA-IIL)	*Pseudomonas aeruginosa*	Fucose, Le^{a}, Le^{x}	Epithelial cells	Infects patients with cystic fibrosis, pneumonia, cancer or injured corneas
Toxins	Toxin A	*Clostridium difficile*	Galα1,3Galβ1,4GlcNAc, (Le^{x}, Le^{y})	Epithelial cells in the intestine	Antibiotic-associated diarrhea
	CT	*Vibrio cholerae*	GM1	Polarized intestinal epithelial cells	Diarrhea
	HT	*Escherichia coli*	GM1, GM2, paragloboside	Intestinal epithelial cells	Diarrhea
	ST	*Escherichia coli*	α1,2-Fucosylated structures	Intestinal epithelial cells	Diarrhea
	Shiga toxin	*Shigella dysenteriae*	Globotriaosylceramide	Intestine	Diarrhea, hemorrhagic colitis, hemolytic uremic syndrome
	Shiga-like toxin	*Escherichia coli*	Globotriaosylceramide, globotetraosylceramide	Intestine	Diarrhea, hemorrhagic colitis, hemolytic uremic syndrome

[a] Globotriaosylceramide (p^{k}), Galα1,4Galβ1,4Glcβ1,1Cer; globotetraosylceramide, GalNAcβ1,3Galα1,4Galβ1,4Glcβ1,1Cer; Forssman antigen, GalNAcα1,3GalNAcβ1,3Galα1,4Galβ1,4Glcβ1,1Cer; GloboA, GalNAcα1,3(Fucα1,2)Galβ1,3GalNAcβ1,3Galα1,4Galβ1,4Glcβ1,1Cer; asialo-GM1, Galβ1,3GalNAcβ1,4Galβ1,4Glcβ1,1Cer; asialo-GM2, GalNAcβ1,4Galβ1,4Glcβ1,1Cer; *N*-acetyllactosamine, Galβ1,4GlcNAc; Le^{b}, Fucα1,2Galβ1,3(Fucα1,4)GlcNAc; H type 1, Fucα1,2Galβ1,3GlcNAc; ALe^{b}, GalNAcα1,3(Fucα1,2)Galβ1,3(Fucα1,4)GlcNAc; SLe^{x}, Siaα2,3Galβ1,4(Fucα1,3)GlcNAc; P1, Galα1,4Galβ1,4GlcNAcβ1,3Galβ1,4Glcβ1,1Cer; blood group B antigen, Galα1,3(Fucα1,2)GlcNAc; Le^{a}, Galβ1,3(Fucα1,4)GlcNAc; Le^{x}, Galβ1,4(Fucα1,3)GlcNAc; Le^{y}, Fucα1,2Galβ1,4(Fucα1,3)GlcNAc; GM1, Galβ1,3GalNAcβ1,4(NeuAcα2,3)Galβ1,4Glcβ1,1Cer; GM2, GalNAcβ1,4(NeuAcα2,3)Galβ1,4Glcβ1,1Cer; paragloboside, Galβ1,4GlcNAcβ1,3Galβ1,4Glcβ1,1Cer.

17.1.1
Fimbriae/Pili

Fimbriae/pili are long surface polymers built up of various numbers of subunits [1]. Usually, one major subunit (for example FimA or PapA of type 1 and type P fimbriae, respectively) forms the bulk of the shaft with minor subunits assisting in fimbriae assembly and lectin integration. The same type of fimbria/pili may be expressed by different types of bacteria. At the same time, the structure of fimbria/pili can vary within subclasses of bacteria based on the subunits supplying the major building blocks. One example is enterotoxigenic *Escherichia coli*. These bacteria express a number of different fimbriae structures as illustrated in Figure 17.1. The structure of the P fimbriae of uropathogenic *E. coli* is shown in the same figure. Each bacterium harbors approximately 200–500 fimbriae.

The lectins are commonly presented at the tip of the fimbriae/pili, but may also be intercalated along the shaft. The carbohydrate specificity of fimbriae lectins differs both within and between species. There are two reasons for this: (i) allelic variants of fimbriae/pili lectins of different bacteria and bacterial strains exhibit minor structural variations; and (ii) the fimbrial shaft itself appears to influence the carbohydrate specificity of the lectins.

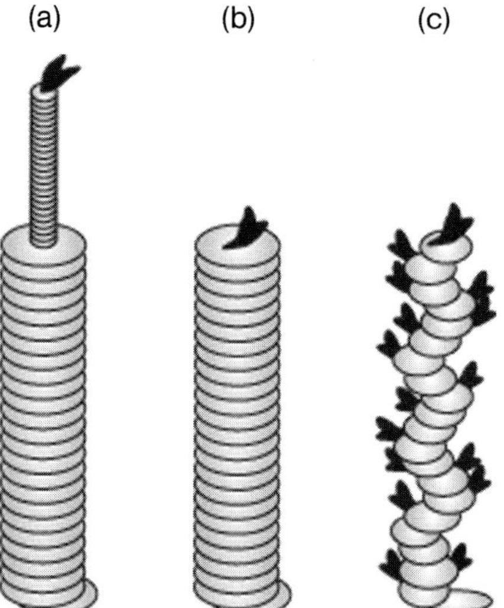

Figure 17.1 A simplified schematic representation of fimbriae/pili of uropathogenic and enterotoxigenic *E. coli*. (a) Fimbriae with the lectin subunit placed at the top of the fimbrial rod (for example type P fimbriae of uropathogenic *E. coli*). (b) Pili with the lectin activity incorporated in the top major subunit (for example CFA/I fimbriae of enterotoxigenic *E. coli*). (c) Fimbriae with the lectin activity intercalated along the shaft of the fimbriae (for example K88 of porcine enterotoxigenic *E. coli*). Adapted from [27].

17.1.1.1 Type 1 Fimbriae

Type 1 fimbriae are cylindrical rods, approximately 7 nm wide and 1 μm long [1]. Their expression is phase variable, that is, the fimbriae are not expressed on the bacterial surface at all times. The carbohydrate specificity of the lectin FimH is for α-mannose moieties (see Chapter 1, especially Figure 1.6, for sugar names, structures and abbreviations).

E. coli expressing type 1 fimbriae are commensal habitants of the intestinal bacterial flora [2, 3]. When entering the urinary tract they bind to glycoproteins carrying high-mannose structures on epithelial cells (see Figure 4.2 and Chapters 6 and 8 for details on *N*-glycans) and cause urinary tract infections. FimH of bacterial isolates from the urinary tract has been shown to have a higher affinity for monomannose as compared to FimH from fecal isolates, which recognizes mainly the trimer. This is thought to be due to different allelic variants of FimH. Recently, it was shown that the affinity of *E. coli* type 1 FimH for monomannose can be increased by shear stress that causes separation of two interacting domains, that is the mannose-binding lectin domain and the fimbriae-incorporating pilin domain, of FimH (for further information on catch bond, please see Chapter 19.2). This may favor attachment to epithelial cell ligands as opposed to free glycoproteins in the urine, which otherwise would facilitate host cell protection.

Klebsiella pneumoniae is an enterobacterium that can colonize the eyes, and the gastrointestinal, respiratory and genitourinary tracts of healthy individuals. It may produce an extended spectrum of β-lactamase enzymes, which cause resistance against a multitude of different antibiotics. The bacterium has therefore emerged as a serious health problem in hospitals, especially in neonatal intensive care units. *K. pneumoniae* can cause transfusion-mediated sepsis, pneumonia, urinary tract infections, surgical site infections and conjunctivitis (eye infections). The mannose-binding specificity of *K. pneumoniae* FimH differs compared to *E. coli* FimH – a difference presumably caused by the fimbrial shaft. This is also the case for FimH of *Salmonella typhimurium*, another member of the Enterobacteriaceae family that exhibits tropism for gut epithelium. Swapping the FimH of an *E. coli* isolate onto the fimbrial shaft of *S. typhimurium* resulted in a switch of the *E. coli* FimH specificity to that of *S. typhimurium* FimH.

17.1.1.2 Type P Fimbriae

Type P fimbriae [1] are common in *E. coli* isolates causing urinary tract infections. They bind to globo series glycolipids which have internal or terminal Galα1,4Gal structures (see Chapters 10 and 30 for detailed information on glycolipid structure and biosynthesis). Four serovariants, *pap*, *fso*, *fst* and *fel*, have been found, all quite homologous to each other. The G subunit, of which three classes have been identified, is responsible for the lectin activity of the fimbriae. Class I corresponds to PapG and binds preferentially to globotriaosylceramide (Gb3) (the carbohydrate sequences of structures referred to in the text are shown in the footnote of Table 17.1 for all bacterial lectins) on uroepithelial cells in the urinary tract. PapG is quite rare in the *E. coli* population, and its clinical effects are consequently poorly known. Class II, corresponding to FsoG and FstG, is common in *E. coli* strains causing upper urinary tract infections and pyelonephritis in man. It binds preferentially to globotetraosylceramide (Gb4), a major glycolipid of human kidney and urether. The third class, present in fimbriae with a P-related sequence (Prs), binds mainly to the Forssman antigen but also to GloboA, and is commonly found in canine *E. coli* isolates. However, this G-protein class has also been associated with human cystitis.

17.1.1.3 **Type S Fimbriae**

The S fimbria [1] is produced only by a limited number of *E. coli* strains. It is strongly associated with meningitis in newborns, with 80% of the isolated strains being S-fimbriated. The lectin specificity is encoded by the SfaS subunit that exists in two forms: SfaS-I and -II. The former has been cloned from uropathogenic *E. coli* isolates, while the latter was cloned from meningitis-associated *E. coli* isolates. SfaS binds to α2,3-sialylated galactose on glycoproteins and possibly also glycolipids. One identified ligand for SfaS-II is fibronectin of interstitial connective tissues and extracellular matrices [4]. The protein carries α2,3-sialylated galactose on *N*-glycans, and is found mainly in developing basement membranes of embryonic tissues and on vascular endothelial cells of adults. It is believed that S fimbriae mediate targeting of meningitis-associated *E. coli* to the epithelium and endothelium of the choroid plexus, which is the portal entry into the cerebrospinal fluid.

17.1.1.4 **Type IV Pili**

This pilin is expressed by a number of different bacteria and exists in two forms: type IVa and IVb [1, 5]. The difference between the two pili is the folding of the β-sheets that surround the α-helices. Type IVa is expressed by *Pseudomonas aeruginosa* (for pathogenicity, see Section 17.1.2.2) and *Neisseria gonorrhoeae*. The carbohydrate binding activity of *P. aeruginosa* type IV pili is encoded in the C-terminal region of the major subunit of the pili, PilA, but it is only exposed for ligand binding at the top of the pilin. This is in contrast to most fimbriae where the lectin activity is encoded by a distinct minor subunit. The carbohydrate ligands of *P. aeruginosa* type IV pili are the glycosphingolipids asialo-GM1 and asialo-GM2 on epithelial cells. *N. gonorrhoeae* shares the specificity for human epithelial cells with *P. aeruginosa*. However, no carbohydrate ligand has been identified. Type IVb pili are expressed by *Vibrio cholerae* and enteropathogenic *E. coli*. The pilin of the former is also known as mannose-sensitive hemagglutinin, while the ligand for enteropathogenic *E. coli* on epithelial cells might be *N*-acetyllactosamine, a common ligand for galectins *in situ* (for information on galectin ligands, please see Table 19.3, on the common fold Figure 13.2).

17.1.2
Bacterial Surface Lectins

Bacterial lectins can also be presented directly on the bacterial surface and as such mediate adhesion to host cells. *Helicobacter pylori* and *P. aeruginosa* are two examples of this as described below.

17.1.2.1 **BabA and SabA**

The adhesins BabA and SabA mediate colonization of *H. pylori* to stomach epithelial cells by binding to Lewis^b (Le^b) and sialyl-Lewis^x (sLe^x), respectively (Table 17.1). The former carbohydrate epitope is naturally expressed on secretor and Lewis-positive individuals, whereas expression of the latter epitope seems to be induced by *H. pylori* infection. It has been shown that BabA exists as allelic variants, of which some, generalists, recognize Le^b-related structures such as H type 1 and ALe^b, while others are pure specialists, binding only to Le^b. Approximately 50% of the world population is infected by *H. pylori* and 10% of these develop peptic ulcer disease, mucosa-associated lymphoid tissue (MALT) lymphoma or gastric cancer.

17.1.2.2 **LecA and LecB**

Formerly known as PA-IL and PA-IIL, LecA and LecB [6] are unusual bacterial lectins in the sense that they require the presence of Ca^{2+} for carbohydrate binding, which is a feature characteristic for several mammalian lectins (see Chapter 16.3 and Figure 16.1h,i,k,l; for details on C-type lectins, see Chapters 19 and 20). LecA binds strongly to human P1 and p^k blood group antigens as well as blood group B, that is, terminal Galα1,4Gal and Galα1,3Gal structures. LecB has an unusually high affinity for fucose, but also recognizes L-galactose, D-mannose and D-arabinose which display the same stereochemical arrangement with one axial and two vicinal equatorial hydroxyl groups. Lewis^a (Le^a) is very well recognized by LecB, while the different orientation of the GlcNAc *N*-acetyl group of Le^x is presumably responsible for its considerably weaker binding to LecB. These lectins also have a toxic function causing damage of epithelial cell primary cultures and blockage of epithelial cell ciliary beating. The two lectins are expressed by *P. aeruginosa*–a pathogen causing serious health problems in patients with cystic fibrosis, pneumonia, cancer or injured corneas. Some of these patients may have a changed cell-surface glycosylation with increased fucosylation and decreased sialylation that appears to promote *P. aeruginosa* binding.

17.1.3
Toxins

Lectins presented on fimbriae/pili or on the cell surface can directly mediate adhesion of a bacterium to the host cell surface and facilitate host infection. Another type of bacterial carbohydrate-binding proteins is constituted by secreted toxins, which are of importance for bacterial pathogenicity.

17.1.3.1 **Toxin A of *Clostridium difficile***

Toxin A is a 308-kDa protein with seven putative binding sites for Galα1,3Galβ1,4GlcNAc (Table 17.1), presumably carried on both lipid and protein carriers. The binding pocket may tolerate some modifications, such as fucosylation, as binding to Le^x and Lewis^y (Le^y) structures (please see Table 27.2) is also possible. Upon binding to the host cell surface, toxin A is endocytosed. It glucosylates Rho proteins in the cytosol thereby disrupting their normal functions including regulation of the epithelial cell barrier [7]. *C. difficile* is an opportunistic pathogen and the most common cause of antibiotic-associated diarrhea. Antibiotics disturb the normal bacterial flora of the intestine, allowing for *C. difficile* overgrowth.

17.1.3.2 **Cholera Toxin**

Vibrio cholerae colonizes the small intestine and cause diarrhea by secretion of a toxin (CT) that disrupts the epithelial cell barrier of the small intestine. CT consists of two subunits. The A subunit exerts the toxic effect of CT, while the B subunits are responsible for its carbohydrate-binding activity. There are five B subunits per toxin, assembled into a pentameric ring, each of them capable of binding with high affinity to a GM1 (see Chapter 25.2 and Info Box 1 in Chapter 30 for the function of GM1 in growth control of human tumor cells and Figure 30.2 for structure) molecule. Studies indicate that this binding takes place at GM1-containing lipid rafts. As GM1 recycles between the cell surface and the Golgi, this may be the toxin's way into the cells [8].

17.1.3.3 **Heat-Labile and Heat-Stable Toxins**

Enterotoxigenic *E. coli* produce heat-labile toxin (LT). It is closely related and very similar to CT both in structure and binding [9]. It too is built up by one A subunit (toxic effect) and five B subunits (lectin activity) that bind to the GM1 glycosphingolipids (for further information on GM1, please see Chapter 10.7–10.10 and Info Box 1 in Chapter 30). However, slight differences do exist as shown by the fact that LT can bind also to desialylated and/or degalactosylated GM1. It also recognizes GM2 and paragloboside which CT does not.

Another toxin produced by enterotoxigenic *E. coli* is heat-stable toxin (ST) [9]. The ligand for this toxin is believed to be α1,2-fucosylated structures (H type 1 and H type 2) as indicated by studies showing that milk oligosaccharides carrying these carbohydrate epitopes protect infants against ST. Enterotoxigenic *E. coli* in contaminated water and food is a common cause of diarrhea in humans as well as pigs. Individuals at high risk for enterotoxigenic *E. coli* infection are travelers, children below the age of 5 years and infants fed milk substitutes. Although related, LT and ST have different mechanisms of action on intestinal epithelial cells. When coexpressed by enterotoxigenic *E. coli* strains, they act synergistically to cause diarrhea.

17.1.3.4 **Shiga and Shiga-Like Toxins**

Similar to CT, the shiga and shiga-like toxins [10, 11] consist of a toxic A subunit and five carbohydrate-binding B subunits. Shiga toxin is produced by *Shigella dysenteriae*. The toxin binds to Gb3-expressing cells and, upon internalization, inhibits protein synthesis leading to diarrhea, hemorrhagic colitis or hemolytic uremic syndrome in infected individuals (for structure of the globo-series triaose, please see also Tables 10.2 and 10.5). It has been shown that cytokines induced by *S. dysenteriae* infection can cause production of Gb3 in some cells. Shiga-like toxin 1 is nearly identical with shiga toxin and also recognizes Gb3. Shiga-like toxin 2 exists in different forms; most of them also recognize Gb3, but one form has been shown to bind to Gb4 as well. Studies indicate that the lipid part of the carbohydrate ligand also plays an important role in recognition. Shiga-like toxins 1 and 2 are produced mainly by enterohemorrhagic *E. coli*, but also by *Aeromononas caviae, Aeromononas hydrophila, Citrobacter freundii* and *Enterobacter cloacae*. Despite the similarity between Shiga toxins and Shiga-like toxins, differences do exist with regard to effects on cells and interactions with the immune system of the host.

The microbe–host cell interactions described above illustrate how lectins in different forms are utilized to achieve bacterial infection. It shows that the surface carbohydrate repertoire of the host is of major importance in determining who will get infected and not. It also implies that the complex interplay between host cell carbohydrate expression and bacterial lectin affinity has evolved over a long time, and it will probably continue to evolve as hosts are challenged with, and forced to respond to, new bacterial strains with changed carbohydrate-binding preferences. As protein–carbohydrate interactions generally are of low affinity, the presence of multiple lectins on each bacterium and the multivalent expression of specific carbohydrate ligands on the cell surface is of major importance for adhesion and subsequent infection (for antibacterial agents, please see below and also Chapter 28.4). Next, we will describe how cell-surface carbohydrate ligands are targets by viruses.

17.2 Virus Binding

The first step in viral infection is the binding of a virus to the cell surface. Viruses bind to a variety of different molecules that can be proteins, lipids or carbohydrates. Some molecules serve only as attachment sites on the host cell surface, while others are true receptors mediating both attachment and uptake of the virus particle. The distribution of molecules mediating viral adhesion determines to a large extent species, tissue and cell tropism [12]. Viral entry into a cell is often a multistep process in which initial attachment is followed by receptor binding and cellular uptake.

Protein–carbohydrate interactions are important in viral adhesion and infection. Some viruses bind to protein- or lipid-carried glycans capped by sialic acid (sialic acid, Sia, *N*-acetylneuraminic acid and Neu5Ac are used interchangeably in the text) and others bind to glycosaminoglycans (see Chapter 11 for details). Both host cell and viral lectins can mediate viral adhesion to host cells. Examples of host cell lectins involved in virus binding are dendritic cell-specific intercellular adhesion molecule-3 grabbing non-integrin (DC-SIGN; see Figure 16.1i and Chapter 19 for X-ray structure and further details as well as Chapter 20 for evolution of the C-type domain) and liver/lymph node (L)-SIGN, which through binding of *N*-linked high-mannose-type structures on virus proteins can mediate adhesion of HIV-1, Sindbis virus, Dengue virus, human cytomegalovirus (HCMV), hepatitis C virus (HCV) and Ebola virus. The mannose-6-phosphate receptor (MPR; see Chapter 19) has been shown to mediate binding of varicella zoster virus (VZV). In Table 17.2, viruses and virus proteins binding carbohydrates are listed together with their

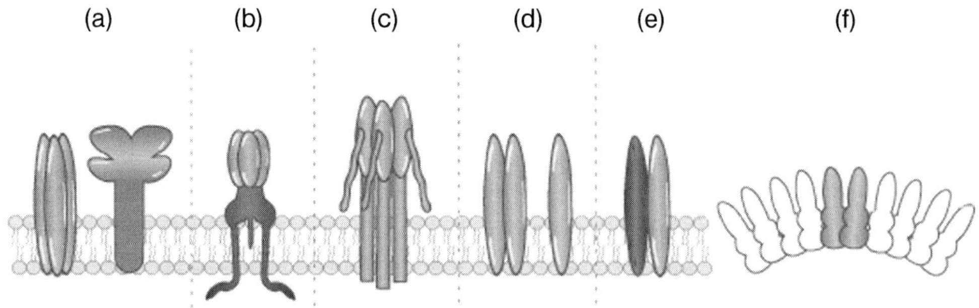

Figure 17.2 Schematic representation of carbohydrate-binding proteins of different viruses. The proteins are not drawn to scale. (a) HA (left) and NA (right) of influenza virus. HA mediates binding to terminal sialic acid. NA is situated in the envelope (env) as a homotetramer and has been referred to as the receptor-destroying enzyme since it hydrolyzes bound sialic acid. It is essential for infection and is the target for several antiviral drugs. (b) Env of HIV-1. Env consists of the trimeric transmembrane gp41 and the trimeric gp120. gp120 can bind to syndecans on macrophages. Adapted from [15]. (c) G-protein of RSV. RSV-G is related to HN of parainfluenza, but does not mediate hemagglutination and lacks NA activity. It is smaller and carries mucin-like regions represented in the figure as protruding spikes. It binds to *N*-sulfated HS. (d) gB and gC of HSV. Both gB and gC bind HS. (e) E1E2 of HCV. E1E2 binds to highly sulfated HS on hepatocytes. (f) VP1 dimer of norovirus. Norovirus lacks an envelope and the VP1 capsid protein forms the virus capsid. VLPs have been shown to bind to human blood group antigens such as H type 1, A, B, Le^a and Le^b.

carbohydrate specificity. In Figure 17.2, a schematic representation of carbohydrate binding proteins of different viruses is shown.

17.2.1 Influenza Virus

17.2.1.1 Influenza Virus Surface Proteins

Influenza virus carries two surface glycoproteins, hemagglutinin (HA) and neuraminidase (NA) (please see also Chapter 28.2). The name hemagglutinin stems from the observation that influenza caused red blood cell aggregation through binding via HA (that is hemagglutination; please see also Table 15.1 for historical account). HA forms a homotrimer, while NA is found as a homotetramer [13]. HA binds specifically to terminal sialic acid (neuraminic acid) on glycoproteins or glycolipids, while NA hydrolyzes terminal sialic acid off glycan chains. Both HA and NA are linkage specific.

The exact cell-surface glycoprotein(s) which mediate(s) uptake of influenza virus is not known (please see also Chapter 26.7). After the virus has been internalized and exposed to the low pH of the endosome, HA conformation changes and mediates fusion of the virus envelope and the endosome membranes. Virus RNA can thus enter the cytosol and start infection.

17.2.1.2 Epidemiology

Influenza virus belongs to the Orthomyxoviridae family and is subdivided into three serotypes: A, B and C. Further classification of influenza is based on the antigenic features of HA and NA. Influenza A is an avian virus that also infects many mammals including humans, while influenza B is mainly found in humans, and influenza C is found in humans and pigs. Influenza type A and B cause recurrent epidemics worldwide. This is due to subtle changes in the virus surface glycoprotein structure that allow the virus to avoid host immunity. The structural changes may be accounted for by new glycosylation patterns on viral surface glycoproteins like HA [13]. Worldwide pandemics such as the Spanish flu 1918–1919, the Asian flu 1957–1958 and the Hong Kong flu 1968–1969 have all been caused by influenza type A. Circulation of influenza type A strains in several species may explain the rapid evolution and emergence of new variants [13]. Emerging influenza virus variants are now closely monitored worldwide due to the potential risk of new pandemics as occured in 2009.

17.2.1.3 **Influenza Virus Species and Tissue Tropism**

Influenza type A does not normally cause disease in its natural host, wild aquatic birds, but the virus is promiscuous in its species tropism causing disease in many mammals. The receptor binding site of influenza A and B HA recognizes sialylated *N*-acetyllactosamine, SiaLacNAc (carbohydrate sequences of structures referred to in the text are shown in the footnote of Table 17.2). The binding specificity varies between different types, subtypes and strains. It has for a long time been known that avian influenza carries HA that preferentially binds sialic acid α2,3-linked to galactose in contrast to human influenza HA, which binds to sialic acid α2,6-linked to galactose. This binding preference correlates with the repertoire of sialosides in the human airways and the bird intestine, respectively.

The HA of the H3 strain of the human influenza type mainly binds to the Neu5Ac residue of SiaLacNAc, while the HA's of the H1 strain of influenza A and influenza type B in addition to binding to Neu5Ac need the GlcNAc and Gal residues of sialylated *N*-acetyllactosamine for binding. The HA of human influenza type C binds to 9-*O*-acetyl neuraminic acid.

Human influenza binds to Siaα2,6Gal on epithelial cells in the nasal mucosa, paranasal sinuses, pharynx, trachea and bronchi. In contrast, avian influenza homes on to Siaα2,3Gal on nonciliated, cuboidal bronchiolar and alveolar type II cells. Of note, Siaα2,3Gal is only found in low abundance at these sites in humans, a factor to explain the low susceptibility of human for avian flu.

17.2.2
Rotavirus

Rotavirus is the most common cause of severe diarrhea in children worldwide. The virus is non-enveloped and has a double-stranded RNA genome. Sialidase treatment of cells renders them resistant to infection by some strains. However, most strains, including human strains, are sialidase-resistant in their infectivity. It is not known whether or not sialidase-resistant strains can bind to sialic acid residues that are not removed by sialidase. Several sialidase-sensitive rotavirus strains have been shown to depend on GM1 and GM3 ganglioside binding for infection. Sialidase-sensitive monkey rotavirus may utilize sialic acid on both glycolipids and glycoproteins.

The monomeric lectin, virus spike protein (VP) 4, mediates binding to sialic acid and integrin $\alpha_2\beta_1$. VP4 is cleaved to form VP5* and VP8*, which interact with integrins and sialic acids, respectively. VP8* has a β-sandwich fold similar to what is found in galectins, even though there is no sequence similarity. The sialic acid binding does not seem to be mediated by the galectin fold (for illustration, see Figure 13.2); rather it is suggested to be localized to a shallow groove which is conserved in sialic acid-dependent viruses [14]. Only VP8* from sialidase-sensitive strains can be cocrystallized with Neu5Ac.

17.2.3 Human Immunodeficiency Virus 1

Carbohydrate-mediated binding plays an important role in HIV-1 infection. The envelope protein of HIV-1, gp120, and its binding to the lymphocyte accessory molecule, CD4, and the chemokine receptor CCR5, have been well characterized. The binding of gp120 to C-type lectins via high-mannose carbohydrates has also been studied extensively. What is less described is the HIV gp120 binding to syndecans–type I transmembrane heparan sulfate (HS) proteoglycans (see Chapter 11.6 for further details of structure). For macrophages, it has been suggested that the binding to long linear HS chains may mediate an initial attachment to the target cell. This is followed by a scan for the entry receptors, that is, CD4 and chemokine receptors [15].

17.2.4 Norovirus

VP1 is the major capsid protein of Norovirus–a non-enveloped, positive-sense single-stranded RNA virus. VP1 can self-aggregate to form virus-like particles (VLPs) consisting of 90 dimers of VP1. It has three domains, where the *N*-terminal faces the interior, the shell domain constitutes the surface and the protruding domain faces outwards. VLPs have been shown to bind to human blood group antigens such as H type 1, A, B, Le^a and Le^b, preferably on mucins or mucin-like proteins [16]. The binding of VP1 to different blood group antigens is strain specific, owing to the high genetic variability of VP1. Noroviruses are a major cause of nonbacterial outbreaks of gastroenteritis. As individuals carry different blood group antigens, natural resistance against some strains of Norovirus are prevalent [17]. This is seen in the case of Norwalk virus strains that cause winter vomiting disease, with approximately 20% of Caucasians being naturally resistant; they are non-secretors lacking the H-type blood group antigens, and consequently A and B antigens also, on epithelial cell surfaces. However, some Norovirus strains can infect non-secretors as well [16].

17.2.5 Herpes viruses

A number of different herpes viruses capable of infecting humans exist. They include *Herpes simplex* virus (HSV)-1 and -2, VZV, Epstein–Barr virus, HCMV, and human herpesvirus-7 and -8 [18]. A shared feature for this group of viruses is their binding to HS, which is found on almost all cells. The herpes viruses have similar sets of envelope glycoproteins involved in cellular attachment and entry. The tropism of the herpes viruses is not conserved to the same degree though.

HSV is an enveloped double-stranded DNA virus. It causes recurrent epidermal/ mucosal lesions, but can also cause encephalitis (HSV-1) and meningitis (HSV-2). At least 12 proteins are present in the envelope of HSV-1, of which glycoprotein B (gB), glycoprotein C (gC) and glycoprotein D (gD) interact with HS in different steps of attachment and fusion. gC binding to HS requires an *N*-sulfated dodecamer carrying at least one 2-*O*- and one 6-*O*-sulfate group [18]. Although similar, the binding to HS is not identical between HSV-1 and -2 gC. The latter displays higher avidity to HS, and the binding is more susceptible to heparin inhibition and HS desulfation [18]. The HS-binding part of HSV-1 gC has been mapped to the *N*-terminal region. gB is a type I membrane glycoprotein, which when mutated in the poly-lysine HS-binding domain reduces the ability of HSV to bind to HS, although to a lesser degree compared with mutated gC. gD is also a type I membrane glycoprotein with three *N*-linked glycosylation sites. gD forms a dimer that requires 3-*O*-sulfated groups to bind HS (please see Figures 1.7d and 11.1 for heparin structure).

HCMV is an opportunistic virus which can infect most human cell types. The broad tropism indicates that the HCMV has several binding molecules in the envelope and that the ligands are found on most cells. HS is required but not sufficient for infection. The complex proteins gM/gN and gB have been shown to bind HS, where gB have several functions besides tethering [19]. gB is a spike protein while gM is a type III glycoprotein with seven membrane-spanning domains. gM is covalently bound to gN.

VZV causes a primary and a secondary disease–varicella (chickenpox) and herpes zoster (shingles). Herpes zoster is caused by a reactivation of dormant viruses residing in nerve ganglia after the primary infection. Infection is inhibited by addition of mannose-6-phosphate. Moreover, cellular downregulation of MPR also stops infection of VZV. This indicates that VZV binds to MPR via mannosylated envelope glycoproteins. gC is responsible for cellular attachment since viruses with mutated gC are unable to bind to target cells. Reports suggest heparin, but not chondroitin sulfate, protects cells from infection and that intact VZV particles bind to heparin, underscoring specificity.

17.2.6 Hepatitis C Virus

HCV is a small, enveloped, positive-stranded RNA virus of the Flaviviridae family that infects the liver. The infection often becomes chronic, and may cause cirrhosis and liver cancer. Recent experiments have revealed that the binding and entry are complex and engage several entry factors. The initial binding involves glycosaminoglycans and lipoprotein lipase (LPL). HCV has two envelope glycoproteins, E1 and E2, which assemble to form a heterodimer and where both subunits are responsible for interactions with HS. E1E2 are type I membrane proteins with a C-terminal transmembrane domain and a larger *N*-terminal ectodomain. Direct binding of E1 and E2 to heparin has been demonstrated in surface plasmon experiments and HCV can be purified using heparin columns. Heparin and heparinase, an enzyme which degrades heparin, inhibit HCV binding. *N*-sulfation, but not 2-*O*- and 6-*O*-sulfation, is required for E1E2–HS interaction. When isolating HS from liver and kidney, only liver HS can inhibit HCV binding [20]. Recently it was shown that HCV associated with lipoproteins can use LPL to grant access to the liver. LPL binds to HS at the cell surface of hepatocytes and act as a bridge to the HCV–lipoprotein complex. Direct binding of HCV to HS on the cell surface may still be of importance for infection yet.

17.2.7 Paramyxoviridae

Attachment to glycans on the cell surface is mediated by two types of glycoproteins in paramyxoviruses, the attachment proteins Hemagglutinin/Neuraminidase (HN) and glycoprotein G. HN is found on several paramyxoviruses including Newcastle disease virus (NDV), human parainfluenza virus (hPIV) 1–4a and parainfluenza virus 5 (PIV 5). The HN lectin binds to, and removes, sialic acid in various linkage positions when present in glycans of glycoproteins and gangliosides. hPIV 1 recognizes *N*-acetyllactosamine with terminal α2,3-linked Neu5Ac, hPIV 3 prefers *N*-acetyllactosamine with terminal α2,3- or α2,6-linked Neu5Ac or α2,3-linked Neu5Gc [21]. Their HN lectin also bind to heparan sulfate. The HN lectin of NDV has many targets, such as linear lacto-series oligosaccharides and gangliosides, such as GM3, GD1a, GT1b, GM2, GM1 or GD1b [21].

The G glycoprotein is expressed by respiratory syncytial virus (RSV) and henipaviruses (Nipah and Hendra). RSV-G binds to HS where a minimal requirement for binding is a decasaccharide with *N*-sulfated residues [18]. It lacks hemagglutination and NA activity. RSV-G is found as an anchored type II membrane protein and as a smaller, soluble form without the *N*-terminal [21]. The protein is highly glycosylated in mucin-like regions, with glycans contributing to 60% of the weight. Sequence and structural homology comparison reveal similarity to a subdomain of tumor necrosis factor receptor.

In the sections above we have learned that cell-surface carbohydrates play an important role also in viral infections. As for bacteria, lectin binding is specific, but the individual receptor–ligand interaction is often of low affinity. However, because viruses, just as bacteria, adhere to host cells via multiple binding points, the overall binding avidity is very high and viral adhesion therefore in principle irreversible.

For many bacterial and viral species, lectin binding to cell-surface carbohydrates is essential for infection to occur. Once the structure–function relationships in the binding is elucidated, specific drugs interfering with the interaction may be developed using structure-based drug design. Examples of such drugs are described below.

17.3
Carbohydrate-Based Antiinfectives

In addition to the sialidase inhibitors Tamiflu and Relenza (see below and Chapter 28.2), glycan and glycoprotein therapies currently on the market include a conjugate glycopeptide vaccine against the bacterium *Haemophilus influenzae* type b, the antibiotics erythromycin (a macrolide antibiotic) and vancomycin (a glycopeptide), and the anticoagulant heparin (please see Chapter 28.5 for clinical applications). The Info Box provides a short description of steps in drug development.

17.3.1
Neuraminidase Inhibitors as Drugs Against Influenza

Influenza virus NA, or sialidase, is like the hemagglutinins anchored to the viral membrane. It is built up by four identical subunits (Figure 17.2) and cleaves α-linked Neu5Ac residues off the glycan chains of various glycoconjugates. NA is believed to facilitate the release of virus particles from infected cells as well as to assist in the movement of the virus through the upper respiratory tract [22]. Thus, sialidase inhibitors are believed to prevent the release of virus particles from cells in infected cells. Virus sialidase inhibitors of low potency were identified before any structural information on the influenza sialidase was available. Structure-based design was then used to structurally optimize the sialidase inhibitors once refined X-ray crystallographic data on the influenza sialidase and its complex with Neu5Ac were obtained. The first compound to be developed into a drug was 4-deoxy-4-guanidino-Neu5Ac-2-en (Figure 17.3a), which under the generic name zanamivir was developed into an inhaled inhibitor of influenza. It was approved in 1999 and is sold under the tradename Relenza (Glaxo) [22]. The next sialidase inhibitor to reach the market, oseltamivir (Tamiflu; Roche), was based on a cyclohexene core template, had a 3-pentyl ether side-chain replacing the alkyl side-chain of Neu5Ac and had sufficient oral bioavailability as the ethyl ester prodrug of oseltamivir carboxylate (Figure 17.3b and Figure 28.6) [22]. Influenza strains resistant to oseltamivir have been identified; strains that have retained their sensitivity to zanamivir. Additional influenza sialidase inhibitors are under clinical development.

17.3.2
Oligosaccharides as Inhibitors of Microbial Adhesion

As evident from the text and tables of this chapter, there is a plethora of microbes that attach to cell-surface carbohydrates as a first and absolutely necessary step in their infectious cycle. It is therefore not a far-fetched idea to assume that

Info Box

The development of a glyco-compound into a successful pharmaceutical to be used in man is a long process and involves several steps. The initial drug discovery phase, during which activity against a particular target molecule is verified, is followed by extensive preclinical and clinical development programs.

The main purpose of preclinical testing is to determine a drug's pharmacodynamics (effects of the drug on the organism), pharmacokinetics (processing of the drug in the organism), absorption, distribution, metabolism and excretion, and toxicity (including carcinogenicity and effects on mammalian reproduction) in animals before testing on humans ensue. Further, a recommended starting dose and dose scheme is established during the preclinical development.

The clinical development program includes phase 0, I, II, III and IV trials. Phase 0, or first-in-man trials, involves administrating single subtherapeutic doses of the drug to a small number (10–15) of subjects with the purpose of acquiring pharmacokinetics and pharmacodynamics data. In phase I studies, small groups of less than 100 healthy subjects are exposed to the drug in order to assess safety, tolerability, pharmacokinetics and pharmacodynamics of the drug. One important purpose of the phase I trial is to find the drug's appropriate therapeutic dose. Most drugs under development fail in phase II, which is set up to assess efficacy and safety in a larger group of between 20 and 300 patients. Phase III studies are usually multicenter trials on thousands of patients with the purpose of comparing the study drug's efficacy with that of the most effective treatment currently used. These studies are the most expensive and time-consuming trials to run. For some drugs at least two successful phase III trials are required in order to get approval from regulatory agencies. Postmarketing surveillance or phase IV trials involves further safety and efficacy assessment on certain patient groups (for example pregnant women or patients with other ongoing pharmacological treatments), and are performed after the drug has received approval and is available on the market.

Drug development is a time-consuming process. Preclinical development may well take at least 5 years before the drug is even tested on humans and clinical development takes on average 8 years from its initiation until the drug is approved for sale.

administration of the sugar ligand in excess in a soluble form should form a complex with the microbe preventing it from binding to the cell surface (please see also Figure 4.3). The idea of a therapy based on an anti-adhesion strategy has been around for several decades and is known to work in breastfed children. Breast milk, with its richness in oligosaccharides and glycosylated proteins, is known to have antimicrobial activity, which is partly explained by glycan-mediated inhibition of microbial adhesion.

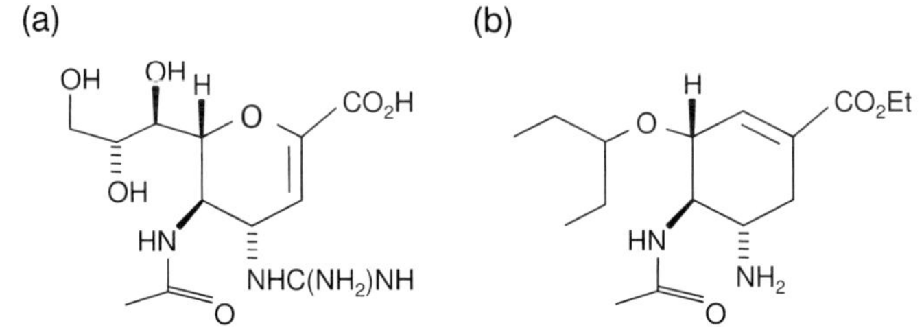

Figure 17.3 Chemical structures of zanamivir (Relenza) and oseltamivir (Tamiflu). (a) 4-Deoxy-4-guanidino-Neu5Ac-2-en (zanamivir) and (b) oseltamivir (GS4104).

The concept of inhibiting bacterial attachment as a means to prevent colonization/infection has been shown to work in a number of animal models, including a rhesus monkey model of *H. pylori*-induced gastritis and a rabbit model of pneumonia caused by *S. pneumonia*. In early studies, it was shown that methyl α-D-mannopyranoside could prevent colonization of the mouse urinary tract by *E. coli* carrying mannose-specific fimbriae [23]. Further, the administration of the tetrasaccharide globotetraose (GalNAcβ1,3Galα1,4Galβ1,4Glc; please see also Table 10.2) decreased colonization of the mouse kidney and bladder by *E. coli* strains with, presumably, the P fimbriae known to mediate binding to cells via globo series glycolipids [24].

In many cases the molecular interaction between the microbial protein and the sugar ligand has been elucidated in great detail using X-ray crystallography, nuclear magnetic resonance and other structural methods. Despite this detailed structural knowledge and animal studies suggesting that the antiadhesive strategy might work, there are no successful human trials in which infection has been prevented or cured by an antiadhesive therapy using monovalent carbohydrate-based inhibitors. In fact, intranasal administration of 3'-sialyllacto-*N*-neotetraose failed to reduce the occurrence of acute otitis media and the nasopharyngeal carriage of *S. pneumoniae, H. influenzae* and *M. catarrhalis* in children [25].

17.3.3
A New Generation of Multivalent, Carbohydrate-Based Inhibitors of Microbial Adhesion

It is clear that the attachment of bacteria and viruses to cell surfaces is mediated by several contact points on the microbe binding several copies of the ligand on the cell surface. For many bacterial toxins having the AB_5 geometry, the pentameric B subunit binds five sugar ligands on the cell surface which contributes to the toxin's overall binding strength. It is because of these so-called multivalent interactions that microbial attachment is so strong and difficult to break by monovalent inhibitors (Figure 17.4). A new generation of carbohydrate-based inhibitors is now being developed and tested in animal models. These include polymers and dendrimers (see Chapter 4) carrying multiple copies of the inhibitory carbohydrate determinant. Two different compounds carrying five (Daisy) and six (SUPER

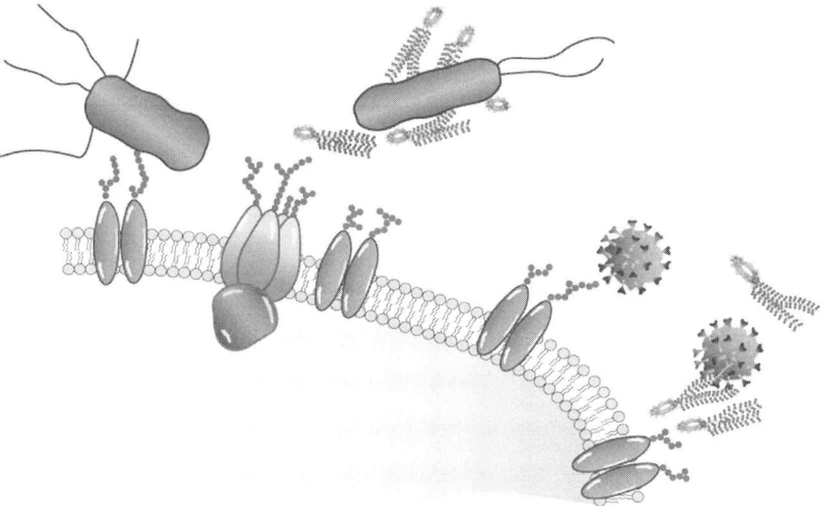

Figure 17.4 Schematic drawing of bacteria (left) and viruses (right) adhering to the cell surface via binding of the glycan chains of glycoproteins. Inhibition of attachment is accomplished by a multivalent glycoconjugate carrying several copies of glycans capable of binding bacterial and viral lectins. Multivalent presentation of the inhibitory carbohydrate is necessary for efficient blocking to occur.

TWIG – a carbosilane dendrimer) copies of the p^k trisaccharide (Galα1,4Galβ1,4Glc), respectively, were effective in protecting mice from the pathogenic effects of shiga-like toxins produced by enterohemorrhagic *E. coli* [26]. A polyacrylamide conjugate substituted with Neu5Acα2,6Galβ1,4GlcNAc protected mice infected by influenza virus. Natural peptide-based scaffolds, such as mucins, carrying multiple substitution of *O*-glycans can be regarded as natural polymers with a multivalent presentation of bioactive glycans [27]. Such molecules carrying the SLe^x or Le^b determinants were shown to be efficient inhibitors of *H. pylori* adhesion (please see Chapter 3.9 for synthetic route to Le^b) [27].

Very few clinical trials have been performed with multivalent, carbohydrate-based inhibitors of microbial adhesion. Orally administrated silicon dioxide particles carrying covalently linked Gb3 failed to diminish the severity of disease in children with diarrhea-associated hemolytic uremic syndrome. However, further development of multivalent inhibitors and more clinical trials are to be expected; innovative developments that will pave the way for tangible advances in the field.

17.4 Conclusions

Bacterial lectins can be found at the tip and/or along the shaft of a fimbriae or pili, or can be expressed directly on the bacterial surface. Some bacterial lectins constitute subunits in bacterial toxins. Their common denominator is to facilitate host cell attachment and to promote infection. Protein–carbohydrate interactions are important also for viral adhesion and infection. Some viruses bind to protein- or lipid-carried glycans capped by sialic acid and others target glycosaminoglycans. Novel carbohydrate-based antiinfective drugs may be developed based on elucidated structure–function relationships in lectin-mediated microbial adhesion. Improved efficacy of carbohydrate-based inhibitors is accomplished by a multivalent presentation of the bioactive carbohydrate determinant.

Summary Box

The sugar code is markedly operative to cause viral and bacterial infections. The insight into the presence and structures of the active lectins and their carbohydrate ligands shapes the concept of rational drug design using carbohydrate derivatives with optimized affinity. Multivalent carbohydrate-based inhibitors of microbial attachment are especially promising and are envisioned to be devoid of the drawbacks associated with the present treatment of bacterial infections, namely antibiotic resistance.

References

1 Smyth CJ *et al.* Fimbrial adhesins: similarities and variations in structure and biogenesis. *FEMS Immunol Med Microbiol* 1996;16:127–39.
2 Gaastra W, Svennerholm AM. Colonization factors of human enterotoxigenic *Escherichia coli* (ETEC). *Trends Microbiol* 1996;4:444–52.
3 Bower JM *et al.* Covert operations of uropathogenic *Escherichia coli* within the urinary tract. *Traffic* 2005;6:18–31.
4 Saren A *et al.* The cellular form of human fibronectin as an adhesion target for the S fimbriae of meningitis-associated *Escherichia coli. Infect Immun* 1999;67: 2671–6.
5 Nudleman E, Kaiser D. Pulling together with type IV pili. *J Mol Microbiol Biotechnol* 2004;7:52–62.
6 Imberty A *et al.* Structures of the lectins from *Pseudomonas aeruginosa*: insight into the molecular basis for host glycan recognition. *Microbes Infect* 2004;6:221–8.
7 Jank T *et al.* Rho-glucosylating *Clostridium difficile* toxins A and B: new insights into structure and function. *Glycobiology* 2007; 17:15R–22R.
8 Chinnapen DJ *et al.* Rafting with cholera toxin: endocytosis and trafficking from plasma membrane to ER. *FEMS Microbiol Lett* 2007;266:129–37.
9 Turner SM *et al.* Weapons of mass destruction: virulence factors of the global killer enterotoxigenic *Escherichia coli. FEMS Microbiol Lett* 2006;263:10–20.
10 Karlsson KA. Microbial recognition of target-cell glycoconjugates. *Curr Opin Struct Biol* 1995;5:622–35.
11 Sandvig K. Shiga toxins. *Toxicon* 2001;39: 1629–35.
12 Smith AE, Helenius A. How viruses enter animal cells. *Science* 2004;304:237–42.
13 Baigent SJ, McCauley JW. Influenza type A in humans, mammals and birds: determinants of virus virulence, host-range and interspecies transmission. *BioEssays* 2003; 25:657–71.
14 Kraschnefsky MJ *et al.* Effects on sialic acid recognition of amino acid mutations in the carbohydrate-binding cleft of the rotavirus spike protein. *Glycobiology* 2009;19:194–200.
15 Roux KH, Taylor KA. AIDS virus envelope spike structure. *Curr Opin Struct Biol* 2007; 17:244–52.
16 Rydell GE *et al.* Human noroviruses recognize sialyl Lewis^x neoglycoprotein. *Glycobiology* 2009;19:309–20.

17 Le Pendu J *et al.* Mendelian resistance to human norovirus infections. *Semin Immunol* 2006;18:375–86.
18 Olofsson S, Bergstrom T. Glycoconjugate glycans as viral receptors. *Ann Med* 2005; 37:154–72.
19 Compton T. Receptors and immune sensors: the complex entry path of human cytomegalovirus. *Trends Cell Biol* 2004; 14:5–8.
20 Helle F, Dubuisson J. Hepatitis C virus entry into host cells. *Cell Mol Life Sci* 2008;65:100–12.
21 Villar E, Barroso IM. Role of sialic acid-containing molecules in paramyxovirus entry into the host cell: a minireview. *Glycoconj J* 2006;23:5–17.
22 von Itzstein M. The war against influenza: discovery and development of sialidase inhibitors. *Nat Rev Drug Discov* 2007;6:967–74.
23 Aronson M *et al.* Prevention of colonization of the urinary tract of mice with *Escherichia coli* by blocking of bacterial adherence with methyl α-D-mannopyranoside. *J Infect Dis* 1979;139:329–32.
24 Svanborg Edén C *et al.* Inhibition of experimental ascending urinary tract infection by an epithelial cell-surface receptor analogue. *Nature* 1982;298:560–2.
25 Ukkonen P *et al.* Treatment of acute otitis media with an antiadhesive oligosaccharide: a randomised, double-blind, placebo-controlled trial. *Lancet* 2000;356:1398–402.
26 Mulvey GL *et al.* Assessment in mice of the therapeutic potential of tailored, multivalent Shiga toxin carbohydrate ligands. *J Infect Dis* 2003;187:640–9.
27 Gustafsson A, Holgersson J. A new generation of carbohydrate-based therapeutics: recombinant mucin-type fusion proteins as versatile inhibitors of protein–carbohydrate interactions. *Expert Opin Drug Discov* 2006;1: 161–78.

18
Plant Lectins

Harold Rüdiger and Hans-Joachim Gabius

Plant material is a notably rich source of lectins and the purification of new lectins continues to be an attractive study aim. Beyond reviewing the natural diversity of plant lectins, their versatile applicability as valuable tools in biochemistry, cell biology and medicine, recently also extended to chip technology (please see Chapter 14), gives further reason to present an account on these proteins. Looking at them in the history of lectinology, Table 15.1 reveals that the first plant lectin detected was a toxin (ricin). As plant lectin-dependent landmarks, a legume protein thereafter became the first member of this superfamily to be crystallized (for original protocol, please see Info Box in Chapter 16) and these proteins' agglutinating activity was crucial for coining the term 'lectin' (for details, please see the historical account in Chapter 15). In this chapter, we have set the priority in the course of covering this topic to first deal with nomenclature, then with the natural diversity of lectin folds and the occurrence of lectins in plants. An outline of the strategy of how to purify them follows, guiding the reader to biomedical applications. Finally, the current concepts for biological functions will be summarized. In other words, we will start by drawing the readers' attention to the historical origin of application of plant lectins in hematology (please see Info Box 1 and Info Box 2 in Chapter 1). As a practical reference source for use in glycophenotyping we present a lectin selection with their target glycans in a survey table. This compilation poses the questions on the structural origin of the lectins' capacity for sugar binding (that is the nature of the folds underlying this activity) and on where lectins can be found in plants. These questions are answered in the next part.

18.1
Nomenclature

Based on hemagglutination, the term 'agglutinin' continues to be common for plant lectins (please see Chapter 15). In fact, it has become an integral part of the abbreviations of lectin names, joining the first letters for genus and species with the 'A' for agglutinin. As listed in Table 18.1, this rule is often obeyed. Only few

The Sugar Code. Edited by Hans-Joachim Gabius

ISBN: 978-3-527-32089-9

Info Box 1

An old rule in chemistry and biochemistry says: 'Never waste pure thinking time on impure preparations'. The following example reveals that – in contrast to this rule – the use of a not yet pure preparation has led to an important discovery. Experimentally, hemagglutination could well be exploited as an efficient means to remove erythrocytes from cell preparations. In 1959, Peter C. Nowell from the University of Philadelphia Medical School planned to prepare a leukocyte suspension from whole blood, devoid of residual erythrocytes. He used a preparation from French beans (*Phaseolus vulgaris*) for removal of erythrocytes by agglutination, at that time commercially available as a freeze-dried crude extract. Expectedly, the cell fractionation worked. What had not been planned was what happened with the leukocyte population: he noted that the number of these cells had increased [P.C. Nowell. Phytohemagglutinin: an initiator of mitosis in cultures of normal human leukocytes. *Cancer Res* 1960; **20**, 462–466]. He hereby discovered the mitogenic activity of lectins. Apparently, the lectin preparation acted both as an agglutinin for red blood cells and as a mitogen for leukocytes. This observation led to further processing of the lectin-containing preparation. As described in 1975, lectin activity underlies the association of two types of subunits, called E and L, to tetramers [R.L. Felsted *et al.* Purification of the phytohemagglutinin family of proteins from red kidney beans (*Phaseolus vulgaris*) by affinity chromatography. *Biochim Biophys Acta* 1975; **405**, 72–81]. If four E-subunits combine to form an E4 complex, a lectin results which exclusively agglutinates red cells (E = erythroagglutinin); if four L-subunits are combined, the resulting L4 complex is inert towards erythrocytes but active on leukocytes as mitogen (L = leukoagglutinin; please see also Table 18.1). The pure PHA-E isolectin, suited to remove the erythrocytes, would have never enabled to discover mitogenicity in this setting.

In a similar way, the preferential agglutination of lymphoma cells by wheat germ agglutinin (WGA) was discovered when a crude wheat germ lipase preparation that also contained WGA was employed as test material (please see introduction of Chapter 25 for further details).

cases deviate from this scheme, that is those which are still referred to by special designations [concanavalin A (ConA), jacalin, peanut agglutinin (PNA), phytohemagglutinin (PHA), and soybean agglutinin (SBA)]. Table 18.1 also clarifies the issue as to whether information on monosaccharide binding covers a plant lectin's carbohydrate specificity entirely. By inspecting the presented sections on mono- and oligosaccharides it becomes evident that plant lectins often have marked preferences for oligosaccharides, with graded selectivity for related structures. For example, complex-type *N*-glycans containing either a bisecting GlcNAc moiety or the β1,6-branch can readily be detected by two different French bean isolectins (for further information on this case, please see Info Box 1; for details on *N*-glycans

Table 18.1 Panel of plant lectins useful for glycophenotyping.

Latin name (common name)	Abbreviation	Monosaccharide specificity	Potent oligosaccharide
Artocarpus integrifolia (jack fruit)	Jacalin (JAC)	Gal/GalNAc	Galβ3GalNAcα, tolerates sialylation of T/T_n antigens
Arachis hypogaea (peanut)	PNA	Gal	Galβ3GalNAcα/β
Canavalia ensiformis (jack bean)	ConA	Man/Glc	GlcNAcβ2Manα6(GlcNAcβ2Manα3)Manβ4GlcNAc
Datura stramonium (thorn apple)	DSA	GlcNAc	$(GlcNAc)_n$, Galβ4GlcNAcβ6(Galβ4GlcNAcβ2)Man, $(Galβ4GlcNAc)_3$
Dolichos biflorus (horse gram)	DBA	GalNAc	GalNAcα3GalNAcα3Galβ4Galβ4Glc > A-tetrasaccharide (Fig. 1.5)
Erythrina cristagalli (coral tree)	ECA	Gal	Galβ4GlcNAcβ6(Galβ4GlcNAcβ2)Man
Galanthus nivalis (snowdrop)	GNA	Man	Manα6(Manα3)ManαR
Glycine max (soybean)	SBA	GalNAc	GalNAcα3Galβ6Glc
Griffonia simplicifolia I	GSA I	GalNAc	GalNAcα3Gal, GalNAcα3GalNAcβ3Galα4Galβ4Glc
Griffonia simplicifolia II	GSA II	GlcNAc	GlcNAcβ4GlcNAc, *N*-glycans with terminal, non-reducing-end GlcNAc
Lens culinaris (lentil)	LCA	Man/Glc	*N*-glycan binding enhanced by core fucosylation
Lycopersicon esculentum (tomato)	LEA	–[a]	Core and stem regions of high-mannose-type *N*-glycans, $(GlcNAcβ3Galβ4GlcNAcβ3Gal)_n$ of complex-type *N*-glycans
Maackia amurensis I (leukoagglutinin)	MAA I	–[a]	Neu5Acα3Galβ4GlcNAc/Glc, 3′-sulfation tolerated

continued

Table 18.1 *Continued*

Latin name (common name)	Abbreviation	Monosaccharide specificity	Potent oligosaccharide
Maackia amurensis II (hemagglutinin)	MAA II	–[a]	Neu5Acα3Galβ3(α6Neu5Ac)GalNAc
Phaseolus vulgaris erythroagglutinin (kidney bean)[b]	PHA-E	–[a]	Bisected complex-type *N*-glycans: Galβ4GlcNAcβ2Manα6(GlcNAcβ2-Manα3)(GlcNAcβ4)Manβ4GlcNAc
Phaseolus vulgaris leukoagglutinin (kidney bean)[b]	PHA-L	–[a]	Tetra- and triantennary *N*-glycans with β6-branching
Pisum sativum (garden pea)	PSA	Man/Glc	*N*-glycan binding enhanced by core fucosylation
Sambucus nigra (elderberry)	SNA	Gal/GalNAc	Neu5Acα6Gal/GalNAc, clustered T_n antigen
Solanum tuberosum (potato)	STA	–[a]	$(GlcNAc)_n$ with preference for high-mannose-type *N*-glycans
Sophora japonica (pagoda tree)	SJA	GalNAc	GalNAcβ6Gal, Galβ3GalNAc
Triticum vulgare (wheat germ)	WGA	GlcNAc/Neu5Ac	$(GlcNAc)_n$, Galβ4GlcNAcβ6Gal
Ulex europaeus I (gorse)	UEA I	Fuc	Fucα2Galβ4GlcNAcβ6R
Vicia villosa (hairy vetch)	VVA	GalNAc	GalNAcα3(6)Gal, GalNAcβ3Gal
Viscum album (mistletoe)	VAA	Gal	Galβ2(3)Gal, Galα3(4)Gal, Galβ3(4)GlcNAc without/with α2,6-sialylation, Fucα2Gal

[a] No monosaccharide known as ligand; experimental example for tumor cell-surface glycophenotyping given in S. André *et al.* Tumor suppressor $p16^{INK4a}$: modulator of glycomic profile and galectin-1 expression to increase susceptibility to carbohydrate-dependent induction of anoikis in pancreatic carcinoma cells. *FEBS J* 2007; **274**, 3233–3256; for further information on specificity of these and other plant lectins, please see I.J. Goldstein and R.D. Poretz. Isolation, physicochemical characterization, and carbohydrate-binding specificity of lectins, in *The Lectins. Properties, Functions, and Applications in Biology and Medicine* (Eds: Liener, I.E., Sharon, N. and Goldstein, I.J.) pp. 33–249, Academic Press, Orlando, FL, 1986; W.J. Peumans and E.J.M. van Damme. Plant lectins: specific tools for the identification, isolation, and characterization of O-linked glycans. *Crit Rev Biochem Mol Biol* 1998; **33**, 209–258.

[b] Source also called French bean.

and glycosylation in plants, please see Chapters 6 and 8). Of note, the introduction of glycan microarrays to mapping glycan specificity (please see Chapter 14) currently extends our knowledge in this respect. Having herewith demonstrated that plant lectins can interact with structures beyond the monosaccharide which defines the basal specificity range of ligand structures, the next section presents information on structural aspects of the receptor side.

18.2 Folding Patterns and Occurrence

After scouring the records for known crystal structures of plant lectins and listing all found entries, it is possible to form categories of folds endowed with carbohydrate-binding activity. The resulting summary of the different folds is presented in Table 18.2. The given structural patterns establish carbohydrate recognition domains in plants (for illustration of the leguminous lectin β-sandwich of ConA, please see Figure 16.1a). The comparison to the respective compilation of folds in animal lectins in Table 19.1 reveals similarities concerning the β-sandwich, β-trefoil, β-prism II and hevein-like domains. In evolutionary terms, they should not immediately be viewed as evidence for a common ancestry. Despite the global similarities between folds in plant and animal lectins, structural details such as positioning of binding sites in the fold, especially β-sandwich, β-trefoil and hevein-like domains, rather point to convergent evolutionary pathways for animal lectins (please see also Chapter 19.1) [1]. In structural terms, especially the hevein-like and β-trefoil domains of plant lectins present telling examples of the interplay between aromatic amino acids and the carbohydrate ligand via CH–π interactions (please see Chapter 13.1 for details) [2, 3]. Clearly, the amino acids tryptophan and tyrosine have conspicuous missions in establishing lectin activity (Figure 13.1). The leguminous lectins illustrate a way how to go beyond amino acids. They recruit a Ca^{2+} as a means to preorganize the contact site for optimal fit (please see Chapter 16.1).

Overall, the compilation in Table 18.2 documents different origins of sugar-binding activity in plants. This result intimates that the presence of lectins is a characteristic which has spread widely. A systematic analysis of the available information is required to verify this assumption. By compiling the status of knowledge (Figure 18.1), it becomes evident that, indeed, lectin presence is widely documented. In quantitative terms, more than half of all known lectins have been described in the legume family, followed by the Araceae, Cucurbitaceae and Liliaceae. Naturally, the available information is based on plants, for which the starting material for purification is readily available. Consequently, mostly cultivated plants have been study objects, making gaps in our knowledge readily apparent. Having presented information on the horizontal spreading of lectin presence in plants, we next turn to the sites in the plants where lectins are known to be present. Plants mostly accumulate lectins in the seeds and other organs that serve for storage purposes, for example bark, rhizomes, bulbs or tubers. Intracellularly, the typical

Table 18.2 Overview of protein folds with lectin activity.

Type of fold	Example for lectin[a]	Example for ligand[a]
β-Sandwich (jelly-roll)	(a) Calnexin, calreticulin (b) Leguminous lectins (ConA, PHA, PNA, PSA, SBA and others)	$Glc_1Man_9GlcNAc_2$ please see Table 18.1 for details
β-Trefoil	(a) Lectin subunit of plant AB-toxins (ricin) (b) Amaranthin (*Amaranthus caudatus*)	Gal T antigen
β-Prism I	Jacalin (*Artocarpus integrifolia*), *Maclura pomifera* agglutinin Artocarpin (*Artocarpus integrifolia*) and *Musa acuminata*, *Calystegia sepium* and *Helianthus tuberosus* agglutinins	T antigen Glc/Man (linear or branched α-mannans/α-glucans)
β-Prism II	GNA, bluebell (*Scilla campanulata*), daffodil (*Narcissus pseudonarcissus*) and garlic (*Allium sativum*) agglutinins	Man
Hevein-like domain	Rubber tree (*Hevea brasiliensis*) and stinging nettle (*Urtica dioica*) agglutinins, WGA	$(GlcNAc)_n$
$(\beta/\alpha)_8$ Barrel (glycoside hydrolase family 18)	Black locust (*Robinia pseudoacacia*) chitinase-related agglutinin; orthologs in *Glycine max*, *Lotus japonicus* and *Medicago truncatula*	Pentasaccharide core of complex-type *N*-glycans
Lysin motif (LysM) domain (βααβ secondary structure)	Receptor-like kinases in rice, *Arabidopsis thaliana*, *Medicago truncatula* and *Lotus japonicus* *CEBiP* (rice)	Lipochitooligosaccharides (Nod factors) Chitin oligomer elicitor

[a] For abbreviations of names of sugars, please see Chapter 1, for those of names of lectins, please see Table 18.1. Fungal lectins have four additional folds, further extending the number of folds with lectin activity: (i) 6-bladed β-propeller (*Aleuria aurantia*), (ii) 7-bladed β-propeller (*Psathyrella velutina*), (iii) pseudo-h-type fold (*Flammulina velutipes*), (iv) actinoporin-like fold (*Agaricus bisporus*).

location of lectins is in the so-called protein bodies – membrane-lined cell organelles which are derived from the vacuole. At this site, the lectins are associated with storage proteins, also enzymes and phytin. The amounts present in seeds and other storage organs vary considerably, ranging from more than 1 g (ConA, PHA) to a few milligrams (lentil, gorse) per 100 g of plant material. Commonly, the lectin contents in plants are thus much higher than for animal lectins, which often require recombinant expression to enable structural studies. As noted above, ConA with its exceptionally high amount in beans could even be purified by crystallization (please see Info Box in Chapter 16). Its biosynthesis encompasses a pathway with a remarkable deviation from the common route (for details, please see Info Box 2). As we have highlighted present gaps in our knowledge on lectin occurrence in plants, with its current preference on cultivated plants, readers might be motivated to enter this field. Thus, we next provide the strategic information on lectin isolation.

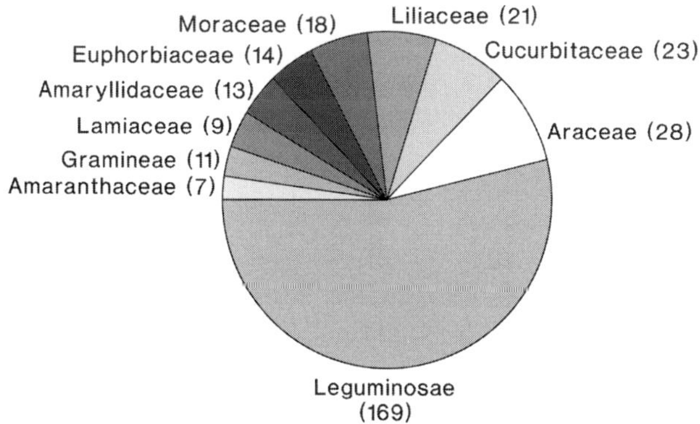

Figure 18.1 Distribution of lectins in the plant kingdom. For each family, the number of lectin-bearing species is given in parentheses.

Info Box 2

It is regarded as a foregone conclusion that alignment of sequences from homologous proteins will be simple, yielding similarity from the first amino acids onward. This is not the case for ConA when compared to other leguminous lectins such as SBA despite their conspicuous homology. Puzzling at first, these sequences reach a maximal sequence similarity only if residue 1 of ConA is aligned to residue 112 in SBA–a phenomenon called circular permutation. As a consequence, the *N*-terminal stretch of SBA and the sequence starting at position 122 in ConA are homologous.

When this–at first encounter–strange relationship was discovered [B.A. Cunningham *et al.* Favin versus concanavalin A: circularly permuted amino acid sequences. *Proc Natl Acad Sci USA* 1979; **76**, 3218–3222; J.J. Hemperly and B.A. Cunningham. Circular permutation of amino acid sequences among legume lectins. *Trends Biochem Sci 1983*; **6**, 100–102], the nature of the underlying process was entirely unclear. In 1985, the mechanism was discovered: after translation, an at that time so far unknown unique rearrangement takes place [D.M. Carrington *et al.* Polypeptide ligation occurs during posttranslational modification of concanavalin A. *Nature* 1985; **313**, 64–67]. The precursor of mature ConA harbors a protein sequence in the expected order, as found in SBA and various other leguminous lectins. This precursor then undergoes processing, first by removing an internal glycopeptide, thereby opening new N- and C-termini, as illustrated below. Instead of simply resealing this cleavage site, it remains open at the expense of peptide bond reformation at another position (118/119 in the mature protein), probably by transpeptidation which displaces a short peptide. Thus, the circular permutation becomes manifest. This reaction is made possible by spatial proximity of the residues at the new connecting point [D.J. Bowles *et al.* Posttranslational processing of concanavalin A precursors in Jack bean cotyledons. *J Cell Biol* 1986; **102**, 1284–1297]. At that time, this feature in ConA biosynthesis was the first example of a protein sequence that is posttranslationally rearranged in this way. It is operative also in other lectins from the subtribe Diocleinae to which the Jack bean (*Canavalia ensiformis*) belongs.

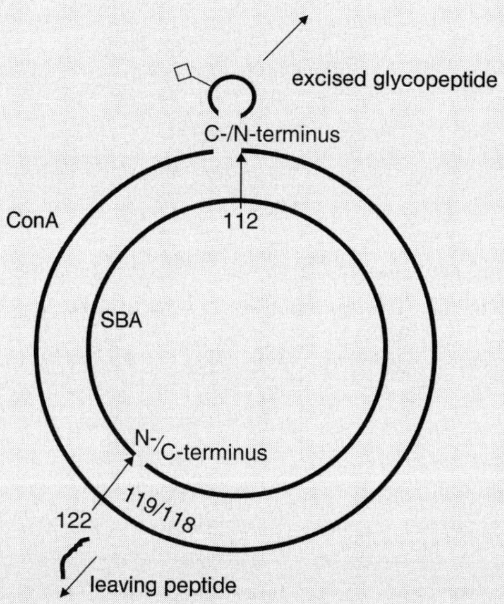

Alignment of the amino acid sequences of concanavalin A (ConA) and soybean agglutinin (SBA) by circular permutation in order to achieve maximal homology. Position 112 of SBA corresponds to the *N*-terminus of ConA, position 122 of ConA to the *N*-terminus of SBA. A new peptide bond is formed between positions 118 and 119 of mature ConA at the expense of a short peptide ('leaving peptide' in analogy to a leaving group in organic chemistry).

18.3
Purification

Initially, extracts – unless crystallization was successful – were processed by rather crude fractionation methods such as precipitation of protein by salts to yield preparations which cannot be considered pure. In his pioneering studies on ricin, H. Stillmark used a 10% NaCl solution for extraction, then filtration and protein precipitation by magnesium sulfate at low temperature [4]. The example of a wheat germ preparation given in Chapter 25 teaches the lesson that remaining minor constituents, here a lectin, can exert the dominant function (please also see Info Box 1). Thus, fractionation schemes needed to be refined. In addition to exploiting size or charge in chromatography, a key to lectin purification would be to take advantage of their intrinsic property, namely sugar binding, and this approach still pays off today. In fact, the breakthrough to obtain pure lectins in a minimum of steps was the introduction of affinity chromatography on immobilized carbohydrates (please see Chapter 15.3). This method is now the central step in lectin purification (Figure 18.2). To optimize yields, the following technical comments can be helpful: careful selection and reliable identification of the plant material is most important in the initial step. Usually, seeds are the source for lectins (please see Info Box 1). If vegetative plant organs are used, the fact that the quality of this material may vary greatly depending on the local environment, season or the year of harvest should be reckoned with. After grinding or chopping the plant material, extraction is performed at a controlled pH (neutral or mildly alkaline). The crude extract is then pre-fractionated by a (classical) precipitation step (salt, low pH) in order to remove compounds of secondary metabolism and reduce the protein content. After dialysis and readjusting the buffer condition to a suitable salt concentration and pH value, the processed extract then is commonly subjected to affinity chromatography.

The choice of the immobilized ligand depends on results from preceding studies, for example inhibition of hemagglutination by saccharides. The binding of neoglycoproteins, derived from chemical conjugation of carbohydrate derivatives to a protein such as albumin, to a lectin in protein mixtures of extracts, presented on a nitrocellulose membrane or a microtiter plate well, also provides evidence for lectin presence [5]. Conveniently, naturally occurring polysaccharides such as dextrans, galactans or chitin (for further information on chitin, please see Chapter 12) can be used at a minimal cost. If they do not satisfy the individual carbohydrate specificity of the lectin under study, tailor-made affinity adsorbents are needed. A versatile procedure to prepare such a resin, enabling to reach high yields, is the use of divinyl sulfone to introduce active sites for the following covalent ligand immobilization (mono-, di- or oligosaccharides, glycoproteins) (for a practical protocol, please see [6] for details). Glycoprotein-loaded resins can even be applied as a matrix for groups of lectins, when using a protein decorated with diverse types of sugar chains, for example ovomucoid or hog gastric mucin, preferably desialylated [7]. Having bound the lectin to the resin and washed out all contaminating

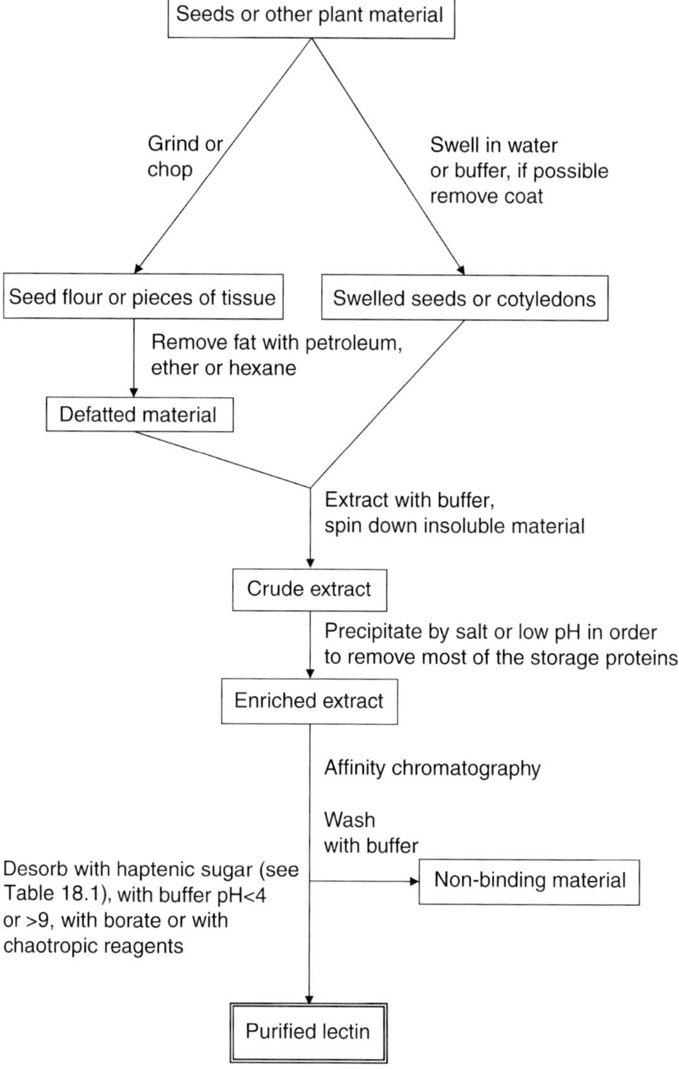

Figure 18.2 General scheme for the purification of plant lectins.

material, the mildest way for desorption of a lectin from an affinity column is to use the haptenic carbohydrate. If this is too expensive, alternatives will have to be devised, for example by testing borate to elute the lectin, prior to shifting the pH value of the elution buffer or having to resort to chaotropic reagents. If applied as a shallow gradient of increasing concentration, it is even possible to resolve closely related isolectins by borate, as demonstrated for the isolectins of French beans (please see Info Box 1) [8]. Following its purification, the new plant lectin may then find its way into laboratories as tool for applications outlined in the next section.

18.4
Applications

A key factor favoring the use of plant lectins in diverse assays is their stability. Also, they can usually be labeled without harming their activities to detect carbohydrate-dependent binding to cells or tissue sections (for detailed protocols, please see [9]). As summarized in Table 18.3, their glycan specificity (for details, please see Table 18.1) makes structural analysis and separation of mixtures of glycans or glycoconjugates possible. Thus, disease-associated alterations of the glycophenotype of cells become readily detectable (please see Chapter 25.2 for examples and Chapter 14.4–6 for technical aspects of arrays). One of the most frequent applications in immunology remains the stimulation of mitosis in lymphocytes, whose detection is recounted in Info Box 1. Mostly T cells are affected, but in certain cases also B cells respond [10]. This cellular activity and also induction of mediator release such as proinflammatory cytokines have nourished the assumption of a therapeutic potential for plant lectins, for example by stimulating antitumor defense mechanisms. However, as it turned out, respective experiments have not validated this assumption, even pointing to growth-stimulatory effects not only on lymphocytes but also on tumor cells *in vitro* and *in vivo* [2, 11]. In general, these results also epitomize the effects of plant lectins on mammalian cells. This widely documented elicitor activity and the widespread occurrence of plant lectins let us expect a series of biological functions in plants.

18.5
Biological Functions

At the outset, the diversity of lectins both in structure and in carbohydrate specificity makes it questionable that a common biological role can be attributed to all lectins. In principle, the main concepts to answer the given question can be grouped into two categories: to assume interaction of a lectin with exogenous ligands or with binding partners in the plant. Table 18.4 is presented accordingly, summarizing the evidence for lectin activities, starting with a defence function. Toxic lectins such as those from castor (*Ricinus communis*) and French beans can apparently protect the nutritious seeds from predators. If insufficiently cooked, consumption of French beans can lead to severe gastrointestinal irritations. However, a lectin/toxin will not necessarily be a biohazard to all animals. Fittingly, the effects of plant lectins on insects are rather specific and can neither be predicted nor generalized [11]. That legume seeds contain lectins in abundant quantities and that legume roots are the site for symbiosis with nitrogen-fixing bacteria led to the idea that lectins may participate in initiating the plant–

Table 18.3 Versatility of plant lectins as research tools (from [11], extended and modified).

Biochemistry
Detection of defined carbohydrate epitopes of glycoconjugates in blots or on thin-layer chromatography plates
Purification of lectin-reactive glycoconjugates by (serial) affinity chromatography
Glycan characterization by serial lectin affinity chromatography (lectin affinity capture)
Glycome analysis by lectin microarrays (glycomics)
Quantification of lectin-reactive glycoconjugates in enzyme-linked lectin-binding assays (ELLA)
Quantification of activities of glycosyltransferases/glycosidases by lectin-based detection of products of enzymatic reaction
Model reagents to assess ligand functionality of carbohydrate-presenting scaffolds (for example glycodendrimers)
Detection of glycomimetic peptides or aptamers
Cell biology
Characterization of intracellular assembly, routing and cell-surface presentation of glycoconjugates in normal and genetically engineered cells (glycomic profiling, spatially defined by cyto- or histochemistry)
Competitive assays with cell extracts on lectin microarrays to define differences in relative glycan abundance between cell populations
Selection of cell variants (mutants, transfectants) with altered lectin-binding properties as models for dissecting the glycosylation machinery and glycan functionality (glycomic profiling, functionally defined)
Fractionation of cell populations by sorting or bead binding
Modulation of proliferation or activation status of cells and dissecting the involved signal pathways
Model substratum for study of cell adhesion, aggregation and migration
Medicine
Analysis of glycosylation in recombinant glycoproteins
Detection of disease-related alterations of glycan synthesis by lectin cyto- and histochemistry
Histo-blood group typing and definition of secretor status
Quantification of disease-associated aberrations of glycan presentation (serum glycoproteins, cell surfaces)
Cell marker for diagnostic purposes including infectious agents (viruses, bacteria, fungi, parasites)
Cell marker for functional assays to pinpoint defects in cell activities such as mediator release

Table 18.4 Functions of plant lectins[a].

	Activity	Example of lectin
	Protection from fungal attack	*Hevea brasiliensis* (rubber tree), *Urtica dioica* (stinging nettle), *Solanum tuberosum* (potato)
	Protection from herbivorous animals	*Phaseolus vulgaris* (French bean), *Ricinus communis* (castor bean), *Galanthus nivalis* (snowdrop), *Triticum vulgare* (wheat)
External	Association of hydrolytic enzyme activity with lectin-like domain	Strawberry β-galactosidase, tomato endo-β1,4-glucanase (cellulase) SlCel9C1[b]
	Involvement in establishing the symbiosis between plants and bacteria	*Pisum sativum* (common pea), *Lotononis bainesii* (miles lotononis), *Arachis hypogaea* (peanut), *Triticum vulgare* (wheat), *Oryza sativa* (rice)
	Biosignaling by a receptor kinase with an extracellular lectin-like domain	LysM domain in Nod-factor perception (*Lotus japonicus, Medicago truncatula*) and leguminous lectin-like domain (*Arabidopsis thaliana, Populus nigra* var. *italica*)
	Storage proteins	Valid for all seed lectins
	Ordered deposition of storage proteins and enzymes in protein bodies and mediation of contact between storage proteins and protein body membranes	*Pisum sativum* (common pea), *Lens culinaris* (lentil), *Glycine max* (soybean), *Oryza sativa* (rice)
Internal	Modulation of enzymatic activities such as phosphatases	*Secale cereale* (rye), *Solanum tuberosum* (potato), *Pleurotus ostreatus* (oyster mushroom), *Glycine max* (soybean), *Dolichos biflorus* (horse gram)
	Participation in growth regulation	*Medicago sativa* (alfalfa), *Cicer arietinum* (chick pea)
	Adjustment to altered environment conditions	*Triticum aestivum* (winter wheat)

[a] For further information on carbohydrate specificities, please see Table 18.1; for recent review, please see [2].

[b] formerly called TomCel8 (tom for tomato; Cel for cellulase), now SlCel9C1 (Sl for *Solanum lycopersicon*, now *Lycopersicon esculentum*).

bacterium symbiosis. These effects may be less direct than initially proposed, with distinct lipochitooligosaccharides (so-called nodulation factors) playing dominant roles in this communication. Sensing of these factors by the plant involves receptor kinases with a lysin motif (LysM) domain [12]. It was originally identified in bacterial peptidoglycan-degrading enzymes (lysins). This domain listed in Table 18.2 is also present in chitinases from the fungus *Kluyveromyces lactis*, the alga *Volvox carteri* and the nematode *Caenorhabditis elegans*, indicating affinity of the module to peptidoglycans and chitin. In a general context, linking receptor-like kinases with a lectin-like domain are also seen in other cases – an attractive means to connect sugar-encoded messages with signaling [13].

As noted above, lectins bind very specifically to components in the plant, for example defined subpopulations of storage proteins [11]. This association depends both on protein–carbohydrate and on protein–protein interactions [14, 15]. *In vivo*, these storage proteins – together with the lectin – are directed to storage organelles, the protein bodies. Further sets of binding partners of lectins are the protein body membranes [16], in line with the hypothesis that lectins might be engaged in routing and packaging the content of the protein bodies (Table 18.4). Intracellular interactions may also have a bearing on properties of the target protein, for example on phosphatases [17]. The detection of lectin modules in proteins with enzymatic activities [in strawberry a β-galactosidase or in tomato an endo-β1,4-glucanase (cellulase)] is evocative of a similar display in modular bacterial/fungal enzymes degrading plant glycopolymers [18] and respective modular proteins in animals (please see Chapter 19). In each case, the enzymatic center might be guided to its substrate by the lectin domain. Modular design also underlies dual-function lectins with enzymatic activity apart from a glycoside hydrolase, for example the *Dolichos biflorus* seed lectin with lipoxygenase activity [19]. Beyond enzymatic centers, plant lectins can also harbor a site with affinity for hydrophobic compounds; physiologically, implications of this reactivity being not yet clear. Finally, the analogy to animal lectins regarding intracellular presence has inspired the concept of plant lectins being engaged in regulatory processes in the cytoplasm and the nucleus (for respective information on animal lectins, please see Chapter 19) [20].

18.6 Conclusions

Lectins are a common constituent of plants, originating from the independent development of sugar-binding activity in different folds. Their ability to select glycan epitopes has turned them into popular laboratory tools for glycophenotyping. Concepts are being developed to define the function of plant lectins, a challenge for this field. Here, research on animal and microbial lectins, for example the detection of the strategic positioning of enzyme activities by a lectin domain in modular proteins (Table 19.2), can advance the status of knowledge.

Summary box

Agglutination of erythrocytes underlies the term 'agglutinin', the 'A' in abbreviations for plant lectins. The diversity of folds with lectin activity and their wide occurrence define the lectins as a common constituent in plants. Due to their selective binding to glycans they have become popular laboratory tools for glycan detection and analysis. Obvious functions *in situ* encompass the role of the lectin part in AB-toxins to initiate cell contact for toxin import into cells and defence against fungi and insects, if the respective lectin targets chitin.

References

1 Loris R. Principles of structures of animal and plant lectins. *Biochim Biophys Acta* 2002;1572:198–208.
2 Gabius H-J *et al.* Chemical biology of the sugar code. *ChemBioChem* 2004;5:740–64.
3 Jiménez M *et al.* AB-type (toxin/agglutinin) from mistletoe: differences in affinity of the two galactoside-binding Trp/Tyr-sites and regulation of their functionality by monomer/dimer equilibrium. *Glycobiology* 2006;16:926–37.
4 Stillmark H. Ueber Ricin, ein giftiges Ferment aus den Samen von *Ricinus comm.* L. und einigen anderen Euphorbiaceen. MD Thesis, Kaiserliche Universität zu Dorpat (now Tartu), Schnakenburg's Buchdruckerei 1888.
5 Gabius S *et al.* Neoglycoenzymes: a versatile tool for lectin detection in solid-phase assays and glycohistochemistry. *Anal Biochem* 1989;282:447–51.
6 Gabius H-J. Influence of type of linkage and spacer on the interaction of β-galactoside-binding proteins with immobilized affinity ligands. *Anal Biochem* 1998;189: 91–4.
7 Freier T *et al.* Affinity chromatography on immobilized hog gastric mucin and ovomucoid. A general method for the isolation of lectins. *Biol Chem Hoppe-Seyler* 1985;366:1023–8.
8 Fleischmann G *et al.* A one-step procedure for the isolation and resolution of *Phaseolus vulgaris* isolectins by affinity chromatography. *Biol Chem Hoppe-Seyler* 1985;366:1029–32.
9 Gabius H-J, Gabius S (Eds.). *Lectins and Glycobiology*. Springer, Heidelberg, 1993.
10 Waxdal MJ *et al.* B- and T-cell stimulatory activities of multiple mitogens from pokeweed. *Nature* 1974;251:163–4.
11 Rüdiger H, Gabius H-J. Plant lectins: occurrence, biochemistry, functions and applications. *Glycoconj J* 2001;18:589–613.
12 Radutoiu S *et al.* LysM domains mediate lipochitin-oligosaccharide recognition and Nfr genes extend the symbiotic host range. *EMBO J* 2007;26:3923–35.
13 André S *et al.* Evidence for lectin activity of a plant receptor-like protein kinase by application of neoglycoproteins and bioinformatic algorithms. *Biochim Biophys Acta* 2005;1725: 222–32.
14 Einhoff W *et al.* Interactions between lectins and other components of Leguminosae protein bodies. *Biol Chem Hoppe-Seyler* 1986; 367:15–25.
15 Freier T, Rüdiger H. *In vivo* binding partners of the *Lens culinaris* lectin. *Biol Chem Hoppe-Seyler* 1987;368:1215–23.
16 Schecher G, Rüdiger H. Interaction of the soybean (*Glycine max*) seed lectin with components of the soybean protein body membrane. *Biol Chem Hoppe-Seyler* 1994; 375:829–32.
17 Conrad F, Rüdiger H. The lectin from *Pleurotus ostreatus*: purification, characterization and interaction with a phosphatase. *Phytochemistry* 1994;36:277–83.
18 Boraston AB *et al.* Carbohydrate-binding modules: fine-tuning polysaccharide recognition. *Biochem J* 2004;382:769–81.
19 Roopashree S *et al.* Dual-function protein in plant defense: seed lectin from *Dolichos biflorus* (horse gram) exhibits lipooxygenase activity. *Biochem J* 2006;395:629–39.
20 van Damme EJM *et al.* Cytoplasmic/nuclear plant lectins: a new story. *Trends Plant Sci* 2004;9:484–9.

19
Animal and Human Lectins

Hans-Joachim Gabius

Previous chapters (Chapters 6–11) have illustrated the enormous structural diversity of glycans. An elaborate synthetic machinery including enzymes for introducing distinct substitutions and remodeling underlies the establishment of the exceptionally high coding capacity of glycans. The question immediately arises as to how this information is turned into biological responses. An intermolecular contact within the decoding process is especially favored at branch ends of mature chains. They are spatially readily accessible (for a primer on the interaction of glycans with proteins, please see Chapter 13). The limited flexibility around glycosidic linkages in glycans accounts for the presentation of only few key-like conformers suited for contact to a lectin (for conformational aspects of glycans, please see Chapters 2 and 13; for definition of the term 'lectin', Chapter 15.3).

Given all these favorable characteristics of glycans at branch ends to enable a role as biochemical signals (structural variability, spatial accessibility and limited flexibility), the level of diversity reached on the sugar side should be matched in numbers by lectins. What held true for bacteria and viruses (please see Chapter 17) and plants (please see Chapter 18), in which numerous lectins had been detected, was questioned for higher animals as late as 1973 [1] (one year before the first mammalian lectin was isolated; please see Chapter 15 for details). The following decades witnessed a remarkable surge in our knowledge about lectins in higher animals, both in number of families and family members [2]. With lectins now finally being accepted as 'a common cell component' [1], lectin–carbohydrate interactions are expected to entail a wide range of functions, a key topic of this chapter. The chapter first provides an overview on protein folds with lectin activity and next on lectin functions. Triggering such responses depends on the binding of distinct glycoconjugates, that is the bioactive lectin ligands, as will be exemplified. Not only their presence but also spatial parameters will be identified to be crucial for bioactivity. As a consequence, six levels, at which the affinity of a glycan for a lectin can be adjusted, will be presented at the end of this chapter. This synopsis can serve as a guideline to interpret structural changes in glycosylation as regulatory events.

The Sugar Code. Edited by Hans-Joachim Gabius

ISBN: 978-3-527-32089-9

19.1
Protein Folds with Lectin Activity

The concept for a complex network of productive protein–carbohydrate interactions would be strongly supported, if several folding patterns were adapted to binding glycans. Then sugar-binding modules are clearly not a singular invention of restricted relevance. The sheer size of the compilation of proven cases, presented in Table 19.1, is compelling evidence for glycan binding to be a rather common aspect of protein evolution (a view on the secondary-structure patterns of C- and P-type lectins, the antiparallel β-sandwich and β-barrel is given in Figure 16.1). Starting from ancestral modules, gene duplications and sequence variations then spurred the process of intrafamily diversification. Already subtle sequence deviations in otherwise closely related C-type lectins (for example a Val/Ser substitution in dendritic and endothelial cell receptors) have a significant bearing on glycan specificity (for details, please see [3]). As recounted in Chapter 20, the C-type domain is present in widely divergent subgroups of respective lectins and has even developed into a module accommodating tyrosine sulfates in addition to carbohydrates as in P-selectin or peptides in natural killer (NK) cell

Table 19.1 Overview of protein folds with lectin activity.

Type of fold	Example for lectin	Example for ligand[a]
C-type	Asialoglycoprotein receptor, collectins, selectins	Fuc, Gal, GalNAc, Man, heparin tetrasaccharide
I-type (Ig fold)	N-CAM, TIM-3, siglecs	$Man_6GlcNAc_2$, HNK-1 epitope, α2,3/6-sialylated glycans
P-type	Mannose-6-phosphate receptors (MR) and proteins with MR homology domain (erlectin, also called XTP3-B [XTP3-transactivated protein], OS-9)	Man-6-phosphate, $Man_{5,8}GlcNAc_2$
β-Sandwich (jelly-roll)	a) Galectins	β-galactosides
	b) Calnexin, calreticulin	$Glc_1Man_9GlcNAc_2$
	c) ERGIC-53	$Man_xGlcNAc_2$
	d) CRD[b] of Fbs1 in SCF E3 ubiquitin ligase and peptide-N-glycanase	$Man_3GlcNAc_2$; mannopentaose
	e) Pentraxins	Glycosaminoglycans, MOβDG, 3-sulfated Gal, GalNAc and GlcA, Man-6-phosphate
	f) G-domains of the LNS family (laminin, agrin)	Heparin

continued

Table 19.1 *Continued*

Type of fold	Example for lectin	Example for ligand[a]
β-Trefoil	a) Fibroblast growth factors b) Cysteine-rich domain of C-type macrophage mannose receptor c) Lectin domain in GalNAc-Ts[c] involved in mucin-type O-glycosylation d) Hemolytic lectin CEL-III of sea cucumber and lectin EW29 of earthworm	Heparan sulfate GalNAc-4-sulfate in LacdiNAc GalNAc Gal
β-Propeller	a) 4-bladed: tachylectin-3 b) 5-bladed: tachylectin-2 c) 6-bladed: tachylectin-1	S-type lipopolysaccharide GlcNAc/GalNAc Gram-negative lipopolysaccharide (2-keto-3 deoxy-octonate: Kdo)
β-Prism II	Pufferfish (fugu) lectin	Man
β-Barrel with jelly-roll topology	Horseshoe crab tachylectin-4, eel (*Anguilla anguilla*) agglutinin, *Xenopus* X-epilectin	Fuc
Fibrinogen-like domain	a) Ficolins b) Intelectins (mammalian, *Xenopus*) c) Tachylectin-5 d) Slug (*Limax flavus*) lectin	GlcNAc Gal*f*, pentoses N-Acetylated sugars Sialic acid
Link module	CD44, TSG-6, LYVE-1, aggregating proteoglycans	Hyaluronic acid
Hevein-like domain	Tachycytin and spider (*Selenocosmia huwena*) neurotoxin; cobra venom cardiotoxin	GlcNAc Heparin-derived disaccharide
$(\beta/\alpha)_8$ Barrel (glycoside hydrolase family 18)	YKL-40 (human cartilage glycoprotein-39, chitinase-like lectin)	$(GlcNAc)_n$
α/β-Fold with two long structured loops	Lectin domain of mouse latrophilin-1, a G-protein-coupled receptor	Rha
Short consensus repeat (complement control protein module)	Factor H (complement regulator)	Glycosaminoglycans, sialic acid

[a] For abbreviations of names for sugars, please see Chapter 1.
[b] Carbohydrate recognition domain.
[c] *N*-acetylgalactosaminyltransferases.

receptors. Equally intriguing, dynamic binding-site evolution holds true for other lectin families such as the galectins or siglecs, fulfilling the structural prerequisite for nonredundant activity profiles [2–5]. There are more, not yet structurally well-defined lectin sites, found in the α_M-integrin (please see also Figure 29.4), α/θ-defensins (please see Chapter 26) or the sea urchin 350-kDa sperm-binding protein and a nucleocytoplasmic GlcNAc-binding protein, the two latter proteins belonging either to the hsp (heat shock protein) 110 and hsp 70 groups, respectively.

When looking at the occurrence of folds in plant lectins (Chapter 18), several motifs are shared among plant and animal lectins such as the β-sandwich fold of leguminous agglutinins, the classical example being concanavalin A (for illustration, please see Figure 16.1a), the β-trefoil fold (first detected in soybean trypsin inhibitor) in the lectin subunit of AB-toxins (ricin and others), and in amaranthin as well as the hevein-like domain in wheat germ agglutinin and also chitinases. Close inspection of positioning of the binding sites and/or the binding mode intimates a convergent rather than divergent course of evolution in the mentioned cases [6]. This aspect should be reckoned with in any future suggestions for a terminology system. With these multiple folds engaged in sugar recognition, the expectation is nourished for multiple functions.

19.2
Functions of Animal and Human Lectins

In concert with the glycans of cellular glycoconjugates, lectins can turn their carbohydrate-binding activity into specific recognition already in the endoplasmic reticulum and then at the cell surface. As already noted in Chapter 6, the nascent and processed *N*-glycans, originating from the precursor common to all Asn-X-Ser/Thr-defined acceptor sites, are signals for quality control by virtue of their ligand capacity. The information at each stage of glycan processing is translated by lectins into efficient monitoring for correct folding of nascent glycoproteins and into ensuing intracellular transport. This intracellular activity profile fills the first part of Table 19.2. Of note, the glycans added in a cotranslational manner may well guide folding pathways and later influence secretory efficiency and protein activity/stability, these processes in principle also involving intramolecular protein–carbohydrate interactions or a switching-off of intermolecular interactions. When glycoproteins with mature glycans finally reach the cell surface, cell adhesion and diverse cellular responses can be attributed to *cis/trans* interactions (Table 19.2). Examples for lectin-elicited intracellular signaling routes leading to responses such as growth control are further explained with illustrations in Chapters 25 and 27 (Figures 25.3 and 27.2).

Toward these ends, a factor different from domain folding comes into play. In addition to direct ligand binding by a lectin's carbohydrate recognition domain (CRD), spatial aspects of CRD arrangement contribute markedly to *in vivo* lectin functionality. They not only increase the affinity by multivalent interactions but

Table 19.2 Functions of animal and human lectins.

Activity	Example of lectin
Recognition of the stem region of *N*-glycans, a signal for ubiquitin conjugation when accessible in incorrectly folded glycoproteins	F-box proteins Fbs1/2 (Fbx2/FBG1, Fbx6b/FBG2) as a ligand-specific part of SCF (Skp1-Cul1-F-box protein) ubiquitin ligase complexes
Molecular chaperones with dual specificity for $Glc_1Man_9GlcNAc_2$ and the protein part of nascent glycoproteins in the endoplasmic reticulum (ER)	Calnexin, calreticulin
Targeting of misfolded glycoproteins with $Man_{8-5}GlcNAc_2$ as carbohydrate ligand to endoplasmic reticulum-associated degradation (ERAD)	EDEM1,2/Mnl1 (Htm1) (lectins or glycosidases?), Yos9 protein (MRH domain) in yeast, erlectin (XTP3-B) and OS-9 in mammals
Intracellular routing of glycoproteins and vesicles and apical delivery	ERGIC-53 and VIP-36 (probably also ERGL and VIPL), P-type lectins, comitin, galectins-3 and -4
Intracellular transport and extracellular assembly	Nonintegrin 67-kDa elastin/laminin-binding protein
Enamel formation and biomineralization	Amelogenin
Inducer of membrane superimposition and zippering (formation of Birbeck granules)	Langerin (CD207)
Cell type-specific endocytosis	Hepatic and macrophage asialoglycoprotein receptors, dendritic cell and macrophage C-type lectins [mannose receptor family members (tandem-repeat type) and single CRD lectins such as trimeric langerin/CD207 or tetrameric DC-SIGN/CD209], cysteine-rich domain (β-trefoil) of the dimeric form of mannose receptor for $GalNAc\text{-}4\text{-}SO_4$-bearing glycoprotein hormones in hepatic endothelial cells, P-type lectins
Recognition of foreign glycans (β1,3-glucans, lipopolysaccharide)	CR3 (CD11b/CD18, Mac-1 antigen), C-type lectins such as DC-SIGN and dectin-1, immulectins, intelectins, *Limulus* coagulation factors C and G, earthworm CCF, tachylectins

continued

Table 19.2 *Continued*

Activity	Example of lectin
Recognition of foreign or aberrant glycosignatures on cells (incl. endocytosis or initiation of opsonization or complement activation)	Collectins, ficolins, C-type macrophage and dendritic cell lectins, CR3 (CD11b/CD18, Mac-1 antigen), α/θ-defensins, pentraxins (CRP, limulin), RegIIIγ (HIP/PAP), siglecs, tachylectins
Targeting of enzymatic activity in multimodular proteins	Acrosin, C-terminal β-sandwich lectin domain of mouse peptide-*N*-glycanase, *Limulus* coagulation factor C, laforin, β-trefoil fold [$(QxW)_3$ domain] of GalNAc-Ts involved in mucin-type *O*-glycosylation (frequent in microbial glycosylhydrolases for plant cell wall polysaccharides)
Bridging of molecules	Galectins, C-type lectins, cerebellar soluble lectin
Induction or suppression of effector release (H_2O_2, cytokines etc.)	Galectins, selectins and other C-type lectins such as CD23, BDCA-2 and dectin-1, I-type lectins [CD33 (siglec-3), siglecs-7 and -9], Toll-like receptor-4
Modulation of enzymatic activities/receptor endocytosis via glycan recognition	Mannan-binding lectin (acting on neprins), galectins (acting on growth factor receptors)
Cell growth control, induction of apoptosis/anoikis and axonal regeneration	Galectins, C-type lectins, amphoterin-like protein, hyaluronic acid-binding proteins, cerebellar soluble lectin, CD22 (siglec-2), MAG (siglec-4)
Cell migration and routing	Galectins, selectins and other C-type lectins, I-type lectins, hyaluronic acid-binding proteins (RHAMM, CD44, hyalectans/lecticans)
Cell–cell interactions	Selectins and other C-type lectins such as DC-SIGN, galectins, I-type lectins (siglecs, N-CAM, P_0 or L1), gliolectin
Cell–matrix interactions	Galectins, heparin- and hyaluronic acid-binding lectins including hyalectans/lecticans, calreticulin
Matrix network assembly	Proteoglycan core proteins (C-type CRD and G1 domain of hyalectans/lecticans), galectins (for example galectin-3/hensin), nonintegrin 67-kDa elastin/laminin-binding protein

also the selectivity for distinct types of ligand display, for example to distinguish self from nonself-glycan signatures [3]. The ways Nature has compensated for the weakness of a single protein–carbohydrate contact are illustrated in Figure 19.1 (please see legend for further details). A firm grip on glycans will be facilitated by the suitably spaced CRDs. This topological arrangement is realized and drawn for galectins, a serum collectin and two other depicted C-type lectins endowed with capacity for endocytic uptake (Figure 19.1). The given arrangement, matched by branched glycans, readily explains the rapid removal of asialoglycoproteins with triantennary *N*-glycans from circulation by the hepatic lectin (please see Chapter 15.4). Figure 19.1 then shows a different and equally effective way to turn sugar binding into a host of cell actions, that is the modular arrangement connecting CRDs with other types of modules. This is encountered in various lectins. As seen for selectins and I-type lectins such as siglecs (Figure 19.1), the presence of spacers ensures the separation of the CRD from the cell surface, a prerequisite for accessibility in cell–cell interactions. In selectins, interdomain contacts and the force-dependent equilibrium between the closed-angle and the extended positioning of the two distal types of module have a strong bearing on the kinetics of binding. The transition to the extended form at low force level prolongs the lifetime of selectin–ligand association. The counterintuitive effect that applied force stabilizes a cell–cell contact leads to regular rolling of leukocytes on a vascular surface (for further details on selectins, please see Chapter 27.3). The so-called catch bonds reduce k_{off}-values, and a structural force-dependent change distant to the CRD, yielding a high-affinity state, underlies the unique and counterintuitive feature of selectins to support rolling and tethering against a hydrodynamic force (please see also Chapter 17.1.1 for example) [7, 8]. In siglecs, ligand binding accounts for covalent alterations by a posttranslational modification far away, that is in the intracellular section of these lectins (please see below).

Modular design has a wide range of functional implications. It aids routing of collectin-opsonized particles, targeting of enzyme activities (for example in laforin, the product of the EMP2A gene, a dual-specificity protein phosphatase with a lectin domain; defects are linked to an autosomal recessive progressive myoclonus epilepsy, the lafora disease, on the clinical level, to lymphoma development in nude mice) or the just mentioned transmembrane signaling reaching intracellular immunoreceptor tyrosine-based activating motifs/immunoreceptor tyrosine-based inhibitory motifs found in several C-type lectins or siglecs (Table 19.2) [3]. Embedded in modular display, the C-type lectin-like domain of aggregating proteoglycans reacts with sugar or peptide motifs and contributes to their role as molecular glue in the matrix (Figure 19.1, upper right part). In cooperation with the selection of the ligand by a CRD, spatial aspects of CRD positioning in the lectin inspect glycosignatures and control the lectins' reactivity to glycans. Implicitly, endogenous lectins are expected to show considerable selectivity for binding partners. Compared to plants as a source for lectins (please see Chapter 18.2 for details), animal tissues contain much less lectin so that recombinant expression has become a valued tool to be able to address this issue.

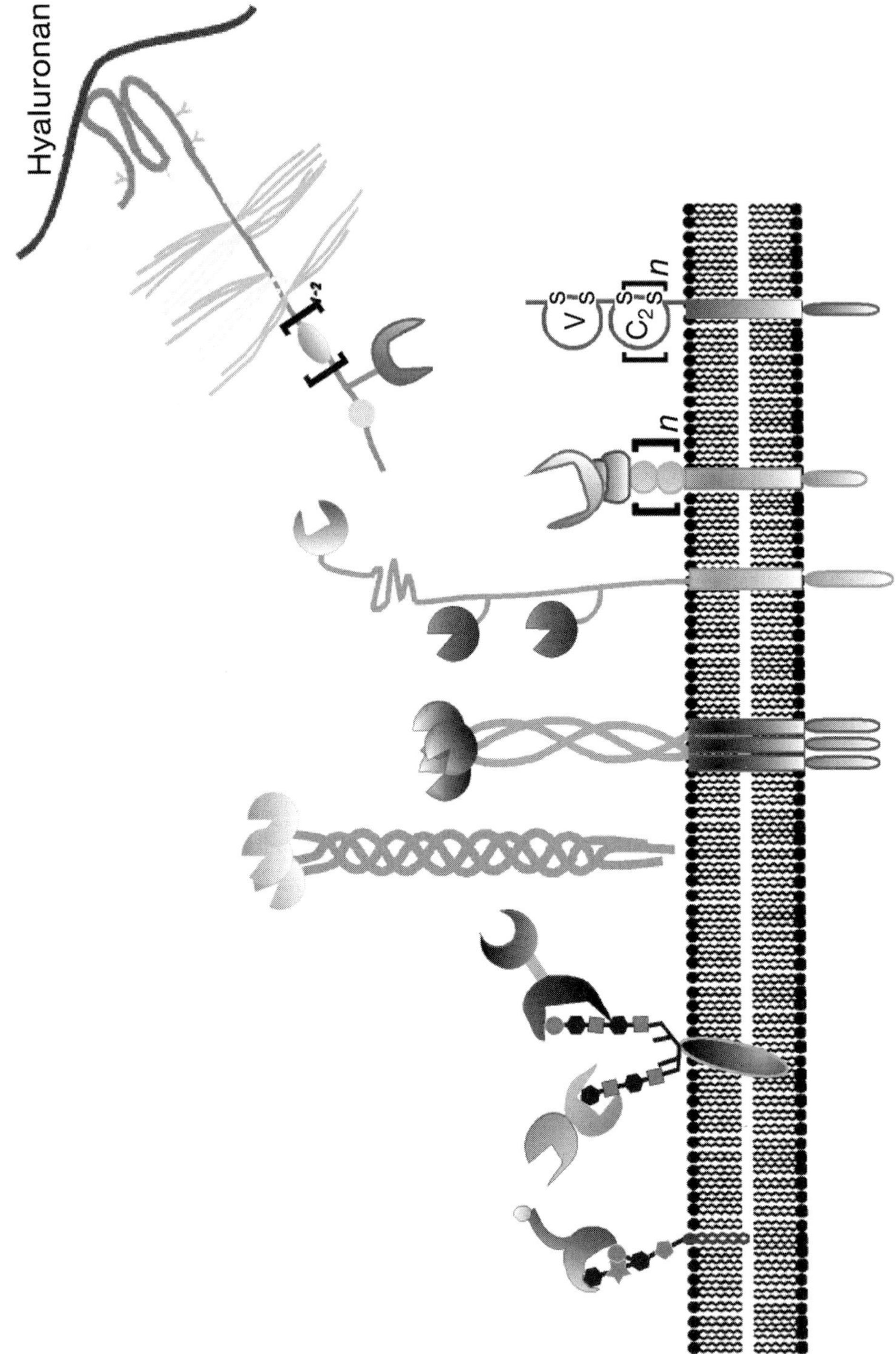
Hyaluronan
V
C2
n
n

Figure 19.1 Illustration of the strategic ways how CRDs in animal lectins are positioned to reach optimal ligand selection (for example to separate self from nonself-glycan profiles in innate immunity) and topological complementarity. From left to right, the CRD display in the three subtypes within the galectin family (chimeric, proto-type and tandem-repeat-type arrangements binding to a ganglioside or a branched complex-type *N*-glycan without or with terminal α2,3-sialylation), the presentation of CRDs (C-type or fibrinogen-like domain) in serum and surfactant collectins or ficolins connected to their collagenous stalks and the noncovalent association of binding sites in transmembrane C-type lectins by α-helical coiled-coil stalks (for example asialoglycoprotein and Kupffer cell receptors, the scavenger receptor C-type lectin, CD23, DC-SIGN or DC-SIGNR) are given. Similar to tandem-repeat-type galectins the C-type family of lectins also has a branch of members with this design, that is, immulectins-1, -2 and -3. Next, the tandem-repeat display in the mannose-specific macrophage receptor (also found on dendritic cells, hepatic endothelial cells, kidney mesangial cells, retinal pigment epithelial cells and tracheal smooth muscle cells) and the related C-type lectin Endo180 with eight domains as well as in the cation-independent P-type lectin with 15 domains is presented. Capacity for sugar binding is confined to only a few domains as depicted. The occurrence of lectin activity for GalNAc-4-SO_4-bearing pituitary glycoprotein hormones in the cysteine-rich domain, a member of the β-trefoil fold family with one $(QxW)_3$ domain in the *N*-terminal section of the macrophage mannose receptor (amino acids 8–128), which is linked via a fibronectin-type-II-repeat-containing module to the tandem-repeat section, is also included into the schematic drawing for these lectins with more than one type of CRD per protein chain. Moving further to the right side, the association of a distal CRD in selectins (attached to an epidermal-growth-factor (EGF)-like domain and two to nine complement-binding consensus repeats) or in the siglec subfamily of I-type lectins using 1–16 C2-set immunoglobulin-like units as spacer equivalents to let the CRD reach out to contact ligands and to modulate capacity to serve in *cis*- or *trans*-interactions on the cell surface is shown. The force-dependent alterations of the topological arrangement of the two distal domains in selectins accounts for catch bonds of selectins (please see text). A canonical immunoreceptor tyrosine-based inhibitory motif (ITIM) together with a putative tyrosine-based signaling motif is frequently present in the intracellular portion of siglecs, especially in the CD33(siglec-3)-related subgroup. C2-Set domains linked to fibronectin-type-III repeats establish the extracellular section of the I-type lectins L1 and neural cell adhesion molecule (N-CAM). In the matrix, the modular proteoglycans (hyalectans/lecticans: aggrecan, brevican, neurocan and versican) interact (i) with hyaluronan (and also link protein) via the link-protein-type modules of the *N*-terminal G1 domain (and an immunoglobulin-like module), (ii) with receptors binding to the glycosaminoglycan chains in the central region and (iii) with carbohydrates or proteins (fibulins-1 and -2 and tenascin-R) via the C-type lectin-like domain flanked by EGF-like and complement-binding consensus repeat modules (kindly provided by H. Kaltner).

19.3
Lectin Ligands and Affinity Regulation

Purification of tissue lectins, often using affinity chromatography as a crucial step (please see Chapter 15.3 for its introduction to lectin research and Chapter 18.3 for further practical details), and also their labeling, as similarly done with plant lectins, were thus crucial to trace endogenous ligands. The range of functions presented in Table 19.2 intimates a correlation between certain lectins and their targets, sometimes already reflected in lectin names. A lysosomal enzyme will present the epitope with mannose-6-phosphate as the postal code for P-type-lectin-mediated transfer to its final destination, and mannan on the cell surface, a characteristic of the mannose-rich glycophenotype of yeast cells, is a high-affinity matrix for the mannan (or mannose)-binding lectin [9, 10]. Functions in cell adhesion and growth or mediator release call for particular sets of glycoligands in the respective context. The summary of currently known binding partners for two multifunctional galectins lists glycoconjugates known for adhesion and cell signaling capacities (Table 19.3). From the wide array of cell-surface glycoconjugates, a tissue lectin is capable of selecting a few binding partners. For example, interaction with the extracellular matrix glycoproteins fibronectin and laminin can modulate adhesion (here a galectin acts as a nonintegrin receptor), binding to CD7 or ganglioside GM1 initiates

Table 19.3 Cellular glycoconjugates and proteins as ligands for endogenous lectins: case study of galectins-1 and -3.

Type of ligand	Galectin-1	Galectin-3
Glycan	Ovarian carcinoma antigen CA125, CD2, CD3, CD4, CD7, CD43, CD45, CD95, carcinoembryonic antigen, fibronectin (tissue), gastrointestinal mucin, hsp90-like glycoprotein, $\alpha_1/\alpha_4/\alpha_5/\alpha_7\beta_1$- and $\alpha_4\beta_7$-integrins, cell adhesion molecule L1, laminin, lamp-1, Mac-2-binding protein, nephrin, neuropilin-1, receptor protein-tyrosine phosphatase (RPTPβ), thrombospondin, Thy-1, tissue plasminogen activator, chondroitin sulfate proteoglycan, distinct neutral glycolipids, ganglioside GM1	CD7, CD11b of CD11b/CD18 (Mac-1 antigen, CR3), CD13 (aminopeptidase N), CD32, CD43, CD45, CD66a,b, CD71, CD95, CD98, carcinoembryonic antigen, colon cancer mucin and MUC1-D (*N*-glycan at Asn36), cubilin, C4.4A (member of Ly6 family), epidermal growth factor receptor, haptoglobin β-subunit (after desialylation), hensin (DMBT-1), glycoform of IgE, β_1-integrin (CD29) and $\alpha_4\beta_1$-integrin, LI-cadherin, laminin, lamp-1/-2, Mac-2-binding protein, Mac-3, MAG, MP20 (tetraspanin), NG2 proteoglycan, T cell receptor complex, tenascin, tissue plasminogen activator, transforming growth factor-β receptor, ganglioside GM1
Protein	Gemin4, oncogenic H-Ras, OCA-B, pre-B cell receptor (human, not murine system)	AGE products, Alix/AIP-1, ATP synthase b-subunit, axin, bcl-2, β-catenin, Cys/His-rich protein, Gemin4, hnRNPQ, mSufu, nucling, oncogenic K-Ras, OCA-B, pCIP, PIAS1, synexin (annexin VII), TTF-1

growth inhibition in activated T cells or in neuroblastoma cells (please see Chapters 25 and 27 for details) [4]. This principle figures as a decisive driving force not only for galectins, but also for other lectins such as selectins (Table 19.2). As further example, intercellular contacts via the dendritic cell lectin DC-SIGN (**d**endritic **c**ell-**s**pecific **I**CAM-3 **g**rabbing **n**onintegrin) similarly hinge on a cell-type-specific selection of the target glycoprotein. The dendritic cell engages the lectin to bind glycans of intercellular adhesion molecule (ICAM)-2 on endothelial cells, of ICAM-3 on T cells and of the α_M-subunit of the Mac-1 antigen (CR3 receptor, $\alpha_M\beta_2$-integrin) on neutrophils and of various pathogens with clustered Man or Lewis epitopes [11]. In addition to glycans, the two galectins, as noted above for the C-type lectin-like NK cell receptors or hyalectans/lecticans (please see legend to Figure 19.1), also bind peptide motifs. This property facilitates intracellular functions in pre-mRNA splicing, placement of oncogenic *ras* or apoptosis regulation (Table 19.3) [4]. Overall, the details emerging teach the following salient lesson: despite the abundance of glycan chains on the cell surface, animal and human lectins have a preference for particular binding partners in a distinct context.

In this sense, shifts in the assembly line for glycans and their substitution pattern can have an impact on affinity toward lectins – not only the presence of a certain epitope counts, but also its topological aspects, starting with the conformation of the lectin-binding determinant and the complete structure of the glycan chain including branching and core substitutions. These parameters have a bearing on lectin binding, although they are not physically involved in the molecular rendezvous [12,13]. It is therefore imperative to systematically profile factors with likely bearing on affinity regulation. If moving in a stepwise manner from the smallest interaction partner of a lectin, a monosaccharide, to the level of microdomains in membranes, six different layers with regulatory potential are then identified (Table 19.4). They give reason to imply an exquisitely tuned interplay between structural glycan tailoring and lectin expression, for example in fertilization (Chapter 24), malignancy (Chapter 25), inflammation (Chapter 27) or immune regulation (Info Box 1 in Chapter 30).

19.4 Conclusions

The assumption that lectin–carbohydrate recognition is a common mode for intermolecular association and biological information transfer is convincingly backed by the large number of folds with lectin activity. Ensuing intrafamily diversification within more than a dozen folds with identified CRDs led to a toolbox of specialized effectors and a broad range of covered functions. Toward this end, covalent and noncovalent CRD clustering opened the way to optimize ligand selection and to distinguish glycosignatures. With lectins thus being also sensors of topological aspects, the affinity of the binding process can swiftly be regulated by altering the local density at the level of individual glycoconjugates and even cell surfaces. Picking functional ligands from the glycomic complexity thus includes context-dependent parameters, making cell specificity of lectin binding possible.

Table 19.4 Six levels of regulation of affinity of glycan binding to a lectin.

1 Mono- and disaccharides (including anomeric position)

2 Oligosaccharides (including branching and substitutions)

3 Shape of oligosaccharide (*differential conformer selection*)

4 Spatial parameters of glycans in natural glycoconjugates
 (a) Shape of glycan chain (for example modulation of conformation by substitutions not acting as lectin ligand such as core fucosylation or introduction of bisecting GlcNAc in *N*-glycans, influence of protein part)
 (b) Cluster effect with bi- to pentaantennary *N*-glycans or branched *O*-glycans [including modulation by substitutions, please see (a)]

5 Cluster effect with different but neighboring glycan chains on the same glycoprotein (for example in mucins)

6 Cluster effect with different glycoconjugates on the cell surface in spatial vicinity forming microdomains

Summary Box

More than a dozen folds of animal and human proteins can accommodate a carbohydrate ligand. Binding entails a broad range of functions. Its affinity is regulated by sequence changes and topological aspects on the levels of the CRD and the glycoepitope. Respective changes on both sides can act in concert to optimize target specificity.

References

1 Roth S. A molecular model for cell interactions. *Q Rev Biol* 1973;48:541–63.

2 Gabius H-J. Animal lectins. *Eur J Biochem* 1997;243:543–76.

3 Gabius H-J. Cell surface glycans: the why and how of their functionality as biochemical signals in lectin-mediated information transfer. *Crit Rev Immunol* 2006;26:43–80.

4 Cooper DNW. Galectinomics: finding themes in complexity. *Biochim Biophys Acta* 2002;1572:209–31.

5 Angata T, Brinkman-van der Linden ECM. I-type lectins. *Biochim Biophys Acta* 2002; 1572:294–316.

6 Loris R. Principles of structures of animal and plant lectins. *Biochim Biophys Acta* 2002; 1572:198–208.

7 Phan UT *et al.* Remodeling of the lectin–EGF-like domain interface in P- and L-selectin increases adhesiveness and shear resistance under hydrodynamic force. *Nat Immunol* 2006;7:883–9.

8 Yago T *et al.* Catch bonds govern adhesion through L-selectin at threshold shear. *J Cell Biol* 2004;166:913–23.

9 Dahms NM, Hancock MK. P-type lectins. *Biochim Biophys Acta* 2002;1572:317–40.

10 Dommett RM *et al.* Mannose-binding lectin in innate immunity: past, present and future. *Tissue Antigens* 2006;68:193–209.

11 Cambi A *et al.* How C-type lectins detect pathogens. *Cell Microbiol* 2005;7:481–8.

12 André S *et al.* Substitutions in the *N*-glycan core as regulators of biorecognition: the case of core-fucose and bisecting GlcNAc moieties. *Biochemistry* 2007;46:6984–95.

13 André *et al.* From structural to functional glycomics: core substitutions as molecular switches for shape and lectin affinity of N-glycans. *Biol Chem* 2009;390:557–65.

20
Routes in Lectin Evolution: Case Study on the C-Type Lectin-Like Domains

Jill E. Gready and Alex N. Zelensky

The preceding chapters have introduced a common protein fold found in animal lectins, that is the C-type lectin domain. As illustrated in Figure 16.1h, which shows the X-ray structure of this fold in human P-selectin (for further details on selectins and their functions, see Chapters 19 and 27), the fold is characterized by a unique combination of individual features suited for stabilizing the structure and the primary sugar-binding site. The term originates from the essential presence of a Ca^{2+} which interacts with sugar ligands via coordination bonds. The domain shows ligand preferences for different types of carbohydrates, but also noncarbohydrate ligands. This has led to general use of the term C-type lectin-like domain (CTLD). As the proteins containing the domain (CTLD-containing proteins (CTLDcps)) comprise a large heterogeneous superfamily with diverse functions, they are an excellent model to study its evolution. Our method of analysis is also relevant to other lectin folds (given in Chapters 16 and listed in Tables 18.2 and 19.1), and to glycosyltransferases and other multimember glycogene groups (given for example in Chapters 6 and 7, with relevance for disease in Tables 22.1 and 23.1).

In this chapter we will show how systematic comparative analysis and data mining have created a strong framework for deciphering the complex functions of CTLDcps and studying their evolution. First, linking of protein sequence with three-dimensional structure provided a framework for interpretation of sugar specificity and binding. Second, tracking how these binding mechanisms have evolved in model organisms such as worm and human has revealed the exceptional capacity of the CTLD fold to evolve new specificities and functions. Definition of the full repertoire of CTLDcps in evolutionary branches by whole-genome sequence analysis has 'laid open' the field, posed unexpected questions and provided novel directions for further study.

20.1
C-Type Lectin (CTL) Evolution as a Case Study

In this chapter we will illustrate how the evolution of C-type lectins [1,2] (see Chapters 16, 19 and 27 for further information) demonstrates key principles of

The Sugar Code. Edited by Hans-Joachim Gabius

ISBN: 978-3-527-32089-9

routes for diversification, using these proteins as an exciting model. The following issues will be addressed in a stepwise manner, emphasizing how these insights have been obtained.

- How a protein scaffold – in this case the CTLD [3, 4] – with superior stability and versatility has the capacity to adapt to bind many types of ligands – not only carbohydrate – specifically. This allows it to be very successful evolutionarily, in the case of the CTLD becoming one of the most abundant protein domains in multicellular animals (Metazoa) [5–8].
- How this abundance of CTLDcps has developed differently in major animal lineages, such as worm [5], fly [6] and vertebrates [7, 8]. This has apparently occurred by independent evolution of novel CTLDcps with lineage-adapted functions, by recruiting the CTLD into the new proteins, usually in combination with other domains.
- How the likely earliest functions of C-type lectins as sugar-binding proteins as part of an innate immune response have been preserved in all animal lineages studied so far.
- How application of systematic whole-genome analyses can provide a sudden revelation of the extent of the CTLDcps repertoire in particular animal lineages, and how this can transform thinking and approaches to defining functions.
- How Nature is always full of surprises, with recent reports of CTLDcps in pre-Metazoa [9], and non-Metazoa, including plants [10], requiring a rethink of the earliest origin of the ‘C-type lectin’ with sugar-binding function.
- How generation of this knowledge has come from scientists working from diverse perspectives – genes and proteins, structure and function, genomics and proteomics, model organisms, and individual species.
- How this richness of acquired knowledge can be integrated into some pithy lessons on the evolution of C-type lectins, which also points the way to how evolution of other lectin domains might be studied.

20.2
CTL Superfamily: Structures and Groups

C-type lectins comprise a large and heterogeneous superfamily of proteins containing CTLDs (Figure 20.1). They have diverse functions and are among the first animal lectins discovered, with conglutinin being detected in 1906, as listed in Table 15.1 which gives a historical overview of lectinology. Until recently, they were regarded as exclusively extracellular proteins that originated in the Metazoan era to fulfill new needs in multicellular organisms for intercellular communication, immune defense and response. Carbohydrate binding is the most common CTLD function in vertebrates, and in this role C-type lectins function either as membrane-anchored or extracellular-matrix proteins in adhesion or immune

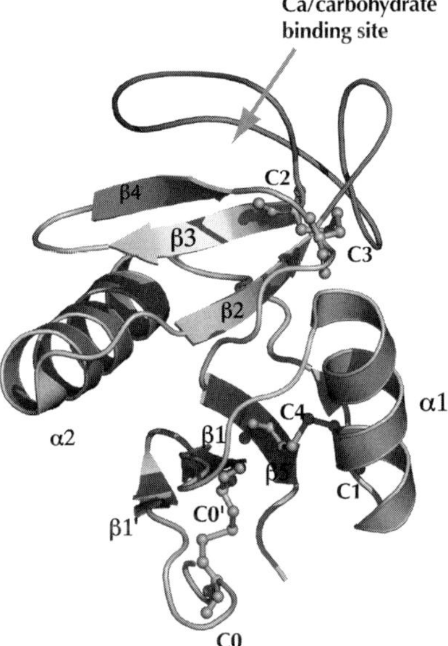

Figure 20.1 A typical CTLD structure shown in cartoon representation. The long-loop region which harbors the primary Ca^{2+}/carbohydrate-binding site (position arrowed) is shown in red. Disulfide bridges are shown as yellow sticks. The archetypical disulfide bridges of the fold are between C1 and C4, and C2 and C3, whereas that between C0 and C0′ is present only in long-form CTLDs which have an N-terminal extension.

defense, intercellular communication and integration, and glycoprotein metabolism (see Chapters 19 and 27 for details). Carbohydrate binding is also thought to be the ancestral function of the superfamily, as evidenced by the many humoral defense CTLDcps characterized in insects and other invertebrates. However, many CTLDs have evolved to recognize ligands other than carbohydrates, particularly proteins, but also lipids and inorganic substances such as ice in fish antifreeze proteins and calcium carbonate in bird egg-shell proteins.

As summarized in Chapters 16, 18 and 19, there are a number of protein folds capable of binding carbohydrate, that is 'lectin domains'. The CTLD is one of them, C-type referring to the presence of a calcium ion in the main carbohydrate binding site and distinguished by its sequence signature from other classes of Ca^{2+}-dependent lectins (for an overview, see Chapter 16). The nomenclature 'C-type lectin-like domain' – rather than just 'C-type lectin domain' – was introduced when it became clear that this domain could bind other ligands [3]. The CTLD, thus, has greater functionality than the lectin domains characteristic of the other lectin types (please see Chapters 18 and 19). They function primarily as carbohydrate recognition domains (CRDs) although I- and P-type lectin-like domains and galectins can also bind proteins (see Chapter 19; protein ligands of galectins are listed in Table 19.3).

In Chapter 19 we saw that a salient feature of many lectins is their modular design (see Figure 19.1). This feature is frequently encountered in CTLDcps. In principle, a domain has a compact three-dimensional structure that may fold independently of the rest of the protein, and may function and evolve semi-independently. The sequence of a protein domain is typically 100–200 amino acids long. A given CTLDcp by definition contains an operative lectin domain, but in addition may contain multiple diverged copies of it with different sugar or other ligand-binding specificity as well as other domains. These features are shown in Figure 20.2.

The development of the C-type lectin field benefited greatly from the early efforts of researchers, notably Kurt Drickamer, to systematize the disparate biological data [1, 2]. This led to their characterization as lectins with CRDs of length 110–140 residues, which bound carbohydrates in a Ca^{2+}-dependent manner. Furthermore, alignment of their sequences showed this protein domain had a characteristic conserved sequence signature. This feature has been of enormous value in advancing C-type lectin research as it has allowed initial identification as a potential CTLDcp directly from analysis of the protein sequence. A further helpful finding from this early sequence analysis was that carbohydrate specificity is often correlated with a particular tripeptide sequence motif within the sequence signature – **EPN** (Glu-Pro-Asn) for mannose-type ligands and **QPD** (Gln-Pro-Asp) for galactose-type ligands. This permits initial functional predictions.

Understanding of the structural basis of the sequence signature came with solution of the first X-ray structure of a CTLD – of rat mannan-binding protein (MBP)-A – in 1991 and a little later by a structure of this CTLD with a bound mannose molecule. Comparison of the residues then identified as the sequence signature – 12 totally conserved and 18 conservatively conserved – against the three-dimensional structure allowed the roles for most of them in Ca^{2+}- and carbohydrate-binding and stabilizing the protein fold to be defined [1].

The basic protein fold is shown in Figure 20.1 and the main details of carbohydrate binding in Figure 20.3. The key characteristics of the CTLD fold are two antiparallel β-sheets and two α-helices, and two disulfide bridges and a hydrophobic core stabilizing the long-loop region which contains the primary sugar-binding site. Examination of the large number of CTLD structures that have been obtained by crystallography shows that this unique loop-in-a-loop structure, in which the large flexible long-loop region is maintained on a stable core, allows the fold to tolerate substantial variation in the shape of the primary ligand-binding site and adjacent regions [4]. This allows specific binding of large multivalent ligands such as complex-type oligosaccharides and mannose-rich structures (triantennary *N*-glycans and mannans are examples given in Chapter 19), non-carbohydrate ligands, and even both. Formation of quaternary complexes, for example in the trimers of the group III vertebrate collectins (for explanation of this term, see Chapter 19), further increases both specificity and affinity of carbohydrate recognition by C-type lectins [3].

In 1993, Drickamer classified mammalian CTLDcps into seven groups (I–VII). This was based primarily on their domain architecture, but the grouping also appeared to reflect evolutionary history as it correlated well with the results of phylogenetic analysis of the CTLD sequences [1]. The classification was revised in

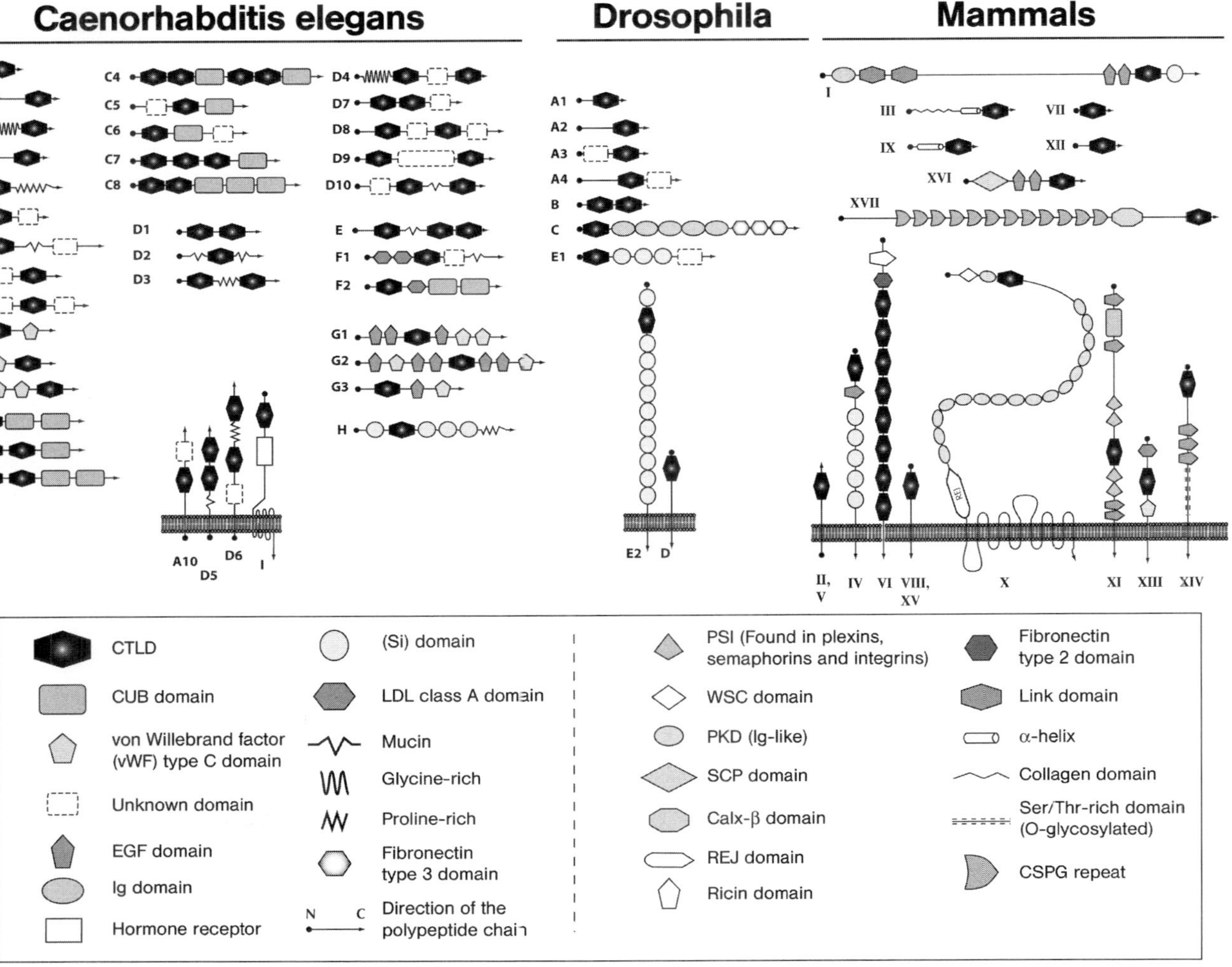

Figure 20.2 Summary of domain architectures in CTLDcps of worm [5], fly [6] and mammals [7, 8] as revealed by whole-genome analysis.

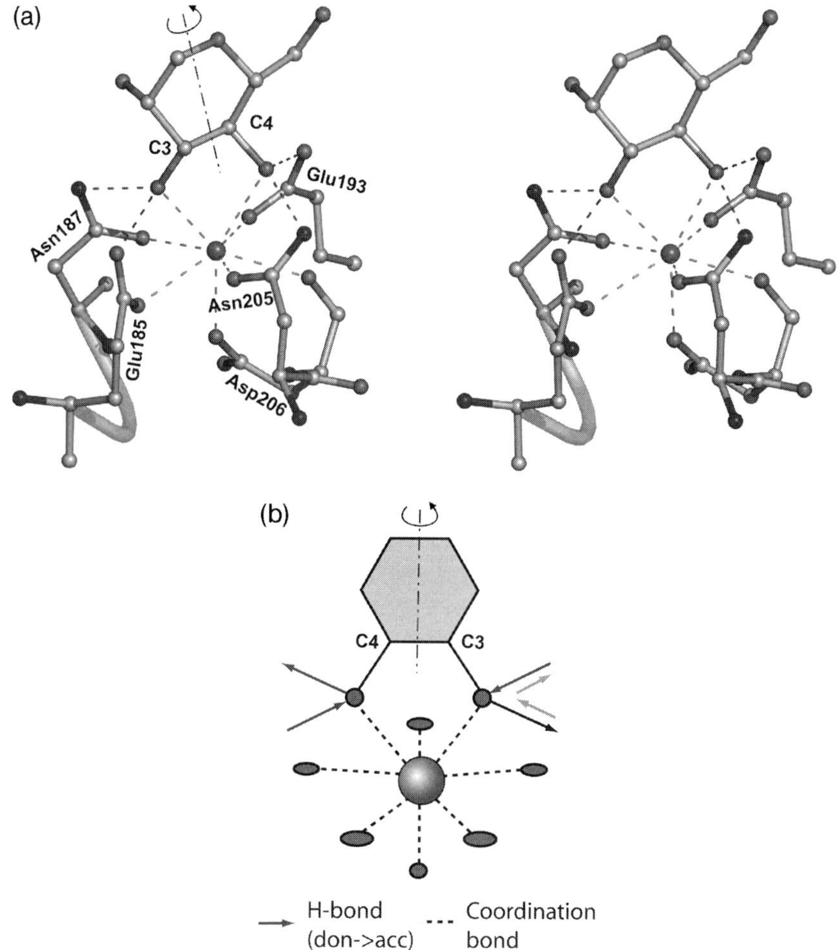

Figure 20.3 Ca^{2+}-dependent monosaccharide binding by CTLDs. (a) Stereo depiction of the structure of rat MBL (formerly called mannose-binding protein, MBP-A) complexed with Ca^{2+} (blue sphere) and mannose. The coordination bonds to Ca^{2+} are shown in orange. Hydrogen bonds where the 3′ and 4′ sugar hydroxyls act as acceptor and donor are shown as red and blue, respectively. (b) Schematic representation of a Ca^{2+}–pyranose–CTLD complex. Two hydroxyl oxygens and the pyranose ring are shown with the Ca^{2+} as a large blue sphere, and oxygens as red circles and ovals. Arrows show the direction of hydrogen bonds in mannose-specific CTLDs, while light-grey arrows indicate the changed directions in galactose-specific CTLDs. The C3 and C4 atoms of the sugar, and the orientation of a rotation axis are shown in both (a) and (b) (please see also Figures 16.1h,i).

2002 with the addition of seven new groups (VIII–XIV) found experimentally in the interim [7]. Subsequently, a further three groups (XV–XVII) have been added based on findings from whole-genome analysis [8]. This classification has facilitated prediction of the oligomerization and ligand-binding properties for newly found members [3]. The domain architectures of the mammalian CTLDcps are summarized in Figure 20.2. This shows the variation in the number and order of CTLD and non-CTLD protein domains in multidomain CTLDcps, as well as whether they are secreted or anchored in the membrane.

20.3 Mechanism of Carbohydrate Binding

In this section we will examine how crystal structures of CTLDs with bound sugar ligands have revealed a general molecular mechanism of carbohydrate binding at the main binding site. Specifically we will look at the roles of the Ca^{2+} and two groups of residues (the '**EPN**'/'**QPD**' and 'WND' motifs) in forming the binding site that interacts with the sugar, and at how the **EPN** and **QPD** motifs discriminate mannose-type and galactose-type sugars, respectively. We confine our attention to this main site but note that some C-type lectins, such as the vertebrate group IV selectins (Chapters 19 and 27), bind additional monosaccharide units of oligosaccharide ligands such as the sialylated and sulfated Lewis^x epitope (see Tables 7.4 and 27.2 for structures) at auxiliary sites of the CTLD (see Figure 16.1h). We also note that it is this site that has been modified in many non-carbohydrate-binding CTLDs to select other ligands (protein, ice, calcium carbonate), reinforcing our developing view that this site is unusually adaptable by evolution.

The architecture of the primary monosaccharide-binding site for mannose-type ligands is illustrated in Figure 20.3a. This is based on the crystal structure for the CTLD of rat MBP-A in complex with the *N*-glycan Man_6-$GlcNAc_2$-Asn; see Chapters 6 and 8 for *N*-glycan structures and Figure 1.6 for structures of monosaccharides. The complex is stabilized by a network of coordination and hydrogen bonds. Oxygen atoms from the 3′ and 4′-hydroxyls of the mannose form two coordination bonds with the Ca^{2+} ion and four hydrogen bonds with residues – Glu185 and Asn187 (**EPN**), Asn205 (WND) and Glu193 – whose carbonyl side-chains coordinate the Ca^{2+}-binding site. This bonding pattern is fundamental for CTLD/Ca^{2+}/monosaccharide complexes, and is observed in all known structures. The **EPN** and WND motifs are in the long-loop region and β4 strand of the CTLD structure, respectively (Figure 20.1). Asp206 of the WND motif contributes another Ca^{2+} coordination bond, while the Trp204 residue is highly conserved and contributes to the hydrophobic core [1, 4].

The arrangement of the hydrogen-bond donors and acceptors and coordination bonds in the binding site, as summarized in Figure 20.3b, has two important features. First, it determines the overall positioning and orientation of the sugar in the binding site. However, as shown in Figure 20.3b, the site has a 2-fold symmetry axis relating the sugar hydroxyls which would allow the sugar to be rotated by 180° without introducing any changes to the bonding scheme. Indeed, examples are now known. The structure of the rat MBP-C complex with mannose showed this hexapyranose bound in the opposite orientation. Also, structures of a galactose-binding mutant of MBP-A and CEL-I, a C-type lectin from the echinoderm sea cucumber, showed galactose bound in the opposite orientation to that observed in the complex of 'TC-14', a lectin found in a tunicate urochordate [2].

Second, constraints imposed by the structure of the Ca^{2+}-coordination site determine the properties of the carbohydrate hydroxyls that the site can accept. This is best demonstrated by the mechanism by which the CRD discriminates between the mannose-type and galactose-type monosaccharides. Crystallographic analysis of the galactose-specific MBP-A mutant in which the **EPN** motif was mutated to **QPD** showed little restructuring of the Ca^{2+}-binding site, suggesting that the key switch for specificity was swapping the hydrogen-bond donor and acceptor across the monosaccharide-binding plane. This changed the hydrogen-bonding pattern from asymmetrical mannose-type (Figure 20.3b; dark-grey arrows) to symmetrical galactose-type (light-grey arrows). The theory is nicely supported by the finding of the same hydrogen-bond distribution in the structure of the TC-14 lectin complex with galactose, even though the details are rather complex as the TC-14 CRD contains an unusual **EPS** (not **QPD**) motif [2].

Although many of the determinants of monosaccharide-binding specificity have been established experimentally by numerous examples, a convincing explanation of the underlying mechanism is still wanting. Although mutual spatial disposition of bonded hydroxyls was initially suggested to be the main determinant of specificity, a growing number of crystal structures of CRDs with the MBP-A-like ('asymmetrical') distribution of hydrogen bonds have shown the binding site is compatible with configurations other than two equatorial hydroxyls (for example 3- and 4-OH of mannose and glucose, 2- and 3-OH of fucose). For example, a combination of axial and equatorial hydroxyls (3- and 4-OH of fucose) have been found in E- and P-selectin structures; Figure 16.1h) [2].

From a comparative study of different lectin–carbohydrate complexes, Elgavish and Shaanan suggested that additional stereochemical factors need to be considered [2] to understand the determinants of specificity. A detailed electronic description of binding of the sugar hydroxyls, which are also coordinated to Ca^{2+}, may be necessary to understand the relative stabilities of possible hydrogen-bonding patterns. Improved understanding is certainly needed as very many sequenced CTLDs – especially the multitude coming from genome sequencing projects – contain atypical or apparently incomplete versions of the carbohydrate-binding motifs we have discussed. Currently, most of these are classified as 'noncarbohydrate binding' but the reliability of these predictions is questionable. A better understanding of specificity would also be useful in assessing whether carbohydrate-binding patterns from known examples, particularly the numerous invertebrate CTLDcps, represent divergent or convergent evolution. The importance of this question will become clearer in the next section where we will see that the repertoire of CTLDcps appears to have been created anew in each major Metazoan branch. This confounds attempts to generate phylogenetic trees, making chemical insights into specificity of great value.

20.4
CTLs in the Genome Era

In the pregenome era, knowledge of C-type lectins/CTLDcps reflected the interests of experimental researchers who found, isolated and characterized them from diverse species, but was heavily skewed to mammalian CTLDcps. While the diversity of CTLDcp domain architectures and ligands the CTLD fold could bind was apparent, the full extent of the variety of the CTLDcp superfamily was unknown. Evolution of the superfamily was a puzzle.

The advent of large-scale DNA sequencing, free public access to sequence databases such as GenBank or EMBL and search tools such as BLAST has changed the way biologists work and the way new genes are discovered. A simplified view of the procedures used to search the sequence of a genome for CTLDcps is shown in Figure 20.4. In the first step, a tool such as the profile-based PSI-BLAST or HMMER (a program using hidden Markov models) is used to look for regions of genome sequence which are similar to that of a supplied query sequence for the CTLD. A simple example is the Drickamer sequence signature discussed above, but this is heavily weighted towards CTLDs which bind carbohydrate by the paradigmatic mechanism discussed in the last section.

The sensitivity of the search to detect weak but significant sequence similarity can be improved by incorporating structural information into the query sequence, as we did in our analysis of CTLDcps in the *Fugu* genome [8]. This also improved the reliability of the search by improving the discrimination between true homologs and spurious sequence similarities. This is critical as it minimizes the chances of invalid findings (false positives) and of missing valid occurrences (false negatives).

Such factors highlight a major problem with the results of automated domain analyses reported in the many-authored papers which announce the availability of a newly sequenced model-organism genome. Such audits using nonoptimized query sequences for each domain are less reliable than studies by researchers with specific knowledge of particular domains. These researchers construct optimum query sequences, and manually check ('curate') the gene identifications and deduced domain architecture of the proteins to make sure they are sensible. This point is well illustrated by the CTLDcps audits of the *Caenorhabditis elegans* and *Fugu rubripes* genomes. The 1998 *Science* paper by the *C. elegans* genome-sequencing consortium reported the exciting result that the CTLD was the seventh most common domain with 120 of them. A subsequent CTLD-specific study a little later produced a larger estimate of at least 125 CTLDcps containing 183 CTLDs [5]. In a recent study the number was updated to 278 CTLDcps [11]. In our study of CTLDcps in the *Fugu* genome we found many instances of mis-prediction of genes. Overall, we verified 32 of the gene structures predicted by the *Fugu* genome sequencing team and available on the public web site (Ensembl) but predicted or revised predictions of a further 63 gene structures [8].

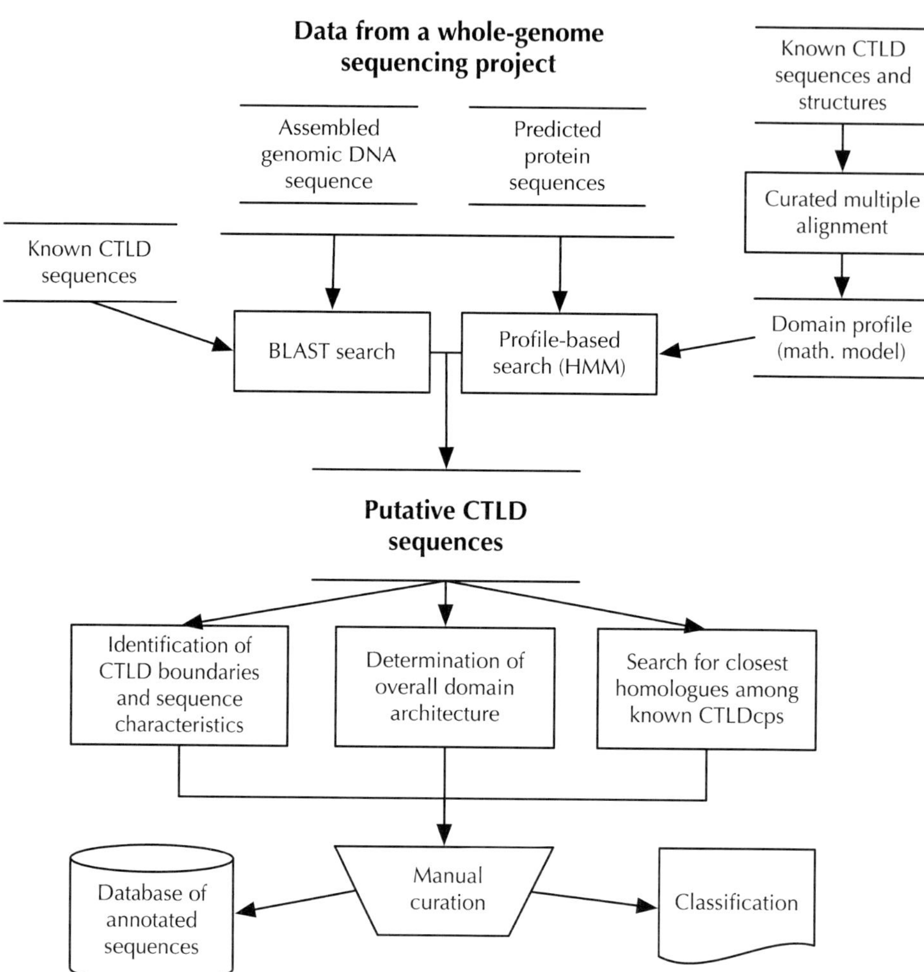

Figure 20.4 Schematic flowchart illustrating the procedure for auditing CTLDcps of a genome. First, the putative CTLD sequences are found by whole-genome analysis. Next, the genes of CTLDcps containing these CTLDs are analyzed to define their domain architectures and closest homologs, and to classify them into groups, illustrated in summary form in Figure 20.2. Further analysis of the sequences of the CTLDs allows initial assessment of whether they may bind carbohydrate and, if so, the likely specificity (**EPN**/**QPD** and W**N**D motifs; see text and Figure 20.3). This information can be included as part of the annotation.

In summary, the biologist needs to exercise caution in assessing published genome statistics. Much careful finding, checking and annotating of genome sequences is necessary before whole-genome CTLDcp statistics and domain architectures can be regarded as sufficiently robust to start drawing evolutionary conclusions. With this caveat that CTLDcp statistics of most genomes should be considered as 'work in progress', we will proceed to consider the main findings of the major genome studies and what they can tell us about the evolution of the superfamily.

20.5 CTL Domain-Containing Proteins (CTLDcps) in Metazoans from Whole-Genome Analysis

In the previous section we have learnt that CTLDcps are abundant in the *C. elegans* (worm) and *F. rubripes* (fish) genomes. In this section we will examine how common this abundance is in the main Metazoan clades by comparing results of whole-genome analyses of model organisms. We will also examine how these proteins are 'put together' (that is their domain architectures). This investigative method is called comparative genomics. These major Metazoan branches are shown in Figure 20.5.

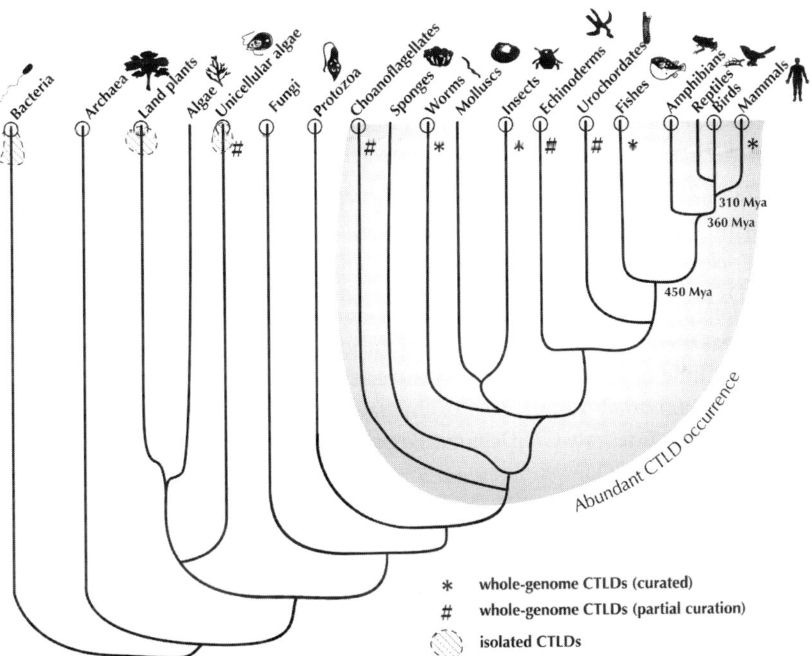

Figure 20.5 The tree of life showing the occurrence of CTLDcps. Branches with sequenced genomes are shown circled. Branches in which CTLDcps have been identified by curated or partially curated analysis are marked by '*' and '#', respectively. The Metazoan and pre-Metazoan (choanoflagellates) branches are shown shaded in grey. The branching region at the Metazoan 'explosion' is tentative only. Non-Metazoan branches in which CTLDcps have been identified are shown by dotted-hashed circles. Evolution times are not drawn to scale.

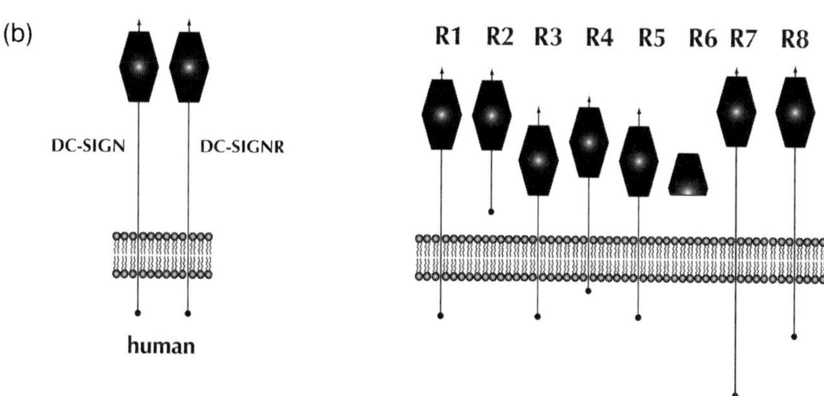

Figure 20.6 Example of a genomic region showing duplicated CTLDcps genes. The eight SIGNR genes with adjacent CTLDcps genes in mouse are shown in (a) gene order and strand orientation, and (b) domain organization and membrane orientation of the proteins. Adapted from [16].

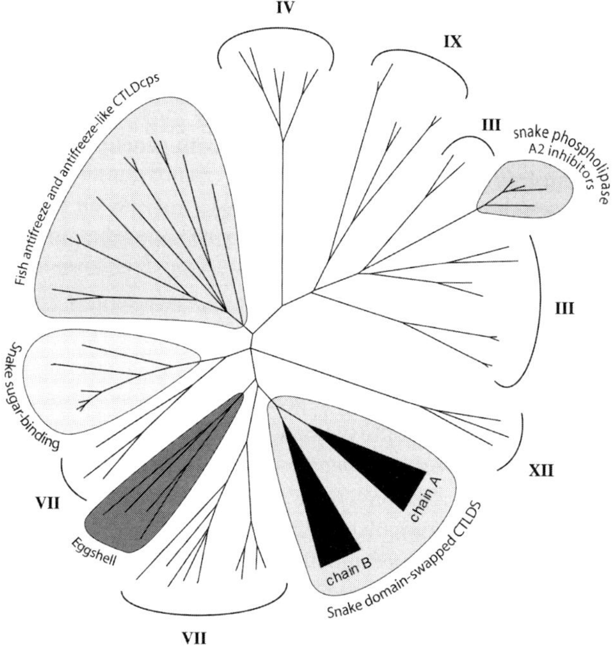

Figure 20.7 Phylogenetic relationships of clade-specific vertebrate CTLDs to the main vertebrate group CTLDs. The positions of the branches of the CTLDs for the three classes of snake venoms, and bird egg-shell and fish antifreeze proteins show their relative sequence similarity to the CTLDs of mammalian group IV, IX, III, XII and VII CTLDcps [2].

20.6 Non-Metazoan CTLDs: From Viruses, Bacteria and Protozoa

To recapitulate what we have learnt so far: it had been thought that the primordial CTLD had originated at the beginning of the Metazoan era and had a carbohydrate-binding function. As shown in Figure 20.5, pre-genome and early-genome studies were consistent with this model showing that CTLDcps were abundant in Metazoa but absent in non-Metazoa such as fungi (genome of the yeast *Saccharomyces cerevisiae*). Notwithstanding these general findings, many interesting examples of non-Metazoan CTLDcps have been reported [2].

Many of these CTLDcps are from parasitic viruses and bacteria which are involved in interactions with the animal host, often as mechanisms to defeat its immune system. The CTLDs of the viral proteins (for example fowlpox, vaccinia and African swine fever viruses) contain sequences similar to mammalian CTLDcps [2]. Well-conserved CTLD sequences are also present in *Trypanosoma*, a parasitic protozoan. The best-characterized bacterial group are toxins (pertussis toxin and proaerolysin) and adhesion proteins (intimin from *Escherichia coli* and invasin from *Yersinia pseudotuberculosis*). Details on bacterial toxins and adhesins are presented in Chapter 18. Their CTLDs cannot be identified by sequence analysis and their structures show a more compact fold [2, 4]. However, several CTLDs (see Info Box 2 for examples) found in sequenced genomes from free-living bacteria contain well-conserved **EPN**/**WND** motifs suggesting they are Ca^{2+}/sugar binding, although some lack conserved cysteines.

How might these CTLDcps in bacteria and viruses have arisen? The most parsimonious explanation of the presence of the viral, protozoan and bacterial CTLDs homologous to those in Metazoans is horizontal gene transfer (that is they are hijacked host proteins in viruses or otherwise acquired) [2]. The high sequence and structural divergence of the CTLD of the parasitic bacterial proteins obscure their origin. They may also have been acquired by horizontal gene transfer or,

Info Box 2

Examples of free-living bacteria with CTLDs found by genome analysis are: *Leeuwenhoekiella blandensis*, a marine flavobacterium; *Synechococcus* sp. RS9917, a marine photosynthetic cyanobacterium; *Marinomonas* sp. MED121, a marine proteobacterium; and *Stigmatella aurantiaca*, a gliding, Gram-negative bacterium. A particularly intriguing example is a putative CTLDcp deduced from the genome sequence of a marine planctomycete *Pirellula* sp. This is the largest protein in the genome (7716 residues) and it contains several CTLD, laminin G and cadherin domains, all of which are domains almost exclusively found in Metazoa (for further information on laminin G domains, see Chapter 16). What could be the function of such a complex protein in a free-living species such as *Pirellula*?

alternatively, may have arisen by convergent evolution, as mimicry of host proteins.

20.7
CTLDcps in Genomes of Pre-Metazoans and Plants

Early findings of CTLD sequences in the genome of the flowering plant *Arabidopsis thaliana*, with a well-conserved ortholog for one of the putative CTLDcps also present in the rice genome, were very perplexing [2]. However, very recent reports of more numerous CTLDcps in genome sequences of a pre-Metazoan, the choanoflagellate *Monosiga brevicollis* (12 CTLDcps) [9], and the unicellular green alga *Chlamydomonas reinhardtii* [10] require a serious reconsideration of the Metazoan hypothesis.

The statistics of the *C. reinhardtii* analysis are compelling, being reminiscent of the general conclusions from the invertebrate and vertebrate genomes we have already outlined. They strongly argue that these CTLDcps have evolved within *Chlamydomonas* for specific roles. Thus, of the 11 CTLDcps containing a total of 67 CTLDs, seven also contain scavenger-receptor cysteine-rich (SRCR) domains, a combination rarely seen in Metazoans and not in CTLDcps for worm, fly and human shown in Figure 20.2. Also, 18 of the CTLDs possess the conserved sequence motifs (**EPN**/**QPD** and W**ND**) associated with Ca^{2+}-dependent carbohydrate binding.

How can the presence of this novel, and likely functional, collection of CTLDcps in a distant non-Metazoan branch be explained? Although the details of the branching pattern around the time of the Metazoan 'explosion' are still an ongoing debate (please see Figure 20.5), such ambiguity cannot explain the apparent absence of CTLDs in fungi, but variable abundance in plants and algae. Although Wheeler *et al.* [10] interpret their findings in terms of loss of CTLDcp genes or divergence of the CTLD beyond recognition as the eukaryotes evolved, a more plausible explanation may be that the plant/algae branch acquired primordial CTLD genes by horizontal transfer, possibly multiple times. Under this model, the patterns of CTLDcps in subbranches such as *Chlamydomonas* simply represent evolution starting at punctuated times to happily build anew groups of CTLDcps to perform novel needed roles, using one of its best domain building blocks – the structurally flexible and functionally versatile CTLD. This explanation is more consistent with the knowledge we have acquired in this chapter of the CTLD as an evolutionarily very successful domain in Metazoans as it does not require an explanation as to why the CTLD was not equally successful in fungi and plants!

20.8
Conclusions

In this chapter we have taken a historical approach to showing how understanding of the structure, function and evolution of C-type lectins has unfolded. We have seen how C-type lectin research has been greatly advanced by application of systematic methods of structural biology and whole-genome sequence analysis, and how such studies have provided a different perspective on the variety and importance of C-type lectins than can be provided by traditional experimental investigative approaches. We have seen how these systematic approaches have provided a strong framework for linking molecular mechanisms to biological functions and have expedited the resolution of many evolutionary questions for the superfamily. So we have learnt some pithy lessons with general relevance, as promised at the beginning of the chapter.

However, new challenges have appeared which follow from some of these lessons, particularly the notion that CTLDcps are an unusually dynamic set of proteins evolutionarily. It appears that we cannot expect to find a paradigmatic set of CTLDcps in widely diverged branches or even for closely related organisms subject to different physiological conditions, such as free-living and parasitic worms. The plenitude of CTLDcps in life highlights the importance of developing an improved understanding of the chemical mechanisms of carbohydrate binding and specificity in order to provide an improved tool for initial deciphering their sugar code using sequence analysis combined with homology modeling. Other questions for future study are whether CTLDcps are present in some fungal branches, and definition of the roles of CTLDcps in plants, especially those with obvious carbohydrate-binding capacity.

Summary Box

Our journey tracking the evolution of the CLTD, the defining protein fold of C-type lectins, has provided many insights. It is one of the most abundant protein domains in Metazoa having been recruited by evolution into proteins, usually with other domains, which carry out the full spectrum of functions essential for multicellular life–cell adhesion, immune defense, intercellular communication and integration, and glycoprotein metabolism. Analyses of genome sequences of model organisms have defined the full extent of this diversity in the major branches of life, including the unexpected presence of CTLDcps with predicted sugar-binding function in non-Metazoa such as plants and algae. The superior stability and versatility of the fold to adapt to bind carbohydrates and many other types of ligands are the keys to its evolutionary success.

References

1 Drickamer K. Evolution of Ca^{2+}-dependent animal lectins. *Prog Nucleic Acid Res Mol Biol* 1993;45:207–32.

2 Zelensky AN, Gready, JE. The C-type lectin-like domain superfamily. *FEBS J* 2005;272:6179–217.

3 Drickamer K. C-type lectin-like domains. *Curr Opin Struct Biol* 1999;9:585–90.

4 Zelensky AN, Gready JE. Comparative analysis of structural properties of the C-type-lectin-like domain (CTLD). *PROTEINS* 2003;52:466–77.

5 Drickamer K, Dodd RB. C-type lectin-like domains in *Caenorhabditis elegans*: predictions from the complete genome sequence. *Glycobiology* 1999;9:1357–69.
6 Dodd RB, Drickamer K. Lectin-like proteins in model organisms: implications for evolution of carbohydrate-binding activity. *Glycobiology* 2001;11:71R–9R.
7 Drickamer K, Fadden AJ. Genomic analysis of C-type lectins. *Biochem Soc Symp* 2002;69:59–72.
8 Zelensky AN, Gready JE. C-type lectin-like domains in *Fugu rubripes*. *BMC Genomics* 2004;5:51.
9 King N *et al*. The genome of the choanoflagellate *Monosiga brevicollis* and the origin of Metazoans. *Nature* 2008;451:783–8.
10 Wheeler GL *et al*. Genome analysis of the unicellular green alga *Chlamydomonas reinhardtii* indicates an ancient evolutionary origin for key pattern recognition and cell-signaling protein families. *Genetics* 2008;179:193–7.
11 Schulenburg H *et al*. Specificity of the innate immune system and diversity of C-type lectin domain (CTLD) proteins in the nematode *Caenorhabditis elegans*. *Immunobiology* 2008;213:237–50.
12 Azumi K *et al*. Genomic analysis of immunity in a Urochordate and the emergence of the vertebrate immune system: 'waiting for Godot'. *Immunogenetics* 2003;55:570–81.
13 Hibino T *et al*. The immune gene repertoire encoded in the purple sea urchin genome. *Dev Biol* 2006;300:349–65.
14 Bottjer DJ *et al*. Paleogenomics of echinoderms. *Science* 2006;314:956–60.
15 Ohta T. Evolution by gene duplication revisited: differentiation of regulatory elements versus proteins. *Genetica* 2003;118:209–16.
16 Powlesland AS *et al*. Widely divergent biochemical properties of the complete set of mouse DC-SIGN-related proteins. *J Biol Chem* 2006;281:20440–9.

21
Carbohydrate–Carbohydrate Interactions

Iwona Bucior, Max M. Burger, and Xavier Fernàndez-Busquets

Carbohydrates are generally considered to act as receptors for complementary binding proteins, typically lectins, in biomolecular recognition (see Chapters 13–19). However, broadening our knowledge of complex carbohydrates warrants including direct carbohydrate–carbohydrate interactions that would trigger initial steps in these multiple and redundant processes. To consider carbohydrate–carbohydrate interactions as important particularly for cell adhesion phenomena stems from the fact that glycans occur on the outermost cell periphery and thereby are likely involved in first intercellular contact.

21.1
Molecular Basis of Carbohydrate–Carbohydrate Interactions

Carbohydrate–carbohydrate interactions have emerged as a novel and highly versatile mechanism for cell adhesion and recognition due to the extraordinary plasticity of glycan chains, to the low affinity and reversibility of individual binding sites, and to the capacity to form multivalent complexes leading to increased association forces (for a review, see [1]). Carbohydrate self-recognition takes place through surfaces determined by the carbohydrate epitopes and is based on non-covalent bonds (see Chapter 13 for the basics of molecular interactions): van der Waals contacts, hydrogen bonds, electrostatic forces and interactions with cations (Figure 21.1a) [2]. On these surfaces, hydrophilic and hydrophobic patches are well defined by the position of functional groups, which determine the three-dimensional (3D) structure adopted. Van der Waals contacts occur through non-polar patches of carbohydrates, providing surface complementarity that contributes to the initial selective formation of carbohydrate complexes. Hydroxyl, carboxyl and amino groups form hydrogen bonds between carbohydrate moieties and with solvating water molecules that increase the stability of the complex (see Figure 13.1 for illustration). Divalent cations such as Mg^{2+} and especially Ca^{2+} not only provide electrostatic forces, but also contribute to the configuration of glycan

The Sugar Code. Edited by Hans-Joachim Gabius

ISBN: 978-3-527-32089-9

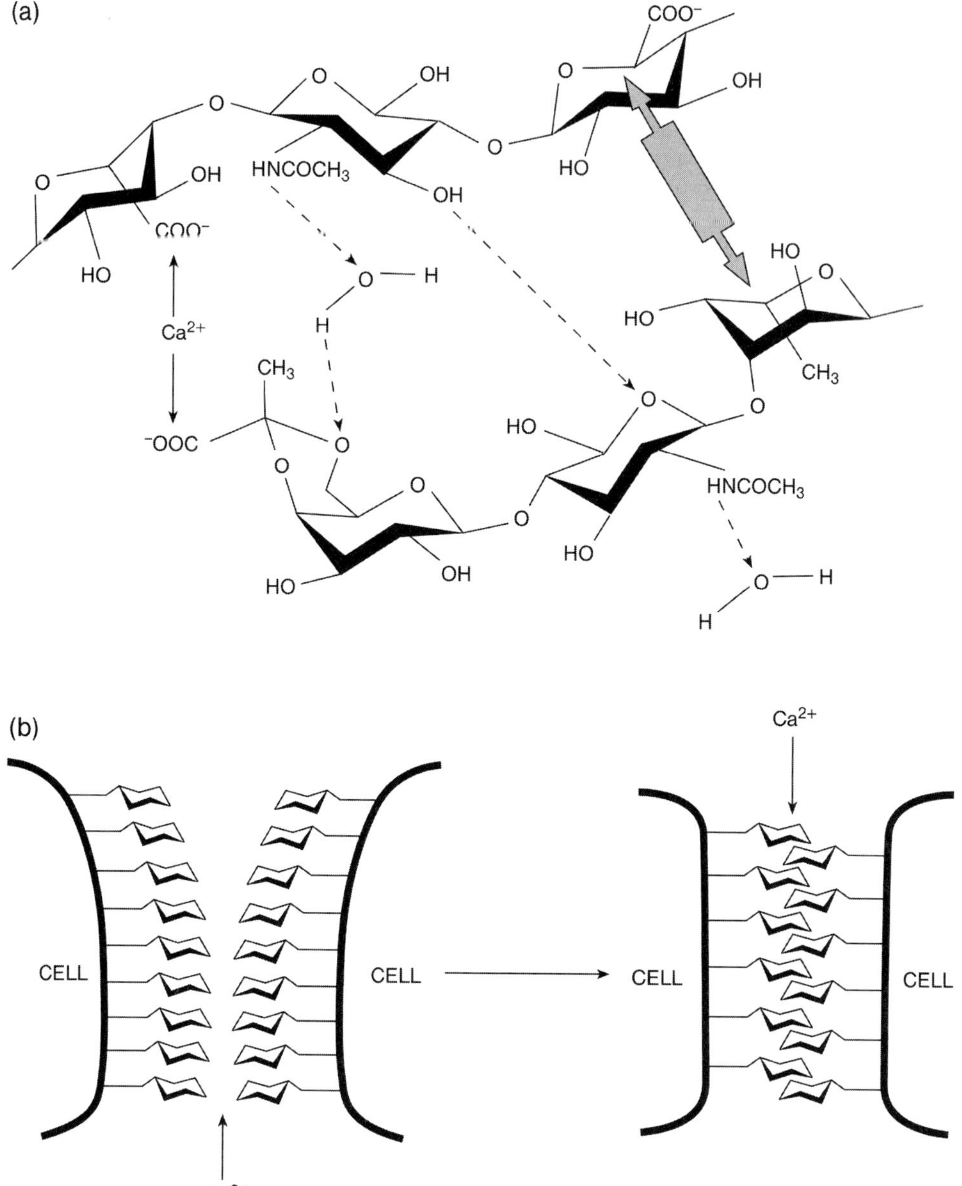

Figure 21.1 Molecular basis of carbohydrate–carbohydrate interactions. (a) Schematic drawing of the different stabilizing forces between two carbohydrate chains (adapted from [2]), showing hydrogen bridges (arrows with dashed lines), hydrophobic surfaces (shaded area) and divalent cation interaction sites (Ca^{2+} between arrows). (b) The zipper adhesion model for carbohydrate–carbohydrate interactions (proposed by [3]).

superstructures and the stability of the complex via coordinative forces, locking sugar chains in an optimal conformation (see Chapter 16 for the role of Ca^{2+} in lectin binding).

Individual carbohydrate–carbohydrate interactions are among the weakest biomolecular binding events, and to generate sufficient affinity, glycoconjugates tend to display a polyvalent configuration at the cell surface. Glycoproteins and proteoglycans present repetitive epitopes on their carbohydrate chains, whereas glycosphingolipids (GSLs) are associated in clusters or patches. A useful analogy to define polyvalent arrays of carbohydrates is to compare them with a velcro pad or a zipper (Figure 21.1b) [3]. In this model, the carbohydrate moieties (the stumps of the zipper) have to be complementary to permit an adequate approach and positioning of chemical groups, whereas Ca^{2+} ions may provide the driving force to maintain the zipper closed. The interacting carbohydrates are arranged on a scaffold – the protein moiety in proteoglycans or the lipid bilayer in GSL clusters – that does not participate in the process, but allows the correct orientation of the groups involved in binding.

21.2 Carbohydrate–Carbohydrate Interactions in Cell Recognition

21.2.1 Proteoglycans

The structure of proteoglycans and their involvement in a number of biological processes are described in detail in Chapter 11. Most proteoglycan-mediated molecular interactions described so far are based on carbohydrate–protein recognition (Chapter 13), but direct carbohydrate–carbohydrate interactions with proteoglycans are also possible in cell adhesion events.

21.2.1.1 Carbohydrate Self-Interactions in Sponge Proteoglycans

Dissociated sponge cells from two different species have the capacity to species-specifically sort out and reaggregate (see Info Box) through cell-surface proteoglycans, termed aggregation factors (AFs) [4], in the same way as mixtures of dissociated embryonic cells from two vertebrate tissues sort out according to their tissue of origin. AFs are large molecules of 2×10^4 to 2×10^7 Da, composed of 30–60% carbohydrates, that have been related to hyalectans [4], which are large, extracellular aggregating modular proteoglycans. However, sponge proteoglycans do not possess the usual glycosaminoglycans (GAGs), but they have instead complex acidic polysaccharides [5]. Atomic force microscopy (AFM) visualization has revealed either a linear or a sunburst-like core structure with 20–25 radiating arms (Figure 21.2a). One of the best-studied sponge proteoglycans carries two *N*-linked glycan molecules (Figure 21.2b): a 6-kDa glycan (g6), present in the arms of the proteoglycan molecule, and a 200-kDa glycan (g200), present in the core structure. A monoclonal antibody that inhibits self-aggregation of AFs is directed against a distinct carbohydrate motif in g200 (reviewed in [1]).

Info Box

In 1907, H.V. Wilson observed a remarkable phenomenon in which dissociated marine sponge cells from two different species reaggregated according to their species of origin to form clones of the parent sponge. This pioneering assay was the first experimental demonstration of cell recognition in the animal kingdom. It is only fitting that the field that focuses on the organization of multicellular organisms was born by studying the oldest multicellular animal on Earth. Aristotle, who first noted that sponges are animals, writes in his *History of Animals* (350 BC): 'Sponges ... get their nutriment in slime: a proof of this statement is the fact that when they are first secured they are found to be full of slime. ... It is said that the sponge is sensitive; and as a proof of this statement they say that if the sponge is made aware of an attempt being made to pluck it from its place of attachment it draws itself together, and it becomes a difficult task to detach it'. More than two millennia later, such strong adhesion was found to be mediated by the carbohydrate structures in sponge AFs. And almost a century after Wilson's experiment, the aggregation specificity of sponge cells was found to reside in the carbohydrate portion of the sponge cell-surface proteoglycans (the *slime* described by Aristotle) – the first experimental demonstration of species-specific carbohydrate–carbohydrate cell recognition.

The first report addressing the species specificity of a carbohydrate–carbohydrate interaction has shown that recognition between g200 glycans on opposite cells mediates species-specific recognition between live sponge cells (reviewed in [1]). According to the current model of carbohydrate-dependent cell–cell recognition, the AF is bound via its arms onto the cell surface through a Ca^{2+}-independent protein–carbohydrate interaction between a cell-surface receptor and the g6 glycan (Figure 21.2c). This structure interacts with an identical counterpart on a different cell through a Ca^{2+}-dependent species-specific carbohydrate–carbohydrate interaction between g200 glycans.

Compositional and possibly architectural differences between carbohydrate chains of sponge proteoglycans contribute to the specificity of the self-interaction (discussed in [1]). Sponge glycans contain Gal, Fuc, Man, GlcNAc, GalNAc and GlcA (see Chapter 1 for abbreviations). The composition of glycans from different individuals of the same species is similar, whereas sponge proteoglycans from different species display major differences in their carbohydrate display. These molecules present novel and species-specific sequences, particularly sulfated and branched pyruvylated structures.

Adhesive forces between individual g200 molecules, as measured by AFM, are in the range of 200–300 pN [1]. These are the strongest forces reported to date for direct carbohydrate–carbohydrate interactions. This study provided the first indication that the strength of specific glycan self-interactions is comparable to protein–protein interactions, e.g. in antibody–antigen recognition. Furthermore, surface

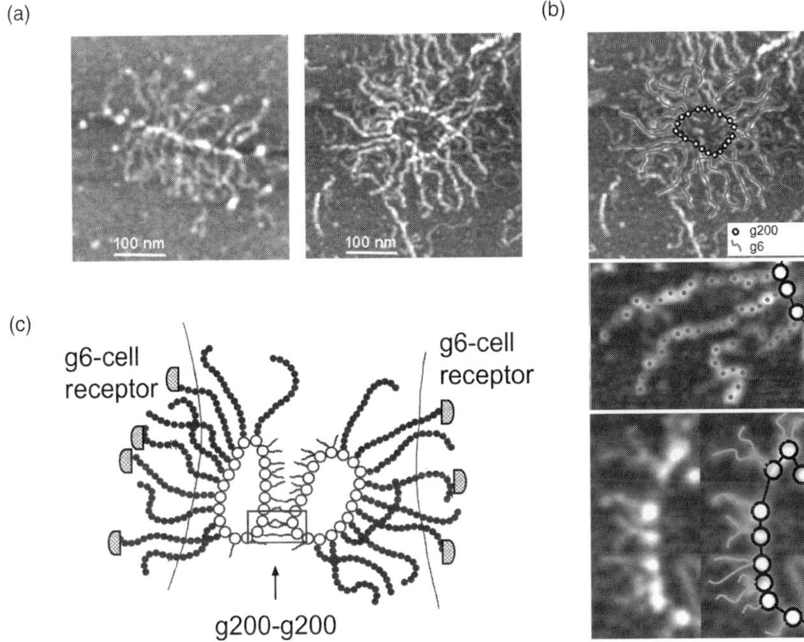

Figure 21.2 Carbohydrate-mediated species-specific intercellular adhesion in sponges. (a) AFM images of linear and circular sponge adhesion proteoglycans with 20–25 radiating arms. (b) AFM images of native *Microciona prolifera* proteoglycan (reprinted from [1]). The grey-coded vertical *z*-scale is 3 nm. The top image shows the localization of g200 glycan in the ring (circles) and g6 glycan in the arms (lines). The enlarged middle image shows 15–16 domains (dots) in each arm in the native structure. The enlarged bottom image shows a detail of the ring structure with short chains protruding, which might represent the g200 glycan (lines). (c) Schematic drawing of the current model depicting the interaction between two sponge cell adhesion proteoglycans (adapted from [7]).

plasmon resonance (SPR) measurements (a technique described later in this chapter) showed that self-recognition of the carbohydrate epitopes in g200 is the major force behind the recognition event [6] and that g200 self-adhesion is much stronger than its binding to other unrelated glycans, such as chondroitin sulfate (CS) [7].

21.2.1.2 **Glycosaminoglycan Self-Interactions**

It has been well established that GAGs located at the cell surface, particularly CS and hyaluronic acid (HA), are involved in cell adhesion, organogenesis and differentiation. However, little is known about the possible role of direct GAG–GAG interactions in these processes. A research line related to the interaction between GAGs demonstrated the specific *in vitro* binding between carbohydrate chains of CS and HA coated onto beads [8]. The authors suggested that specific binding between heterologous GAG chains can represent one mechanism controlling adhesive behavior and morphogenetic phenomena.

21.2.2 Glycolipids

The organization of GSLs into microdomains on the cell surface and their role in cellular recognition and adhesion is described in Chapter 10. Here, we focus on the role of carbohydrate–carbohydrate interactions in such GSL-mediated phenomena (for reviews, see [9, 10] and [1]; for synthesis and structures, see Chapters 10 and 30; for functional aspects of gangliosides, see Chapter 25).

21.2.2.1 Le^x–Le^x Interactions

It was far more difficult to determine specific carbohydrate–carbohydrate interactions between glycoconjugates of proteoglycans or glycoproteins than of glycolipids because, in the former case, the protein portion may have acted as a lectin. Thus, specific carbohydrate–carbohydrate interaction was first shown for glycolipids (Le^x–Le^x) and only later for sponge proteoglycans (see Table 7.4 and Table 27.2 for structures). Embryonal stem cells and embryonal carcinoma cells expressing Le^x at the cell surface, but not Le^x-negative cells, have been shown to autoaggregate through a Ca^{2+}-dependent interaction independent of E-cadherin [11]. The autoaggregation mimics compaction of morula-stage mouse embryos – the first in a series of adhesion events during embryogenesis. E-cadherin is coexpressed with Le^x during compaction but the process proceeds well in an E-cadherin knockout mouse. These findings suggest that embryonal compaction and cell autoaggregation depend on a cooperative effect of Le^x- and E-cadherin-dependent adhesion. It supports the notion that cell adhesion consists of multiple redundant steps initiated by flexible, reversible carbohydrate–carbohydrate recognition, followed by reinforcing protein–carbohydrate and protein–protein interactions (Figure 21.3).

Binding affinity studies by SPR spectroscopy revealed that Le^x trisaccharides linked to self-assembled monolayers of alkanethiolate on gold films strongly bound soluble Le^x trisaccharides ($K_d = 5.4 \times 10^{-7}$ M) [12]. The interaction force between individual Le^x molecules measured by AFM is 20 pN, one of the lowest biomolecular adhesion forces reported to date [10]. This example is consistent with the view that weak binding forces between single molecules can become biologically relevant due to the presence of large amounts of oligosaccharide chains on cell surfaces and of multiple binding sites on each chain, assembling a powerful recognition system based on versatile polyvalent interactions.

21.2.2.2 Gb4-Dependent Adhesion

Autoaggregation of human embryonal carcinoma 2102 or TERA-2 cells constitutes a model of primate embryonic compaction process. Le^x is highly down-regulated on these cells, but they abundantly express lactoneotetraosylceramide (nLc_4), globoside (Gb4, see Table 10.2 for structures), and extended globo series GSLs termed stage-specific embryonic antigen (SSEA)-3 and -4 (for structures and further details, see Chapter 10.6 and Table 24.3) that are maximally expressed at morula stage and possibly mediate the compaction process of the primate (presumably

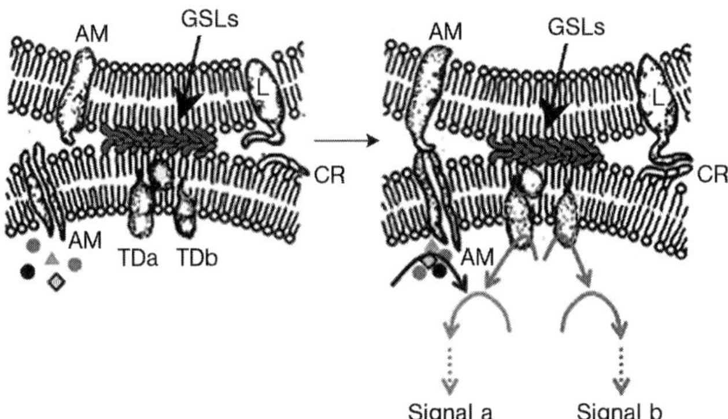

Figure 21.3 Model of intercellular recognition mediated by GSLs (adapted from [9]). The early event in cell recognition consists of highly specific, low-affinity and polyvalent interactions between the carbohydrate moieties of GSLs at the cell surface (left panel). The polyvalency is obtained through organization of GSLs into microdomains (see Chapter 10). This initial step is subsequently reinforced by protein–carbohydrate interactions between lectins and their carbohydrate receptors (see Chapters 13–19), and protein–protein interactions between different adhesion molecules (right panel). The binding is associated with signal transduction. AM, adhesion molecule; CR, carbohydrate receptor; GSLs, glycosphingolipids; L, lectin; TD, signal transducer.

including human) embryo. Homotypic aggregation of 2102 cells is based on interactions between Gb4 and nLc_4 (see Chapter 10 for structures), or between Gb4 and GalGb4 (the major SSEA-3 epitope) [9, 10]. This Gb4-dependent adhesion is thought to be characteristic of the human/primate compaction process that leads to activation of the transcription factors cAMP-responsive element binding protein and activation protein 1. In this adhesion system, the possibility of protein–carbohydrate interaction through Gb4- or GalGb4-binding proteins remains to be studied.

21.2.2.3 **GM3-Dependent Adhesion**

Tumor cell adhesion to endothelial cells (ECs) through carbohydrate–carbohydrate interaction promotes tumor cell metastasis (reviewed in [9, 10], see Chapters 10.8, 25.2 and 30.5 for functional aspects of gangliosides). Metastatic and invasive abilities of mouse melanoma B16 cells are closely correlated with the level of sialosyllactosylceramide (GM3, see Table 10.3 and Figure 30.3 for structures) surface expression, and also with their degree of adhesion to cultured ECs. Such adhesion is not selectin- or integrin/intercellular adhesion molecule-dependent, but the process is based on the heterotypic interaction of GM3 with lactosylceramide and Gb4 expressed on ECs, where both adhesion and motility are strongly enhanced. This mechanism is synergistic with integrin-dependent adhesion and motility (Figure 21.3).

Gangliotriaosylceramide (Gg3; expressed on mouse lung ECs, see Table 10.3 for structure) is a strong ligand of GM3, and Gg3 trisaccharide-carrying polystyrene adsorbs onto a GM3 monolayer strongly and specifically with an apparent affinity constant of $K_a = 2.5 \times 10^6 M^{-1}$ [13]. The GM3–Gg3 heterotypic interaction may account for B16 metastasis to lung, and liposomes containing Gg3 or GM3 inhibit this metastasis [9, 10]. Furthermore, adhesion of melanoma cells through GM3–Gg3 interaction enhanced tyrosine phosphorylation by c-Src (cellular Src from the Src family of tyrosine kinases) and focal adhesion kinase (FAK), and enhanced GTP binding to RhoA and Ras (GTPase proteins). Increased motility of melanoma cells mediated by GM3-dependent adhesion to ECs has been regarded as the initial step in melanoma cell metastasis, where up-regulation of c-Src, RhoA, Ras and FAK may provide a pathophysiological foundation for this initial step in the metastatic process.

21.3 Carbohydrates as DNA-Binding Motifs

A very important aspect of carbohydrate recognition, although much less studied than the examples described above, is the interaction between carbohydrates and DNA. DNA-binding drugs that contain carbohydrates have been known for over 50 years, but the role of the carbohydrate portion of these drugs has been unclear for a long time. The first studies on the importance of the carbohydrate moiety for the biological activity of DNA binders were carried out with the family of anthracycline antibiotics [14, 15]. The number and the substitution pattern of carbohydrate chains have a profound effect on biological activity, and removal of the sugar side-chain results in inactivation of the antibiotic. Structural studies by X-ray crystallography and nuclear magnetic resonance have revealed that the carbohydrate makes van der Waals contacts with the minor groove of the DNA.

There is enough evidence to support the importance of the saccharide moiety of anthracyclines for their interaction with DNA, but this is not as clear for other DNA-binding drugs that contain carbohydrates. However, the discovery of the enediyne antitumor antibiotic calicheamicin opened a new view on the carbohydrates as DNA binders [16]. Calicheamicin binds to DNA in a sequence-selective manner and causes site-specific double-stranded cleavage. The major DNA contact surface of the antibiotic is an aryltetrasaccharide moiety that interacts with the minor groove of the DNA duplex. Derivatives lacking two or more carbohydrate units exhibit less efficient and nonselective cleavage, indicating that the oligosaccharide portion of calicheamicin is the principal DNA-binding element and is largely responsible for the oligopyrimidine selectivity. The preferred target sequences are TCCT, TCTC and TTTT, and the carbohydrate tail of the antibiotic is oriented toward the 3′-end of the tetranucleotide surface [17]. The aryltetrasaccharide segment binds to the DNA minor groove in an induced-fit mechanism, where the extended oligosaccharide conformation spans the recognition sequence of the duplex. This interaction has obvious implications for the functions of DNA since the carbohydrate portion interferes with the binding of transcription factors and with the activation of transcription, foreseeing new contributions to the understanding of the chemical control of genetic information by regulating genes or entire genetic programs of cells.

The realization that the paired bases of DNA are on the inside of the double helix structure whereas the sugar, 2-deoxyribose, is facing outwards, makes it very likely that some DNA–carbohydrate interactions are actually carbohydrate–carbohydrate interactions. This opens a whole new perspective on the importance of carbohydrate self-recognition in the binding of DNA drugs (see Chapter 28 for sugars as pharmaceuticals).

21.4 New Strategies to Study Multivalent Carbohydrate–Carbohydrate Interactions

Characteristic features of carbohydrate–carbohydrate interactions are a strong dependency on calcium, high specificity and low affinity for the individual binding site that is compensated in nature by a multivalent presentation of the carbohydrate units. These features make the process difficult to study and quantify with monovalent ligands due to limitations in the resolution of many biochemical techniques. Current approaches to study polyvalent carbohydrate–carbohydrate interactions include (i) analytical ultracentrifugation, (ii) the construction of two-dimensional (2D) and 3D models to generate polyvalency, amenable to evaluation by new surface analytical techniques such as SPR and the quartz crystal microbalance (QCM), and (iii) the development of methodologies sufficiently sensitive for the study of single molecule interactions such as single-molecule force spectroscopy (SMFS).

21.4.1 Analytical Ultracentrifugation

Hydrodynamic methods are optimal tools for looking at weak interactions, although these techniques have only recently started being applied to the study of carbohydrate–carbohydrate associations [18]. Analytical ultracentrifugation allowed determining the sedimentation velocity of different bioactive heteroxylan species stabilized by weak self-interactions of the monomer–dimer type frequently found in protein systems. This association was shown to be temperature dependent and likely to be hydrophobic in nature, with higher temperature increasing the strength of the interaction and vice versa. Such hydrophobic interactions are not unexpected in polysaccharides since they can be amphiphilic, possessing both hydrophobic (carbon/hydrogen atoms) and hydrophilic (oxygen-containing) regions.

21.4.2 2D/3D Polyvalent Model Systems

To generate polyvalent 2D model systems neoglycoconjugates of the natural epitopes are deposited on gold surfaces to form self-assembled monolayers (SAMs) [19] that can mimic, for instance, the multivalent arrays of GSL patches in the cell membrane [20]. Polyvalent 3D model systems are mainly based on carbohydrate-modified gold nanoparticles [19, 21]. Such glyconanoparticles (GNPs) provide a glycocalix-like surface with globular carbohydrate display and chemically well-defined composition optimally suited to study cooperative carbohydrate interactions [22]. GNPs can be easily prepared by *in situ* reduction of a gold salt in the presence of an excess of the corresponding thiol-derivatized carbohydrate [21]. These GNPs can have a core down to only 2 nm across, and are highly water-soluble and stable for months without flocculation under physiological conditions [20]. Manipulation of the ratio of the ligands permits preparation of GNPs with differing carbohydrate density at the surface, providing a versatile model for investigating the effect on molecular recognition events of epitope clustering and presentation.

Functionalized liposomal nanoparticles have also been applied to the analysis of carbohydrate–carbohydrate associations. Lipid vesicle micromanipulation with the micropipette aspiration technique was used to study Le^x interactions (see Section 21.2.3) [23]. The results obtained revealed a Ca^{2+}-dependent self-binding of monomeric Le^x, whereas dimeric Le^x exhibited a repulsive behavior in the presence of Ca^{2+}. The lactose–GM3 interaction involved in B16 melanoma cell adhesion and signaling processes (see Section 21.2.2) was investigated using micelles of a lactosyl lipid and monolayers of GM3 [24]. In the absence of divalent cations the lactose–GM3 binding was strengthened at higher NaCl concentrations in the subphase of the monolayer, whereas when divalent cations were present the process was not as sensitive to ionic strength. These results suggested a role for both cation-dependent as well as -independent lactose–GM3 interactions.

21.4.2.1 Surface Plasmon Resonance (SPR)

SPR spectroscopy is an evanescent-wave biosensor technology that monitors the interaction of two or more molecules or molecular assemblies in real-time [25]. In SPR, a ligand is immobilized on the surface of a gold-coated sensor chip, whereas the corresponding binding partner, or analyte, is carried in a flow of buffer solution along a miniature flow cell. Light from a laser source arriving through a prism at the angle of total internal reflection induces a nonpropagative evanescent wave that penetrates into the cell opposite the prism. At a given angle dependent upon the refractive index of the solution, resonance between the evanescent wave and free electrons in the gold layer results in a reduction in the intensity of reflected light. The change in angle of reduced intensity reflects changes in the refractive index of the solution in the flow cell immediately adjacent to the gold layer. A dextran layer coupled to the gold surface allows immobilization of ligands within the evanescent field. Any binding event on the surface of the sensor chip leads to a change in refractive index at the surface layer and is continuously monitored by a detector (for example diode array). Time-dependent changes in the refractive index are recorded as sensorgrams that provide information about bound mass, binding kinetics and strength of the interaction.

SPR detection was used to mimic the role of a sulfated disaccharide and of a pyruvylated trisaccharide implicated in Ca^{2+}-dependent marine sponge cell adhesion [6]. The results showed self-recognition of the sulfated disaccharide to be a major force behind the Ca^{2+}-dependent event. The interaction was not simply based on electrostatic interactions, as other sulfated carbohydrates did not self-associate. Ca^{2+} specificity was confirmed by the complete eradication of the self-binding when Ca^{2+} was substituted by other divalent cations such as Mg^{2+} or Mn^{2+}.

The kinetics of the Le^x–Le^x interaction in the presence of Ca^{2+} was studied with a combination of SAMs formed by the neoglycoconjugates on a biosensor gold surface as ligand and gold GNPs as the analyte [20]. The SPR sensorgrams obtained for the binding of Le^x-GNPs to the Le^x-SAMs indicated a slow association and a gradual dissociation phase in the presence of Ca^{2+} ($K_d = 5.4 \times 10^{-3}$ M). These results indicated that Le^x self-aggregation is a selective, multivalent and Ca^{2+}-dependent event consistent with the proposal that carbohydrate *trans*-associations are part of the intercellular interaction machinery.

SPR has been used to investigate carbohydrate–carbohydrate associations between clustered GM3 on phospholipid monolayers and clustered Gg3 trisaccharide along a polystyrene chain (see Section 21.2.2) [26]. The results obtained revealed that the *N*-acetyl groups of *N*-acetylneuraminic acid in GM3 and of GalNAc in Gg3 play an important role in the interaction and that polystyrene(Gg3) recognizes not only some specific portions of GM3, but also the trisaccharide as a whole.

Finally, SPR has recently been applied to investigate the self-binding of the 200-kDa g200 glycan responsible for the Ca^{2+}-dependent, species-specific aggregation of sponge cells (see Section 21.2.1) [7]. SPR measurements showed that g200 self-adhesion is much stronger than its binding to other unrelated glycans such as chondroitin sulfate. The data obtained in artificial sea water containing only 2 mM Ca^{2+} (that is low Ca^{2+} conditions) showed a steep decrease in AF self-binding compared with AF–g200 binding, suggesting the existence in low calcium of g200 adhesive properties different from those observed at the physiological sea water concentration of 10 mM Ca^{2+}(for further SPR applications in lectin research, please see Chapter 14).

21.4.2.2 **Quartz Crystal Microbalance (QCM)**

The QCM technique relies upon the piezoelectric effect in quartz crystals [27], whose frequency of oscillation changes in proportion to the amount of mass adsorbed onto their surface. This mass sensing technique eliminates the need for any specific labeling step to be part of the signal transduction mechanism, which operates well in complex media. The QCM has a wide detection range: at the low mass end, it can detect monolayer surface coverage by small molecules on polymer films; at the upper end, it is capable of detecting much larger masses bound to the surface, up to multilayers above 100 nm deep. Another important and unique feature of this methodology is the ability to measure mass and energy dissipation properties of biopolymer films while simultaneously carrying out electrochemistry on solution species or upon film systems bound to the upper electrode on the oscillating quartz crystal surface [27].

The QCM has been used to characterize the structural growth of layers of the polysaccharides sodium alginate and polygalacturonic acid [28]. Monitoring the variations in frequency and dissipation energy provided information on the average bound mass and the viscoelastic properties of the adsorbed layer of polyelectrolytes along with the associated ions and water molecules. Different swelling behaviors between both polysaccharides indicated differences in their complexation modes.

21.4.3
Single-Molecule Detection and Manipulation

Despite the successful application of multivalent model systems to mimic native polyvalency, study of the molecular details of carbohydrate–carbohydrate recognition requires new models and techniques to probe the interaction of individual motifs. A promising line of work contemplates binding carbohydrates onto a conformationally defined helical scaffold (a so-called foldamer) to allow carbohydrate–carbohydrate interactions to be studied within a controlled and nonmultivalent environment [29]. Spectroscopic methods such as circular dichroism can be used to directly evaluate the interactions, which may also be associated within (and reportable via) the scaffold core. However, most of the few studies to date on individual carbohydrate–carbohydrate interactions have been performed by SMFS.

21.4.3.1 Single-Molecule Force Spectroscopy (SMFS)

During the last decade SMFS has developed into a highly sensitive tool for the investigation of single biomolecule interactions [30]. Most SMFS experiments use either optical tweezers or AFM to measure dissociation forces of single ligand-receptor complexes in the picoNewton range.

Optical tweezers use forces exerted by a strongly focused beam of light to trap small objects (for a review, see [31]). Optical tweezers can trap particles as small as 5 nm and can measure forces exceeding 100 pN with resolutions down to 100 aN. This is an optimal range for applying forces on biomolecular systems and for recording their responses. Optical tweezers have been used to determine for the first time the interaction between decorin GAG chains [32]. The rupture force between single GAG–GAG bonds was found to be 16.5 ± 5.1 pN. Such valuable quantitative data will undoubtedly contribute to understanding the mechanical properties of connective tissues at the molecular level.

The adhesion forces between carbohydrates have also been evaluated with AFM-based SMFS (reviewed in [1, 9, 10]). After functionalizing the AFM tip and surface with the corresponding glycans, force–distance curves can be recorded under physiological conditions. In studies of the Le^x–Le^x interaction [20], analysis of several hundred force curves measured in different areas of the sample indicated that the specific interaction force between two single Le^x molecules was 20 ± 4 pN. This value indicates that only five pairs of Le^x molecules would be necessary to reach the binding strength (100 pN) between neural retina cells of embryonic chicken, and 16 Le^x pairs (320 pN) would be sufficient to hold T and B lymphocytes together in the absence of antigen stimulation.

SMFS studies performed with sponge-derived proteoglycans (see Section 21.2.1) [1, 7] of their calcium-dependent interactions have revealed functional intermolecular domains that can contribute to many adhesive and elastic extracellular matrix interactions. Force peaks resulting from g200–g200 glycan interactions in AFs are separated by an average distance of 20 nm (Figure 21.4), which corresponds to the spacing between g200 glycan anchoring points in the AF. These data are consistent with a self-adhesion mechanism where each peak in *M. prolifera* AF force–extension curves corresponds to the breaking of a single g200–g200 interaction. Fitting to a worm-like chain model [33] of the individual peaks (Figure 21.4) indicates that there is an elastic component involved in the process, probably resulting from a molecular stretching, before each break of the glycan–glycan interaction, of the *M. prolifera* AF protein to which g200 is covalently linked. The multiplicity of binding sites confers a high degree of modulability as required in most biological interactions, in contrast to the higher stability of a single, strong bond.

21.5 Conclusions

This chapter describes noncovalent, versatile and polyvalent carbohydrate–carbohydrate interactions in molecular processes. Cell-surface glycoconjugates are involved in carbohydrate self-interactions in cell recognition processes where specificity and flexibility are essential for proper social behavior of cells. Cell recognition based on carbohydrate–carbohydrate interactions is unique in that it displays (i) high variability depending on the degree of molecule clustering and/or repetition of the binding motif along the carbohydrate chain, and (ii) faster reactivity than that of protein–protein interactions that typically occur through integrins, cadherins and other cell adhesion molecules. Therefore, carbohydrate–carbohydrate interactions are not alternative or supplemental to protein–protein and protein–carbohydrate interactions. Rather, they represent initial steps leading to multiple and redundant mechanisms fundamental to cell recognition and adhesion events. A system, yet to be explored in detail, where carbohydrate–carbohydrate recognition may play an important role is the binding between carbohydrate moieties of drugs and the DNA minor groove. Understanding the molecular basis of this interaction and its importance for the biological activity of DNA-binding drugs is essential for designing more specific and effective

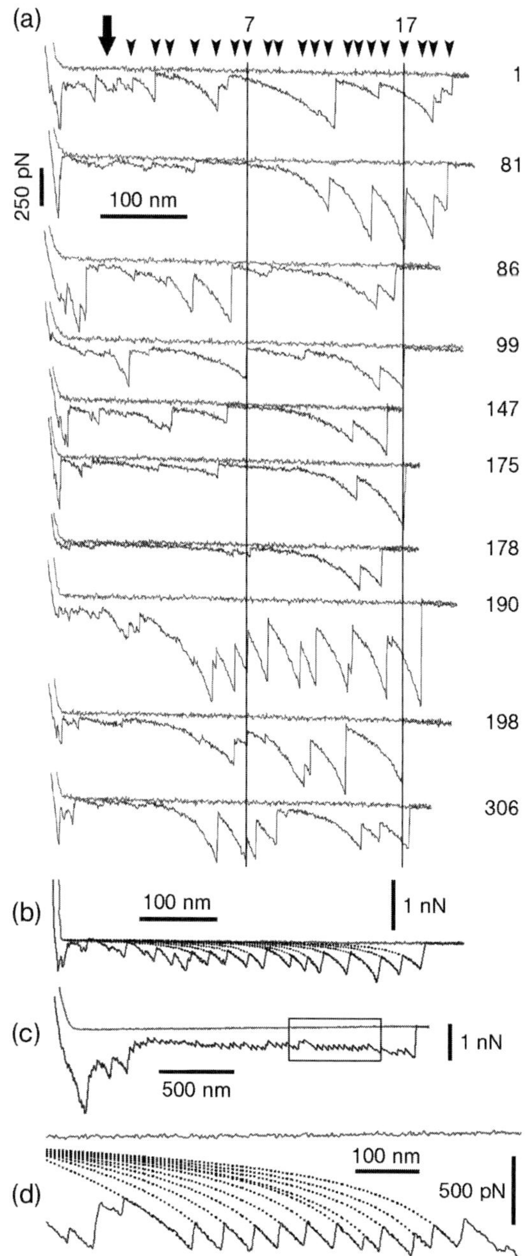

Figure 21.4 SMFS analysis of carbohydrate-based sponge proteoglycan adhesion. (a) Alignment of 10 force–extension curves selected from several hundred consecutive approach–retract cycles in a single SMFS experiment. The numbers on the right indicate recording order. Every curve reflects the interaction of a single *M. prolifera* AF molecule on the AFM cantilever tip. Piling up the adhesion curves of the retracting phase permits an alignment of the force peaks into discrete positions separated by an average distance of 20 nm (arrowheads). Aligned peaks in curves 99, 190 and 306 for position 7, and in curves 81, 99, 175 and 198 for position 17 are indicated by vertical lines to illustrate the repetitive on/off binding of discrete individual g200 glycans. Peaks in the first 50 nm of the retracting phase (arrow) are not considered due to possible unspecific short range tip-surface binding artifacts. (b) Worm-like chain fit of a *M. prolifera* AF force–extension curve in a typical single-molecule interaction. (c) Force–extension curve corresponding to a concatamer of several *M. prolifera* AF molecules. (d) Blow-up of the retraction curve section boxed in (c), showing the worm-like chain fit.

therapeutic compounds. The study and evaluation of carbohydrate–carbohydrate interactions go hand in hand with the development of novel and more sensitive experimental approaches to better understand the nature and biological roles of these low-affinity interactions that govern essential cellular processes.

Summary Box

Carbohydrate–carbohydrate interactions provide a highly versatile mechanism based on the low affinity and polyvalent configuration of carbohydrate chains. Carbohydrates offer many combinatorial possibilities for finely tuned cellular adhesion and recognition processes, as well as for interactions between drugs or natural ligands with DNA. Precision and sensitivity of novel experimental approaches allow more thorough investigation of these interactions at the single-molecule level.

References

1 Bucior I, Burger MM. Carbohydrate–carbohydrate interactions in cell recognition. *Curr Opin Struct Biol* 2004;14:631–7.

2 Spillmann D, Burger MM. Carbohydrate–carbohydrate interactions in adhesion. *J Cell Biochem* 1996;61:562–8.

3 Spillmann D. Carbohydrates in cellular recognition: from leucine-zipper to sugar-zipper? *Glycoconj J* 1994;11:169–71.

4 Fernàndez-Busquets X, Burger MM. Circular proteoglycans from sponges: first members of the spongican family. *Cell Mol Life Sci* 2003;60:88–112.

5 Guerardel Y *et al.* Molecular fingerprinting of carbohydrate structure phenotypes of three porifera proteoglycan-like glyconectins. *J Biol Chem* 2004;279:15591–603.

6 Haseley SR *et al.* Carbohydrate self-recognition mediates marine sponge cellular adhesion. *Proc Natl Acad Sci USA* 2001; 98:9419–24.

7 Garcia-Manyes S *et al.* Proteoglycan mechanics studied by single-molecule force spectroscopy of allotypic cell adhesion glycans. *J Biol Chem* 2006;281:5992–9.

8 Turley EA, Roth S. Interactions between the carbohydrate chains of hyaluronate and chondroitin sulphate. *Nature* 1980; 283:268–71.

9 Hakomori S. Carbohydrate-to-carbohydrate interaction, through glycosynapse, as a basis of cell recognition and membrane organization. *Glycoconj J* 2004;21:125–37.

10 Regina Todeschini A, Hakomori SI. Functional role of glycosphingolipids and gangliosides in control of cell adhesion, motility, and growth, through glycosynaptic microdomains. *Biochim Biophys Acta* 2008;1780: 421–33.

11 Handa K *et al.* Le^x glycan mediates homotypic adhesion of embryonal cells independently from E-cadherin: a preliminary note. *Biochem Biophys Res Commun* 2007;358: 247–52.

12 Hernaiz MJ *et al.* A model system mimicking glycosphingolipid clusters to quantify carbohydrate self-interactions by surface plasmon resonance. *Angew Chem Int Ed Engl* 2002;41:1554–7.

13 Matsuura K *et al.* Surface plasmon resonance study of carbohydrate–carbohydrate interaction between various gangliosides and Gg3-carrying polystyrene. *Biomacromolecules* 2004;5:937–41.

14 Smith CK *et al.* DNA–nogalamycin interactions: the crystal structure of d(TGATCA) complexed with nogalamycin. *Biochemistry* 1995;34:415–25.

15 Temperini C *et al.* Role of the amino sugar in the DNA binding of disaccharide anthracyclines: crystal structure of the complex MAR70/d(CGATCG). *Bioorg Med Chem* 2005;13:1673–9.

16 Lauria A *et al.* DNA minor groove binders: an overview on molecular modeling and QSAR approaches. *Curr Med Chem* 2007; 14:2136–60.

17 Sissi C *et al.* Interaction of calicheamicin γ1(I) and its related carbohydrates with DNA–protein complexes. *Proc Natl Acad Sci USA* 1999;96:10643–8.

18 Patel TR *et al.* Weak self-association in a carbohydrate system. *Biophys J* 2007;93: 741–9.

19 Carvalho de Souza A, Kamerling JP. Analysis of carbohydrate–carbohydrate interactions using gold glyconanoparticles and oligosaccharide self-assembling monolayers. *Methods Enzymol* 2006;417:221–43.

20 de la Fuente JM, Penadés S. Understanding carbohydrate–carbohydrate interactions by means of glyconanotechnology. *Glycoconj J* 2004;21:149–63.

21 Barrientos AG *et al.* Gold glyconanoparticles: synthetic polyvalent ligands mimicking glycocalyx-like surfaces as tools for glycobiological studies. *Chem Eur J* 2003; 9:1909–21.

22 Rojo J *et al.* Gold glyconanoparticles as new tools in antiadhesive therapy. *ChemBioChem* 2004;5:291–7.

23 Gourier C *et al.* Specific and non specific interactions involving Le^X determinant quantified by lipid vesicle micromanipulation. *Glycoconj J* 2004;21:165–74.

24 Santacroce PV, Basu A. Studies of the carbohydrate–carbohydrate interaction between lactose and GM3 using Langmuir monolayers and glycolipid micelles. *Glycoconj J* 2004;21:89–95.

25 Jonsson U *et al.* Real-time biospecific interaction analysis using surface plasmon resonance and a sensor chip technology. *Biotechniques* 1991;11:620–7.

26 Matsuura K, Kobayashi K. Analysis of GM3–Gg3 interaction using clustered glycoconjugate models constructed from glycolipid monolayers and artificial glycoconjugate polymers. *Glycoconj J* 2004;21:139–48.

27 Marx KA. Quartz crystal microbalance: a useful tool for studying thin polymer films and complex biomolecular systems at the solution–surface interface. *Biomacromolecules* 2003;4:1099–120.

28 de Kerchove AJ, Elimelech M. Formation of polysaccharide gel layers in the presence of Ca^{2+} and K^+ ions: measurements and mechanisms. *Biomacromolecules* 2007;8:113–21.

29 Simpson GL *et al.* Glycosylated foldamers to probe the carbohydrate–carbohydrate interaction. *J Am Chem Soc* 2006;128:10638–9.

30 Zlatanova J *et al.* Single molecule force spectroscopy in biology using the atomic force microscope. *Prog Biophys Mol Biol* 2000; 74:37–61.

31 Grier DG. A revolution in optical manipulation. *Nature* 2003;424:810–6.

32 Liu X *et al.* Direct measurement of the rupture force of single pair of decorin interactions. *Biochem Biophys Res Commun* 2005; 338:1342–5.

33 Janshoff A *et al.* Force spectroscopy of molecular systems – single molecule spectroscopy of polymers and biomolecules. *Angew Chem Int Ed Engl* 2000;39:3212–37.

Part Five
Biomedical Aspects and Case Studies

The Sugar Code. Edited by Hans-Joachim Gabius

ISBN: 978-3-527-32089-9

22
Diseases of Glycosylation

Thierry Hennet

Can incorrect glycosylation make us sick? This question has not always received a clear-cut answer. Knowledge of diseases of glycosylation has long remained limited to mild hematological disorders. These few diseases with cryptic names, like Tn syndrome and paroxysmal nocturnal hemoglobinuria (PNH), gave the impression that glycosylation is probably so redundant that single-gene defects can be dispensable to normal health. This misconception has definitively disappeared over the last decade, which has seen the description of close to 40 diseases of glycosylation, encompassing neurological disorders, congenital muscular dystrophies, connective tissue disorders, immune deficiencies and coagulopathies [1].

This recent progress has been made possible by the development of sensitive analytic techniques, and especially thanks to the coordinated efforts of clinicians and basic researchers. However, despite the recent advances, the identification of diseases of glycosylation remains a challenging venture. Glycosylation is ubiquitous and the tremendous structural complexity of glycans makes it quite difficult to predict the biological importance of individual structures. For example, it is impossible to guess the biological role of a given glycan structure based on its cellular distribution or its presence on specific proteins. Some diseases of glycosylation have been identified by detecting changes in the binding patterns of lectins and carbohydrate-specific antibodies. Changes in the molecular weight of known glycoproteins have also pinpointed glycosylation defects in previously untyped diseases. Pure genetic approaches like linkage analysis and homozygosity mapping continue to extend the list of known inherited diseases of glycosylation. Finally, the use of animal models may also help to delineate glycan functionality (please see Chapter 23).

Diseases of glycosylation have been described in nearly all glycosylation pathways. The clinical features associated with these diseases reflect the multiple contributions of glycans in human development and physiology. However, beyond this general statement, what have we learnt so far from diseases of glycosylation? First, by comparing the clinical features of various diseases of glycosylation, it became clear that mutations in genes acting in the same glycosylation pathway

The Sugar Code. Edited by Hans-Joachim Gabius

ISBN: 978-3-527-32089-9

yield diseases with similar symptoms. This similarity is often used to isolate new gene defects within a glycosylation pathway. Secondly, some diseases of glycosylation present the same clinical features as diseases caused by the loss of specific glycoproteins, thus suggesting that these glycosylation defects are restricted to given glycoproteins. This level of specificity could certainly not be inferred from the *in vitro* study of substrate specificity of glycosyltransferase enzymes. To provide some examples, the present chapter groups diseases of glycosylation according to glycosylation pathways. Since an extensive discussion of these diseases would exceed the scope of this chapter, readers are invited to consult the online database Online Mendelian Inheritance in Man (OMIM) of genetic diseases (http://www.ncbi.nlm.nih.gov/Omim), which covers in depth aspects such as clinical features, diagnosis, molecular and population genetics, and animal models. The OMIM reference numbers of each disease outlined below are given to simplify the retrieval of the corresponding reports in the database.

22.1 *N*-Glycosylation

N-Glycosylation is unique among the classes of glycosylation as it involves the assembly of the oligosaccharide $GlcNAc_2Man_9Glc_3$ (see Chapter 1 for the nomenclature) on the carrier lipid dolichol at the endoplasmic reticulum (ER) membrane. This lipid-linked oligosaccharide (LLO) is then transferred 'en bloc' to selected asparagine residues of nascent proteins during protein translation. Once bound to proteins, *N*-glycans undergo additional trimming and elongation steps in the Golgi apparatus as described in Chapter 6. Defects of LLO assembly in the ER lead to limited availability of the oligosaccharide substrate and thus to limited occupancy of *N*-glycosylation sites on proteins. This results in a quantitative defect of glycosylation, although the quality of the remaining *N*-glycans is unaffected. By contrast, altered trimming and elongation yield qualitative defects of *N*-glycosylation, while the degree of *N*-glycosylation site occupancy remains unchanged. Quantitative defects of *N*-glycosylation are classified as congenital disorders of glycosylation (CDG) type I. Similarly, qualitative defects of *N*-glycosylation are called CDG type II [2].

Both types of *N*-glycosylation defects can be detected by isoelectric focusing of serum glycoproteins such as transferrin (Figure 22.1). The absence of *N*-glycans or their defective elongation decreases the number of sialic acid residues on *N*-glycans, thus changing the number of negative charges and hence the isoelectric point of the glycoprotein examined. This simple test, which is routinely used in clinical chemistry laboratories, can be rapidly performed on few microliters of blood (please see also Info Box 1). A more detailed analysis of *N*-glycosylation defects necessitates more specialized tests, such as liquid chromatography and mass spectrometry, requiring more expertise and infrastructure.

To date, 16 genetic defects have been identified as causing CDG-I (Table 22.1). The corresponding genes encode proteins involved in the synthesis of GDP-Man,

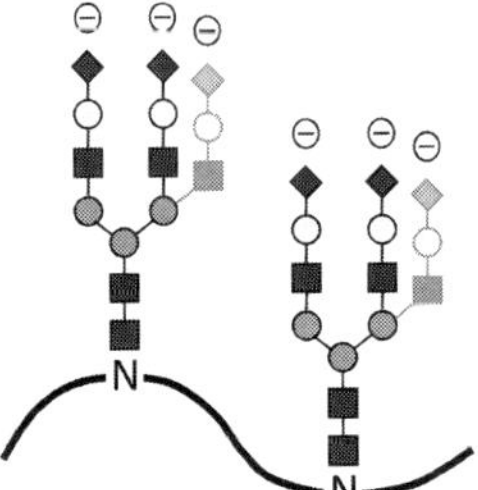

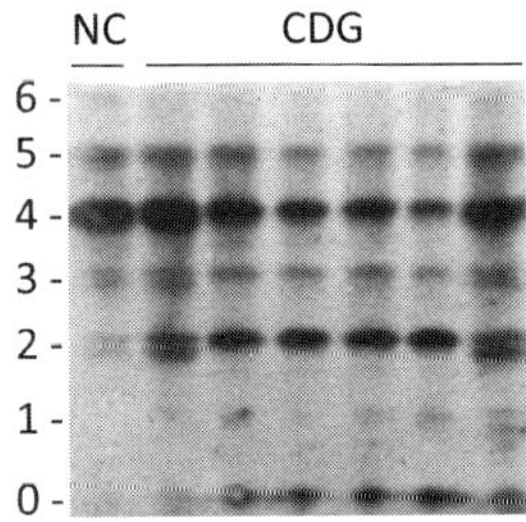

Figure 22.1 Detection of *N*-glycosylation defects by isoelectric focusing of serum transferrin. Transferrin carries two *N*-glycans, which are normally terminated by Sia, thereby introducing four to six negative charges (left panel). Isoelectric focusing separates the transferrin glycoforms according to their amount of negative charges. Normal control (NC) transferrin mainly shows a glycoform carrying four negative charges (right panel). Transferrin from CDG samples accumulates glycoforms with 3, 2, 1 or even 0 negative charges, which indicates a loss of glycan structures. The double-band patterns seen in lanes 1, 2 and 7 are caused by an amino acid polymorphism often found in transferrin.

Info Box 1

In 1976, the Swedish neurologist Helena Stibler described structural abnormalities in the serum glycoprotein transferrin in cases of alcoholism. The cause of the anomaly was shown to be related to the absence of glycan chains on transferrin, which normally carries two *N*-glycan chains. The decreased glycosylation, as evidenced by isoelectric focusing analysis, matched with episodes of alcohol abuse since transferrin glycosylation normalized in periods of abstinence. Because of its high sensitivity and specificity, carbohydrate-deficient transferrin (CDT) became a standard diagnostic marker of alcohol abuse. As the CDT test became broadly adopted, a few cases of false-positivity were noticed in patients affected by a rare inherited disorder of protein glycosylation, which was then called carbohydrate-deficient glycoprotein syndrome. This syndrome was later found to represent several forms of CDG-I and -II. The pioneer work of Helena Stibler and of the Belgian pediatrician Jaak Jaeken was instrumental in the first description of diseases of *N*-glycosylation. The isoelectric focusing test of serum transferrin has paved the way for the identification of many types of *N*-glycosylation disorders. Whereas the importance of the test in the diagnosis of glycosylation disorders is undisputed, the biochemical mechanisms underlying the effect of ethanol exposure on protein glycosylation are still enigmatic.

Dol-P-Man, in the assembly of LLO in the ER and in the transfer of oligosaccharides to proteins (Figure 22.2). The first gene associated with CDG-I, and by far the most frequent defective one, is the phosphomannomutase-2 (*PMM2*, OMIM 212065) gene. The estimated incidence of *PMM2* deficiency averages 1:50 000, meaning that, fortunately, it is a rare disease. Remarkably, the most common

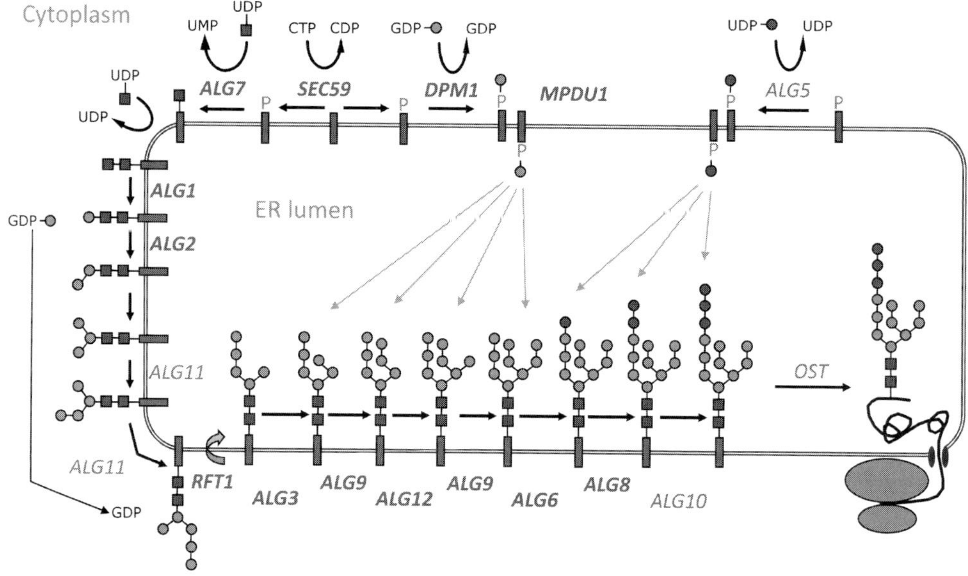

Figure 22.2 Pathway of LLO assembly in the ER. Monosaccharides are added sequentially onto the lipid carrier dolichol (dark rectangle) by glycosyltransferase enzymes. The genes that have been related to a form of CDG-I are marked in dark grey.

effect of mutations in human ER glycosyltransferases genes by expression in yeast stains that are deficient for the orthologous glycosyltransferases gene. These yeast strains are engineered so that the glycosylation defects impair cell growth. The expression of the human orthologous genes in yeasts corrects the glycosylation defect and thereby normalizes cell growth. However, the expression of human genes with mutations sometimes does not restore yeast growth, thus proving the pathogenically of the human mutation in question.

Whereas most steps of the LLO assembly pathway have been associated to a disease of glycosylation, defects of the oligosaccharyltransferase (OST) complex (see Chapter 6) have just been discovered recently with the description of mutations in the subunits *TUSC3/OST3* and *IAP/OST6* in a form of non-syndromic mental retardation. Surprisingly, these defects only impair the *N*-glycosylation of selected proteins, meaning that they cannot be diagnosed by transferrin isoelectric focusing like most types of CDG. The identification of restricted *N*-glycosylation deficiency as a cause of mental retardation suggests that the disease may still be largely undiagnosed. Along this line, the lack of sensitivity to the transferrin assay indicates the need for the development of novel diagnostic tools.

Qualitative defects of *N*-glycosylation, i.e. various types of CDG-II, are less frequent than CDG-I. To date, only deficiency of *MGAT2* β1,2-GlcNAc-transferase-2, defined as CDG-IIa (OMIM 212066), and of ER glucosidase-I, defined as CDG-IIb

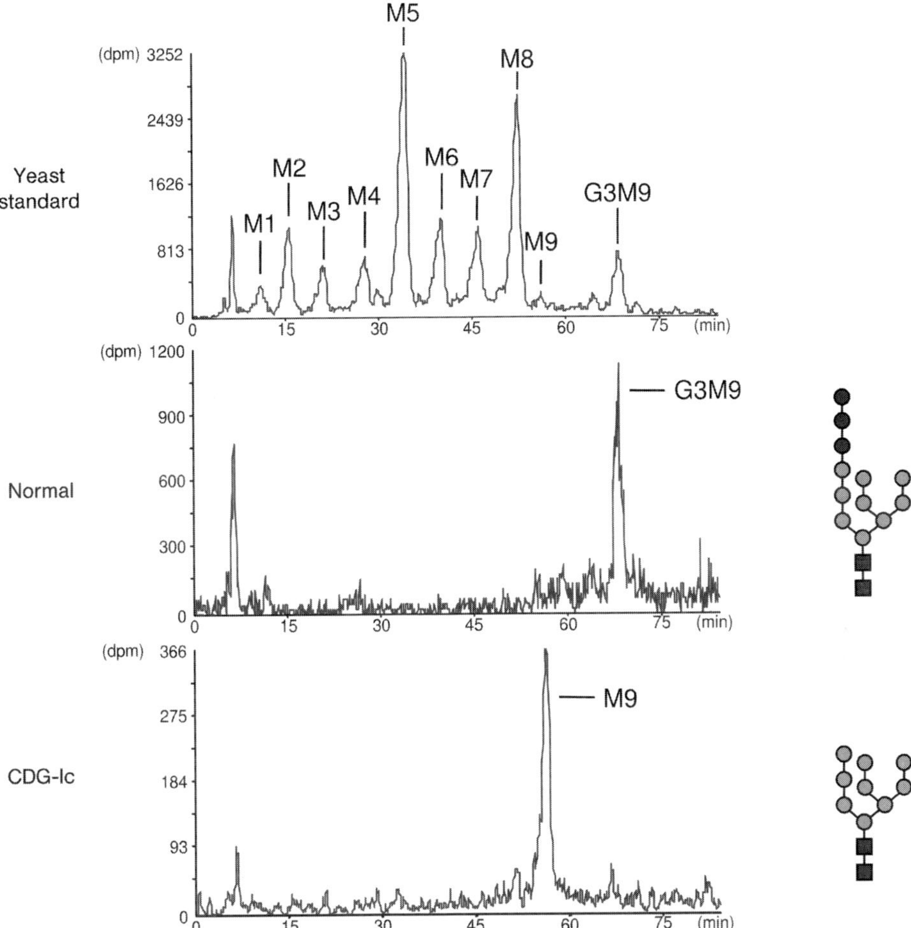

Figure 22.3 Profiling of LLO by high-pressure liquid chromatography (HPLC). Fibroblast cells isolated from a skin biopsy are incubated with radiolabeled [^{3}H]mannose, which is incorporated into LLO. After extraction from the cells, LLOs are separated by HPLC and the resulting profiles are compared to a yeast LLO standard to identify the type of LLO possibly accumulating in the cells investigated. In the case of CDG-Ic (lower panel), the cells accumulate the incomplete lipid-linked $GlcNAc_2Man_9$ (M9) structure.

(OMIM 606056), represent true defects of *N*-glycosylation (Table 22.1). Deficiencies of GDP-Fuc transporter (CDG-IIc, OMIM 266265), *B4GALT1* β1,4-Gal-transferase-1 (CDG-IId, OMIM 607091) and CMP-sialic acid (Sia) transporter (CDG-IIf, OMIM 603585) also affect the formation of *O*-glycans and glycolipids. Defects in the subunits of the COG complex (see below, Section 22.10), which is involved in vesicular trafficking, have also been referred to as types of CDG-II (Table 22.1), since these defects affect *N*-glycosylation. However, multiple glycosylation pathways and possibly other cellular processes are also disturbed in COG deficiency so that these disorders of vesicular transport should not count solely as diseases of glycosylation.

The severity of *N*-glycosylation disorders largely depends on the genes affected and on the degree of inactivation achieved by the underlying mutations. Most types of CDG are associated with symptoms such as psychomotor retardation, dysmorphic features, hypotonia, cerebellar hypoplasia, hormonal disorders, coagulopathies and stroke-like episodes. By contrast, deficiency of phosphomannose isomerase, known as CDG-Ib (OMIM 602579), is associated with normal psychomotor development, but presents with hypoglycemia, vomiting and diarrhea. Deficiency of the GDP-Fuc transporter (CDG-IIc) has also been called leukocyte adhesion deficiency (LAD)-II, because the lack of fucosylated glycans impairs leukocyte rolling mediated by P-, E- and L-selectins (see Chapters 19 and 27). Consequently, CDG-IIc/LAD-II patients are immune compromised and thereby prone to infections. Deficiency of the CMP-Sia transporter, assigned as CDG-IIf, affects the sialylation of many glycans classes. The single CDG-IIf patient described to date died a few months after birth of severe bleeding related to a blood platelet disorder. It is unclear whether other organ functions were affected by the generalized sialylation deficiency.

Unfortunately, successful therapies are limited to CDG-Ib and CDG-IIc/LAD-II, which can be corrected by oral supplementation with mannose and fucose, respectively (please see Info Box 2). Other forms of CDG are insensitive to carbohydrate supplementation. Various derivatives of sugar metabolites like mannose-1-phosphate have been tested in cell culture with little success. Such approaches are questionable since manipulation of sugar-phosphate levels can have drastic effects on glycolysis and on the cellular ATP pool.

22.2 *O*-Glycosylation

Some forms of *O*-glycosylation occur nearly on any kind of glycoproteins (for details, please see Chapter 7). This is the case for mucin-type *O*-GalNAc glycosylation. By contrast, *O*-Man and *O*-Fuc glycans are mainly limited to few classes of glycoproteins. Accordingly, defects in the assembly of *O*-GalNAc, *O*-Man and *O*-Fuc glycans impair distinct biological processes and hence cause different types of diseases.

22.2.1 *O*-GalNAc Glycosylation

As discussed in Chapter 7, *O*-GalNAc glycosylation is initiated by a family of more than 15 polypeptide GalNAc-transferase enzymes. The various polypeptide GalNAc-transferase isoforms show varying degrees of acceptor substrate specificity and the level of redundancy between these isoforms makes it difficult to predict the pathological outcome of a given enzyme deficiency. Probably for this reason,

Info Box 2

Two forms of CDG, namely CDG-Ib and CDG-IIc/LAD-II, can be treated by dietary uptake of mannose and fucose, respectively. The simplicity of the treatment and the good tolerance to the carbohydrates lead to the belief that other CDG patients might also profit from a supplementation with mannose or fucose. Although these carbohydrates are not toxic, the long-term effects of such supplementation regimens are largely unclear. Carbohydrates can react with the free amino groups of proteins in a process called nonenzymatic glycation, thereby altering the properties of the affected proteins. The level of protein glycation becomes critical in situations of elevated serum carbohydrate concentrations, such as found in diabetes mellitus, where increased protein glycation has been related to chronic complications like atherosclerosis and retinopathy. Glucosamine is another carbohydrate that is commonly taken up as dietary supplement. Glucosamine is a precursor of GAG chains and it is thought that the uptake of glucosamine favors the regeneration of cartilage tissue, thus preventing osteoarthritis. While the benefit of glucosamine against osteoarthritis is controversial, some studies even claimed that increased glucosamine levels promote insulin resistance by elevating the intracellular *O*-GlcNAc modification of signaling proteins. The long-term effects of carbohydrate supplementations are certainly difficult to resolve due to the multiple involvements of carbohydrates, as energy molecules, structural units and information signals. So, what should we recommend at the present stage? Probably, like all good things, carbohydrates should be enjoyed with moderation.

the first association of a polypeptide GalNAc transferase defect to a disease came from a genetic study. Using genetic linkage analysis, mutations in the *GALNT3* encoding polypeptide GalNAc transferase-3 have been identified in cases of familial tumoral calcinosis (OMIM 211900), a disease characterized by extreme subcutaneous deposition of calcium and by elevated phosphate levels in blood. Another cause of familial tumoral calcinosis had been linked to deficiency of the protein fibroblast growth factor (FGF) 23, a glycoprotein hormone that regulates vitamin D metabolism and phosphate homeostasis. Remarkably, FGF23 carries *O*-GalNAc glycans, which are initiated by the *GALNT3* polypeptide GalNAc transferase-3 enzyme. These *O*-GalNAc glycans are necessary for the proper secretion of the active hormone. Interestingly, the binding of FGF23 to its receptor requires another protein, named Klotho, which is a secreted glycosidase. Whether Klotho enables the binding of FGF23 by affecting the glycosylation of the hormone or of its receptor is still unclear [3].

The first disease of *O*-GalNAc glycosylation described was Tn syndrome (OMIM 300622), a rare autoimmune disorder characterized by a mild hemolytic anemia. A portion of the erythrocytes of Tn patients express the Tn antigen, i.e. bare *O*-GalNAc on glycoproteins, due to a defect of core-1 β1,3-Gal-transferase activity.

These Tn-erythrocytes are recognized and eliminated by the immune system, thereby explaining the mild anemia. Although the core-1 β1,3-Gal-transferase activity is deficient in some hematopoietic cells of patients with Tn syndrome, no mutations are found in the *C1GALT* core-1 β1,3-Gal-transferase gene. Rather, the defect is localized at the level of the *COSMC* gene, which encodes a chaperone protein required for the proper folding of the *C1GALT* core-1 β1,3-Gal-transferase enzyme. Somatic mutations in the X-linked *COSMC* gene also arise in tumor cells, thereby exposing the Tn antigen, which is recognized by circulating antibodies and contributes to the removal of the tumor cells by immune surveillance mechanisms. Finally, *COSMC* mutations in hematopoietic stem cells are probably at the origin of IgA nephropathy (OMIM 161950), the most common form of glomerulonephritis worldwide. IgA nephropathy is characterized by the mesengial deposition of IgA1 aggregates, which leads to an inflammatory destruction of glomeruli. The IgA1 protein contains 10 *O*-GalNAc glycans in its hinge region. The reduction of these *O*-glycans to bare *O*-GalNAc decreases the stability of the IgA1 protein and thereby promotes its deposition in kidneys.

By analogy to the transferrin isoelectric focusing assay, through which disorders of *N*-glycosylation can be rapidly identified, isoelectric focusing of the *O*-glycoprotein ApoCIII has been described as a test to identify possible defect of *O*-GalNAc glycosylation. However, ApoCIII glycosylation is simple and does not reflect the diversity of *O*-glycosylation sufficiently well to represent a comprehensive test for *O*-GalNAc glycosylation defects.

22.2.2
O-Man Glycosylation (*O*-Mannosylation)

O-Man glycans are mainly found on the α-dystroglycan protein [4]. These *O*-Man glycans mediate essential interactions between α-dystroglycan and extracellular matrix proteins like laminin-2, thereby promoting cell adhesion and considerably stabilizing muscle fibers (Figure 22.4). Deficiencies at the level of the dystroglycan complex, such as caused by deficiency of *O*-mannosylation, lead to degeneration of muscle tissue – a condition defined as muscular dystrophy. The main form of congenital muscular dystrophy is Duchenne muscular dystrophy, which is caused by mutations in the dystrophin gene that encodes a cytoskeleton protein binding the intracellular side of the dystroglycan complex. Several forms of congenital muscular dystrophy have been related to defects of *O*-mannosylation (Table 22.2). Mutations in the *POMT1*/*POMT2* protein *O*-mannosyltransferase-1 and -2 genes and in the *POMGNT1* β1,2-GlcNAc-transferase disable the transfer of the first two monosaccharides of *O*-Man glycans. Mutations in the fukutin, fukutin-related protein and *LARGE1* genes also lead to congenital muscular dystrophy. These genes encode potential glycosyltransferases, although their substrate specificities have not yet been determined.

Abnormal glycosylation of the dystroglycan complex is primarily associated with congenital muscular dystrophy, but other organ dysfunctions are also often encountered. Dystroglycan is involved in the control of neuronal migration in the

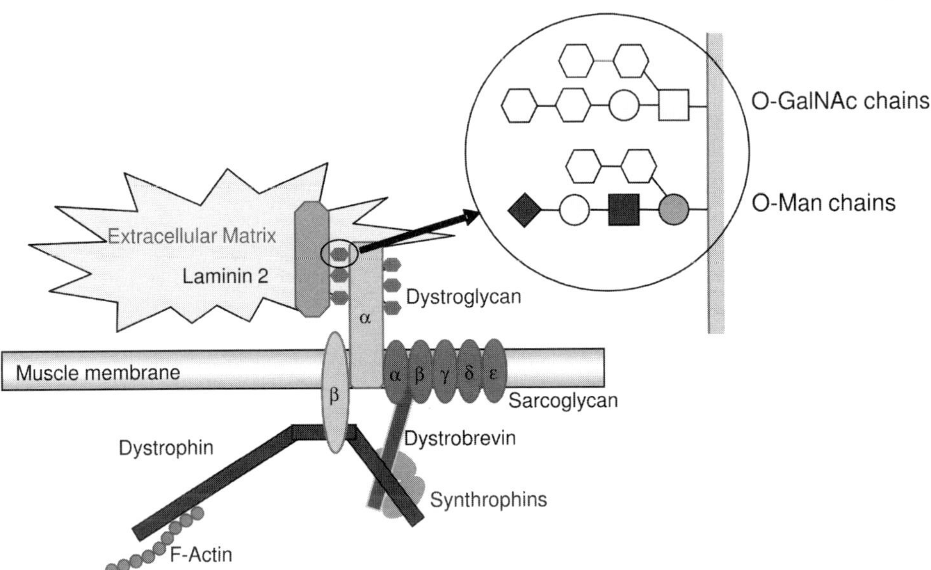

Figure 22.4 Dystroglycan *O*-mannosylation. α-Dystroglycan interacts at the cell membrane with β-dystroglycan and with the sarcoglycan complex. The interactions of α-dystroglycan with the extracellular protein laminin-2 are mediated by the numerous *O*-Man and *O*-GalNAc glycans chains decorating α-dystroglycan. However, the exact structure of these glycan chains is largely unclear.

Table 22.2 Diseases of *O*-mannosylation (adapted from a figure made by Jane Hewitt, University of Nottingham).

Name	Gene
Walker–Warburg	POMT1/POMT2
Muscle–eye–brain	POMGNT1
Fukuyama congenital muscular dystrophy	FKTN
Congenital muscular dystrophy (type 1D)	LARGE
Congenital muscular dystrophy (type 1C)	FKRP
Limb–girdle muscular dystrophy	

brain cortex, and in the organization and function of the retina. Depending on the severity of the *O*-mannosylation defect, the degree of muscular dystrophy, neuronal involvement and retinal degeneration will be more or less pronounced. This variable severity and organ involvement leads to the definition of different clinical entities, i.e., different diseases with distinct names. From the most to the least severe one, these diseases are: Walker–Warburg syndrome (OMIM 236670), muscle–eye–brain disease (OMIM 253280), Fukuyama congenital muscular dystrophy (OMIM 253800), congenital muscular dystrophy 1C (OMIM 606612), congenital muscular dystrophy 1D (OMIM 608840) and limb–girdle muscular dystrophy type 2I (OMIM 607155). Although each of these forms of congenital muscular dystrophies are mainly linked to a specific gene defect, the discrimination is not clear-cut since mutations in the same glycosyltransferase gene, for example in *POMT1*, have been found in Walker–Warburg syndrome, in muscle–eye–brain disease, and even limb–girdle muscular dystrophy (Table 22.2).

Defects of dystroglycan glycosylation are detected using antibodies recognizing the glycans moiety of dystroglycan, using either Western blotting or *in situ* histochemistry. These tests, which are far from being trivial, are usually performed in research or specialized clinical laboratories. The envelope protein of the lymphocytic choriomeningitis virus (LCMV) has been shown to bind to *O*-Man glycans on α-dystroglycan. Therefore, this viral protein can represent an alternative probe to glycosylation-specific antibodies for the recognition of *O*-mannosylation defects. However, as for many genetic diseases, the identification of a congenital muscular dystrophy in a patient does not open the door to any therapy. No treatment is available to date, although the overexpression of the *LARGE1* gene in skeletal muscle of dystrophic mice, achieved through gene transfer, leads to stabilization of the dystroglycan complex.

22.2.3
O-Fuc Glycosylation (*O*-Fucosylation)

Epidermal growth factor (EGF)-like repeats and thrombospondin domains often carry *O*-Fuc glycans. As compiled in Table 7.9, these *O*-Fuc glycans mediate important functions, e.g., by regulating the interactions between the Notch receptor and its ligands [5]. Apart from the regulation of Notch signaling, little is known about the biological significance of *O*-fucosylation. As for other glycosylation pathways, the identification of diseases of *O*-fucosylation expands the scope of the developmental and physiological processes known to depend on this type of glycosylation. Two diseases of *O*-fucosylation have been identified to date. The first one, a form of skeletal deformity known as spondylocostal dysostosis (OMIM 277300), has been mapped to mutations of the Lunatic Fringe *LFNG* gene that encodes a β1,3-GlcNAc-transferase adding GlcNAc to the *O*-Fuc core on EGF-like repeats. Notably, another form of spondylocostal dysostosis is associated to mutations in the *DLL3* gene, which codes for a ligand to the Notch receptor. The second disease of *O*-fucosylation has been related to a defect of the β1,3 Glc-transferase, which also elongates the *O*-Fuc core, but this time on thrombospondin domains. The corresponding disease, known as Peters-Plus syndrome (OMIM 261540), is characterized by a defect of eye development resulting in the clouding of the cornea and an incomplete cleavage of the lens from the cornea. In addition, Peters-Plus patients have short limbs and hands and are psychomotor retarded. Peters-Plus syndrome has not been associated to any protein defect, meaning that the identity of the proteins requiring glucosylated O-Fuc glycans is still unknown.

The detection of O-Fuc glycans relies on complex analytic methods based on liquid chromatography and mass spectrometry, which are only available in specialized research laboratories. So far only a few proteins, like the complement protein properdin, have been shown to be aberrantly O-fucosylated in Peters-Plus patients, and no biochemical analysis has been undertaken yet to quantify the glycosylation defect found in Lunatic Fringe-associated spondylocostal dysostosis.

22.3 Glycosaminoglycans

The functional investigation of glycosaminoglycan (GAG) chains in model organisms has demonstrated the importance of these glycans in the localization and activation of various proteins (please see Chapter 11.1 structural aspects). Therefore, it is somehow surprising that only few diseases of GAG chain assembly have been documented so far. The main disorder related to GAG chain defects is multiple hereditary exostoses (OMIM 133700) – an autosomal dominant disease caused by mutations in the *EXT1* and *EXT2* genes [6]. Exostoses represent protrusions of bone material mainly occurring on long bones and originating from benign cartilaginous tumors. Exostoses are quite frequent with an estimated prevalence of 1 : 50 000 in most ethnic groups. The *EXT1* and *EXT2* genes encode glycosyltransferases building heterodimers that catalyze the polymerization of heparan sulfate (HS) chains. The inclusion of a mutant EXT1 or EXT2 subunit to the heterodimer inactivates the enzyme, thereby explaining the dominant-negative effect. Theoretically, this situation causes a decreased enzymatic activity to 50% of normal levels. This decreased activity leads to a significant reduction of HS chain formation in tissues, in which the amount of such GAG chains is tightly regulated. This is the case for bone tissue, where HS chains are required for the proper localization of various factors regulating endochondral ossification, a process characterized by a directional sequence of chondrocyte proliferation and differentiation (Figure 22.5). Notably, the complete loss of the EXT activity is not compatible with life, explaining why homozygous or combined heterozygous mutations in the *EXT* genes have not been reported. Exostoses can transform, with a frequency of about 5%, into malignant chondrosarcomas. This increased predisposition to cancer accounts for the designation of *EXT1* and *EXT2* as tumor suppressor genes (for phenotype of KO mice, please see Table 23.1).

Another defect of GAG chain biosynthesis is caused by mutations in the xylosylprotein β1,4-Gal-transferase *XGPT1*, which catalyzes the transfer of the first Gal residue of the GAG core tetrasaccharide GlcAβ1,3Galβ1,3Galβ1,4Xyl(β1-O)Ser (please see Chapter 11.1). The resulting disease, which is called the progeroid form of the Ehlers–Danlos syndrome (OMIM 130070), is characterized by mental retardation, short stature and connective tissues abnormalities, such as loose skin, wrinkled facies, osteopenia and joint hypermobility. The disease is very rare and

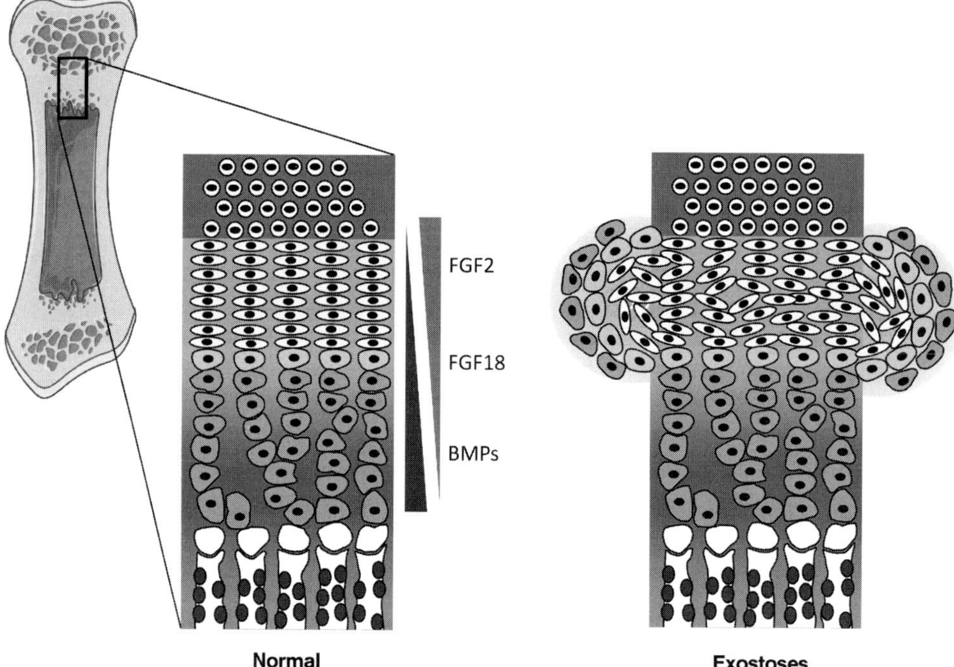

Figure 22.5 Endochondral ossification and exostoses. The formation of bone tissue from cartilage relies on the ordered proliferation and differentiation of chondrocyte cells. This process is controlled by gradients of growth factors, like FGF2, FGF18 and bone morphogenetic proteins (BMPs), which are maintained by HS GAG chains. The decreased expression of HS GAG chains caused by *EXT1* and *EXT2* mutations leads to disordered chondrocyte proliferation and thus to the formation of exostoses.

less than 10 cases have been reported to date. Little is known about the extent of the GAG chain biosynthetic defect in the affected patients. Analysis of skin fibroblasts from patients indicated that some proteoglycan proteins are secreted free of GAG chains.

GAG chains carry many sulfate groups, which are essential to the properties and functions of these glycans [7]. GAG chain sulfation is mediated by a large family of carbohydrate sulfotransferases as outlined in Chapter 11. Two diseases have been shown to be caused by deficiencies of carbohydrate sulfotransferases. The first, Omani type spondylepiphyseal dysplasia (OMIM 608637), is caused by a deficiency of the chondroitin 6-*O*-sulfotransferase-1 enzyme that catalyzes the 6-*O*-sulfation of GalNAc found in chondroitin and dermatan sulfate GAG chains. The symptoms of Omani type spondylepiphyseal dysplasia are typical of connective tissue disorders, i.e., short stature, skeletal dysplasia, kyphoscoliosis and arthritic joints. Patients show a normal intelligence. The second sulfation defect affects the GlcNAc-6-sulfotransferase *CHST6* gene, which is involved in the sulfation of keratan sulfate chains. The *CHST6* gene is expressed in many cell types but the clinical features of GlcNAc-6-sulfotransferase deficiency are limited to the eye, resulting in a progressive opacity of the cornea, a condition referred to as macular corneal dystrophy (OMIM 217800). Keratan sulfate chains are important for the solubility and water retention of corneal proteoglycans. The decreased level of sulfation caused by CHST6 deficiency leads to aggregation of corneal proteoglycans and thereby to increased opacity of the cornea. Reduced sulfation of keratan sulfate chains is also found in other tissues like cartilage but this sulfation defect has no apparent effect on the integrity and functions of these tissues. Considering the number of GAG-specific sulfotransferase genes in the human genome, one can confidently expect additional diseases to be related to defects of carbohydrate sulfation.

22.4 Glycosphingolipids

Whereas several diseases of glycosphingolipid degradation have been described over recent decades (please see Chapter 10.14 and Table 30.1), only a single defect of glycosphingolipid assembly is known to date. Recently, a genetic study on a form of infantile epilepsy occurring in the Amish community has lead to the identification of a sialyltransferase deficiency. The sialyltransferase *ST3GAL5* gene encodes the enzyme catalyzing the formation of the ganglioside GM3. The loss of the ganglioside GM3 was associated with symptoms like epilepsy, failure to thrive, psychomotor retardation, poor feeding, vomiting, muscular weakness and blindness. The disease of GM3 deficiency is also known as Amish infantile epilepsy syndrome (OMIM 609056). Notably, most of these features are also encountered in CDG-I and in many metabolic disorders, which explains the difficulty in recognizing the disease clinically. Ganglioside GM3 deficiency can be confirmed by structural analysis of glycosphingolipids isolated from blood serum. However, this analysis is not a routine procedure, thus rendering the search for additional defects of glycosphingolipid biosynthesis in candidate patients a difficult endeavor.

GM3 deficiency is not to be mistaken with disorders of ganglioside degradation, known as gangliosidosis, which belong to the family of lysosomal storage diseases. The latter diseases are caused by defects of hexosaminidases, sialidases (neuraminidases), galactosidases and glucosidases and are also known as Tay-Sachs, Sandhoff, Fabry and Gaucher diseases. These lysosomal storage diseases are characterized by the intracellular accumulation of various glycosphingolipids (please see also Table 30.1).

22.5 Glycosylphosphatidylinositol Anchor

The assembly of the glycosylphosphatidylinositol (GPI) anchor is initiated by the transfer of GlcNAc to phosphatidylinositol (see Chapter 9). This step is mediated by a multiprotein complex containing the PIGA, PIGC, PIGH and PIGQ subunits. The systemic loss of the GPI anchor is not compatible with life, but a GPI anchor deficiency limited to hematopoietic cells has been associated with a form of hemolytic anemia called paroxysmal nocturnal hemoglobinuria (PNH, OMIM 311770) [8]. This complex and somewhat intimidating name indicates that the affected patients notice dark-colored urine in the morning after sleep. Red blood cells lacking GPI-anchored proteins are sensitive to complement-mediated lysis, thus explaining the hemolysis and the accumulation of hemoglobin in the urine. PNH is an acquired genetic disease mainly caused by mutation in the X-linked *PIGA* gene arising in hemopoietic stem cell clones. PNH patients are usually diagnosed at about 40 years of age. The disease also features an increased incidence of leukemia and increased thrombotic episodes. These complications usually lead to the death of the patients within 10–20 years after diagnosis. Platelet transfusion and anticoagulant therapies are commonly used to improve the survival rate.

22.6 Defects Affecting Multiple Classes of Glycosylation

Defects of nucleotide-sugar biosynthesis and transport affect multiple classes of glycosylation. Accordingly, the phenotypes of such defects are expected to be severe considering the multiple biological processes are involved. Two nucleotide-sugar transporter defects have been identified to date in humans. Mutations in the Golgi GDP-Fuc transporter gene have been found to cause a deficiency of protein fucosylation (CDG-IIc/LAD-II) [9]. In fact, CDG-IIc/LAD-II patients present with recurrent infections and increased levels of circulating granulocytes due to the defective trafficking of leukocytes outside of the blood stream. The fucosylation defect does not only affect leukocyte functions, since psychomotor retardation, short stature and mild dysmorphism are also encountered in the affected patients. It is presently unclear whether *O*-fucosylation is also impaired in CDG-IIc/LAD-II. However, the absence of severe skeletal deformity, as seen in the Lunatic Fringe defect (see Section 22.5), would suggest that the ER-localized *O*-fucosylation is not affected by the Golgi GDP-Fuc transporter defect. As the mutations detected in CDG-IIc/LAD-II do not completely inactivate the transporter activity, the fucosylation disorder can be successfully corrected by dietary supplementation with fucose. This treatment has been shown to reduce the occurrence of infections and lead to a normalization of blood leukocyte levels in the patients.

Mutations in the CMP-Sia transporter have been found in an infant presenting with a severe bleeding disorder, characterized by decreased blood platelet and neutrophil levels (CDG-IIf). The CMP-Sia transporter defect leads to an apparent loss of protein sialylation on hematopoietic cells, while several serum proteins, like transferrin, were surprisingly unaffected. Furthermore, the impact of the CMP-Sia transporter defect on ganglioside biosynthesis remains unknown. The description of additional cases encompassing additional mutations is required to better define the scope of organ functions and processes altered by reduced sialylation.

The defect of CMP-Sia transport demonstrates the importance of sialylation in human physiology. Accordingly, defects of Sia or CMP-Sia biosynthesis would also be expected to cause severe bleeding disorders and probably other organ dysfunctions. The UDP-GlcNAc 2-epimerase/*N*-acetylmannosamine kinase (GNE) protein is rate limiting in the synthesis of Sia. Accordingly, GNE activity also correlates with the level of sialylation on hematopoietic cells. Therefore, the association of GNE mutations to a mild form of a neuromuscular disorder called hereditary inclusion body myopathy (IBM, OMIM 600737) came as a surprise. IBM is characterized by a late-onset myopathy mostly limited to leg muscles. Histological examination of IBM tissue reveals typical intramuscular vacuoles and filamentous inclusions of unknown origin. Glycan analysis of IBM muscle biopsies revealed decreased levels of sialylated *O*-glycans, whereas sialylation of *N*-glycans remained unchanged. The rather weak phenotype of GNE deficiency raises several questions concerning the activity of possible salvage pathways that could compensate for the loss of endogenous Sia biosynthesis. However, the targeted disruption of the *GNE* gene in mice leads to embryonic lethality, suggesting that salvage pathways cannot fully compensate for the loss of the GNE activity. Additional work is required to understand the relationship between the pathogenesis of IBM, the level of residual GNE activity and the contribution of alternate sources of Sia. Interestingly, GNE mutations in the allosteric site of the epimerase domain cause another disease, sialuria (OMIM 269921), a dominant inherited disorder characterized by increased cellular and urinary Sia levels.

Other defects of sugar metabolism, such as galactosemia (OMIM 230400) and fructosemia (OMIM 229600), lead to limited availability of specific nucleotide sugars and thus to abnormal glycosylation. However, because these defects impair several pathways linked to the energy metabolism, they are not considered as true diseases of glycosylation and are therefore not discussed here.

22.7 Trafficking Disorders

Proper glycosylation not only requires adequate donor substrates and enzyme levels, the latter must also be correctly localized along the secretion pathway. The localization of glycosylation reactions is a dynamic process that is tightly regulated together with the flow of secreted proteins and the shaping of the ER, ER–Golgi intermediate compartment and Golgi apparatus. Recently, defects of glycosylation have been related to the abnormal assembly of the COG complex, which is part of the vesicular transport machinery. COG, which stands for 'conserved oligomeric Golgi', is a complex of eight subunits found in eukaryotes from yeast up to human cells [10]. Although the functions of COG are still being investigated, it is clear that COG is required for the retrograde transport of glycosyltransferases between

Golgi cisternae. To date, mutations have been found in the COG1, COG7 and COG8 subunits of the complex. These COG mutations have been identified in patients classified as CDG-IIx, who were tested positive for *N*-glycosylation disorder based on an abnormal glycosylation of the serum protein transferrin (Figure 22.1). Therefore, COG defects follow the CDG classification, and have been designated as CDG-IIe (OMIM 608779), -IIg (OMIM 611209) and -IIh (OMIM 611182) for the COG7, COG1 and COG8 deficiencies, respectively. At the cellular level, COG defects can be recognized by measuring the delayed collapse of the Golgi apparatus into the ER induced by the toxin brefeldin A. The clinical features associated with COG defects are similar to those found in CDG patients, i.e., a neuromuscular involvement, dysmorphia, psychomotor retardation and coagulopathy. However, the severity of these symptoms varies greatly between the COG subunit defects. COG7 deficiency is associated with high mortality, whereas COG1 deficiency leads to a relatively mild disease.

Abnormal *N*-glycosylation of serum transferrin and *O*-glycosylation of serum ApoCIII were also recognized in a group of patients presenting with wrinkled skin, osteopenia, dysmorphia, hypotonia and developmental delay. The cells of these patients showed the same phenotype to brefeldin A treatment as those of COG-deficient patients, i.e., a delayed collapse of the Golgi apparatus, thus suggesting a possible defect of Golgi vesicular transport. The gene to blame for this defect, named *ATP6V0A2*, was identified by genetic linkage analysis and found to encode a subunit of a H^+-ATPase pump. The subunit A2 of the H^+-ATPase pump is localized in the Golgi apparatus and in early endosomes, where it is possibly involved in pH regulation. The mechanisms relating an H^+-ATPase pump to the glycosylation machinery is presently unclear. In any case, the *ATP6V0A2* defect provides fascinating clues about the complexity underlying the homeostasis of glycosylation.

22.8 Conclusions

The last 10 years have seen a large number of diseases being classified as disorders of glycosylation. It is likely that the increased awareness for the medical relevance of glycosylation combined with the progress made in human genetics and in analytical methods will lead to the discovery of novel diseases of glycosylation. Beyond the identification of pathogenic mutations, scientists are also beginning to investigate relationship between nucleotide polymorphisms in glycosylation genes and common multigenic diseases. Along this line, a first study demonstrated a link between an increased activity for a given glycosyltransferase and susceptibility to diseases such as multiple sclerosis. Further work in this direction will certainly shed new light on the biological importance of glycosylation and thereby find how much glycosylation contributes in this 'Pandora's box'.

Summary Box

Many diseases of glycosylation have been described over the last decade. These diseases represent different clinical entities, which reflect the diverse involvements of glycans in regulating organ functions. Most diseases of glycosylation are caused by mutations in glycosyltransferase genes, but defects in nucleotide-sugar transporters and in vesicular transport proteins also lead to disorders of glycosylation.

References

1 Freeze HH. Genetic defects in the human glycome. *Nat Rev Gen* 2006;7:537–51.
2 Jaeken J, Matthijs G. Congenital disorders of glycosylation: a rapidly expanding disease family. *Annu Rev Genomics Hum Genet* 2007;8:261–78.
3 Razzaque MS, Lanske B. The emerging role of the fibroblast growth factor-23–klotho axis in renal regulation of phosphate homeostasis. *J Endocrinol* 2007;194:1–10.
4 Barresi R, Campbell KP. Dystroglycan: from biosynthesis to pathogenesis of human disease. *J Cell Sci* 2006;119:199–207.
5 Haines N, Irvine KD. Glycosylation regulates Notch signalling. *Nat Rev Mol Cell Biol* 2003;4:786–97.
6 Duncan G *et al.* The link between heparan sulfate and hereditary bone disease: finding a function for the EXT family of putative tumor suppressor proteins. *J Clin Invest* 2001;108:511–6.
7 Kusche-Gullberg M, Kjellen L. Sulfotransferases in glycosaminoglycan biosynthesis. *Curr Opin Struct Biol* 2003;13:605–11.
8 Luzzatto L *et al.* Somatic mutations in paroxysmal nocturnal hemoglobinuria: a blessing in disguise? *Cell* 1997;88:1–4.
9 Becker DJ, Lowe JB. Fucose: biosynthesis and biological function in mammals. *Glycobiology* 2003;13:41R–53R.
10 Ungar D *et al.* Retrograde transport on the COG railway. *Trends Cell Biol* 2006;16:113–20.

Info Box 1

Isozymes, also known as isoenzymes, are enzymes that are encoded by different genes, but catalyze the same chemical reaction. These enzymes display different kinetic parameters, different regulatory properties or different expression patterns. The existence of isozymes permits the fine tuning of metabolism to meet the particular needs of a given tissue or developmental stage.

although the mucin-type *O*-glycan plays a critical role in the immune system (please see Section 23.2.2 and Chapter 27.4). This problem may be overcome by the production of 'double-KO' or 'triple-KO' mice, in which two or three genes that are functionally associated are simultaneously inactivated. We are able to produce these multiple KO mice by mating the individual KO mice that have an inactive allele at a different gene locus.

When KO mice are developmentally lethal, it is impossible to investigate the gene function in mature mice. 'Conditional KO' mice, in which gene disruption can be induced at specific times or in particular tissues using a special technique such as the Cre-loxP system, may overcome this problem.

Occasionally, the phenotypes of independently generated KO mice are different from each other even if the same gene is targeted. This is the case for KO mice for β1,4-GlcNAc transferase (GnT-III; Mgat-3; please see Chapter 6.8.1 for this part of the biosynthetic pathways of *N*-glycans) and double-KO mice for GD3 synthase and GM2/GD2 synthase (please see Figures 10.2 and 30.3 for biosynthetic pathways of glycolipids). This discrepancy may be explained by the difference in the targeting construct or the genetic background of the mice. In the case of GnT-III KO mice, for example, one mutated allele lacked the whole coding region, and the other mutated allele retained a part of the coding region. KO mice with the former allele showed no abnormal phenotype, while KO mice with the latter allele exhibited a subtle neurological disorder, suggesting that the phenotype may arise from the presence of truncated, inactive GnT-III.

The preceding chapter (Chapter 22) provides insights into the connection of defects in glycosylation to human diseases. Since humans share many genes with mice, the phenotype observed in KO mice may give hints for understanding how the orthologs cause diseases in humans (for definition, please see Info Box 2). However, it should be taken into account that the phenotype of KO mice is occasionally different from that observed in humans in which the ortholog gene is inactivated.

Organisms are hierarchically integrated with multiple levels that are populated by molecules, supramolecular assemblies, organelle, cells, tissues and organs from the lower to the upper level. Upper levels are composed of elements with greater integrity than lower levels. According to the theory of complex systems, an

Info Box 2

Orthologs and paralogs are two types of homologous sequences, homologs. Orthologs are genes in different species that evolved from a common ancestral gene by speciation. Usually, orthologs retain the same function. Paralogs are homologous genes within a single species that diverged by gene duplication.

upper-level element emerges from interactions between elements at the immediately lower level via nonlinear and large-scale interactions, called self-organization. It is not easy to elucidate the whole mechanism of pathogenesis, even if disruption of a gene function displays some abnormal phenotype at the higher level of organism formation. To comprehend the biological processes, we cannot skip the intermediate levels. The same is true for the study of human diseases.

Information on individual gene KO mice is available in the Mouse Genome Informatics (MGI) database of the Jackson Laboratory (http://www.informatics.jax.org/).

23.2 Specific Features of Glycogene KO

23.2.1 Relationship between Glycogenes and Related Glycans

Glycans represent the secondary gene products formed by the reactions of glycan-synthesizing enzymes. The biosynthesis of glycans is regulated by the activity of the glycan-synthesizing enzymes, their substrate specificity and their localization in specific tissue sites. Genes that encode the glycan-synthesizing enzymes, including glycosyltransferases, sulfotransferases, nucleotide-sugar transporters and related enzymes, are called 'glycogenes' (please see Chapters 6–11 for activity profiles of these enzymes). More than 180 glycogenes have been cloned at present. Information on individual glycogenes is available in the GGDB GlycoGene DataBase of the National Institute of Advanced Industrial Science and Technology, Japan (http://riodb.ibase.aist.go.jp/rcmg/ggdb/). This database also links to the MGI database.

Since carbohydrate chains are basic components of organisms, they are thought to function in a wide variety of biological processes. In fact, the inactivation of specific glycogenes in KO mice has proved that glycogenes are essential for development, the immune system, the nervous system, the reproductive system, cancer progression, the establishment of infection, and so on (Table 23.1) [1–4]; however, the actual functions of most of the carbohydrate chains in the glycoconjugates remain unknown (for human diseases of glycosylation, please see Table 22.1; please see also Table 27.4).

Table 23.1 Mice of knockout of glycan-synthesizing enzymes.

Glycogenes and enzymes (abbreviations)	Phenotypes of KO mice	Molecular functions (section number)
(A) *N*-Glycans		
GlcNAcT-I (GnT-I, Mgat-1)	embryonic lethality	unknown, multiple (23.2.4)
GlcNAcT-II (GnT-II, Mgat-2)	neonatal death	unknown, multiple (23.2.4)
GlcNAcT-III (GnT-III, Mgat-3)	no apparent change	unknown (23.1)
GlcNAcT-IVa (GnT-IVa, Mgat-4a)	type 2 diabetes	modification of glucose transporter 2 to restrain their endocytosis (23.2.6)
GlcNAcT-V (GnT-V, Mgat-5)	suppression of tumorigenesis, enhancement of T cell response	modification of cytokine receptors to restrain their endocytosis (23.2.6)
α6-FucT (FucT-VIII, Fut8)	neonatal semilethality, growth retardation, emphysema	modification of cytokine receptors for their ligand binding (23.2.4)
α-mannosidase II (MII)	anemia corresponding to human HEMPAS	unknown, multiple (23.2.5)
α-mannosidase IIx (MX)	male sterility	adhesion between germ cells and Sertoli cells (23.2.5)
β4GalTI	neonatal semilethality, growth retardation, IgA nephritis	unknown, multiple (23.2.5)
βGalα2,6-sialyltransferase (ST6Gal-I)	impaired B cell function	production of CD22 ligand to mask CD22 (23.2.4)
(B) *O*-Glycans		
(B.1) *O*-GalNAc (mucin type)		
polypeptide GalNAcT	no apparent change (single KO)	unknown, multiple (23.1)
core 2 GlcNAcTI (C2GnT-I)	disorder of leukocyte trafficking	production of selectin ligand (23.2.2)
α3-FucTIV (FucT-IV, Fut4)	disorder of leukocyte trafficking	production of selectin ligand (23.2.2)
α3-FucTVII (FucT-VII, Fut7)	disorder of leukocyte trafficking	production of selectin ligand (23.2.2)
GlcNAc 6-sulfotransferase-I	disorder of leukocyte trafficking	production of selectin ligand (23.2.2)
HEC-GlcNAc 6-sulfotransferase (LSST)	disorder of leukocyte trafficking	production of selectin ligand (23.2.2)
Galβ3GalNAc sialyltransferase (ST3Gal-I)	decrease of $CD8^+$ T cells caused by apoptosis	modification of CD8 to regulate the binding to MHC class I
(B.2) *O*-Fuc (Notch signaling system)		
protein *O*-FucT 1 (PoFucT 1, pofut 1)	embryonic lethality	chaperone and modification of Notch
β3-GlcNAcT (Lunatic fringe)	abnormality in somite formation	modification of Notch
β3-GalT (Brainiac)	embryonic lethality	unknown

continued

Table 23.1 *Continued*

Glycogenes and enzymes (abbreviations)	Phenotypes of KO mice	Molecular functions (section number)
(B.3) *O*-GlcNAc		
polypeptide *O*-β-GlcNAcT (PoGlcNAcT, Ogt)	embryonic stem cell death	unknown, multiple
(C) Glycolipid		
GlcCer synthase	embryonic lethality	unknown (23.2.4)
GalCer synthase (Cgt)	tremor, ataxia, paralysis, male sterility	myelin formation and maintenance, spermatogenesis
sulfatide synthase (Cst)	tremor, ataxia, paralysis, male sterility	myelin maintenance, spermatogenesis
GM3 synthase	enhanced insulin sensitivity, deaf	regulation of the insulin receptor signal (23.2.6)
GD3 synthase	reduced nerve regeneration	unknown (23.1)
GM2/GD2 synthase	late-onset neuronal degeneration, male sterility	maintenance of neurons, transportation of testosterone (23.2.4)
Gb3 (CD77) synthase	normal, resistant to verotoxin	production of the verotoxin receptor
(D) Proteoglycans		
(D.1) HS/heparin chain		
polymerase EXT-1	embryonic lethality	modulation of signaling pathways (23.2.4)
polymerase EXT-2	embryonic lethality (null mutant), ectopic bone growth (heterozygous)	modulation of signaling pathways (23.2.4)
N-deacetylase/*N*-sulfotransferase (NDST-1)	neonatal death, pulmonary atelectasis, defects of the forebrain	modulation of signaling pathways (23.2.4)
N-deacetylase/*N*-sulfotransferase (NDST-1)	defects in mast cells	production of sulfated heparin (23.2.5)
C5-epimerase	neonatal death, renal agenesis, lung defects, skeletal malformation	modulation of signaling pathways (23.2.4)
heparan sulfate 2-*O*-sulfotransferase (2-OST)	neonatal death, renal agenesis, defects of the eye and skeleton	modulation of signaling pathways (23.2.4)
heparan sulfate 6-*O*-sulfotransferase (6-OST-1)	embryonic or perinatal lethality	modulation of signaling pathways
heparan sulfate 3-*O*-sulfotransferase (3-OST-1)	normal hemostasis	(23.2.5)
(D.2) chondroitin sulfate chain		
chondroitin 6-sulfotransferase (C6ST-1)	decreased naive T lymphocytes in the spleen	unknown
chondroitin 4-sulfotransferase (C4ST-1)	neonatal death, multiple skeletal abnormalities	modulation of signaling pathways, cartilage growth plate morphogenesis

As mentioned in the preceding section, organisms are integrated by a hierarchical structure in which the basal level is populated by the genome. When it comes to glycogenes, the following levels are presumed from the basal level: glycogenes, glycan-synthesizing enzymes, glycans (glycoconjugates), supramolecular assemblies (cell formation), intercellular networks (tissue formation), intertissual networks (organ formation) and interorganic networks (organism formation) (Figure 23.1). The relationship between a glycogene and its related glycan-synthesizing enzyme is linear, and has one-to-one correspondence. In contrast, the relationship between a glycan-synthesizing enzyme and its related glycan is nonlinear, and has one-to-many correspondence. In other words, glycans are molecules formed via self-organization. Therefore, inactivation of a glycogene will affect a number of glycans, which will display complex phenotypes.

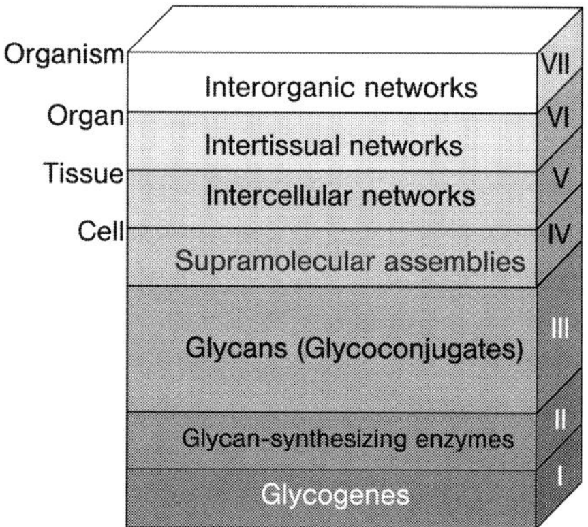

Figure 23.1 Hierarchical structure to integrate the organism.

23.2.2 Functional Association of Glycogenes

Although glycans are actual players in the biological events, their remodeling by means of gene targeting of glycogenes is the best approach to determine their *in vivo* functions. Several enzymes participate in the biosynthesis of an oligosaccharide, meaning that more than one glycogene is responsible for the formation of an oligosaccharide. These genes are functionally associated with each other.

This issue is easier to understand if we look at the example of a glycan ligand involved in lymphocyte homing (please see Chapter 27, especially Figure 27.1, and also Chapter 29, especially Figure 29.5). Lymphocytes patrol the entire body, protecting it against infection by circulating between the blood vessels and the lymphatic organs. The entry of lymphocytes from the blood vessels to the lymph nodes is called lymphocyte homing. This homing process is initially mediated by the interaction between L-selectin on the surface of lymphocytes and its glycan ligand on the surface of high endothelial venules (HEVs) of lymph nodes. L-Selectin belongs to the family of C-type lectins (for further information, please see Chapters 19 and 20 and also Figure 16.1h for crystal structure). The structural analysis of an L-selectin ligand GlyCAM-1 and histochemical studies of HEVs with carbohydrate-specific antibodies suggest that the physiological L-selectin ligand requires a 6-sulfo sialyl-Lewis^x (Le^x) structure (Figure 23.2) (please see Chapter 27.4 for selectin ligands). The key enzymes in the biosynthesis of the 6-sulfo sialyl Le^x are α1,3-fucosyltransferase (FucT) and GlcNAc 6-sulfotransferase (GlcNAc6ST). There are multiple isozymes of these enzymes, six α1,3-FucTs (FucT-III–VII and IX) and five GlcNAc6STs (GlcNAc6ST-1, HEC-GlcNAc6ST, I-GlcNAc6ST, GlcNAc6ST-4 and C-GlcNAc6ST). The contribution of individual enzymes to the lymphocyte homing assessed using KO mice, in which these isozymes are singly or doubly knocked out, is summarized in Figure 23.3 [2]. This graph indicates that both α1,3-FucT and GlcNAc6ST are needed for lymphocyte homing. As to α1,3-FucTs, FucT-VII is mainly responsible for the homing activity and FucT-IV helps to some extent. For GlcNAc6STs, HEC-GlcNAc6ST and GlcNAc6ST-1 contribute partially in this order.

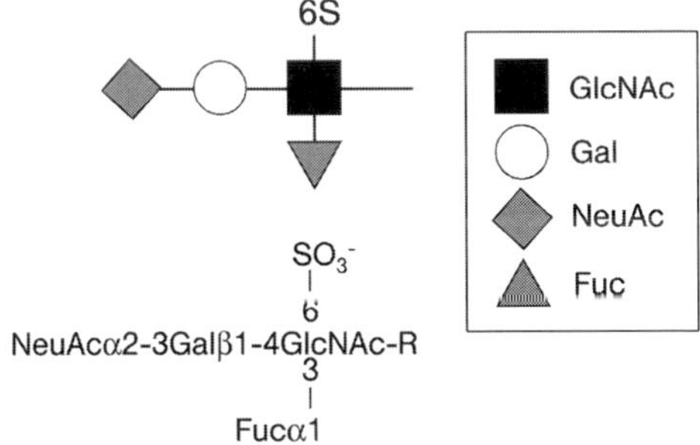

Figure 23.2 Structure of 6-sulfo sialyl-Le^x determinant.

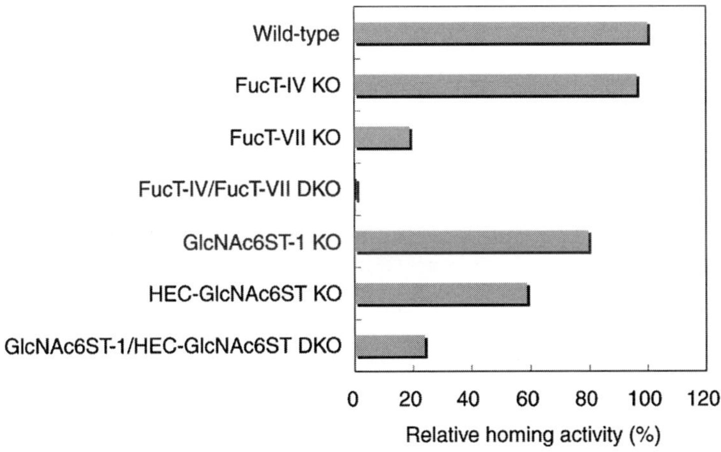

Figure 23.3 Lymphocyte homing activities to peripheral lymph nodes in α1,3-FucT and GlcNAc6ST KO mice.

23.2.3
Which Are Essential – Glycans or Their Carriers?

Although all selectins bind a common carbohydrate structure, sialyl-Le^x, their binding affinities toward carbohydrate chains are very weak. Specific aglycons presenting the sialyl-Le^x determinant are required for high-affinity interactions. The HEV L-selectin ligands are carried by mucin-type glycoproteins such as GlyCAM-1, CD34 and MAdCAM1. Neither CD34 nor GlyCAM-1 KO mice show abnormality in the lymphocyte homing activity, indicating that L-selectin ligand activity is not dependent on carrier proteins. In this case, glycans play a pivotal role and their scaffold proteins are replaceable [2].

P-Selectin glycoprotein ligand (PSGL)-1 is a mucin-type glycoprotein expressed on most white blood cells and plays an important role in their recruitment into inflamed tissues, where cell adhesion molecules such as P-selectin and E-selectin are induced on the surface of endothelial cells (please see also Chapters 27.4 and 29.5). Leukocytes flowing in the blood vessel interact with such adhesion molecules. The first step in this process is carried out by the interaction between PSGL-1 and P-selectin and/or E-selectin. This interaction results in rolling of leukocytes on the endothelial cell surface followed by stable adhesion and transmigration into the inflamed tissue. T helper (T_h) 1 cells from PSGL-1 KO mice do not bind to P-selectin and migrate less efficiently into the inflamed skin than those in the wild type. Furthermore, PSGL-1 null T_h1 cells do not migrate into the inflamed skin of E-selectin-deficient mice, indicating that PSGL-1 on T_h1 cells is the sole ligand for P-selectin *in vivo*. In contrast, PSGL-1 null T_h1 cells migrate into the inflamed skin of P-selectin-deficient mice less efficiently than wild-type T_h1 cells, indicating that PSGL-1 on T_h1 cells functions as one of the E-selectin ligands *in vivo*. The PSGL-1 case demonstrates the necessity of collaboration between glycans and their carrier proteins.

In the case of heparan sulfate (HS) proteoglycans, gene targeting can be applied to both carrier proteins (also called core proteins) and glycan-synthesizing enzymes [3]. There are three common types of HS proteoglycans (Figure 23.4) (please see

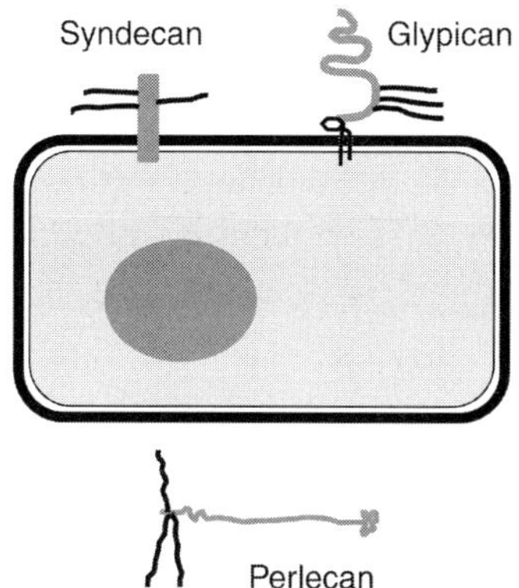

Figure 23.4 Three types of HS proteoglycans.

also Chapters 11.5 and 11.6 for details on proteoglycans). Syndecans (four members identified) are cell-surface transmembrane proteins with a single transmembrane domain. Glypicans (six members identified) are attached to the plasma membrane via a glycosylphosphatidylinositol (GPI) anchor at the C-terminal end (please see Chapter 9 for GPI anchor). Perlecan is a secretory protein and present in the basement membranes. HS chains of different types of proteoglycans synthesized in the same cell have a similar structure. Gene targeting of HS chain-synthesizing enzymes yields a more severe and complex phenotype than that of core proteins (see Section 23.2.4). Syndecan-1, -3 and -4 KO mice are all viable and fertile, although they show some specific phenotypes: reduced Wnt-1-induced tumorigenesis in mammary glands in syndecan-1 KO mice, altered feeding behavior in syndecan-3 null mice, and delayed wound repair and reduced angiogenesis in syndecan-4-deficient mice. Mice lacking exon 3 in perlecan, without HS chain attachment sites, show early degeneration of the lens, but otherwise develop normally. Mice deficient in glypican-3 display several clinical symptoms similar to Simpson-Golabi-Bemel syndrome patients, such as developmental overgrowth, abnormal lung development and perinatal death (please see also Chapter 11.6 for information these two types of cell surface proteoglycans).

23.2.4
Effects of Elimination of the Core and Terminal Structures of Glycans

The biological functions of glycans are often attributed to their unique terminal structural features, which are recognized by specific glycan receptors or lectins. Glycosyltransferases and glycan-modifying enzymes successively synthesize carbohydrate chains from the reducing end connected to aglycon moieties. Since the core structure (backbone) of sugar chains is a prerequisite for the terminal structures, the inactivation of enzymes forming the backbone abolishes the formation of the terminal structures. KO of enzymes involved in the formation of backbone yields mice with systemic, serious and complex phenotypes, and which are difficult to analyze. In contrast, KO mice in which the enzymes forming the terminal structure are lacking could define a molecular mechanism involving glycans at a specific tissue.

Asparagine-linked glycosylation (i.e., *N*-glycan) is the most common modification of glycoproteins (please see Chapters 6 and 8 for survey on *N*-glycans). Many types of membrane proteins and secretory proteins undergo this type glycosylation. Dolichol phosphate GlcNAc-1 phosphate transferase (GPT) is required for the first step in the formation of the dolichol oligosaccharide precursor, which is essential for the production of all types of *N*-glycans (please see Figure 22.2 for pathway of its assembly). GPT KO mouse embryos completed preimplantation development and implantation, but died shortly thereafter, between 4 and 5 days postfertilization, indicating essential roles of *N*-glycans during early embryogenesis and postimplantation development (please also see Chapter 24 for glycans in fertilization and early embryogenesis) [2].

GnT-I (Mgat-1) transfers βGlcNAc residue to the $Man_5GlcNAc_2$ core structure of *N*-glycans (Figure 23.5 and also Figure 6.5). GnT-I plays a key role in the

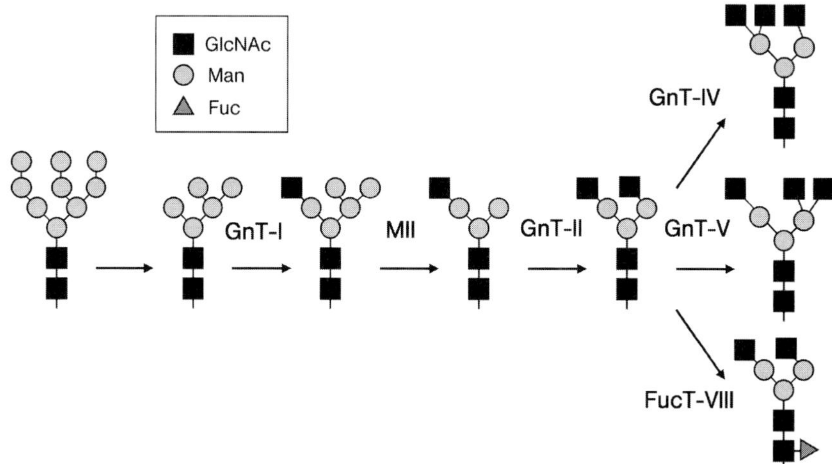

Figure 23.5 Processing of *N*-glycans.

biosynthesis of the complex- or hybrid-type *N*-glycan, as the GnT-I product is a prerequisite for all other modifications (Chapter 6.7). Knockout of GnT-I is embryonically lethal, and GnT-I KO mice die by day 10.5 after gestation (please see also Table 8.1) [2]. Abnormality is found, among other sites, in vascular endothelial cells, somite numbers, neural tube formation, and body axis. These observations indicate that complex-type or hybrid-type *N*-glycans are essential for a variety of developmental processes. Conditional KO mice with GnT-I KO in neuronal tissue are born apparently normally. However, most of them die within 8 weeks due to neuronal apoptosis.

GnT-II (Mgat-2) transfers βGlcNAc residue to the $GlcNAcMan_3GlcNAc_2$ core structure (Figure 23.5). This step is essential for the formation of complex-type *N*-glycans. GnT-II KO mice lack complex-type glycans, but possess hybrid-type glycans [2]. Most GnT-II KO mice die during the neonatal period. Null pups exhibit severe gastrointestinal, hematological and osteogenic abnormalities. These observations indicate that complex-type *N*-glycans are essential for early postnatal life. Conditional KO mice with GnT-II KO in neuronal tissue yield no apparent abnormality in contrast to that of GnT-I.

Sialylation is a common terminal decoration of carbohydrate chains in glycoproteins (please see Chapter 6.8.4 and Figure 7.4 for further information) as well as glycolipids (please see Chapters 10.7 and 30.3 for glycosphingolipids). Galβ1-4GlcNAc α2,6-sialyltransferase-I (ST6Gal-I) transfers sialic acid at the C6 position of the terminal Gal residue on *N*-glycans. A Siaα2-6Galβ1-4GlcNAc terminus on *N*-glycans is a common structure and present in a variety of tissues. ST6Gal-I KO mice are apparently healthy, but exhibit severe immunosuppression, reduced serum IgM levels and impaired B cell proliferation in response to the B cell receptor (BCR) activation [1, 2]. As shown in Figure 23.6, CD22 (also known as siglec-2), an accessory molecule of the BCR, binds to the Siaα2-6Galβ1-4GlcNAc structure

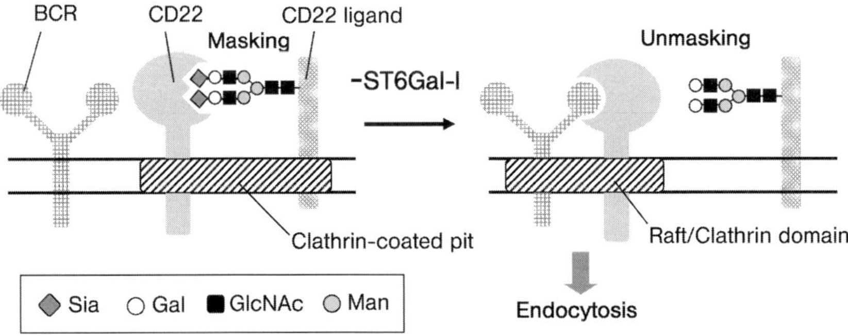

Figure 23.6 CD22-dependent BCR endocytosis in ST6Gal-I KO mice.

on neighboring CD22 ligands (please see Chapter 27.6 for CD22 and siglecs). CD22 is predominantly associated with clathrin-coated pits in resting B cells, whereas the BCR is excluded from the clathrin domains. In ST6Gal-I KO mice, CD22 colocalizes with the BCR in fused raft–clathrin domains due to the dissociation of endogenous ligand. Consequently, the BCR is downregulated by the clathrin-mediated endocytosis (please see Table 27.4 for phenotype of CD22 KO mice).

Another example of terminal decoration of *N*-glycan is α1,6-fucosylation of its innermost GlcNAc residue (Figure 23.5). This fucosylation is catalyzed by FucT-VIII (please see Chapters 2.7 and 6.7 and Table 25.2 for further information). Knockout of FucT-VIII is semilethal during the neonatal period [2]. The surviving mice eventually show emphysema-like degeneration in the lung tissue. The abnormal production of matrix metalloproteinases (MMP)-1, -12 and -13 has been implicated in the induction of emphysema. Signaling through transforming growth factor (TGF)-β receptors, which are α1,6-fucosylated proteins, negatively regulates the MMP expression. This is a key factor in regulation of extracellular matrix (ECM) proteins (Figure 23.7). Signaling through the TGF-β1 receptor is downregulated in FucT-VIII null mice as the ligand affinity to the receptor is less potent. Therefore, MMPs are highly expressed and consequently the degradation overwhelmed the synthesis of ECM, which causes emphysema in the mutant mice. Furthermore, the generation of FucT-VIII KO mice has revealed that α1,6-fucosylation of *N*-glycans attached to the epidermal growth factor (EGF) receptor is required for the binding of EGF. The EGF-induced phosphorylation of the EGF receptor and subsequent signal transduction are suppressed in the embryonic fibroblasts established from FucT-VIII KO mice. These results indicate that the TGF-β1 and EGF receptors are functional target proteins for FucT-VIII. These findings show that it is possible to link a terminal structure of glycans with a specific biological process.

The same is true for glycolipids and proteoglycans. The inactivation of GlcCer synthase, which acts most upstream in the biosynthetic pathway of glycosphingolipid (please see also Figures 10.2 and 30.3), results in the disappearance of all the distal structures, even though downstream enzymes are active (shaded area in Figure 23.8a; please see also Chapters 10.7 and 30.3) [4]. The phenotype of GlcCer

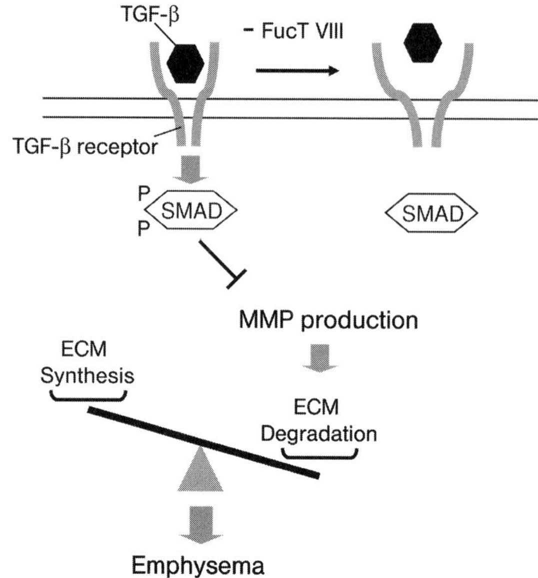

Figure 23.7 Dysregulation of the TGF-β signal leads to emphysema in FucT-VIII KO mice.

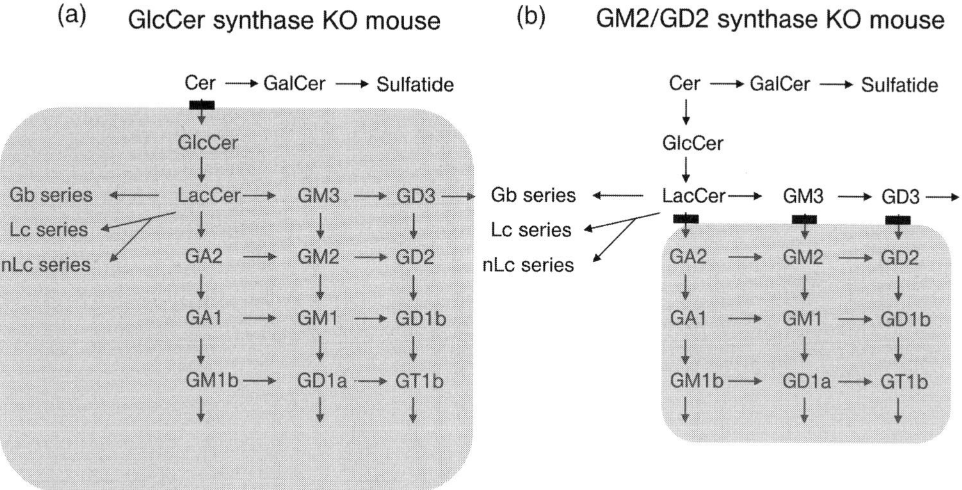

Figure 23.8 Glycolipids absent in glycolipid synthase KO mice.

KO mice represents the sum of the loss of all glycosphingolipids except galactolipids, resulting in embryonic death. In contrast, mice with KO of GM2/GD2 synthase, which is a key enzyme in the synthesis of complex gangliosides (Figure 23.8b), are born and develop apparently normally (please see also Chapter 30.3). They are, however, sterile and show aspermatogenesis due to a disorder of testosterone transport. Although complex gangliosides completely disappear in the brain, GM2/GD2 synthase null mice have apparently normal brain morphology. With age, the KO mice show marked degeneration and demyelination of nerves, and poor regeneration of resected nerves, indicating that gangliosides are essential for the maintenance and repair of nerve tissues.

The biosynthesis of the HS chain is divided into three phases: chain initiation, polymerization and modification (please see Chapter 11.1 for biosynthetic pathway of HS chain). In the chain-initiation phase, linkage tetrasaccharides (GlcAβ1-3Galβ1-3Galβ1-4Xylβ1-O-Ser) are assembled on selected serine residues in the core protein. Following addition of GlcNAc to the linkage tetrasaccharides, polymerization of the chain is carried out through alternating transfer of GlcA and GlcNAc units to the nonreducing end of the growing polysaccharide. As the chain grows, modifying enzymes introduce sulfate groups at various positions and some of the D-glucuronic acid residues are converted into L-iduronic acid (please see Chapter 1.3 for special features of L-iduronic acid). The obligatory initiation step of this modification is catalyzed by GlcNAc *N*-deacetylase/*N*-sulfotransferase (NDST). Depending on the targeted enzyme, HS proteoglycans are remodeled in a different way (Figure 23.9) [5]. If polymerases (EXT1 and EXT2, called after hereditary multiple exostoses, please see Chapter 22.3) are knocked out, only short stubs containing the linkage tetrasaccharides are attached. If NDST is inactivated, nonsulfated polysaccharide chains are formed. If other specific *O*-sulfotransferases are knocked out, a more subtle change is induced.

A number of important signaling molecules, such as fibroblast growth factors (FGFs), Wingless (Wg/Wnt), TGF-β and Hedgehog, need HS proteoglycans to display their proper function during development. Therefore, mice lacking the HS chain would be expected to present developmental disorders. Indeed, mouse embryos lacking EXT1 or EXT2 die around the time of gastrulation [3]. NDST1 KO mice die early after birth due to lung failure, and with skull and brain defects, while NDST2 KO mice survive until adulthood and are fertile. The NDST2 null

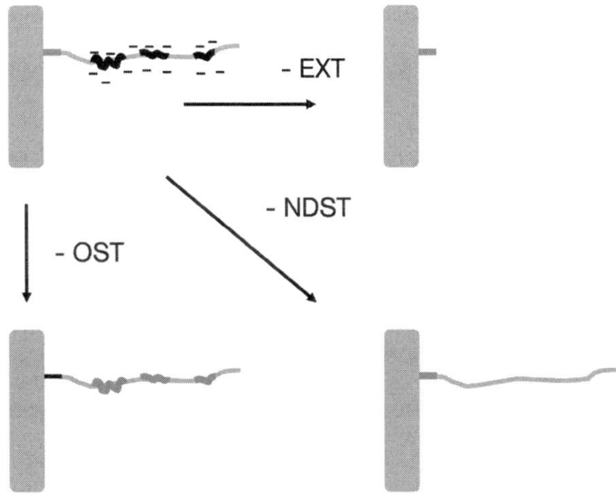

Figure 23.9 HS proteoglycans in HS chain-synthesizing enzymes KO mice.

mice, however, display only a phenotype with abnormal mast cells containing no heparin (see Section 23.2.5). Mice lacking both NDST1 and NDST2 die during early embryonic development. Phenotypes of mice with KO of 2-*O*-sulfotransferase and C5-epimerase are similar to each other. They die during neonatal period and lack kidneys. C5-epimerase KO mice also display a lung abnormality that is not present in the 2-OST-deficient mice but similar to the phenotype of NDST1-null mice.

23.2.5 Unexpected Findings Provide New Insights into Glycan Functions

A number of unexpected phenotypes have been revealed by generation of glycogene KO mice. These findings have provided a series of new insights into the functions of glycans.

β1,4-Gal transferase (β4GalT-I), which was the first cloned glycogene, forms the Galβ1-4GlcNAc structure that is quite common in complex-type *N*-glycans, *O*-glycans and glycolipids (please see Chapters 6–8, 10 and 30 for oligosaccharide structures in *N*-glycans, *O*-glycans and glycolipids, respectively). Therefore, inactivation of β4GalT-I had been expected to show severe systemic phenotypes. On the contrary, β4GalT-I KO mice were born healthy and fertile, but they exhibited growth retardation and semilethality during the neonatal period [2]. This finding suggested that other isozymes compensate for the loss of β4GalT-I function. In fact, a number of β1,4-Gal transferases (β4GalT-II~VII) have subsequently been found (please see also Table 22.1 on CDG-IId).

α-Mannosidase-II (MII) trims two Man residues to yield Man_3 after GlcNAc transfer by GnT-I, which serves as the precursor for GnT-II in the biosynthesis of complex-type *N*-glycans (Figure 23.5) (please see also Chapter 6.6 for the biosynthetic pathway of *N*-glycans). It is therefore expected that MII KO mice should display a more severe phenotype than GnT-II KO mice. However, it was found that complex-type *N*-glycans were retained in most tissues other than erythrocytes in MII KO mice [2]. Associated with the loss of complex-type *N*-glycans in erythrocytes, null mice exhibit dyserythropoiesis that correlates with human congenital dyserythropoietic anemia type II (HEMPAS). α-Mannosidase IIx (MX) was cloned as an MII-related gene. MX KO mice were born and grew normally, but male mice were found to be infertile. To elucidate the relationship between MII and MX function *in vivo*, MII/MX double-KO mice were produced. The double-KO mice completely lack complex-type *N*-glycans and display perinatal death. Thus, MII and MX compensate for each other in most tissues except for erythrocytes and testes.

Heparin and HS proteoglycan are well-known anticoagulants (please see Chapter 28.5 for a pharmaceutical role of heparin). 3-*O*-Sulfotransferase-1 has been thought to be a key enzyme in producing the pentasaccharide domain interacting with antithrombin (Figure 1.7d). However, 3-*O*-sulfotransferase-1 KO mice did not show a procoagulant phenotype [3]. The NDST-2 KO mice completely lack sulfated heparin in the secretory granules of mast cells. As NDST-2 null mice show no signs of thrombosis, endogenous heparin is not involved in the regulation of coagulation. It instead plays a role in regulating the mast cell mediators, such as histamine and the mast cell-specific proteases (chymases and tryptases).

23.2.6 Insights into Human Diseases

The phenotype of KO mice may provide a symptomatic clue in a search for human deficiencies that are poorly understood at present (please see Chapter 22 for more details on human disorders of glycosylation). KO mice also may serve as a test animal model for the newly developed treatment.

Fucosylation is a common modification of glycoproteins and is catalyzed by fucosyltransferases, all of which require GDP-Fuc as the donor substrate (please see Chapters 6 and 7 for biosynthesis of glycoproteins). Fucosylations in mammals are classified into four groups based on the specific linkages formed and the substrates to which Fuc is added: α1,2-fucosylation, α1,3/4-fucosylation, α1,6-fucosylation and *O*-fucosylation. The biological significance of fucosylation in humans has been proved by studies on an autosomal recessive disease called leukocyte adhesion deficiency (LAD)-II/congenital disorder of glycosylation (CDG)-IIc, which is characterized by leukocyte adhesion deficiency as well as severe neurological and developmental abnormalities (please see Chapter 22.6 for LAD and CDG and Table 27.4). Patients with LAD-II/CDG-IIc have a mutation in the Golgi GDP-Fuc transporter, resulting in decreased GDP-Fuc levels in the Golgi lumen and, hence, reduction of fucosylation in many types of glycoproteins. The LAD may be ascribed to the loss of sialyl-Le^x determinant containing α1,3-fucosylation. It had been assumed that the neurological and developmental symptoms of LAD-II/CDG-IIc patients might be linked to impaired Notch signaling mediated by *O*-fucosylation. However, *O*-fucosylation does not occur in the Golgi, but in the ER and, hence, is generally unaffected in LAD-II/CDG-IIc patients. Alternatively, the neurological and developmental symptoms might be related to the loss of α1,6-fucosylation. This possibility should be assessed by studies on FucT-VIII KO mice in the future (for emerging functions of *O*-fucosylation, please see Table 7.9).

The analysis of glycogene KO mice sometimes provides an insight into common diseases such as diabetes mellitus and cancer. Glucose transporter 2 (Glut-2) on the pancreatic β cell plasma membranes plays a critical role in glucose-stimulated insulin secretion, thereby controlling blood glucose homeostasis in response to dietary intake. The GnT-IVa (Mgat-4a) KO mice (please see Chapter 6.8 for this part of the biosynthetic pathway of *N*-glycans) exhibit attenuation of β cell Glut-2 glycosylation [2]. The attenuated glycosylation leads to endocytosis of Glut-2 and thereby the cell-surface expression of Glut-2 is reduced. Thus, GnT-IVa KO mice develop type 2 diabetes. The induction of diabetes by chronic ingestion of a high-fat diet is associated with reduced GnT-IV expression and attenuated Glut-2 glycosylation coincident with Glut-2 endocytosis. The lesson from the GnT-IVa KO mice is that the fluctuation of β cell Glut-2 glycosylation can be the pathogenesis of type 2 diabetes (see also Table 25.2). On the other hand, insulin sensitivity is enhanced in skeletal muscle of GM3 synthase (lactosylceramide α2,3-sialyltransferase, ST3Gal-V, SAT-1) KO mice [4], consistent with the finding that increased GM3 decouples the functional association between the insulin receptor and the insulin receptor substrate in a type 2 diabetes animal model. These results suggest that inhibition of GM3 synthesis could prevent the development of insulin-resistant type 2 diabetes. In fact, pharmacological inhibition of GlcCer synthase is found to enhance insulin sensitivity.

Many tumor cells have increased GlcNAcβ6 branching in their *N*-glycans (please see Chapter 6.7). GnT-V (Mgat-5) catalyzes the addition of this branching GlcNAc, leading to poly-*N*-acetyllactosamine extension and increased sialylation. The functional significance of the GlcNAcβ6 branching in tumorigenesis has been studied using GnT-V KO mice [2]. The polyomavirus middle T (*PyMT*) oncogene induces multifocal breast tumors in mice when expressed from a transgene in mammary epithelium. *PyMT*-induced tumor growth and metastasis are markedly suppressed in GnT-V KO mice compared with their *PyMT* transgenic littermates expressing GnT-V. Consistent with these phenotypes, tumor cells from GnT-V KO mice are insensitive to multiple cytokines, including EGF, FGF, platelet-derived growth factor, insulin-like growth factor 1 and TGF-β. Galectin-3 binds to poly-*N*-acetyllactosamine extended on the GlcNAcβ6 branches (for further information on galectin-3, please see Chapter 19.3). The binding of galectin-3 to the glycans attached to the growth factor receptors suppresses their endocytosis and maintains their number on the cell surface. As galectin-3 may not bind to their glycans in GnT-V KO mice, the growth factor receptors undergo endocytosis and their cell-surface expression is downregulated. Of note, however, absence of galectin-3 did not reduce *PyMT*-induced tumor formation [6]. These findings have suggested that the GlcNAcβ6 branching promotes tumor development and metastasis, and related molecules including galectins can become a promising target for molecular therapy of cancer (please see also Chapter 25.2 for galectins in growth regulation of tumors).

23.3 Other Gene Manipulation Techniques

Other gene manipulations techniques such as RNA interaction and overexpression systems are also applicable to define the function of glycans. These techniques would be particularly effective in elucidating the detailed molecular mechanisms of glycans if they could be used in combination with cultivated cells established from KO mice (please see Table 25.2).

Gene knockout in lower model organisms such as *Caenorhabditis elegans*, *Drosophila melanogaster* and zebrafish also provides valuable information on glycan functions (please see Chapter 8 for glycosylation in lower model organisms), and their comparative studies with KO mice will be helpful in understanding the glycan functions from an evolutionary point of view.

23.4 Conclusions

The functions of glycans are intrinsically entangled, and our knowledge and ideas regarding glycan functionality are restricted. Glycogene KO mice are valuable resources to untangle the complex functions of glycans, although it will require a great deal of time and labor to fully understand the roles of individual carbohydrate chains using KO mice.

> **Summary Box**
>
> **Glycogene KO mice have proved that glycans are essential for a variety of biological events and that some glycogenes are functionally associated. KO mice eliminating the core structure yield a more systemic, serious and complex phenotype than those eliminating the terminal structure. Unexpected findings obtained in KO mice provide a new insight into glycan functions. Finally, KO mice may provide insights into human diseases.**

References

1 Lowe JB, Marth JD. A genetic approach to mammalian glycan function. *Ann Rev Biochem* 2003;72:643–91.

2 Muramatsu T. Knockout mice and glycoproteins. In: *Comprehensive Glycoscience from Chemistry to Systems Biology* (Ed.: Kamerling JP), pp. 121–47. Elsevier, Oxford, 2007.

3 Kimata K *et al.* Knockout mice and proteoglycans. In: *Comprehensive Glycoscience from Chemistry to Systems Biology* (Ed.: Kamerling JP), pp. 159–89. Elsevier, Oxford, 2007.

4 Furukawa K *et al.* Knockout mice and glycolipids. In: *Comprehensive Glycoscience from Chemistry to Systems Biology* (Ed.: Kamerling JP), pp. 149–57. Elsevier, Oxford, 2007.

5 Forsberg E, Kjellen L. Heparan sulfate: lessons from knockout mice. *J Clin Invest* 2001;108:175–80.

6 Eude-Le Parco *et al.* Genetic assessment of the importance of galectin-3 in cancer initiation, progression and dissemination in mice. *Glycobiology* 2009;19:68–75.

24
Glycobiology of Fertilization and Early Embryonic Development

Felix A. Habermann and Fred Sinowatz

The preceding chapters have provided detailed information on the glycan structures of cellular glycoconjugates as well as on the presence and biochemical nature of glycan receptors (lectins). Biomedical aspects were already covered by dealing with diseases caused by defects in glycan synthesis/degradation or lectin expression (for example, Chapter 11 on proteoglycans or Chapter 19 on human lectins). The emerging connection between glycosylation and disease was further described in Chapters 22 and 23. This chapter opens the final section of the book, where *in vivo* aspects of the sugar code are presented. Naturally, fertilization is the starting point, thus warranting intense investigation of this process and ensuing development. We will start with a bird's eye view on mammalian fertilization and the description of the major components defining the glycophenotype of oocytes.

24.1
Primer to Mammalian Fertilization

In mammals, oocytes are fertilized in the ampulla of the oviduct. At ovulation, the oocyte is surrounded by the zona pellucida (ZP) – a transparent, porous and viscoelastic extracellular glycoprotein matrix (please see below). The ZP itself is enclosed by several layers of cells of the cumulus oophorus of the ovarian follicle (Figure 24.1a) [1]. These are embedded in an extracellular matrix primarily composed of hyaluronic acid (for further information on glycosaminoglycans, please see Chapter 11). The oocyte completes the first meiotic division – the pairs of homologous chromosomes are separated. A haploid (single) set of chromosomes, each still comprised of two identical sister chromatids, is retained in the oocyte, while the other one is discarded by extrusion of a small cell, the first polar body. Subsequently, the oocyte is arrested at the metaphase stage of the second meiotic division, ready to fuse with a sperm (Figure 24.1a). *In vivo*, mammalian sperm acquire competence to fertilize after ejaculation. In the female reproductive tract, a final maturation process is initiated, termed capacitation. It denotes a series of changes on the cell surface and in the cell, which enable the sperm to fertilize the oocyte.

The Sugar Code. Edited by Hans-Joachim Gabius

ISBN: 978-3-527-32089-9

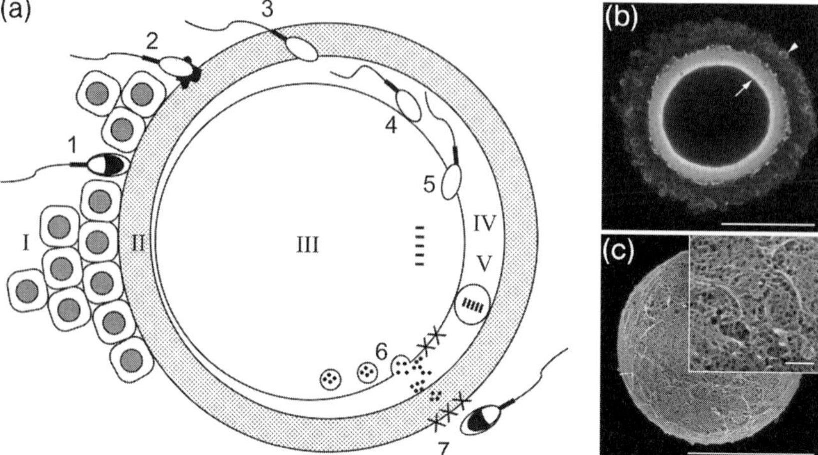

Figure 24.1 Fertilization in mammals. (a) Schematic diagram of the stages of sperm–oocyte interaction: (1) binding of a sperm to the ZP, (2) acrosome reaction, (3) transit of the sperm through the ZP, (4) sperm–oocyte fusion, (5) sperm entry and oocyte activation, (6) cortical reaction, and (7) zona reaction blocking polyspermy; (I) cells of the cumulus oophorus from the ovarian follicle, (II) ZP, (III) mature fertilizable oocyte arrested at the metaphase of the second meiotic division, (IV) perivitelline space and (V) first polar body. The image panel on the right side illustrates how the structure of the ZP and the presence of glycan epitopes can be visualized by staining with plant lectins and confocal laser scanning microscopy. (b) Optical section through the center of a canine ZP, which is intensely stained with wheat germ agglutinin (WGA; for lectin acronyms and sugar specificity, please see Table 18.1). In comparison, the extracellular matrix between the cumulus cells surrounding the ZP is weakly stained. Please note that the oocyte and cumulus cells themselves are not visible. Scale bar = 100 μm. (c) The surface of a bovine ZP stained with *Sambucus nigra* agglutinin (SNA). The maximum intensity projection was computed from a stack of 60 serial optical sections. The cumulus cells have been mechanically removed prior to staining. The insert shows the lattice-like structure and the pores at higher magnification. Scale bars = 100 and 10 μm (insert).

What happens next is that the sperm have to pass through the layers of cumulus cells and the ZP. Towards this end, sperm are equipped with appropriate biochemical means. In detail, they have a cap-like secretory vesicle at the tip of their head, called the acrosome. This organelle is filled with inactive hydrolases (proenzymes) such as proacrosin (for information on acrosin, a protease with a lectin site to target enzymatic activity, please see Table 19.2) and hyaluronidase. The content of the acrosome is released by the so-called acrosome reaction only when the sperm head specifically recognizes and binds to the ZP. Acrosomal proenzymes are now activated and enable the sperm to cut its way through the ZP to the oocyte. Necessarily, specific contact formation between sperm and the ZP depends on structural complementarity and, in particular, on sugar-encoded information. After entering the perivitelline space beneath the ZP, a sperm eventually succeeds in binding to and fusing with the oocyte plasma membrane (Figure 24.1a).

Thereby, oocyte activation is triggered – a complex cascade of events. Herein, the second meiotic division is completed and the two sister chromatids of each chromosome are separated. One haploid set of single-chromatid chromosomes remains in the oocyte, while the other one is discarded by extrusion of the second polar body. Moreover, in the cortical reaction, the content of peripheral secretory vesicles, the cortical granules, is released by exocytosis into the perivitelline space. The cocktail secreted by the oocyte includes proteases as for instance tissue-type plasminogen activator (tPA) and initiates an elaborate set of mechanisms to prevent polyspermy, that is fertilization of an oocyte by more than one sperm. The ensuing zona reaction alters the ZP in such a way that it can no longer be penetrated by sperm. In conclusion, the ZP plays a key role in mammalian fertilization by restricting the access of sperm to the oocyte, and this in a species-specific manner. Later on, after fertilization, the ZP protects and supports the developing embryo while moving down the oviduct. Having provided an overview on the basic steps, we now direct our attention to the morphology and the molecular structure of the ZP, which will guide us into the realm of the glycobiology of fertilization.

24.2 The Functional Morphology of the Zona Pellucida (ZP)

The ZP is formed during follicle and oocyte growth in the ovary, spatially between the oocyte and the innermost layer of cumulus cells (corona radiata), which extend fine cytoplasmic processes through the ZP and form gap junctions with the oocyte. At first glance, the ZP appears to be a rather unspectacular structure. Depending on the species, the ZP is constituted by only three or four different glycoproteins, which build up a fibrogranular structure by noncovalent interactions. The most obvious morphological difference between species is the thickness of the ZP. It varies from only 1–2 μm in opossums, 5 μm in mice, 15 μm in humans and pigs to 27 μm in cattle. In many species, the ZP is arranged in concentric layers, which differ in their glycoprotein and carbohydrate composition. Composition and structures of the glycans of the ZP glycoproteins can be determined by mass spectrometry or other biophysical separation techniques (see Chapter 5 for further details). The functional morphology of the ZP as well as the presence and localization of distinct glycan epitopes within the ZP has to be investigated by microscopic techniques, for instance using antibodies and plant lectins (for further information on plant lectins, please see Chapter 18) [1].

When applying scanning electron microscopy or confocal laser scanning microscopy, the mammalian ZP is visualized as a very intricate three-dimensional network structure with many pores (Figure 24.1b and c). Thereby, the outer surface has a fenestrated lattice-like appearance, while the inner surface is more evenly structured. The pores mentioned above result from the cytoplasmic processes of the cumulus cells, which surround the oocyte in the ovarian follicle. Typically, the pore diameter centripetally decreases in size. Diameter, density and shape of the pores as well as the fineness of the network structure vary from species to species. When the ZP surface is compared among species, the smallest pores are found in cattle, the largest ones in mice and cats. The ZP structure allows the penetration of relatively large molecules such as immunoglobulins. On the other hand, and surprisingly, smaller sized molecules such as the proteoglycan heparin (please see Chapter 11) are precluded from passing through the ZP. Obviously, the ability of molecules to move through this extracellular matrix depends on the size of the penetrating molecule and further on biochemical or physicochemical properties like surface charge. Since embryo transfer has become increasingly popular, serious questions have arisen concerning the risk of virus transmission and epidemiological consequences. Therefore, it was a reassuring observation that an intact ZP of *in vitro*-produced bovine embryos acts perfectly as a protective barrier against bovine herpes virus-1 and bovine viral diarrhea virus – the two most important viral pathogens in cattle. Having examined the architecture of the ZP, we next turn to a fine-structural analysis of the ZP glycoproteins.

24.3 The Glycoproteins of the ZP and Their Encoding Genes

The genome projects have catalogued and sequenced the ZP genes in numerous species and provided the basis for a consistent nomenclature across all species. Moreover, DNA sequence similarity and chromosomal position disclosed the phylogeny of the ZP genes in mammals and other classes of animals [2]. Details of the current nomenclature of ZP genes including the nomenclature used in this chapter are given in Table 24.1. In general, four different ZP genes can be found in mammals. In fact, species such as humans, chimpanzees, macaques and rats have the full set of four functional ZP genes. In other species such as dogs and cattle, the ZP contains no ZP1. In mice, the ZP contains ZP1, but no ZP4. This suggests that the mechanism of sperm–ZP interaction requires the presence of both ZP2 and ZP3 as well as one or both of ZP1 and ZP4. The latter share a rather recent ancestor and are more similar to each other than ZP2 and ZP3. The peptide sequences of ZP2 and ZP3 are rather well conserved across the mammalian species studied (65–98% amino acid sequence identity) [3]. Functionally important and highly conserved are the number and position of cysteine residues and, notably, of *N*-glycosylation sequons (the Asn-X-Ser/Thr sequence; for details on *N*-glycosylation, please see Chapters 6, 8 and 22.1).

Overall, the four mammalian ZP proteins share a modular architecture (Figure 24.2). In detail, the following highly conserved domains are present: a *N*-terminal signal peptide, a ZP domain, a consensus furin cleavage site and a C-terminal transmembrane domain [4]. The secreted ZP proteins no longer contain a *N*-terminal signal peptide and are cleaved at the C-terminal furin site. Particularly important is the ZP domain, which is located relatively close to the C-terminus. ZP domains consist of around 260 amino acids with eight strictly conserved cysteine residues, implying similar patterns of formation of disulfide bonds. Further constant features are similar hydrophobicity profiles and the frequent occurrence

Table 24.1 The nomenclature of mammalian ZP genes[a].

HUGO[b] gene symbol	Alias gene symbols/Entrez Gene GeneIDs[a]				
	Human	Mouse	Rat	Pig	Cattle
ZP1	ZP1 22917	Zp1 22786	Zp1 85271	–[d]	–[d]
ZP2	ZP2, ZPA 7783	Zp2 22787	Zp2 81828	ZP2, ZPA 396846	ZP2 280963
ZP3	ZP3A, ZP3B, ZPC 7784	Zp3 22788	Zp3 114639	ZP3β, ZPC 396927	ZP3 280964
ZP4	ZP1, ZPB 57829	*Zp4*[c] 664793	Zp4, Zpb 282833	ZP3α, ZPB 397111	ZP4 280965

[a] According to the Entrez Gene database of the National Center for Biotechnology Information (NCBI; www.ncbi.nlm.nih.gov).
[b] Human Genome Organization (HUGO; www.hugo-international.org).
[c] Pseudogene.
[d] Not found (NCBI; www.ncbi.nlm.nih.gov).

of turn-forming motifs. ZP domains are part of numerous extracellular proteins with widely varying functions, found in many organisms ranging from nematodes to mammals [3]. ZP1 and ZP4 additionally share a trefoil domain (for relevance of β-trefoil domains in plant and mammalian lectins, please see Table 18.2 and Table 19.1). Having herewith characterized the protein parts of ZP glycoproteins, we next provide information on their glycan parts.

24.4 Glycan Structures of ZP Glycoproteins

Whereas the sequences of the protein backbone of the four mammalian ZP glycoproteins including the sites for potential *N*- and *O*-glycosylation are remarkably well conserved across species, the carbohydrate composition differs greatly. The diversity between species concerns both *N*- and *O*-glycosylation patterns (for structural details on *N*- and *O*-glycans, please see Chapters 6 and 7). Additionally, ZP oligosaccharides can be both sulfated and sialylated [1]. These substitutions further substantially increase the structural diversity and establish the low isoelectric point (for structural examples of sialylated and sulfated glycans and their role for enhancing the coding capacity of glycans, please see Chapter 1.3).

The structural analysis of the ZP glycans requires sophisticated techniques and equipment such as two-dimensional chromatography and mass spectrometry (as illustrated in Chapter 5). For a simple reason, sequences of the carbohydrate chains have been most intensely studied in pigs and to a somewhat lesser degree in cattle [5]. In these two species, it is feasible to collect sufficient amounts of ZP glycoproteins. To give an idea on actual quantities, a single porcine ZP contains about 30 ng of glycoproteins and a bovine ZP about 20 ng, while the total yield of ZP glycoproteins per ovary is 10 times higher in pigs. Efficient large-scale isolation methods greatly facilitated the elucidation of glycan composition and structures for porcine ZP glycoproteins. By gel electrophoresis, porcine ZP glycoproteins can be resolved into two components, with molecular masses of about 55 and 90 kDa. The 55-kDa component represents about 80% of the total glycoprotein content and consists of ZP3 (ZPC) and ZP4 (ZPB), which can be separated only after partial deglycosylation with endo-β-galactosidase. Each of the polypeptides has five sequons for *N*-glycosylation (positions of *N*-glycan sequons of the human ZP glycoproteins are shown for comparison in Figure 24.2). Carbohydrate composition analysis of the glycoproteins detected the presence of four *N*-glycans and six *O*-glycans on ZP3 (ZPC) as well as three *N*-glycans and three *O*-glycans on ZP4 (ZPB). For further analysis, the *N*-linked glycans were released by hydrazinolysis

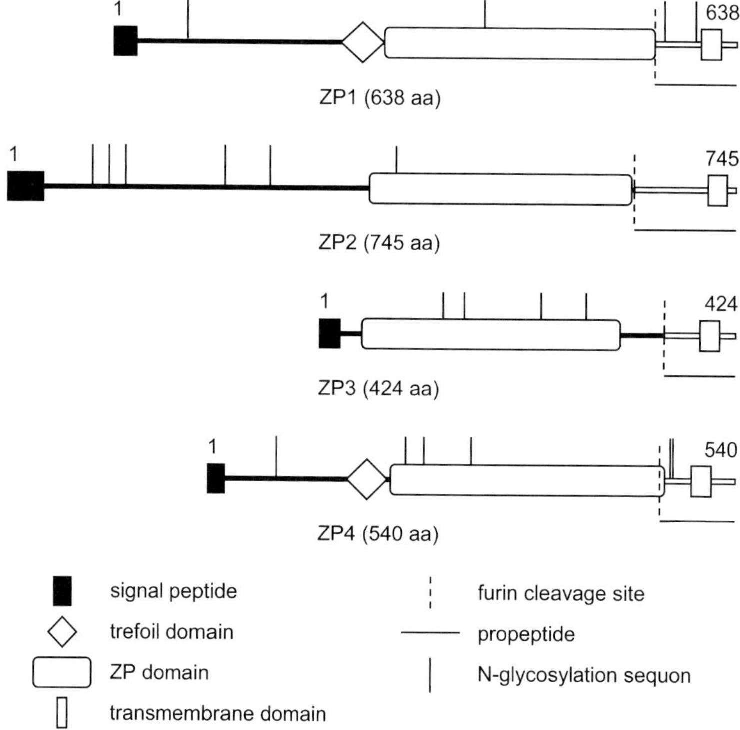

Figure 24.2 The modular architecture of the human ZP genes (aa = amino acids). Source: UniProtKB/SwissProt.

from the 55-kDa glycoprotein fraction and then separated into neutral (28%) and acidic (72%) carbohydrate chains by anion exchange high-performance liquid chromatography (for chromatographic techniques in glycan separation, please see Chapter 5). Both fractions comprise complex-type bi-, tri- and tetraantennary compounds with core fucosylation (for structural illustration of triantennary glycans, please see Figure 22.1; for the role of core fucosylation on glycan conformation, see Chapter 2.6; for different forms of this substitution in animals, see Chapter 8; on the relation to disease, see Chapter 23). Moreover, fucosylation can occur at the non-reducing end [5]. Having so far focused on *N*-glycans, we now turn to the *O*-glycans.

In pigs, the molecular structure of eight types of neutral *O*-linked chains and 23 acidic *O*-linked chains from ZP glycoproteins was determined [1]. Poly-*N*-acetyllactosamine extensions of *O*-glycans have anionic substitutions in common with branch ends of *N*-glycans. Interestingly, extended chains predominantly terminate with α2,3-linked sialic acids Neu5Ac/Gc. Chains with no or only one *N*-acetyllactosamine unit can additionally be α2,6-sialylated at the proximal GalNAc residue (sialyl-Tn), whereas terminal α-Gal or β-GalNAc residues are also found [1]. The number of *N*-acetyllactosamine repeats and the degree of sialylation and sulfation contribute considerably to the enormous structural complexity and diversity of the carbohydrate portion of the ZP glycoproteins, as already referred to above.

24.5 The Synthesis of ZP Glycoproteins

Depending on the species and the thickness of their ZP, mammalian ZP glycoproteins are synthesized by the oocyte and/or the ovarian follicle cells surrounding the oocyte [1]. Thus, in mice forming a thin ZP (around 5 μm), the ZP glycoproteins are exclusively synthesized by the oocyte. Notably, this species is particularly suited for gene knockout models (see Chapter 23 for technical comments), but can only provide limited information on the situation in other mammals: investigations in species that develop a thick ZP, such as humans, rabbits, dogs, pigs and cattle, revealed that both the oocyte and the follicle cells produce ZP glycoproteins. Moreover, studies on bovine ovaries using *in situ* hybridization and immunostaining have shown that both ZP gene transcription and glycoprotein synthesis by the oocyte and/or the cumulus cells change during folliculogenesis and oogenesis [1]. Thereby, the oocyte and the cumulus cells may glycosylate the ZP protein backbones differently. In conclusion, the spatially and temporally regulated glycosylation of the individual ZP glycoproteins by the oocyte alone and/or the cumulus cells during follicle and oocyte development is able to generate an enormous structural complexity of the ZP. Having so far analyzed the structure of ZP glycans, the issues of whether they serve as ligands need to be addressed next.

24.6
Ligand Properties of ZP Glycans

In principle, it has been demonstrated in several mammalian species that carbohydrates of the ZP play a crucial role in sperm–ZP interactions [1]. Our knowledge of the actually functionally relevant ZP glycan determinants and their respective sperm receptors is still rather limited. A step on the way to identify distinct carbohydrate ligands for sperm–ZP interactions is to detect and localize defined carbohydrate epitopes within the ZP *in situ*, for example in whole-mount preparations and in tissue sections. As illustrated in Figure 24.3 and Figure 24.4, this can be done using specific sugar-binding proteins (lectins; for definition, please see Chapter 15; for information on plant lectins used as tools, please see Chapter 18; Table 18.1 presents a compilation of respective reagents with detailed information on their glycan specificity; for information on mammalian lectins as tools, please see Chapter 25; for information on how to measure lectin specificity, please see Chapter 14). Comparative analyses using lectins revealed species-specific differ-

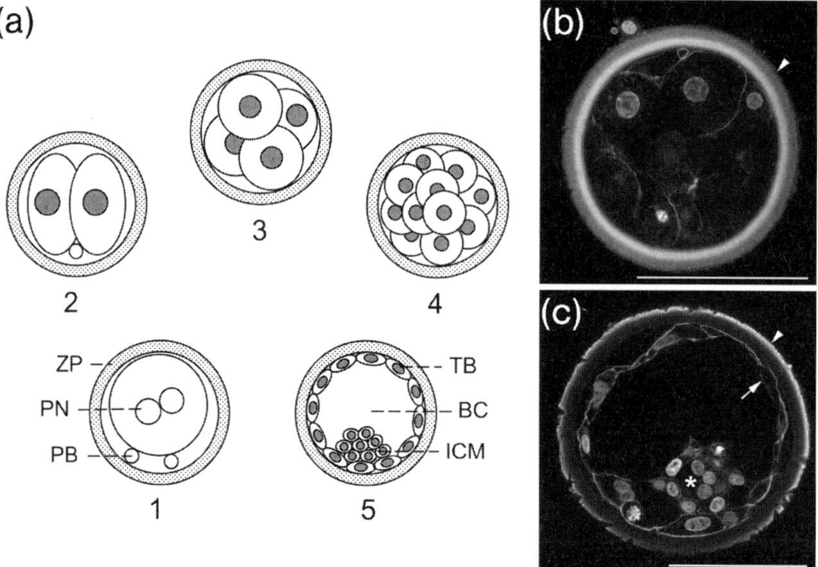

Figure 24.3 Mammalian development from the fertilized oocyte to the blastocyst stage. (a) Schematic diagram (for further details, see Info Box): (1) pronuclear zygote, (2) two-cell stage, (3) four-cell stage, (4) morula and (5) blastocyst (PN = pronuclei, PB = polar bodies, TB = trophoblast, BC = blastocoel, ICM = inner cell mass). The image panel on the right side shows a bovine morula (b) and a bovine blastocyst (c) surrounded by their ZP, as seen by confocal laser scanning microscopy. (b) WGA binds across the entire thickness of the ZP (arrowhead). (c) In contrast, SNA intensely stains the outer ZP surface (arrowhead); the ICM is marked by an asterisk, the trophoblast by an arrow (for information on the two mentioned plant lectins and the reactive glycan epitopes, please see Chapter 18). Both embryos were additionally stained with the DNA-binding dye DAPI and fluorescent phalloidin to visualize cell nuclei and the F-actin microfilament cytoskeleton. Scale bars = 100 μm.

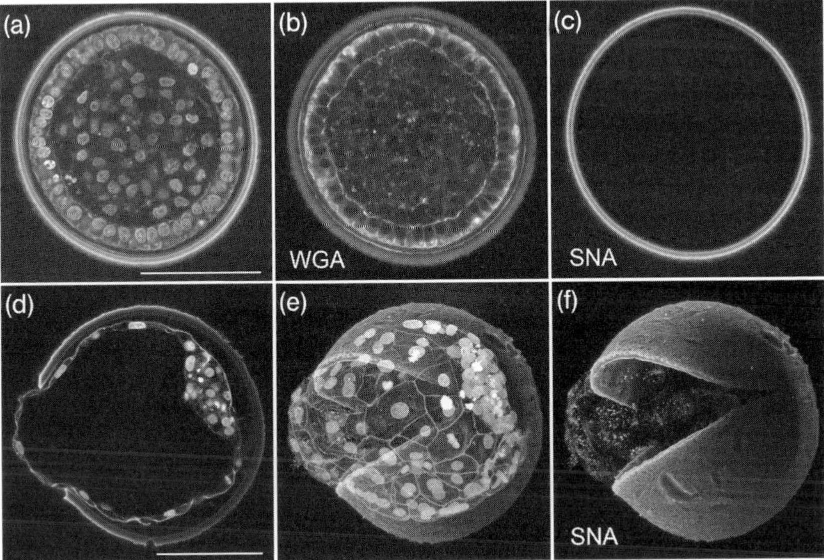

Figure 24.4 Glycan profiling of mammalian blastocysts using lectins and confocal laser scanning microscopy. The upper panel shows a horse blastocyst prior to hatching. In this species, the trophoblast cells produce a glycoprotein capsule replacing the ZP and the blastocyst remains spherical during further growth. The specimen was stained with WGA and SNA and in addition with the DNA-binding dye DAPI to visualize cell nuclei (for information on the two mentioned plant lectins, please see Chapter 18). (a–c) Optical sections through the center of the embryo: (a) overlay image of the three fluorescent stainings, and (b) WGA and (c) SNA alone, respectively. The lower panel shows a hatching bovine blastocyst, which will elongate to a long filamentous conceptus. The specimen was stained with SNA, DAPI (see above) and fluorescent phalloidin marking the F-actin microfilament cytoskeleton. (d) Central optical section, (e) and (f) maximum intensity projections computed from 170 serial optical sections; (d) and (e) overlay images of the three fluorescent stainings, (f) SNA alone. Scale bars = 100 μm.

ences in the morphological structure and in the carbohydrate signature across the ZP [1]. The differences in lectin-binding patterns between species appear to correlate with their phylogenetic distance. Of particular note in terms of spatial access to a receptor is that most variations appear to reside in the nonreducing ends of the ZP glycans. Rather small structural differences at strategic positions can well be important for establishing species-dependent sperm–ZP interactions. This concept holds similarly true for inflammatory or metastatic tumor cells homing to distinct endothelia (please see Chapters 25 and 27). In terms of methodology, these studies exemplify the use of plant lectins as versatile laboratory reagents referred to in Chapter 18.4.

On the way to identify ZP ligands and sperm receptors, the actual ZP–sperm interactions in several species were examined by monitoring sperm binding to the ZP and/or defined putative ligands *in vitro*. The results point to different oligosaccharides of the ZP as key molecules in species-specific sperm–ZP interaction. As an example, in mice a Lewis^x (Le^x)-containing glycan (Galβ1,4[α1,3Fuc]GlcNAc-R)

Table 24.2 Examples for proteins on sperm with binding properties to ZP glycans.

Localization	Name	Species	Carbohydrate ligand
Sperm head surface	α-mannosidase	Mouse	Man
	β1,4GalT-I[a]	Mouse	GlcNAc
	Asialoglycoprotein receptor 2	Rat	Gal
	Spermadhesins	Pig	Galβ1,3GalNAc, Galβ1,4GlcNAc, sulfated glycosaminoglycans
Acrosome	Acrosin[b]	Mouse	Sulfated glycans
	Bindin	See urchin	350-kDa glycoprotein with sulfated and sialylated *O*-glycans, sulfated fucans
	P-Selectin	Pig	Not yet characterized[c]
	Sp17	Mouse	Heparin
	Sp38	Pig	Sulfated glycoproteins
	Sp56	Mouse	*O*-Linked oligosaccharides of ZP3
	Zonadhesin	Pig	Sulfated glycans

[a] β1,4-galactosyltransferase.
[b] Protease with a lectin site (see Table 19.2).
[c] For information on P-selectin ligands in the immune system, please see Chapter 27.

turned out to be a potent competitor of sperm binding to the ZP (for Lewis carbohydrate structures, please see Table 7.4) [6]. Studies in cattle strongly suggest that α-D-Man residues at the nonreducing ends of high-mannose-type *N*-glycans play an essential role in ZP binding and fertilization [5]. In humans and rats sperm penetration through the ZP was inhibited by pretreatment of the sperm with D-mannose. In guinea pigs, hamsters, rats and humans, L-Fuc and fucoidin, a sulfated fucan from algae, also active to block selectin binding in inflammation (please see Chapter 27 for details), were shown to be potent inhibitors of sperm–ZP interaction. The latter findings intimate that sperm binding to the ZP engages a 'selectin-like' interaction [1] (for information on selectins, please see Chapters 16, 19, 23 and 27). From looking at carbohydrate ligands of the ZP we next turn to the question of the corresponding carbohydrate receptors of sperm.

Table 24.2 presents a survey of prominent sperm head surface proteins and acrosomal proteins with binding properties to glycans [7, 8]. The latter are only released or exposed after induction of the acrosome reaction (see above). Apparently, the heterogeneity, complexity and high coding capacity of mammalian ZP glycans is paralleled by the evolution of multiple sperm receptors for the ZP (see also Table 18.2 and Table 19.1 for the diversity of structural folds with lectin activity). In each species, sperm–ZP interaction appears to be based on the cooperation of multiple ligand–receptor systems with inherent redundancy to compensate a loss (a similar redundancy is also seen in bacterial adhesion, see Chapter 17). Due to the high degree of evolutionary divergence the key ligands and receptors will have to be identified individually in each species of interest. In the following, we move on from the ZP and its relevance for mammalian fertilization to early embryo development.

24.7 The Glycoprotein Shell of Mammalian Embryos

Early mammalian embryos are surrounded by a glycoprotein shell, that is still the ZP and/or other glycoprotein layers formed by the embryo and/or the oviductal and/or uterine epithelium. This shell has essential functions listed as follows, while the fertilized oocyte, the zygote, develops to the blastocyst (see Info Box and Figure 24.3) and travels through the oviduct into the uterus [9]. The glycoprotein shell keeps the early blastomeres together before they form intercellular connections, and hinders the early embryo to prematurely adhere to the oviductal and uterine epithelium. Ensuring an appropriate protective microenvironment, it shields the embryo from mechanical damage, toxins and xenobiotics, bacteria, large viruses, maternal phagocytes and other immune cells. Finally, the glycoprotein shell is a selective permeability barrier, which plays a central role in the communication between the embryo and its mother: not only nutrients and metabolic waste, but also all messenger molecules transmitting information between embryo and its mother have to pass through this shell.

Both the embryonic needs and the maternal environment change dramatically during the development from the zygote to blastocyst stage. Accordingly, the glycoprotein shell undergoes dynamic changes. Depending on the species, the ZP is altered and supplemented by secretions from the embryo and/or the oviductal and/or uterine epithelium. Finally, the early blastocyst must dissociate from the ZP to allow for further growth and implantation in the uterus. Hatching of the blastocyst from the ZP (Figure 24.4) is a central event of early mammalian development and occurs in most domestic mammals in the uterus four to eight days after fertilization. Hatching failure due to abnormal embryo development and improper trophoblast function and/or ZP anomalies is a serious obstacle for *in vitro* fertilization in humans and other species [9]. Notably, the course and duration of mammalian preimplantation development differ greatly from species to species.

Info Box

The fertilized oocyte is called a zygote and initially contains two separate sets of haploid chromosomes, which form the female and the male pronucleus. Both replicate their DNA and gradually approach each other (for illustration, see Figure 24.3). The two pronuclei fuse and the zygote is divided by a series of mitotic cell divisions, called cleavage divisions, into smaller and smaller cells, referred to as blastomeres. The morula, a solid cluster of (around 16 to 64) blastomeres develops to the blastocyst, which is constituted by a single layer of cells, the trophoblast, surrounding a fluid-filled cavity, the blastocoel and the inner cell mass (ICM) or embryoblast. The trophoblast mediates the implantation into the uterine wall and gives rise to the fetal part of the placenta, while the fetus originates from the ICM.

In parallel, the size and shape of the implanting conceptus, embryo–maternal signaling and the mode of implantation in the uterus vary amazingly. In ruminants and pigs, the blastocyst develops to a long filamentous structure. In contrast, in horses, rabbits and carnivores, the implanting conceptus is by and large spherical. Depending on the species, embryonic secretions and/or oviduct and/or uterine secretions form additional glycoprotein layers covering the inner and/or outer surface of the ZP. In rabbits, for example, the ZP is covered by a thick layer of highly sulfated mucoproteins produced by the oviduct epithelium. In horses, the blastocyst (Figure 24.4) produces a glycoprotein capsule mainly composed of mucin-like glycoproteins beneath the ZP as a prerequisite for further development. This capsule replaces the ZP to enclose the conceptus during the second and third weeks of pregnancy.

Notably, glycans are not only a central component of the ZP and additional glycoprotein coats around the embryo. Later on, cell surface glycans are crucial for proper embryo implantation in the uterus, in the right place at the right time. Both the trophoblast of the conceptus and the uterine luminal epithelium (ULE) are covered by a glycocalyx. The surface of the nonreceptive ULE has a high density of transmembrane mucins such as MUC1 precluding implantation (for information on mucins, please see Chapter 7.5). Uterine receptivity appears to go along with a loss of MUC1 and a remodeling of the glycocalyx of the ULE [10]. At least equally important and an arising matter is that cell-surface glycans appear to be critically involved in the maintenance and development of embryonic stem cells.

24.8
Surface Glycans of Stem Cells

Stem cells are the 'master cells' of the organisms, since they can both renew themselves through mitotic cell division and generate a progeny of specialized cells. They constitute the basis for tissue formation, regeneration and repair. Strategically presented at the cell surface, glycans are suitably positioned to be involved in the control of stem cell maintenance, proliferation and differentiation, for instance by modulating structure-activity profiles of signaling molecules (for examples how glycosylation changes growth factor receptor parameters, please see Table 25.3) [11]. Regulation of growth and differentiation factors can, for instance, be exerted through the following effector routes: (i) Fibroblast growth factor (FGF) 2 is a key player in regulating self-renewal and proliferation of stem cells. FGFs bind to both high-affinity FGF receptors and heparan sulfate proteoglycans at the cell surface, whose affinity is regulated by the pattern of substitutions such as sulfation (for information on proteoglycans, please see Chapter 11 and Chapter 22.6). (ii) Wnt proteins are secreted glycoproteins associated with the extracellular matrix, which induce intracellular signaling. Matrix contact is mediated by heparan sulfate proteoglycans. (iii) Notch is a large single-pass transmembrane glycoprotein essential for development by its regulatory function in stem cell fate determination. Notch signaling depends on glycosylation of extracellular epidermal growth factor

(EGF) domains by the *O*-fucosyltransferase *O*-FucT1 and the β1,3-N-acetylglucosaminyltransferase Fringe (for further details on *O*-fucosylation, please see Chapter 7; for the relationship between *O*-fucosylation or glycosaminoglycan parameters and disease, see Chapter 22) [12]. As analyzed by applying both carbohydrate-specific antibodies and plant lectins as probes, each stem cell type appears to display a distinct glycan signature at the cell surface [11]. This property might pave the way for the characterization of stem cell lineages – a step toward envisioned applications in regenerative medicine. Examples of emerging stem cell glycan markers are shown in Table 24.3 and serve to illustrate the significance of glycomic profiling for cell typing. The ensuing question on examples for a functional role of glycan markers will be answered for different types of tumor cells in Chapter 25 and for inflammatory cells in Chapter 27.

Table 24.3 Examples for cell-surface glycan markers used for identification and enrichment of stem cells[a].

Stem cell lineage	Species	Glycan marker	Description/structure
Embryonic stem cells	Mouse	SSEA-1 (Le^x, CD15)[b]	Galβ1,4[α1,3Fuc]GlcNAc on glycoproteins and glycosphingolipids
	Human	SSEA-3	Pentasaccharide of a globo-series glycosphingolipid
	Human	SSEA-4	Hexasaccharide of a globo-series glycosphingolipid
	Human	TRA-1-60	Neuraminidase-sensitive and keratan sulfate-presenting proteoglycan
	Human	TRA-1-82	Neuraminidase-insensitive keratan sulfate-presenting proteoglycan
Hematopoietic stem cells	Human	CD34	Type 1 transmembrane sialomucin (glycoforms)
Hematopoietic progenitor cells	Human	CD164	Transmembrane sialomucin
Hematopoietic and endothelial stem cells, neural stem cells	Mouse, human	CD133 (prominin 1, PROM1)[b]	Five-transmembrane-domain glycoprotein
Neural stem cells	Human	PSA-NCAM	α2,8-Linked oligosialic acid chains as part of a complex-type *N*-glycan on a type 1 transmembrane glycoprotein

[a] SSEA = stage-specific embryonic antigen; CD = cluster of differentiation (see Info Box in Chapter 27); TRA = tumor rejecting antigen; PSA-NCAM = polysialic acid-presenting neural cell adhesion molecule.

[b] Alias protein names; for glycan nomenclature, please see Chapter 1; for details on *N*-glycans, see Chapters 6 and 8; on mucins, see Chapter 7.5; on glycolipids, see Chapters 10 and 30; on proteoglycans Chapter 11; on the glycoprotein neural cell adhesion molecule (NCAM), please see Chapters 6 and 30.

24.9 Conclusions

This chapter provides insights into salient aspects of the glycobiology of fertilization and early embryonic development in mammals. To start with, sperm–oocyte contact is mediated in a species-dependent manner by the ZP, a glycoprotein matrix that encapsulates the oocyte. Penetration of the ZP by a sperm requires a complex and intimately regulated set of sperm–ZP interactions. The ZP is constituted by only three or four ZP glycoproteins, which share a modular design highly conserved across species. Structural versatility of *N*- and *O*-glycans of the ZP encompassing particular patterns of sulfation and sialylation allows for the evolution of multiple and partially redundant ligand–receptor systems and great diversity between species. Later on in development, the glycocalices of the trophoblast and the uterine luminal epithelium are crucially involved in embryo implantation. The elucidation of the role of particular carbohydrate epitopes in the control of stem cell maintenance, proliferation and differentiation, especially as signals [13], is an emerging challenge in biomedical research with the long-term perspective for respective therapeutic approaches.

Summary Box

Glycans of the ZP harbor information required for controlling sperm–oocyte interaction in a species-specific manner. Also depending on the species, the ZP and further glycoprotein layers produced by the embryo or the oviductal and uterine epithelia are then essential for the development of the blastocyst. They protect the embryo and mediate the communication between the embryo and its mother. Later on, the glycocalices of the embryonic trophoblast and the uterine epithelium are involved in ensuring proper embryo implantation in the uterus. Of note, cell-surface glycans are connected to the development and maintenance of embryonic stem cell lineages.

References

1 Sinowatz F *et al.* Functional morphology of the zona pellucida. *Anat Histol Embryol* 2001;30:257–63.

2 Goudet G *et al.* Phylogenetic analysis and identification of pseudogenes reveal a progressive loss of zona pellucida genes during evolution of vertebrates. *Biol Reprod* 2008;78:796–806.

3 Jovine L *et al.* Zona pellucida domain proteins. *Annu Rev Biochem* 2005;74:83–114.

4 Gupta SK *et al.* Structural and functional attributes of zona pellucida glycoproteins. *Soc Reprod Fertil Suppl* 2007;63:203–16.

5 Yonezawa N *et al.* Structural significance of *N*-glycans of the zona pellucida on species-selective recognition of spermatozoa between pig and cattle. *Soc Reprod Fertil Suppl* 2007;63:217–28.

6 Kerr CL *et al.* Lewis^x-containing glycans are specific and potent competitive inhibitors of

the binding of ZP3 to complementary sites on capacitated, acrosome-intact mouse sperm. *Biol Reprod* 2004;71:770–7.

7 Tanphaichitr N *et al.* New insights into sperm–zona pellucida interaction: involvement of sperm lipid rafts. *Front Biosci* 2007;12:1748–66.

8 Mengerink KJ, Vacquier VD. Glycobiology of sperm–egg interactions in deuterostomes. *Glycobiology* 2001;11:R37–43.

9 Herrler A, Beier HM. Early embryonic coats: morphology, function, practical applications. An overview. *Cells Tissues Organs* 2000;166:233–46.

10 Carson DD. The glycobiology of implantation. *Front Biosci* 2002;7:d1535–44.

11 Lanctot PM *et al.* The glycans of stem cells. *Curr Opin Chem Biol* 2007;11:373–80.

12 Haltiwanger RS, Lowe JB. Role of glycosylation in development. *Annu Rev Biochem* 2004;73:491–537.

13 Gabius HJ. Glycans: bioactive signals decoded by lectins. *Biochem Soc Trans* 2008; 36:1491–6.

25
Glycans as Functional Markers in Malignancy?

Sabine André, Jürgen Kopitz, Herbert Kaltner, Antonio Villalobo, and Hans-Joachim Gabius

'Within the last few years many scientists have focused their attention on the nature of the cell membrane. It is well established that cellular adhesions and interactions are largely dependent on the surface properties of cells. With the observation that neoplastic cells differ from normal cells in the nature of these reactions, the cell surface was implicated as an important factor in the assumption of a neoplastic state' [1]. These sentences introduce a classic report from 1963 on aggregate formation of tumor cells (mouse lymphoma and Ehrlich ascites cells). The agglutination was triggered by supernatants of a preparation of wheat germ lipase treated at 65 °C [1]. Normal control cells 'remain almost completely isolated' [1]. Thus, a difference in cell surface properties sensed by a plant protein separates the tested normal from the tumor cells, and, in this special case, it concerns cell surface glycosylation (for detailed structure information on cell surface glycans, please see Chapters 6–11 and 30). In fact, the active compound binding and cross-linking the tumor cells was identified four years later as a lectin, the wheat germ agglutinin (WGA), by one of the authors of Chapter 21 [2]. As characteristic for a lectin (for a current definition of this term, please see Chapter 15.3; for further details on lectins in plants and also their purification, please see Chapter 18), tumor cell agglutination was reversibly inhibited by a sugar, that is *N*-acetylglucosamine [2].

Soon after detecting this first evidence for lectin reactivity as a feature of tumor cells, biochemical analysis uncovered structural aspects of glycans associated with the malignant phenotype. Gel filtration of detergent-solubilized membrane fractions from 3T3 fibroblasts [controls and cells after viral (SV40) transformation for oncogenesis] and monitoring elution of [^{3}H]/[^{14}C]glucosamine-labeled glycoconjugates were performed. Glycoprotein/glycopeptide fractionation revealed profile changes in size and monosaccharide composition [3]. In general, the glycome of tumor cells shows quantitative changes with over- and underexpression relative to the situation in normal cells, sometimes reminiscent of that during embryogenesis (for further information on glycophenotypes of stem cells and cells in early embryogenesis, please see Chapter 24.7/8). With glycans also appearing in the list

The Sugar Code. Edited by Hans-Joachim Gabius

ISBN: 978-3-527-32089-9

of established tumor markers such as CA19-9, it is tempting to answer the question given in this chapter's title positively. Examining the historical course and present status of tumor glycosciences teaches the lesson that this is indeed justified, in a defined context and without guarantee for *a priori* generalization. Initial work in this area focused on pursuing comparative analysis between normal and tumor cells.

25.1 The Past

On the grounds of the pioneering work with WGA, plant lectins recommended themselves as tools for the comparative mapping of tumor cell glycomes (for information on the fine specificity of a panel of plant lectins commonly used for this purpose, please see Table 18.1). Such type of analysis with lectins as sensors for glycan epitopes was performed at the level of cells and tissues (Table 25.1). Due to the influence of fixation and sample processing on cellular reactivity (organic solvents, for instance, can extract glycolipids and thus deplete cells of respective glycoconjugates), technical details are to be kept strictly constant in a comparative study and their consequences on the specimen noted to ensure reliable results. Although a large array of glycans is covered by plant lectins, typical branch-end structures, especially sialylated Lewis epitopes (for structural diagrams, please see Tables 7.4 and 27.2), are not. It was thus necessary to supplement the toolbox for glycan detection (Table 25.1).

Toward this end, the range of carbohydrate structures detectable by plant lectins was extended by adding monoclonal antibodies, opening the way, for instance, to

Table 25.1 The four classes of reagents used in glycocyto- and histochemical analysis.

Experimental aim	Type of reagent	Example
Detect certain aspects of glycosylation	Plant lectin/carbohydrate-specific antibody	Monitoring the presence of β1,6-branching in *N*-glycans or of sialylated Lewis epitopes
Detect accessible sites for binding a distinct carbohydrate epitope	Neoglycoconjugate	Detecting binding sites for sialylated Lewis epitopes in colon cancer
Detect distinct lectins *in situ*	Antibody specific for endogenous lectin	Performing immunohistochemical galectin fingerprinting in colon cancer (with prognostic relevance for galectins)
Detect accessible ligands (glycan/peptide) for an endogenous lectin	Tissue lectin	Delineating prognostic relevance for galectin-3 binding in sections of head and neck cancer specimens

studies on the mentioned epitopes (for details on their role in inflammation and on receptors of the selectin group of C-type lectins, please see Chapters 19 and 27.3; for structural details of a low-energy conformation of the sialyl-Lewis^x tetrasaccharide in complex with P-selectin, please see Figure 16.1h). With these two classes of reagents listed in Table 25.1 at hand, the evidence from early biochemical analysis given above was confirmed. Several aspects of glycan presentation in tumors deviate from normal cells, among them relatively increased degrees of branching and sialylation of *N*-glycans, enhanced presentation of short-chain mucin-type *O*-glycans with sialylation, and alterations in the abundance of histo-blood group AB0 and Lewis epitopes (for details on these structures, please see Chapters 1, 7 and 27) [4, 5]. The nature of these changes can vary with the tumor type so that the detected malignancy-associated deviations from controls should always be interpreted in the context of the analyzed tumor histology and never be extrapolated to other tumor classes.

At the molecular level, the discovery of oncogenes afforded the opportunity to further test the hypothesis that there is a connection between factors causing malignancy and glycosylation. Transforming cells with oncogenes should then induce aberrant glycan production. Indeed, this was the case, and a connection between oncogenes such as *ras* or *src* and transcriptional regulation of glycosyltransferases, especially *N*-acetylglucosaminyltransferase-V responsible for adding the β1,6-branch to complex-type *N*-glycans and α2,6-sialyltransferase-I adding sialic acid to branch ends (for details on *N*-glycan structure, please see Chapters 6 and 8, on KO models in Tables 23.1 and 27.4), was inferred to underlie increases in the respective substitutions [4, 5]. While surely contributing in a valuable manner to histopathological characterization and even delivering prognostic assessments in certain cases, the quest for a truly tumor-specific carbohydrate marker still continues. However, this search must not necessarily be successful to prove the importance of tumor cell glycosylation. Already quantitative changes in a few key aspects of global glycosylation may be able to make their presence felt as biologically potent features of the malignant phenotype. Moving from initial detection of glycome differences to ascribing functional relevance guides us from the past to the present. By doing so we will encounter two main routes glycans can take to exert an influence on cell properties.

25.2 The Present

The first mode by which tumor-associated glycans can have an impact on the establishment of the malignant phenotype is by modulating the functionality of the protein part of a glycoprotein. Evidently, if altered glycosylation shifts the distribution of charge, sialylation being a major factor, and conformational preferences of glycan chains (for details on modeling glycan parameters and the role of substitutions in the *N*-glycan core as modulators of conformational preferences, please see Chapter 2), cell surface glycoproteins such as integrins or growth factor receptors may no longer maintain their regular activity, as though being exposed to a blocking antibody. In other words, a deviation in glycan structure from the normal phenotype accounts for functional consequences ranging from modulating binding activity of a ligand to growth retardation of mice (Table 25.2). Indeed, the examples given in Table 25.2 from the study of models underscore that tumor-associated changes in glycan display bring about changes in cell behavior. Without requiring a tumor-specific event, shifts in glycosylation are thus not merely phenomenological aberrations (for further information on the relationship between glycosylation and disease *in vivo*, see Chapters 22 and 23; on therapeutic approaches by interfering with glycosylation, see Chapter 28.1). Obviously, glycan chains can act as molecular switches on protein activities in a glycoprotein by virtue of adopting distinct conformations [6]. Even seemingly subtle modifications, for example presence of core fucosylation conspicuously distant from branch-end sugars in extended antennae, will be effective (for details, please see Table 25.2), and there is a second functional dimension alluded to above. In fact, the same sensitivity of

Table 25.2 Examples for the impact of structural changes in the glycan part of cell surface glycoproteins on their functionality.

Protein	Experimental system	Altered glycosylation	Functional consequence
EGFR	Overexpression of GnT-III (Mgat3) in transfected human cervical adenocarcinoma HeLaS3 cells	Increased level of bisecting GlcNAc	Reduced EGF binding (low-affinity sites), increased internalization and ERK phosphorylation
EGFR	Downregulation of GnT-V (Mgat5) by small interfering RNA in metastatic and invasive human breast carcinoma MDA-MB231 cells and breast carcinoma cells from GnT-V (Mgat5) null mice	Reduced level of *N*-linked β1,6-branching	Suppression of FAK signaling despite unaltered EGF binding, increased receptor internalization, reduced signaling, tumor cell invasiveness and growth
EGFR	FucT-VIII overexpression in transfected human embryonic kidney HEK293 cells	Enhanced core fucosylation	Enhanced EGF-dependent cell growth and sensitivity to gefitinib (an EGFR inhibitor)
EGFR	Stable FucT-VIII knock-down by short hairpin RNA in human embryonic kidney HEK293 cells and human lung cancer A549 cells	Reduced core fucosylation	Decreased EGF-dependent cell growth and sensitivity to gefitinib

continued

Table 25.2 *Continued*

Protein	Experimental system	Altered glycosylation	Functional consequence
Fibronectin receptor ($\alpha_5\beta_1$-integrin)	Enzymatic de-*N*-glycosylation in human chronic erythroleukemia K562 cells	Reduced extent of *N*-glycosylation	Dissociation or altered association of $\alpha_5\beta_1$-subunits, inhibition of cell binding to fibronectin
Fibronectin receptor ($\alpha_5\beta_1$-integrin)	Inhibition of *N*-glycan processing (1-deoxymannojirimycin; swainsonine) in human fibroblasts and breast carcinoma cells from GnT-V (Mgat5) null mice	Reduced extent of complex-type *N*-glycosylation	Reduced cell and receptor binding to fibronectin; reduced fibronectin fibrillogenesis
Fibronectin receptor ($\alpha_5\beta_1$-integrin)	Differential desialylation affecting α2,8-linked oligosialic acid chains in human melanoma G361 cells	Cleavage of α2,8-linked sialic acid chains	Inhibition of integrin binding to fibronectin
Fibronectin receptor ($\alpha_5\beta_1$-integrin)	Reconstitution of expression of tumor suppressor p16^{INK4a} in human pancreatic carcinoma Capan-1 cells	Altered galactosylation and sialylation	Increased cell surface presentation and susceptibility to anoikis
β_1-Integrin in the fibronectin receptor ($\alpha_5\beta_1$-integrin)	Overexpression of GnT-V (Mgat5) in transfected human fibrosarcoma HT1080 cells	Increased level of *N*-linked β1,6-branching	Inhibition of integrin clustering, cell adhesion and spreading, increased cell migration
Glut-2	GnT-IVa (Mgat4a) null mice	Reduced *N*-glycan branching	Reduced Glut-2 half-life at cell surface, insulin resistance, type 2 diabetes
TGF-βR	FucT-VIII null mice	Reduced core fucosylation of *N*-glycans	Growth retardation *in vivo*, emphysema-like lungs, MMP overexpression and ECM downregulation
TrkA/NGFR	Overexpression of GnT-III (Mgat3) in transfected rat pheochromocytoma PC12 cells	Increased levels of *N*-linked β1,6-branching	Absence of NGF-induced TrkA phosphorylation and cell differentiation

The abbreviations used are: ECM, extracellular matrix; EGFR, epidermal growth factor receptor; ERK, extracellular signal-regulated kinase; FAK, focal adhesion kinase; FucT-VIII, α1,6-fucosyltransferase-VIII; GnT-III/IVa/V, *N*-acetylglucosaminyltransferases; Glut-2, glucose transporter 2; MMP, matrix metalloproteinase; NGF, nerve growth factor; NGFR, NGF receptor; TGF-βR, transforming growth factor-β receptor; TrkA, tropomyosin-related kinase A. For information on animal models, please see Chapter 23, especially Table 23.1.

glycan features to structural alterations holds true for the other mechanism as well, that is sugar epitopes as docking points for endogenous lectins, a process governing, for example, serum clearance of glycoproteins, a topic for rational therapeutic glycoengineering [6]. The readers are also already familiar with the interaction between tumor cell glycans and plant lectins (please see introduction). In view of the information on lectins in quality control during glycan remodeling and human lectins in general, given in Chapters 6 and 19, aberrant glycosylation may very well play into this mode of cellular communication.

Biochemical purification using affinity chromatography with immobilized sugar ligands demonstrated lectin expression in tumor cells [7] and, of particular relevance for tumor spread and immunosurveillance, by endothelial and inflammatory cells (for details on introduction of affinity chromatography to this field and its application, please see Chapters 15.3 and 18.3). This work guided the development of a histochemical method for detecting the presence of lectin activity in tissue sections. When conjugated at the anomeric position via a linker to an inert carrier such as albumin or a synthetic backbone such as substituted polyacrylamide for this purpose, the carbohydrate structures will maintain ligand capacity and thus be operative to trace lectin sites in any type of biological material (for information on neoglycoproteins and glycodendrimers, please see Chapter 4). This experimental approach to detect binding of carbohydrate is thus complementary to using plant lectins or monoclonal antibodies (Table 25.1). Explicitly, the screening for carbohydrate-binding activities is therefore readily feasible with these carrier-immobilized carbohydrates (neoglycoconjugates) in extracts, cells and tissue sections [8]. With the help of synthetic chemistry (please see Chapter 3) the ligand part can be tailored to meet any biological requirements, for example to visualize binding activity for a tumor-associated oligosaccharide, and the practice of this so-called reverse lectin histochemistry has delineated the presence of tissue lectins and tumor-associated changes in their expression [8]. With these data in hand, the experimental guideline is established for the ensuing purification of detected lectins and then the production of lectin-specific antibodies. When available, the neoglycoconjugates and lectin-specific antibodies are suitable tools to characterize the protein side of protein–carbohydrate interactions in tumors (Table 25.1). As noted above for plant lectin histochemistry, respective investigations (lectin fingerprinting) have led to insights ranging from phenotypic characterization to prognostic relevance [8]. These results imply the fundamental strategy of a coregulation of glycan and lectin for a functional interplay. Is there experimental evidence for the validity of this far-reaching assumption?

In the mentioned functional terms, pro- or antitumoral consequences of increased appearance of distinct glycans can indeed already be attributed to their ligand properties. In colon cancer, rich decoration of the glycoprotein carcinoembryonic antigen (CEA, also a tumor marker) by Lewis$^{x/y}$ epitopes appears to be capable of downregulating the maturation of dendritic cells. Here, their recognition through these cells' C-type lectin DC-SIGN has a bearing on their defence activity and sialylated Lewis$^{a/x}$ epitopes on glycoproteins promote carcinoma metastasis by interaction with endothelial selectins, while clusters of Lewis$^{a/b}$ epitope-containing termini of *N*-glycans can serve as targets for a defence factor of innate immunity – the serum collectin mannan-binding lectin [9] (for details about these human lectins, please see Chapter 19). This lectin 'reads' aberrant glycosignatures and also foreign surface displays of infectious organisms. Culturing colon cancer cells in a hypoxic environment also promotes expression of selectin ligands, with an assumed effect on hematogenous metastasis. In this case, transcriptional upregulation through the hypoxia-inducible factor is documented for genes of enzymes responsible for fucosylation and sialylation in the course of synthesis of selectin ligands [10]. The following three examples illustrate how glycan and lectin presentation are orchestrated to cause negative growth regulation:

a) Cell density-dependent growth inhibition of neuroblastoma cells (SK-N-MC) in culture rests on a remodeling of sugar-encoded signals. In detail, an activity increase of the cell surface ganglioside sialidase activity is central to initiating cell responses (Info Box). Desialylation of gangliosides elevates cell surface

Info Box

Gangliosides of the plasma membrane are potent modulators of diverse cellular functions including cell adhesion, cellular interactions, growth and differentiation as well as neuronal repair (please see Chapters 10 and 30). Thus, control of a cell's ganglioside signature is of profound functional relevance. Major sites of ganglioside synthesis are the endoplasmic reticulum and Golgi apparatus, and their degradation commonly takes place in lysosomes. Is it possible to remodel the ganglioside patterns on the cell surface? Indeed, the detection of a plasma membrane-bound ganglioside-specific sialidase helps shape the concept that desialylation on the cell surface is not simply a degradation, but a transition between sugar-encoded signals of different meaning. Its enzymatic activity produces a shift from higher sialylated species to ganglioside GM1 and a conversion of ganglioside GM3 to lactosylceramide (for details on ganglioside structure, please see Chapters 10 and 30). The activity of this enzyme is markedly induced during logarithmic growth of the neuroblastoma cells. To underline its role in cell growth, the presence of a specific enzyme inhibitor in the culture medium led to a release of the cells from contact inhibition and to a loss of differentiation markers. Further investigation of the underlying mechanism of decoding the sugar-encoded information revealed that binding of galectin-1 to ganglioside GM1 is the link between an altered ganglioside profile and the intracellular response (please see above and also Info Box 1 in Chapter 30). In structural terms, the lectin accommodates a distinct low-energy conformer of the pentasaccharide different from that bound by cholera toxin, an example for differential conformer selection by sugar receptors with relevance for drug design [H.-C. Siebert *et al.* Unique conformer selection of human growth-regulatory lectin galectin-1 for ganglioside GM1 versus bacterial toxins. *Biochemistry* 2003; **42**, 14762–14773] (for further details on combining experimental and computational techniques for defining receptor-bound conformations of glycans in solution, please see Chapters 2 and 13).

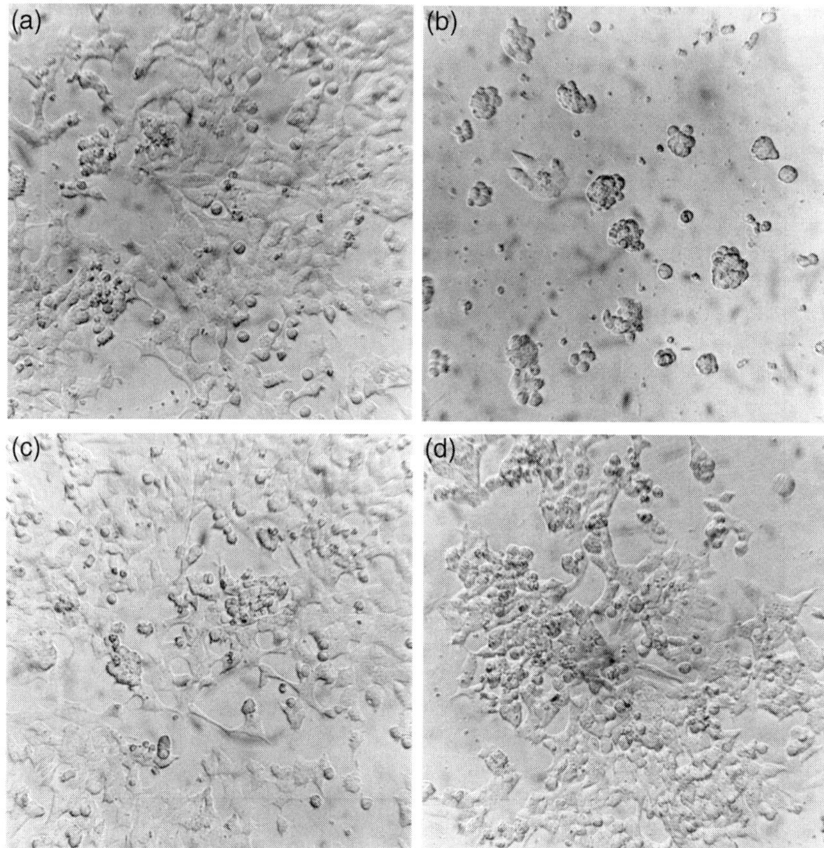

Figure 25.1 Illustration of effect of ganglioside GM1-binding human lectins on cell growth and morphology of human neuroblastoma cells (SK-N-MC) in culture. Untreated controls (a) are compared with cell populations exposed to galectins-1 and -3 at 125 μg/ml (b and c) and galectin-1 in the presence of a 10-fold excess of galectin-3 (D) for 48 h; magnification: ×125.

presentation of ganglioside GM1 (for details on glycosphingolipids, please see Chapters 10 and 30). Will there be an endogenous protein reading this sugar-encoded message? Figure 25.1 illustrates the effect that a cross-linking human lectin (galectin-1) has on cell cultures (for crystal structure of this lectin, please see Figure 13.2; a survey of its natural ligands is presented in Table 19.3). Importantly, this lectin is expressed by the neuroblastoma cells and its cell surface presentation follows the course of ligand generation – a convincing case of coregulation [11]. The same players can also originate from two different cell types, as is the case in cross-talk between T regulatory cells, the source of galectin-1, and T effector cells presenting the ganglioside [12]. The knowledge of the presence of other galectins (please see Chapter 19 and also Chapter 27.5) inspired us to probe into the possibility of functional antagonism. Indeed, the detection of competitive inhibition by another member of this lectin family, the

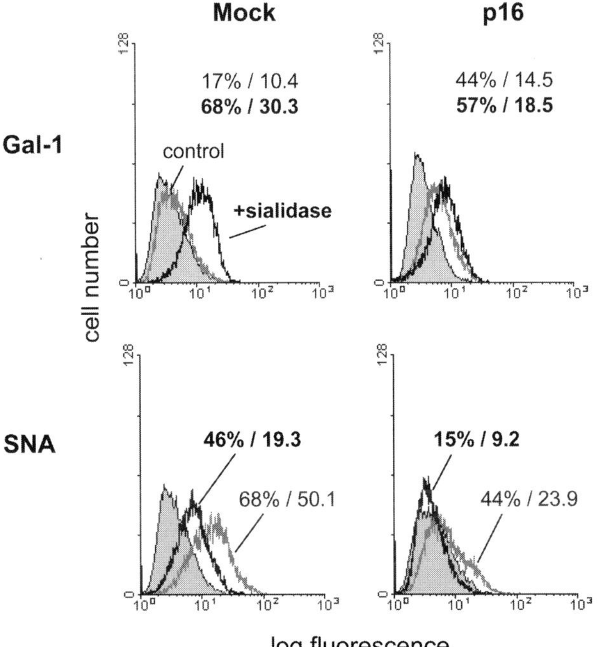

Figure 25.2 Illustration of lectin-binding properties of tumor suppressor-negative (Mock) and tumor suppressor-expressing ($p16^{INK4a}$) pancreatic carcinoma cells (Capan-1) by cytofluorimetric analysis. Quantitative data on percentage of positive cells and fluorescence intensity are given in each panel. The gray area represents the control in the absence of lectin, the gray line represents carbohydrate-dependent staining without removal of cell surface sialic acids by sialidase and the black line represents staining after enzymatic desialylation. The growth regulator galectin-1 (Gal-1) and the plant lectin *Sambucus nigra* agglutinin (SNA) recognizing α2,6-sialylated Gal/GalNAc epitopes were used as probes.

monomeric galectin-3, teaches the lessons that binding and ensuing cross-linking are required for the biological effect, and that functional antagonism among members of a lectin family is operative (Figure 25.1) [11]. That lectin pathways of growth regulation can be activated by master regulators of tumor growth, especially a tumor suppressor, is illustrated in the next paragraph.

b) Loss of integrity of the tumor suppressor $p16^{INK4a}$ is observed frequently in pancreatic cancer. In a model of human pancreatic carcinoma cells (Capan-1) the restoration of suppressor activity renders the cells susceptible to enter programmed cell death when detached from a matrix (anoikis). Figure 25.2 presents a clue that the activity of this suppressor encompasses an effect on cell surface glycosylation: the cell-staining profiles obtained with labeled human lectin revealed that suppressor-positive cells are strongly reactive with galectin-1, whereas control cells (Mock) must be subjected to a prior desialylation step to acquire reactivity. The differing extent of sialylation is documented using a plant lectin (*Sambucus nigra* agglutinin; for details on its sugar specificity,

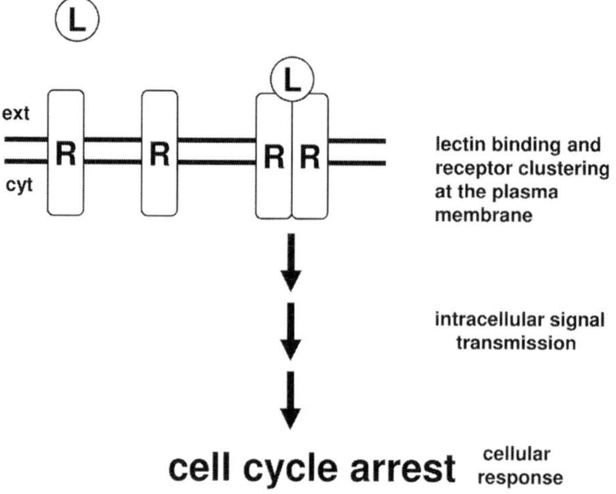

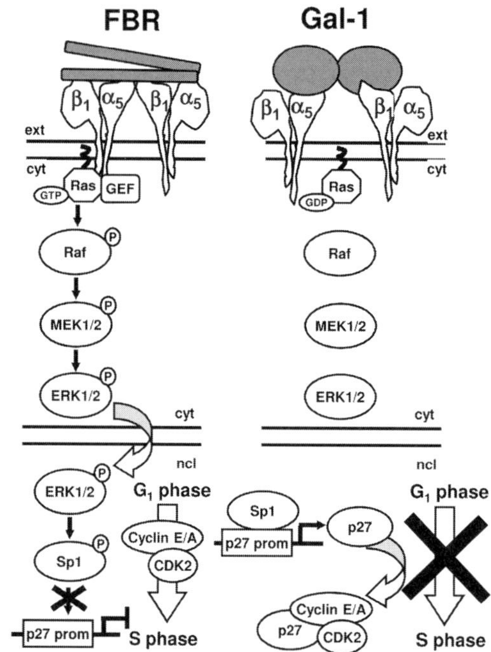

Figure 25.3 Illustration of signaling for cell cycle arrest by a lectin. (Top panel) Binding of a lectin (L), for example galectin-1, to a cell surface receptor (R) induces receptor clustering and activation. The ensuing signaling route results in a cellular response, for example cell cycle arrest. (Bottom panel) Illustration of signal transmission and cellular response to binding of the extracellular matrix protein [fibronectin (FBR)] or a lectin [galectin-1 (Gal-1)] to the fibronectin receptor ($\alpha_5\beta_1$ integrin). The specific interaction involves different sites of the integrin: peptide motifs for fibronectin and glycan motifs for the lectin. (Bottom panel, left column) The binding process induces integrin clustering and activation. This leads to recruitment of plasma-membrane-anchored Ras protein and its activation upon exchange of GDP by GTP, a process catalyzed by a guanine nucleotide exchange factor (GEF). The GTP-loaded Ras then initiates the sequential activation of a chain of protein kinases including Raf, mitogen-activated protein kinase (MAPK)/ERK kinases 1 and 2 (MEK1/2) and extracellular-signal-

please see Table 18.1) (Figure 25.2). As galectin-1 expression is at the same time upregulated, similarly noted in the studies on neuroblastoma, the status of sialylation acts as molecular switch-on signal for galectin-1 binding also in this case [12]. In detail, the tumor suppressor works via transcriptional upregulation of lectin/enzymes of the glycosylation machinery, and–at the same time–the cell surface presence of a likely effector target glycoprotein of galectin-1 (the fibronectin receptor) is elevated, thereby completing an orchestrated interplay between glycan remodeling and lectin expression [13]. How such a functional interplay is clinically manifested during tumor progression is presented in the third example.

c) Cutaneous T cell lymphoma cells in the Sézary syndrome are known to lose distinct surface markers in the course of disease progression, especially the CD7 antigen. Its absence renders these tumor cells resistant to the induction of apoptosis with galectin-1 as effector [14]. In contrast, $CD7^+$ cells undergo lectin-induced apoptosis so that the differential sensitivity to the tissue lectin may well be tied to the shift in subpopulations of this tumor type [14].

These three examples highlight the exquisite target specificity of an endogenous lectin for functional cell surface ligands (for further examples of orchestration of lectin–glycan expression in inflammation, please see Chapter 27). Despite sharing monosaccharide specificity, plant and human lectins cannot be expected to have identical fine-specificity profiles due to their structural disparities. Practically, the experimental data shown in Figure 25.2 illustrate that the binding characteristics of cell populations for an endogenous lectin can entail functional implications, in this case ascertained by measuring galectin-1-induced anoikis [13]. Thus, application of endogenous lectins as tool (Table 25.1) can be a salient step to track down lectin-triggered effector mechanisms. They will comprise a series of intracellular signaling steps leading from the initial binding step to the cellular response. An example for the intracellular signaling pathway following cell surface binding is shown in Figure 25.3 for the galectin-1-dependent cell cycle

regulated protein kinases 1 and 2 (ERK1/2) by sequential transphosphorylation. Active (phosphorylated) form of proteins is marked by a small-encircled P. The phosphorylated ERK1/2 translocates from the cytosol to the nucleus and here phosphorylates the transcription factor signaling protein 1 (Sp1). As a consequence, the binding of this protein to the promoter of the $p27^{KIP1}$ gene (p27 prom) is abrogated. The protein $p27^{KIP1}$ is a cyclin-dependent kinase (CDK) inhibitor. It acts on cyclin E/A-cyclin-dependent kinase 2 (CDK2) complexes blocking cell cycle progression. As a result of the downregulation of its expression, cells can now enter the S phase. (Bottom panel, right column) In contrast, the interaction of the galectin-1 homodimer with the integrin does not trigger engagement of Ras, which remains in its inactive (GDP-loaded) form. This prevents the activation of the MAPK pathway, its components remain inactive (unphosphorylated). Without being phosphorylated, Sp1 induces transcription of the $p27^{KIP1}$ gene after binding to its promoter (p27 prom). The produced $p27^{KIP1}$ protein (p27) then associates with cyclin E/A–CDK2 complexes, causing G_1 arrest. The abbreviations 'ext', 'cyt' and 'ncl', respectively, denote the extracellular, cytosolic and nuclear compartments. The two crosses signify that the respective process is blocked (for experimental details, please see [15]).

arrest of susceptible carcinoma cells [15] (for dissecting signaling routes in galectin-dependent induction of apoptosis in activated T cells as referred to in the third example given above, please see Figure 27.2; for further examples, please see [16]). The apparent involvement of an integrin by virtue of its glycans gives reason to revise a hitherto held paradigm. Integrins thus exploit not only peptide determinants in intermolecular recognition by peptide-peptide recognition. Because it is responsive to changes in expression profiles of genes in glycan production and tailoring, integrin glycosylation is a versatile biochemical signal, and, as Chapter 29 reveals, a case is even known for an integrin acting as lectin. At this stage, it is thus justified to conclude that glycans are functional ligands in malignancy (for further examples in other biological contexts, please see following chapters and also Table 19.2 on lectin functions; for the ligand activity of glycans in carbohydrate–carbohydrate interactions, please see Chapter 21). This clear answer to the question in this chapter's title gives reason to proceed to sketch future directions.

25.3 The Future

If altered glycosylation in malignancy is not simply an aberration without any functional dimension, then its different aspects will deserve to be rigorously examined in diverse classes of tumors: by glycomic profiling to define changes structurally, by glycoproteomics to define target proteins and by activity or lectin-binding studies to relate structure to function (for analytical aspects, please see Chapters 5 and 14). To add clinical relevance, case study of microsatellite-unstable colon cancer cells has even delineated impact of compensation of loss of genes not directly associated with the machinery for glycan generation on cell surface glycosylation [17]. Microarray technology combining the monitoring of the synthetic machinery for glycans and of lectins as well as of ligand properties of tumor-associated glycans will be instrumental in this research line to spot relevant changes and coregulation. The results emerging will likely provide practical guidelines for innovative therapeutic approaches by rational manipulation of glycan/lectin expression, for example targeting sialylation and rendering tumor cells vulnerable to lectin-mediated defence mechanisms (for further strategic aspects, please see Table 28.1). In principle, the concept to exploit natural growth-regulatory mechanisms via glycans has considerable appeal, giving respective efforts a clear and promising direction.

25.4 Conclusions

In the quest to define tumor features of functional importance, glycosylation is attracting increasing interest. The corresponding changes in malignancy are no longer considered as being purely phenomenological. In a broader biological context, the definition of certain glycan epitopes as oncofetal antigens by application of plant lectins/monoclonal antibodies means that they can embody signals equally relevant during normal development (please see previous chapter and also Chapters 22 and 23). Instead of constituting seemingly random deviations, the emerging evidence on tumor-associated glycans signifies that they – by (i) modulating protein activities and (ii) acting as docking sites for tissue lectins – are an integral part of the tumor phenotype. As the examples given above teach us, the accrued insights justify the perspective for devising innovative therapeutic approaches. Hereby, we are already moving beyond positively answering the question of this chapter's title.

Summary Box

Malignancy engenders multiple changes in glycosylation. The tumor-associated glycan remodeling can gain a functional dimension via (i) glycan-triggered activity changes of distinct glycoproteins such as growth factor receptors or integrins and (ii) lectin-mediated responses. Orchestration of glycan and lectin expression with functional consequences is documented. The insights drawn from model studies in tissue culture and from histopathology establish a perspective for innovative therapeutic approaches by rational manipulation of glycan/lectin expression.

References

1 Aub JC *et al.* Reactions of normal and tumor cell surfaces to enzymes I. Wheat-germ lipase and associated mucopolysaccharides. *Proc Natl Acad Sci USA* 1963;50:613–9.

2 Burger MM, Goldberg AR. Identification of a tumor-specific determinant on neoplastic cell surfaces. *Proc Natl Acad Sci USA* 1967;57:359–66.

3 Meezan E *et al.* Comparative studies on the carbohydrate-containing membrane components of normal and virus-transformed mouse fibroblasts. II. Separation of glycoproteins and glycopeptides by Sephadex chromatography. *Biochemistry* 1969;8:2518–24.

4 Hakomori S-I. Aberrant glycosylation in tumors and tumor-associated carbohydrate antigens. *Adv Cancer Res* 1989;52:257–331.

5 Brockhausen I *et al.* Glycoproteins and their relationship to human disease. *Acta Anat* 1998;161:36–78.

6 André S *et al.* From structural to functional glycomics: core substitutions as molecular switches for shape and lectin affinity of N-glycans. *Biol Chem* 2009;390:557–66.

7 Gabius H-J *et al.* Biochemical characterization of endogenous carbohydrate-binding proteins from spontaneous murine rhabdomyosarcoma, mammary adenocarcinoma, and ovarian teratoma. *J Natl Cancer Inst* 1984;73:1349–57.

8 Gabius H-J. Glycohistochemistry: the why and how of detection and localization of endogenous lectins. *Anat Histol Embryol* 2001;30:3–31.

9 Aarnoudse CA *et al.* Recognition of tumor glycans by antigen-presenting cells. *Curr Opin Immunol* 2006;18:105–11.

10 Koike T *et al.* Hypoxia induces adhesion molecules on cancer cells: a missing link between Warburg effect and induction of selectin-ligand carbohydrates. *Proc Natl Acad Sci USA* 2004;101:8132–7.

11 Kopitz J *et al.* Negative regulation of neuroblastoma cell growth by carbohydrate-dependent surface binding of galectin-1 and functional divergence from galectin-3. *J Biol Chem* 2001;276:35917–23.

12 Wang J *et al.* Cross-linking of GM1 ganglioside by galectin-1 mediates regulatory T cell activity involving TRPC5 channel activation: possible role in suppressing experimental autoimmune encephalomyelitis. *J Immunol* 2009;182:4036–45.

13 André S *et al.* Tumor suppressor $p16^{INK4a}$: modulator of glycomic profile and galectin-1 expression to increase susceptibility to carbohydrate-dependent induction of anoikis in pancreatic carcinoma cells. *FEBS J* 2007;274:3233–56.

14 Rappl G *et al.* $CD4^{+}CD7^{-}$ leukemic T cells from patients with Sézary syndrome are protected from galectin-1-triggered T cell death. *Leukemia* 2002;16:840–5.

15 Fischer C *et al.* Galectin-1 interacts with the $\alpha_5\beta_1$ fibronectin receptor to restrict carcinoma cell growth via induction of p21 and p27. *J Biol Chem* 2005;280:37266–77.

16 Villalobo A *et al.* A guide to signaling pathways connecting protein-glycan interaction with the emerging versatile effector functionality of mammalian lectins. *Trends Glycosci Glycotechnol* 2006;18:1–37.

17 Patsos G *et al.* Compensation of loss of protein function in microsatellite-unstable colon cancer cells (HCT116): a gene-dependent effect on the cell surface glycan profile. *Glycobiology* 2009;19:726–34.

26
Small Is Beautiful: Mini-Lectins in Host Defense

Robert I. Lehrer

Having already illustrated the emerging roles of glycans as biochemical signals in fertilization, early embryogenesis and malignancy, we will next consider how some antimicrobial peptides – an ancient means of host defense – interact with glycans. Table 19.2 lists endogenous lectins that recognize foreign glycosignatures, among them the defensins. α-Defensins, with 29–35 amino acid residues, and θ-defensins, with only 18 residues, are considerably smaller than hevein, a mini-lectin found in plants, as described in Chapter 18 (see also Chapter 15 for definition). This chapter will focus on θ-defensins – among the smallest known lectins and the only cyclic peptides of animal origin. However, before you meet them, we will introduce their older relatives.

26.1
Meet the Families

The defensin peptides of vertebrate animals comprise three subfamilies, called α-, β- and θ-defensins. The sequences of selected α- and β-defensin peptides appear in Figure 26.1. Human α-defensin (*DEFA*) genes cluster on the short arm of chromosome 8 (8p21), with many β-defensin (*DEFB*) genes located nearby. This subtelomeric location favors gene reduplication, and several *DEFA* and *DEFB* genes show marked copy number polymorphism. *DEFA* genes encode pre-pro-peptides containing a negatively charged pro-domain and positively charged C-terminal defensin domain. After posttranslational processing, mature α-defensin peptides have largely β-sheet structures that are stabilized by three intramolecular disulfide bonds. The α- and β-defensins have similarly folded peptide backbones, but show limited amino-acid identity.

Three of the six human α-defensin peptides (HNP-1–3) have identical sequences, except for their N-terminal residue. Collectively, these three peptides comprise 5–7% of total protein of a human neutrophil [polymorphic neutrophils (PMNs)]. For any readers who are curious about the origin of the word 'neutrophil' and its abbreviation as 'PMN', both terms can be traced to the pioneering studies of Paul

The Sugar Code. Edited by Hans-Joachim Gabius

ISBN: 978-3-527-32089-9

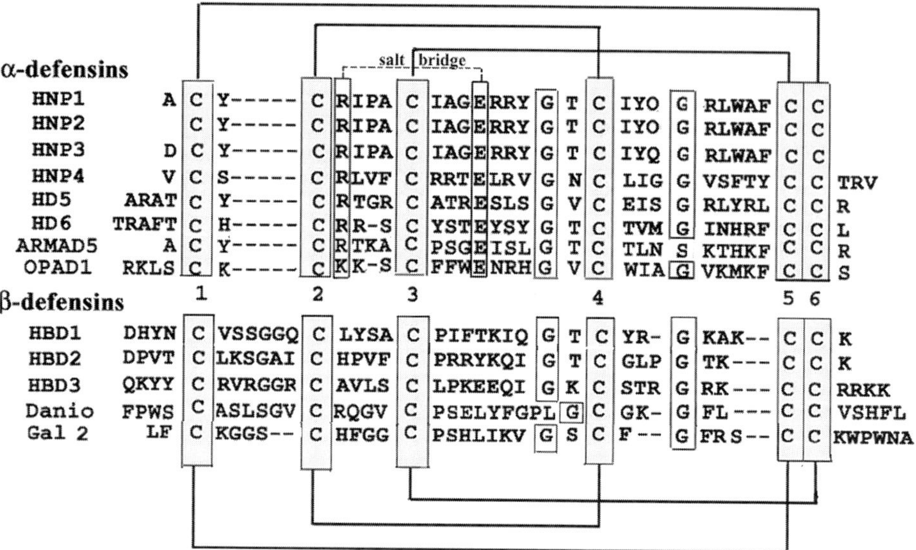

Figure 26.1 α- and β-Defensins. Six human α-defensins (HNP-1-4 and HD-5 and -6) and three human β-defensins (HBD-1–3) are shown. Conserved residues are boxed. Cysteine residues are numbered, and their pairing is shown by solid lines. A dashed line indicates a salt bridge in HNP-1–3. ARMAD5 is from the nine-banded armadillo, a basal mammal, and OPAD1 is from the opossum, a marsupial. Gaps (–) were introduced to maximize the alignments.

Ehrlich, who called attention to the polymorphic shape of the cell's nucleus and the neutrophilic (neither acidophilic nor basophilic) staining properties of its granules. The six α-defensins of rabbit PMNs are even more abundant, accounting for over 15% of the cell's total protein. The PMNs of cattle and fowl lack α-defensins, but contain multiple β-defensins. Mice (but not rats), pigs, sheep and horses have PMNs that lack α- and β-defensins. Instead, these PMNs contain other antimicrobial peptides, including many that belong to the cathelicidin family.

Figure 26.1 shows three human β-defensins (HBD-1–3), plus one each from the zebrafish (*Danio*) and the chicken (Gal 2). The existence of defensins in a basal mammal, the armadillo, and the opossum, a marsupial, indicates that the lineage of α-defensins extends back at least around 135 million years. β-Defensins are at least 300 million years older than α-defensins, judging from their presence in zebrafish. The α- and β-defensin peptides have several structural differences. Whereas Cys1 pairs with Cys6 in α-defensins, it pairs with Cys5 in β-defensins. Their inter-cysteine spacing also differs, with a single residue between Cys1 and Cys2 in α-defensins, and five or six intervening residues in β-defensins. In contrast, Cys4 and Cys5 are more widely separated in α-defensins. These reciprocal deletions and insertions permit Cys1–Cys6 pairing in α-defensins, but eliminate the short, α-helical segment found in many β-defensins. Much more ancient defensin-like peptides also exist (see Info Box 1); however, their genetic kinship to the defensins mentioned above is uncertain.

Info Box 1

Defensin-like antimicrobial peptides (DLPs) are expressed by fungi, plants and many invertebrates, including ticks, spiders, scorpions, mosquitos, dragonflies, crustaceans and mollusks. These DLPs contain six cysteines and have a conserved structural motif consisting of two antiparallel β-strands, one of which is attached to an α-helical domain. Such peptides are called cysteine-stabilized (CS) αβ-defensins. Plectasin, a CSαβ-defensin isolated from a forest fungus, shares considerable structural and sequence identity with several invertebrate CSαβ-defensins. Data base searches have uncovered additional families of fungal DLPs, and even found CSαβ-defensin-like peptides in a few myxobacteria.

26.2 Where Do α- and β-Defensins Reside?

Clues to the functions of defensins come from the cells that contain them, the stimuli that promote their synthesis and secretion, their *in vitro* properties, and their *in vivo* behavior. HNP-1–3 are most prominent in leukocytes, including PMNs, natural killer cells and certain T lymphocytes, such as γδ T cells. PMNs are phagocytic cells that can ingest and kill bacteria and fungi, especially after the microbial surfaces are modified (opsonized) by the attachment of antibody, complement or certain lectins, including the collectin mannan-binding lectin (MBL; a C-type lectin, please see Chapter 19 and Figure 20.3). Human PMNs store fully processed α-defensin peptides in cytoplasmic structures called 'azurophil granules', organelles whose contents also include other antimicrobial proteins. These granules can be delivered to vacuoles that contain ingested microbes or their contents can be released into the blood or tissues. Human PMNs are remarkably short-lived cells that circulate for less than a day ($t_{1/2} < 7\,h$) in the blood. Consequently, humans must produce at least 10–15 mg HNP-1–3/kg body weight/day just to equip new circulating PMNs under basal conditions and then increase this production rate when microbial infection occurs. When we first began studying human α-defensins, we wondered if their small size (less than 3.5 kDa) might result in their renal excretion. Accordingly, we collected large amounts of urine from patients with chronic myelogenous leukemia only to be disappointed by the absence of HNP-1–3. Serum levels of HNP-1–3 are very low (less than 0.1 μg/ml) and what happens to 'yesterday's HNPs' is unknown. *In vitro*, epithelial cells and macrophages acquire α-defensins from the medium, so if similar uptake of α-defensins occurs *in vivo*, it could reinforce the ability of such cells to resist infection. Human PMNs contain a fourth α-defensin, HNP-4, that is present in much smaller amounts than HNP-1–3. Although HNP-4 is also antimicrobial, its raison d'être is not yet understood.

this stop codon would abort translation, its presence explained why human PMNs lacked θ-defensin peptides. Later we found that the *DEFT* genes of chimpanzees and gorillas carried an identical silencing mutation, but most orangutan *DEFT* genes were intact [2]. This time-line suggests that our ancestors had lost the ability to make θ-defensins at least 7.5 million years ago.

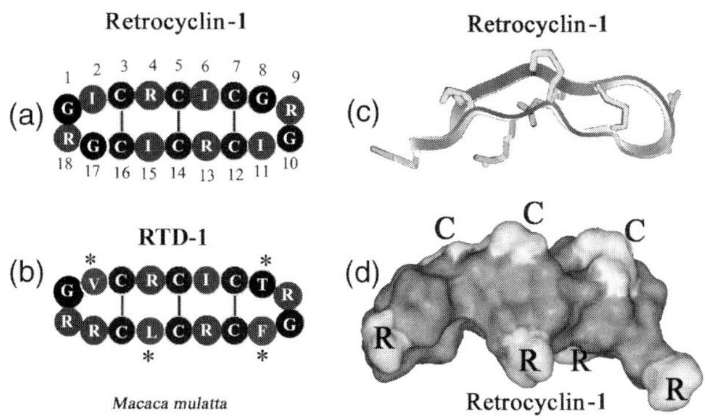

Figure 26.3 θ-Defensins. RC-1 appears in panels (a), (c) and (d), and RTD-1 in panel (b). The four RTD-1 residues that differ from those in RC-1 are marked with an asterisk. Structural hallmarks of θ-defensins include a cyclic backbone, a tridisulfide ladder that connects their antiparallel β-sheets, at least four arginines and no anionic residues.

From the RTD studies, we learned which pro-peptide residues end up in the mature peptide. Our human bone marrow cloning studies had exhumed an ancestral θ-defensin mRNA [3] and we later found additional human *DEFT* pseudogenes by database searching [2]. The retrieved sequence information allowed us to synthesize θ-defensins perhaps last produced by an ancestral protohominid. We named the new/old peptides retrocyclins (RCs) to commemorate their antiquity (retro = backwards or behind) and their cyclic backbone. RC-1 and RTD-1 are compared in Figure 26.3. The antimicrobial activity of RC-1 resembled that of HNP-1–3. However, when we found that RC peptides protected human cells from HIV-1 infection this became a focus of our work and led us to recognize their lectin-like properties, which are illustrated below in an old-fashioned way.

26.5 Hemagglutination

The historical survey in Chapter 15 reminds us of the important role of hemagglutination assays in the classical era of lectin research. Although we now have many more ways to examine protein–carbohydrate interactions (please see Chapter 14), hemagglutination assays retain the virtue of simplicity. Accordingly, we tested the ability of 22 different defensins to agglutinate human red blood cells. Mindful of the century-old advice of Silas Weir Mitchell, a distinguished neurologist, scientist and writer (see the table and first paragraphs in Chapter 15 for his prominent role in the history of lectinology), we used washed red cells and performed these assays in serum-free phosphate-buffered saline. Under these conditions, all six θ-defensins (RC-1–3 and RTD-1–3) caused hemagglutination (Figure 26.4). The addition of 30 mM cellobiose or lactose (for the structures of cellobiose/lactose, please see Chapter 1) had no effect, but adding 30 mM chitobiose or 10% normal human serum completely prevented hemagglutination. Chitobiose (see also Chapter 18 for chitobiose-binding plant lectins) is a characteristic component of chitin (see Chapter 12) which occurs in fungal cell walls, insect and arthropod cuticles, and *N*-linked glycans. Among the other defensins, only rabbit α-defensin NP-1 (net charge +10), human β-defensins HBD-2 (net charge +5) and human HBD-3 (net charge +13) caused hemagglutination. In later surface plasmon resonance experiments, HBD-2 and -3 showed little or no ability to bind carbohydrates, suggesting that hemagglutination resulted from their high net positive charge. The jury remains undecided about NP-1, awaiting further experimental results. To exclude the possibility that RC-1 acted through a charge-related mechanism, we compared RC-1 with its *enantio* (RC-112) and *inverso* (RC-111) analogs, all of which had a net charge of +4. RC-1 and -112 caused hemagglutination, but RC-111 did not (Figure 26.5).

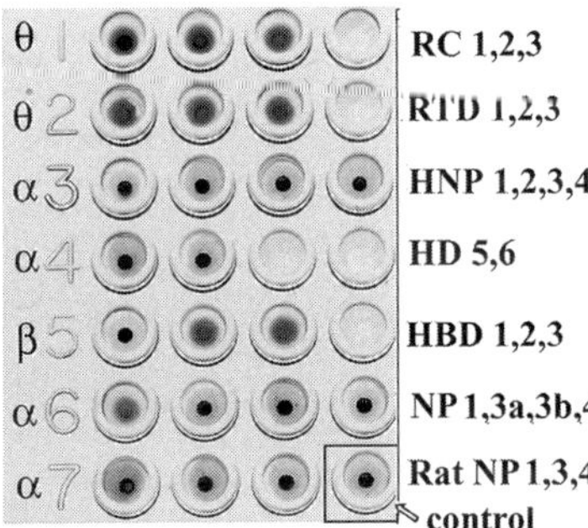

Figure 26.4 Hemagglutination by defensins. Washed normal human erythrocytes in phosphate-buffered saline were mixed with 50 μg/ml of defensin and allowed to settle. Agglutinated red cells covered a large area at the bottom of the microwell. Nonagglutinated red cells formed a tight button at the center of the well, as seen in the peptide-free control (arrow). The defensin subfamily (α, β or θ) is indicated next to the row numbers. *Abbreviations:* RC, retrocyclin; RTD, rhesus θ-defensin; HNPs and HDs, human α-defensins; HBDs, human β-defensins; NP, rabbit α-defensin; RatNP, rat α-defensin. The rabbit and rat defensins were purified from leukocytes and the other defensin peptides were synthetic.

Although a red blood cell's surface looks smooth and featureless through a regular microscope, it is covered by a dense thicket of glycoproteins called a glycocalyx [4]. Glycophorin A (GpA), with over 1 million copies on each erythrocyte, is among the most abundant molecules in this carbohydrate-rich surface coat. GpA, a transmembrane protein that is around 60% carbohydrate by weight, is bound by θ-defensins. RCs also self-associate and this property should allow them

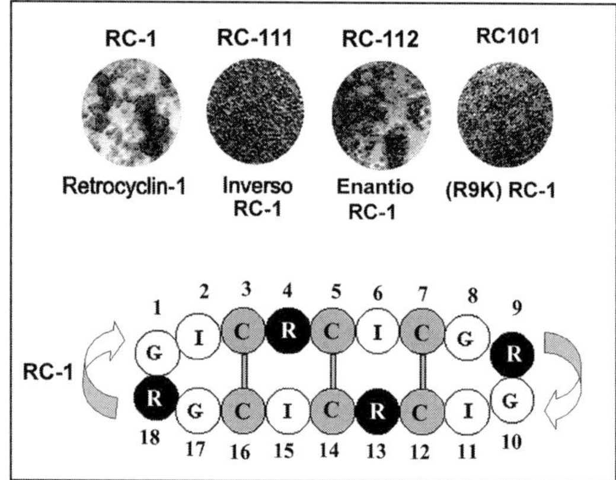

Figure 26.5 Effect of net charge, polarity and structure. RC-1 and -112 are enantiomers. RC-111, the *inverso* analog of RC-1, has the following sequence: 1-RGCICRCIGRGCICRCIG-18. RC101 and -1 are identical, except that residue 9 of RC101 is lysine. The rabbit erythrocytes used in this study were photographed through a microscope. Mouse and rat red cells were also hemagglutinated by RC-1. RC101 does agglutinate human RBC. Hemagglutination of human erythrocytes was independent of the presence or absence of common blood groups, such as ABO, Lewis, Kell, Rh, Duffy and others. Residue numbering is based on the synthetic peptide precursor of RC-1.

to cross-link other θ-defensin molecules that are bound to GpA. The ability of chitobiose to block θ-defensin-mediated hemagglutination is noteworthy, since more ancient, defensin-like antifungal peptides of plants, such as hevein and Ac-AMP2, also have an affinity for chitobiose and chitin – molecules that are prominent components of fungal cell walls and insect exoskeletons (see Chapter 18.5 for information on the potential roles of plant lectins in host defense).

26.6 How HIV-1 Enters Target Cells

To initiate infection, viruses must enter a permissive host cell, subvert its biosynthetic machinery, and force it to make and release new viruses to infect additional cells. HIV-1 usually infects cells bearing a surface receptor called CD4, that otherwise mediates interactions between T cells and antigen-presenting cells (please see Chapter 17.2.3). To be susceptible to infection by HIV-1, $CD4^+$ cells must also display a coreceptor: either CXCR4 or CCR5 [3]. As RC-1 protected $CD4^+$ human lymphocytes from infection by HIV-1 strains requiring either coreceptor, we assumed it would bind either CD4 or gp120, the surface glycoprotein of HIV-1 that interacts with CD4. To our surprise, RC-1 bound gp120, CD4 and galactosylceramide (a cell-surface glycolipid) with very high and almost equal affinity [5].

Since gp120, CD4 and galactosylceramide are glycosylated, we wondered if RC-1 was binding to the carbohydrate moieties of these structurally distinct molecules.

Binding of RC-1 to gp120 and CD4 was greatly reduced once the glycoproteins were enzymatically deglycosylated. The converse was also true, since θ-defensins bound neoglycoproteins (for information on neoglycoconjugates, please see Chapters 4 and 27) more than unmodified albumin. Cyclic analogs of RC-1 lacking its internal tridisulfide ladder had little ability to bind glycoproteins or neoglycoproteins. Noncyclic RC-1 analogs with the tridisulfide ladder and a β-hairpin structure were also relatively ineffective. From these observations, we concluded that RCs had lectin-like activity, and that both their cyclic backbone and their disulfide ladder contributed to this property [5]. Since other lectins, most notably cyanovirin, had been shown to afford protection against HIV-1 (and other viruses) by preventing viral entry, we felt confident that we were on the right path to showing how θ-defensins prevented HIV-1 entry.

We tested many RC-1 analogs, including its *enantio*, *retro* and *retroenantio* counterparts, and variants in which noncysteine residues of RC-1 were replaced one-by-one or in groups by other residues. The ability of these analogs to inhibit entry of HIV-1 correlated well with their ability to bind gp120 and/or CD4 [5].

26.7 Studies with Influenza A Virus

RC-2 potently inhibits infection by influenza A virus (IAV), again acting as an entry inhibitor [6]. Unlike HIV-1, the entry of IAV is initiated by binding interactions between the hemagglutinin of the virus and cell-surface sialic acid receptors of the target cell, and viral internalization takes place by subsequent endocytosis (please see also Chapter 17.2.1). Fusion between the viral envelope and the endosomal membrane allows IAV to enter the cell. RC-2 did not change viral binding, inhibit endocytosis or directly inactivate the virions. Intact disulfide bonds were necessary for activity against IAV, as they were for activity against HIV-1. How then was fusion inhibited?

Fusion mediated by viral glycoproteins is a multistep process that depends on close contact between the viral and cell membranes. Normally, the virus–cell contact zone is crowded with membrane proteins that cover both of the membranes. To consummate the fusion process, these membrane proteins must be displaced from the future fusion site. Binding to and then cross-linking the surface glycoproteins by RC-2 frustrates this process by preventing the displacement of proteins from the fusion site. A similar mechanism allows RC-2 to inhibit fusion mediated by the glycoproteins of Sindbis virus and baculovirus, suggesting that the antiviral effects of RCs do not arise from interactions specific to any one class of viral or cellular (co)receptors or viral fusion proteins. This mechanism is not unique to RC-2, since both HBD-3 and MBL also use this mechanism to block viral fusion (for further information on the collectin MBL, please see Chapter 19).

Any reader who has remained with this chapter has no doubt noticed the author's circumforaneous (you can look it up) tendencies and will not be surprised that he is about to take you on another side trip. There is a plaque on his office door that states 'Not all who wander are lost'. If this observation is correct, the excursion may be relevant. If not, at least the excursion will be short.

26.8 A Toxic Side-Trip

In addition to their antimicrobial, antiviral and lectin-like properties, θ-defensins (and α-defensins) inhibit several bacterial exotoxins, none of which is glycosylated. For example, anthrax lethal factor and botulinum neurotoxin A are zinc-dependent proteases. The α- and θ-defensins bind them, and neutralize their toxicity by inhibiting their enzymatic activity in a noncompetitive manner [7]. They also inactivate listeriolysin-O and pneumolysin, which are nonenzymatic, pore-forming toxins. Low micromolar concentrations of RC-1 protect cells by binding and cross-linking the toxin monomers, and preventing their assembly into pores (unpublished data). Since none of the aforementioned toxins are glycosylated, the protective effects of θ-defensins cannot be attributed to their lectin-like behavior.

26.9 And Now, the Surprise

As mentioned above, the ability of RCs to prevent HIV-1 entry and their lectin-like properties were strongly correlated. One might assume (as we did) that the protective mechanism delineated in our studies with IAV would apply to HIV-1. However, RC-1 inhibited HIV-1 envelope-mediated fusion without either cross-linking membrane glycoproteins, as in IAV, or inhibiting gp120–CD4 interactions [8]. Instead, RC-1 acted late in the HIV-1 Env fusion cascade, just prior to 6-helix bundle formation, by binding to the C-terminal heptad repeat of gp41 and preventing 6-helix bundle formation – an integral component of the entry process. Binding of RC-1 to the heptad repeat was selective, showed high affinity and (here comes the surprise) it was glycan independent.

Why were the carbohydrate-dependent binding of RCs and their ability to prevent viral entry strongly correlated? Perhaps the molecular features of RCs that promote high affinity binding to anionic residues in heptad repeat-2 also contribute substantially to its carbohydrate-binding properties. The θ-defensins contain only 18 residues. Of these, 10 (six cysteines and the four arginines common to RC-1 and RTD-1) were also conserved in the *DEFT* genes found in other primate species [2]. The importance of the cysteine disulfide bonds for carbohydrate binding and activity against HIV-1 has already been mentioned [3, 5]. Molecular modeling and studies with RC analogs indicated that the four conserved arginines were strategically positioned to interact with highly conserved anionic residues found in HIV-1 [8]. In HIV-2, these anionic residues are not conserved and this retrovirus is, in fact, considerably more resistant to RCs. The highly mutable genome of HIV-1 favors the emergence of resistant mutants when selection pressure is exerted by antiretroviral agents. Passaging HIV-1(BAL) serially under pressure by RC-101 caused relatively minor changes in viral susceptibility. Emergent viral isolates had three amino acid changes in their envelope glycoprotein, one was in a CD4-binding region of gp120, and two in heptad repeat domains of gp41. Each mutation replaced a neutral or electronegative residue with one that was positively charged. Single-site mutations in heptad repeat-2 replacing anionic residues with cationic ones, often generated viruses with impaired infectivity [9].

26.10
It Takes Two to Tangle

Can one very small molecule use different mechanisms to prevent HIV-1 and IAV? Our answer will begin with an easily overlooked technical detail. The studies with IAV [6] were performed under serum-free conditions and the HIV-1 studies were done in medium containing fetal bovine serum. These conditions were chosen because IAV infects host cells on serum-free surfaces in the upper or lower airways and because the HIV-1 targets required serum-containing media for optimal growth and survival (for influenza hemagglutinins and therapy by sialidase inhibitors, please see Chapters 17.3.1 and 28.2).

Fetuin is a heavily glycosylated protein that is one of the most abundant proteins in fetal bovine serum. It harbors three sites for *N*-glycosylation (complex-type, bi- and triantennary *N*-glycans) and also *O*-glycosylation (core 1) (for details, please see Chapters 6 and 7). Fetuin binds to RCs with high affinity and competes for its binding to cell-surface glycoproteins [5]. Under the serum-free conditions, RCs formed a tangled barrier of surface glycoproteins whose presence barred the entry of IAV virus [6]. RC-mediated hemagglutination (Figures 26.4 and 26.5) probably resulted from entangled surface glycoproteins and was also inhibited by serum. The reversible cross-linking of surface glycoproteins by θ-defensins is a form of polymerization in which the monomers (glycoproteins) cross-linked by θ-defensins form a lattice. By diverting θ-defensins from the cell surface or otherwise acting as lattice breakers, free-floating glycoproteins can prevent barrier formation. Thus, in a serum-containing environment, a mechanism based on electrostatic binding to heptad repeat-2 of gp41 will predominate, whereas a barrier mechanism can exist in other environments.

26.11
Which Human α- and θ-Defensins Are Lectins?

The author has investigated α-defensins for three decades, and lived with them personally for seven, so he will not feign indifference to these peptides. Perhaps because α-defensins are expressed in humans *and* mice, whereas θ-defensins are expressed only by non-human primates, α-defensins have received considerably more attention. Human α-defensins have antimicrobial, antiviral, cytokine-like and antitoxic properties. Some of them also have lectin-like properties. We could have recognized their carbohydrate-binding properties over two decades ago, had we stopped to think about the curious behavior of HNP-1–3 when subjected to size-exclusion chromatography on a Sephacryl S-200 column. Sephacryl is a tradename for a cross-linked copolymer of allyl dextran and *N,N*-methylenebisacrylamide, and dextran is a branched polysaccharide composed of glucose molecules. HNP-1–3 eluted from our S200 column very late, with an elution volume considerably larger than the column's bed volume. In contrast, HNP-4 eluted earlier, and within the column's bed volume. In describing this behavior, we attributed the delayed elution of HNP-1–3 to 'nonspecific interactions between these peptides and the gels matrix'. Neoglycoproteins as tools (please see Chapter 4.3) document HD-5's capacity to home in on carbohydrates in multivalent interactions [10]. Among the human α-defensins, only HNP-1–3 and HD-5 are lectins. Info Box 2 lists five reasons why an ability to bind carbohydrates could be useful for α-defensins. The other human α-defensins, HNP-4 and HD-6, had little or no affinity for carbohydrates in our studies.

Although HBDs 1–3 did not exhibit carbohydrate-binding properties in our studies, they represent fewer than 10% of all the β-defensins in the human genome. In rodents, many β-defensins are expressed in the male reproductive tract, especially in the epididymis. Alternative splicing has enabled one rat epididymal β-defensin, called 'E3', to acquire a lectin domain that resembles one found in wheat germ agglutinin [11]. Rat spermatozoa are coated with this hybrid lectin–

Info Box 2

Five possible reasons why HNP-1–3 bind carbohydrates are:

(i) Since they are stored in azurophil granules that have a peptidoglycan-rich matrix, electrostatic and lectin-like interactions with this matrix may prevent their premature release.

(ii) The affinity of secreted HNP-1–3 for cell-surface glycoproteins and glycolipids should concentrate these peptides at a cell's surface, positioning them optimally to intercept incoming bacteria, viruses or toxins.

(iii) HNP-1–3 is chemotactic for monocytes, immature dendritic cells and certain T cells. The gradual elution of defensins from cell surface glycomolecules can establish a longer lasting chemotactic gradient.

(iv) The small size of defensins makes them subject to rapid clearance by renal glomerular filtration, unless they bind larger serum glycoproteins.

(v) Immunoglobulins (IgG, IgM and IgA) are glycoproteins that bind HNP-1–3 reversibly and with high affinity (unpublished). Perhaps defensin-carrying immunoglobulins are a natural embodiment of Ehrlich's century-old 'magic bullet' concept.

defensin molecule, which may protect sperm against microbes encountered during their journey to an ovum.

Although a detailed description of the sugar specificity of α- and θ-defensins cannot be provided at this time, it clearly conforms more to the 'Hello sailor, new in town?' model than to a highly selective one. Since these defensins must be able to interact with a myriad of different pathogens, their ability to recognize multiple different pathogen-associated molecular patterns is advantageous. Highly refined information on the sugar specificity of lectins larger than defensins will be presented in the next chapter.

26.12 Conclusions

Defensins are host defense peptides that have been around, in one form or another, for over a billion years. This chapter deals with θ-defensins, cyclic octadecapeptides whose antiparallel β-sheets are cross-connected by a tridisulfide ladder. θ-Defensins, which are expressed only by nonhuman primates, have antimicrobial, antitoxic and antiviral properties. Being among the smallest currently known lectins (please see also Chapter 19 on hevein-like domains) [12, 13] allowed θ-defensins to be included in this textbook. Their small size and simple structure suggests that these miniature lectins could also serve as molecular templates for developing novel therapeutic and prophylactic agents.

Summary Box

The θ-defensins – until recently, the smallest known lectins – are cyclic octadecapeptides expressed by the leukocytes of non-human primates. Ancestral hominid θ-defensins known as RCs can protect human cells from infection by HIV-1 and IAV viruses, acting via different mechanisms that directly or indirectly use their carbohydrate-binding properties. Their small size and unique structural characteristics make θ-defensins intriguing templates for designing novel antimicrobial, antiviral and antitoxic agents.

References

1 Tang YQ *et al.* A cyclic antimicrobial peptide produced in primate leukocytes by the ligation of two truncated α-defensins. *Science* 1999;286:498–502.
2 Nguyen TX *et al.* Evolution of primate θ-defensins: a serpentine path to a sweet tooth. *Peptides* 2003;24:1647–54.
3 Cole AM *et al.* Retrocyclin: a primate peptide that protects cells from infection by T- and M-tropic strains of HIV-1. *Proc Natl Acad Sci USA* 2002;99:1813–8.
4 Roseman S. Reflections on glycobiology. *J Biol Chem* 2001;276:41527–42.
5 Wang W *et al.* Retrocyclin, an antiretroviral θ-defensin, is a lectin. *J Immunol* 2003;170:4708–16.
6 Leikina E *et al.* Carbohydrate-binding molecules inhibit viral fusion and entry by cross-

linking membrane glycoproteins. *Nat Immunol* 2005;6:995–1001.

7 Wang W *et al.* Retrocyclins kill bacilli and germinating spores of *Bacillus Anthracis* and inactivate anthrax lethal toxin. *J Biol Chem* 2006;281:32755–64.

8 Gallo SA *et al.* θ-Defensins prevent HIV-1 Env-mediated fusion by binding gp41 and blocking 6-helix bundle formation. *J Biol Chem* 2006;281:18787–92.

9 Fuhrman CA *et al.* Retrocyclin RC-101 overcomes cationic mutations on the heptad repeat 2 region of HIV-1 gp41. *FEBS J* 2007;274:6477–87.

10 Lehrer RJ *et al.* Multivalent binding of carbohydrates by the human α-defensin, HD5. *J Immunol* 2009;183:480–90.

11 Rao J *et al.* Cloning and characterization of a novel sperm-associated isoantigen (E-3) with defensin- and lectin-like motifs expressed in rat epididymis. *Biol Reprod* 2003; 68:290 301.

12 Gabius HJ *et al.* The chemical biology of the sugar code. *ChemBioChem* 2004;5:740–64.

13 Li J *et al.* Odorrana lectin is a small peptide lectin with potential for drug delivery and targeting. *PLoS ONE* 2008;11:e2381.

27
Inflammation and Glycosciences

Reinhard Schwartz-Albiez

Carbohydrate–lectin interactions play a decisive role in many immunological reactions. In the preceding chapter the protective role of α/θ-defensins as part of the innate immune system was described. These are smallest lectins known with target selectivity for glycans of pathogens. Functions of animal and human lectins and structural requirements for their binding behavior are explained in Chapter 19 (summarized in Table 19.2, see also the X-ray structures of two C-type lectins and their carbohydrate ligands in Figure 16.1h,i and Figure 20.3). Here, we see how these carbohydrate–lectin interactions are integrated into the complex network of immune reactions during inflammation and against infectious pathogens (see also Chapter 17 for bacterial/viral lectins) and also against tumor cells (see also Chapters 19 and 25). Deficiencies in the glycosylation apparatus or in lectin expression can lead to severe defects in immune responses and finally to a disease status.

One of the first reactions of the immune system against invaders from outside is inflammation. In general, we define inflammation as the cooperative response of the immune and vascular system to injury caused by invading pathogens (viruses, bacteria, fungi or parasites), damaged or malignant cells, or irritating chemical or physical agents. Its ultimate achievement, if successful, is to remove the deleterious stimuli and to initiate the healing process. As a rule, the acute inflammatory process is tightly regulated. Macrophages and neutrophils, as the inflammatory cells of the first line of immune defense, recognize the intruding pathogen by direct cell contact. They then release chemokines and cytokines, initiating inflammation. In a second step, the process of adaptive immune response involving T and B lymphocytes comes into action. Local inflammation may also be triggered by activation of the complement system. Under certain circumstances, the timely and spatially orchestrated process can run out of control. This is the case when, for instance, an infection by Gram-negative bacteria producing the pyrogenic lipopolysaccharide (LPS) cannot be restricted or abolished by immune cells (see Chapter 10.3 for information on LPS) and spreads over the entire body causing the dangerous stage of sepsis. Sepsis may cause the so-called septic shock syndrome with failure of vital organs such as lung and kidney, finally leading to

The Sugar Code. Edited by Hans-Joachim Gabius

ISBN: 978-3-527-32089-9

death. Uncontrolled inflammation may also be symptomatic for a large variety of diseases, most of them induced by allergic or autoimmune dysregulation, such as rheumatoid arthritis, atherosclerosis, multiple sclerosis, chronic obstructive pulmonary disease or inflammatory bowel diseases (IBDs), to name just a few. These diseases eventually become chronic and have a poor prognosis since treatment modalities today are still very limited. Chronic inflammation is often a precancerous stage of tumors such as hepatocellular carcinoma (induced by hepatitis viruses), lung cancer (as a final stage of a long history of chronic inflammation induced, for instance, by heavy tobacco smoking), cervical cancer (induced by papilloma viruses) and colorectal cancer (possibly as a late stage of IBDs such as Crohn's disease or colitis ulcerosa). In the following we take a closer look at the different steps leading to inflammatory reactions.

27.1 Sequence of Events

Leukocytes circulating in the bloodstream must be directed to, and migrate into, the inflamed tissue in order to attack the invading microorganisms or to eliminate the irritating substance. This occurs either by direct contact and subsequent phagocytosis or by release of an array of cytokines and inflammatory mediators that promote cross-talk with other immune cells and maintain the inflammatory process, or of enzymes that can damage the infecting organisms. As a rule, granulocytes (neutrophils) and monocytes are the predominant immune cells involved in acute inflammation, whereas monocytes, T and B lymphocytes prevail in chronic inflammations. To reach the site of inflammation the lymphocytes and other leukocytes must receive signals from the vascular endothelium adjacent to the inflamed tissue. These endothelial cells are triggered by soluble factors like histamine or cytokines, such as tumor necrosis factor or interleukin-1β (TNF or IL-1β), to switch into an activation mode and sequentially express receptors that are recognized by leukocytes. At the molecular level, the first contact between vascular endothelial cells and leukocytes is mediated via interactions between E-selectin on endothelial cells (for information on selectins, see Chapter 19 and below) and distinct oligosaccharides, preferentially of the Lewis^x (Le^x) histo-blood group expressed on the cell surface of leukocytes [the Le carbohydrate sequences are given below in either the biochemical nomenclature, i.e., sialyl-Le^x, or the cluster of differentiation (CD) nomenclature, e.g., CD15s; see also Figure 7.4 for structures]. For many proteins including lectins and also carbohydrate sequences known to be involved in immune reactions the CD nomenclature has been established based on interactions of these protein and carbohydrate structures with distinct monoclonal antibodies. Wherever it is possible, we refer here to the CD nomenclature (Figure 27.1a; for further information, see Info Box.) Due to the low affinity of this carbohydrate–protein interaction, contacting leukocytes start to roll along the endothelial surface until they come to a firm halt. This halt is mediated by protein–protein interactions, i.e., leukocyte integrins binding to endothelial

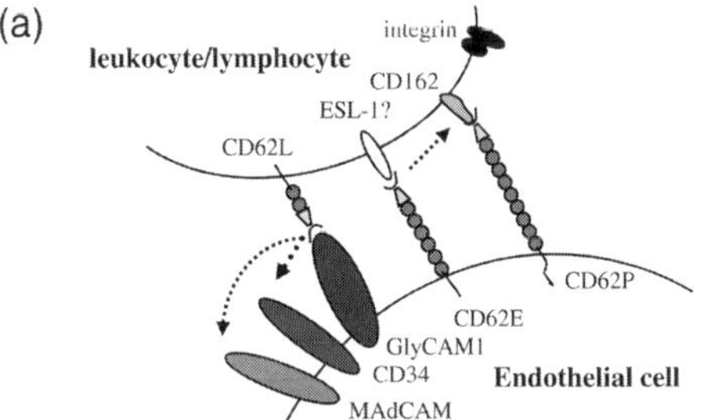

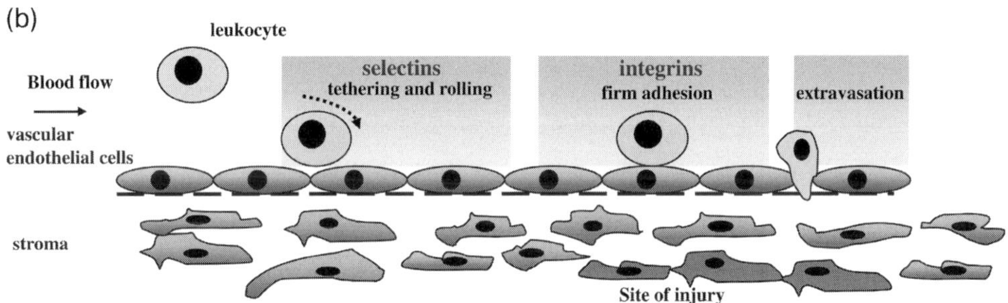

L-selectin (CD62L): lymphocytes/leukocytes E-and P-selectin (CD62E/CD62P): activated endothelial cells

Figure 27.1 (a) Interaction partners of selectin-mediated cell–cell contacts between leukocytes/lymphocytes and vascular endothelial cells. (b) Mechanism of selectin interactions during vascular adhesion and extravasation of leukocytes to inflammatory tissue.

receptors (Figure 27.1b). During the next step leukocytes/lymphocytes have to extravasate by transmigration induced by chemokine gradients (diapedesis), and then move along a gradient of chemoattractants to the source of inflammation where they finally fulfill their task of anti-pathogen immune attack (please see also Figure 29.5).

27.2 Where Do Carbohydrate–Lectin Interactions Play a Role During Acute Inflammation?

We focus here on the most pertinent effectors involved in mechanisms of inflammation and lymphocyte activation. Components of the innate immune system like the complement system, in which carbohydrate interactions play a role both for initiating the complement cascade (lectin pathway) and for controlling the effects of complement-mediated cell destruction (like inhibition of C1q, the first element of the classical complement pathway, by chondroitin sulfates) are not dealt with here (please see Chapter 19 for lectin in innate immunity). There are three major families of lectins that play an important role in immune reactions: the selectins, the galectins and the siglecs.

Info Box

A convenient way to explore glycan structures on cell surfaces is the application of anti-carbohydrate monoclonal antibodies. Indeed, antibodies have the ability to detect subtle differences in oligosaccharide structures like differences in anomericity, terminal monosaccharides and even the number of repeats of distinct sugar motifs. The best example for exact detection of defined oligosaccharide structures is the recognition by natural blood group antibodies of blood groups A and B. Monoclonal antibody technology has provided a powerful tool for detecting a large variety of oligosaccharide sequences. To organize the array of monoclonal antibodies produced in various laboratories, the integration and classification of anti-carbohydrate monoclonal antibodies within the CD nomenclature was initiated during the course of the Human Leukocyte Differentiation Antigen (HLDA) workshops and conferences. Most of these structures recognized by monoclonal antibodies are expressed on cells of the hematopoietic system, but also to varying degrees on endothelial and epithelial cells. Today we have several established carbohydrate CD groups:

CD	Carbohydrate sequence
CD15	Le^x
CD15s	Sialyl-Le^x
CD15u	Sialyl-Le^x
CD15su	6-sulfo-sialyl-Le^x
CD17	Lactosylceramide (LacCer)
CD60a	GD3 and related structures
CD60b	9-O-acetyl GD3
CD60c	7-O-acetyl GD3
CD65	Fucosylated ganglioside with four lactosaminyl repeats
CD65s	Sialylated form of CD65
CD75	N-Acetyllactosaminyl structures with several repeats
CD75s	Sialylated CD75 sequences
CD77	Globo series neutral glycosphingolipids, (preferentially Gb3, p^k blood group)
CD173	Blood group H
CD174	Le^y
CD175	Tn
CD175s	Sialylated CD175
CD176	Thomsen–Friedenreich (TF) antigen

Monoclonal antibodies are commercially available for many of these carbohydrate CDs. Many lectins are also included in the CD nomenclature, such as CD22 (siglec-2), CD62 (selectins), CD162 [P-selectin glycoprotein ligand (PSGL)-1], CD169 (sialoadhesin, siglec-1).

27.3
Selectins

Selectins are a family of three type 1 transmembrane glycoproteins. E-selectin (CD62E) is expressed on the surface of endothelial cells and L-selectin (CD62L) on the surface of leukocytes. P-selectin (CD62P), the largest selectin, is stored in α-granules of platelets and in Weibel-Palade bodies of platelets and endothelial cells, and is transferred to the cell surface upon activation (for further description of selectins, please see Figure 16.1h, Chapter 19 and Chapter 29). Selectins are involved in capture, rolling and first contacts of leukocytes to the vascular endothelial cell layer. In addition to their function in leukocyte migration and adhesion to the vascular endothelium during the course of inflammatory processes, selectins are also important adhesion molecules in the migration of naïve lymphocytes to lymphoid organs. In 1964, James Gowan and his colleagues at Oxford University discovered that lymphocytes migrate from the blood circulation into lymphoid organs. This process was later termed 'homing' [1]. L-Selectins of lymphocytes interact with their carbohydrate ligands expressed on a distinct type of endothelial cells within postcapillary venules in the cortex of lymph nodes.

These vessels are commonly named high endothelial venules (HEVs) because of the characteristic form of their endothelial cells. Similar to the rolling adhesion to vascular endothelium at inflammatory sites, lymphocytes also form L-selectin-mediated contacts with HEVs by tethering and rolling along the apical site of HEVs. This interaction with HEVs of lymphoid organs is an important early step during the process of T lymphocyte-dependent B lymphocyte activation. Therefore, selectins have vital functions both for the innate immune system (leukocyte trafficking to inflammatory sites) as well as for the adaptive immune response (lymphocyte homing). Further features of L-selectin are in leukocyte–leukocyte interactions during inflammation and leukocyte–tumor cell interactions during the formation of metastasis. The interaction of selectins on endothelial cells with ligands on tumor cells is described in Chapter 28.1.

27.4
Selectin Carbohydrate Ligands and Their Carrier Glycoproteins

Selectin–ligand interactions during various immune reactions are summarized in Table 27.1. Three different structural features characterize selectin carbohydrate ligands to varying degrees: fucosylation, sialylation and sulfation. Carriers for the carbohydrate selectin ligands are surface-expressed glycoproteins of the mucin type as described in more detail below (see also Figure 27.1a) [2].

Selectins belonging to the group of C-type lectins have a carbohydrate recognition domain (CRD) in their N-terminal domain (see Figure 16.1h and Chapter 19, and also Chapter 29). Sialyl-Le^x also fits well in the complex with the CRD of CD62E. CD62E and CD62P have been shown to have a much higher affinity for sialyl-Le^x (CD15s) than for Le^x (CD15). Structural details of a series of selectin

Table 27.1 Selectin–ligand interactions during inflammation and lymphocyte homing.

Selectin type	CD	Selectin expressed on	Ligand molecules	Ligand carrier	Ligand expressed on	Biological function
L-Selectin	CD62L	B, T cells, monocytes, granulocytes, NK cells	CD15su ≫ CD15s > CD15	GlyCAM 1, CD34, MAdCAM	Vascular endothelial cells, HEVs, Peyer's patches	Lymphocyte homing
P-Selectin	CD62P	Platelets, endothelial cells, megakaryocytes	CD15s > CD15 tyrosine sulfate	PSGL-1 (CD162)	T cells, leukocytes, monocytes	Inflammation
E-Selectin	CD62E	Endothelial cells	CD15s > CD15 tyrosine sulfate	PSGL-1 (CD162), CD44, ESL-1	T cells, leukocytes, monocytes	Inflammation

ligands, including sialylated and sulfated structures of the Le blood group family, are depicted in Table 27.2. The carbohydrate specificity of selectins may vary depending on their specific tasks. For instance, the ligand for L-selectin (CD62L) of naïve T lymphocytes on human HEVs has been identified as sialyl-6-sulfo-Le^x (CD15su). CD15su is strongly expressed on HEVs and the binding of CD62L to this oligosaccharide can be inhibited by a specific monoclonal antibody. CD62L apparently prefers carbohydrate 6-sulfation in contrast to CD62P and CD62E that, in addition to CD15s, require tyrosine sulfate for efficient binding. The accurate expression of specific ligands of the Le family affords the coordinated activity of various glycosyl- and sulfotransferases (Table 27.3). Sulfated tyrosine residues at the N-terminus of the respective mucin-type glycoprotein ligands may also be involved in the binding of selectins.

It has been proposed that there are significant physiological differences of selectin interactions between sulfated and non-sulfated Le^x groups (Table 27.1). It is postulated that – as a rule – CD15s and CD15su on leukocytes/lymphocytes serve as ligands for CD62E and CD62P. Further, CD15s plays a key role in the recruitment and piloting of leukocytes to sites of inflammation. CD15su is involved in the homing of naïve $CD62L^+$ T lymphocytes to peripheral lymph nodes [3]. Natural killer (NK) cells of the peripheral blood system appear to use CD15su interactions. CD15su is not only expressed on HEVs of peripheral lymph nodes, but also on HEVs of Peyer's patches of the gastrointestinal immune system. In the mucosal immune system of the gut, mature $CD62L^+$ T helper memory lymphocytes bind to $CD15su^+$ HEVs. These T cells are necessary to maintain immunological homeostasis in the intestine.

Table 27.2 Ligands for selectin interactions.

CD	Structure name	Structure
CD15	Le^x	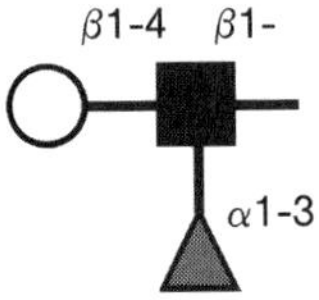 Galβ1,4(Fucα1,3)GlcNAcβ1–
CD15s	Sialyl-Le^x	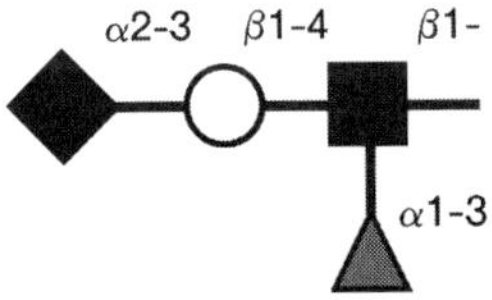 NeuAcα2,3Galβ1,4(Fucα1,3)GlcNAcβ1–
CD15su	6-sulfo-sialyl-Le^x	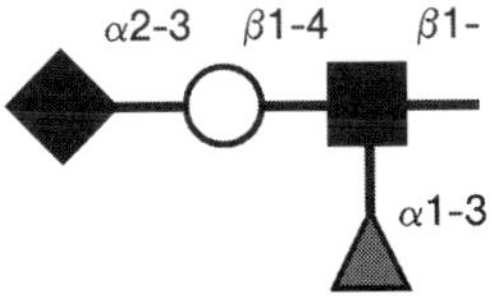 NeuAcα2,3Galβ1,4(SO^{3-}-6)(Fucα1,3)GlcNAcβ1,3Gal–
No CD	Le^y	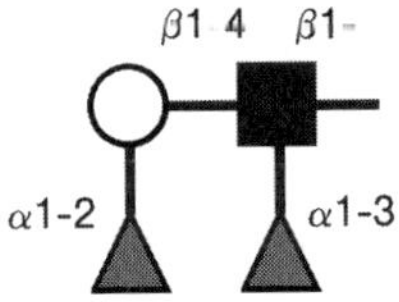 (Fucα1,2)Galβ1,4(Fucα1,3)GlcNAcβ1–
No CD	Le^a	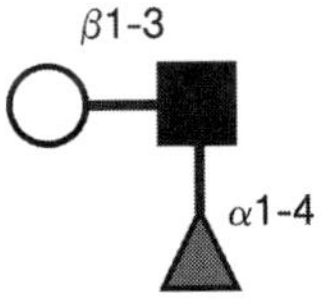 Galβ1,3(Fucα1,4)GlcNAcβ1–
No CD	Sialyl-Le^a	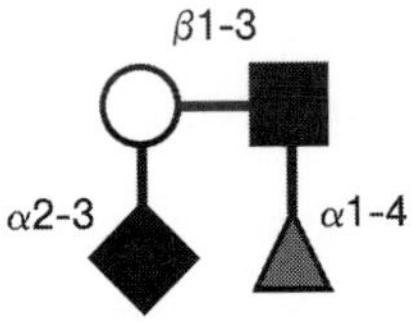 NeuAcα2,3Galβ1,3(Fucα1,4)GlcNAcβ1–
No CD	Le^b	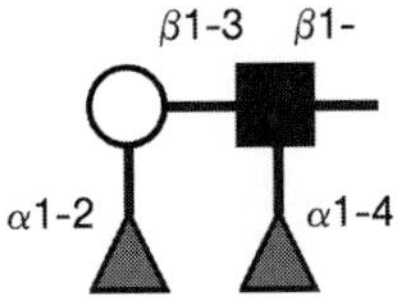 (Fucα1,2)Galβ1,3(Fucα1,4)GlcNAcβ1,3Galβ1–

Please see Figure 28.3 for sialyl-Le^x with sugars drawn as chairs.

Table 27.3 Examples for enzymatic regulation of selectin ligands.

Key enzymes	Product
Core 2 GlcNAcT + FUT7 (FUT4)[a]	Le^x (CD15)
Core 2 GlcNAcT + FUT7 (FUT4) + ST3Gal-IV	sialyl-Le^x (CD15s)
Core 2 GlcNAcT + FUT7 (FUT4) + ST3Gal-IV + sulfotransferase	6-sulfo-sialyl-Le^x (CD15su)
Core 2 GlcNAcT + FUT7 (FUT4) + FUT1	Le^y (CD174)[b]

[a] FUT4 and FUT7 are the most relevant fucosyltransferases for synthesis of selectin ligands of the six fucosyltransferases responsible for α1,3 fucosylation in the human system (FUT3–7 and 9).

[b] There seems to be a differential regulation of α1-2 fucosyltransferases with regard to cellular Le^y (FUT1) and the secreted form of Le^y (FUT2).

On the other hand, carbohydrate ligands for CD62E and CD62P are expressed on leukocytes and lymphocytes. CD15s is constitutively expressed on granulocytes and monocytes, and is suppressed in most resting lymphocytes. Upon activation, CD15s expression is enhanced in T lymphocytes and CD15su seems to decrease at the same time due to diminished sulfation.

With regard to which selectin ligands are the 'right ones' under physiological conditions, it has to be considered that *O*-glycans of glycosylation-dependent cell adhesion molecule (GlyCAM)-1 – one of the mucin-type glycoprotein carriers of L-selectin (CD62L) ligands – possess multiple sulfation modifications. In addition, and as described above, both lectins and carbohydrate ligands may be expressed on leukocytes, lymphocytes and the reactive vascular endothelium in different combinations. This further complicates functional studies of single components in the sterical context of live cell–cell interactions. There is also an organ selectivity with regard to selectin function. For instance, neutrophil migration to liver and lung is largely CD62 independent, whereas for recruitment to skin and mucosa CD62P and CD62E are vital.

The major carrier glycoprotein of ligands for CD62P-mediated interactions is PSGL-1 (CD162) [4]. Activated platelets expressing CD62P can bind to CD162 on leukocytes and monocytes, which may be an important interaction for the recruitment of inflammatory leukocytes to thrombi. In the context of atherosclerosis, platelet CD62P is needed for interactions with monocytes and endothelial cells where activated platelets deposit proinflammatory chemokines on the surface of endothelial cells and monocytes.

There are three major glycoprotein ligands known for CD62L: GlyCAM-1, CD34 and mucosal addressin cell adhesion molecule (MAdCAM). Glycosylation of CD34 on vascular endothelium is differentially regulated because in non-inflamed vascular endothelium it does not react with CD62L. For CD62E the situation in terms of glycoprotein carriers for carbohydrate ligands is not yet clear; CD162, CD44 and E-selectin ligand (ESL)-1 are possible candidates (Table 27.1 and Figure 27.1a).

The importance of selectins for effective leukocyte recruitment during the course of inflammation becomes evident in distinct genetic defects such as leukocyte adhesion deficiency (LAD)-II (see also Chapter 22.1). Patients suffering from this rare disease have a mutation in the gene encoding a fucose transporter and thus incorporation of fucose into selectin ligands is hampered. Consequently, leukocytes are unable to bind to endothelial selectins, and bacterial infections preferentially of the mucosa and skin cannot be counteracted effectively. Targeted deficiencies in murine knockout model systems can also give valuable information about the functional relevance of distinct lectins or oligosaccharides (examples are summarized in Table 27.4).

Whether the extravasation of tumor cells as an early, decisive step of metastasis formation follows mechanisms similar to those described above for leukocytes is still a matter of discussion, although a large body of evidence speaks in favor of

Table 27.4 Biological consequences of genetic deficiencies affecting the glycosylation machinery or lectin synthesis.

Deficiency	Organism	Cell type	Functional aberrations
CD22	Mouse, gene knockout	B lymphocyte	Elevated Ca^{2+} mobilization after BCR stimulation, reduced surface IgM on peripheral B lymphocytes, impaired response to T lymphocyte independent antigens
ST6Gal-I	Mouse, gene knockout	B lymphocyte	Reduced Ca^{2+} mobilization after BCR stimulation, reduced proliferation in response to CD40, BCR and LPS signaling, reduced surface IgM
FX	Mouse, gene knockout	Leukocytes and other cells	Extreme neutrophilia, disturbed myeloproliferation, absence of selectin ligand expression
Galectin-1	Mouse, gene knockout	Endothelium	Decreased microvessel formation
Galectin-3	Mouse, gene knockout	Monocytes, macrophages, mast cells	Defects in phagocytosis, reduced IgE-mediated response, reduced airway allergic and inflammatory reactions
GFTP	LAD-II patients (see Chapter 22.6)	Neutrophils, lymphocytes, nerve system	Increased peripheral neutrophil counts, (reduced homing and adhesion (rolling) to vascular endothelium?), recurrent infections, defects in mental development

ST6Gal-I: the sialyltransferase that is responsible for α2,6-sialylation of LacNAc sequences; FX: GDP-4-keto-6-deoxymannose-epimerase/reductase that is required for *de novo* synthesis of GDP-Fuc from GPD-Man; GFTP: the Golgi GDP-Fuc transporter, which is an antiporter translocating GDP-Fuc into the Golgi lumen in exchange for GMP; please see also Table 23.1.

it. It has been convincingly demonstrated that tumor cells exhibit altered surface expression of carbohydrate moieties compared to their normal counterparts. For instance, surface expression of CD15s is positively correlated with tumor progression and the capacity to form metastases. In addition, other forms of the Le histo-blood group family like Le^a and Le^y are frequently expressed on human carcinomas. These oligosaccharide sequences may also serve as selectin ligands, primarily for CD62E and CD62P, although this has not yet been extensively studied.

Selectins preferentially bind to carbohydrate structures of the Le blood group family and have vital functions both in directing leukocytes to sites of inflammation and in homing of lymphocytes to lymph nodes. We next explain how the complex group of galectins is involved in diverse immune reactions.

27.5 Galectins

Galectins are an ancient family of lectins that do not contain a transmembrane-spanning domain like selectins (for fold, please see Figure 13.2). They are soluble molecules found in the outer environment, on the cell surface in dimeric form, in the cytoplasm and even in the nucleus. This makes their functional analysis more difficult than of membrane-anchored lectins (see Chapters 15, 16 and 19). Galectins are defined by their affinity for the β-galactoside core; however, they differ in their affinity for substituted β-galactosides complex carbohydrate sequences (see Chapters 13, 19 and 25). A multitude of functions has been ascribed to galectins (please see Table 19.2 for respective entries). Here, we con-

Figure 27.2 Illustration of lectin (galectin) signaling during T cell apoptosis. (Top panel) Binding of a lectin (L) to a cell surface receptor (R) induces receptor clustering and initiation of signaling. Hereby a cellular response is generated, in this case apoptosis. (Bottom panel) Galectin-1 homes in on distinct cell-surface glycoproteins as receptors, such as CD7, CD43 or CD45. This interaction triggers apoptosis because the activation of sphingomyelinase induces increased levels of ceramide. The ensuing activation of the initiator proteases caspase-8 (casp8) and caspase-9 (casp9) and their effectors caspase-3, -6 or -7 results in the degradation of intracellular proteins. Active caspase-9 is produced from its inactive precursor by the apoptosis-activating factor-1 (Apaf-1) when cytochrome *c* (cyt c) is available in the cytosol. This requirement is met by the galectin-dependent upregulated expression of the pro-apoptotic protein Bax and the downregulation of the anti-apoptotic protein Bcl-2. Overall, these two proteins regulate the permeability of the outer mitochondrial membrane in an antagonistic manner. Tilting the balance in favor of Bax lets cytochrome *c* enter the cytosol. In addition to protein fragmentation, the effector caspases also facilitate DNA degradation via caspase-activated DNases (CAD). Prior to their nuclear translocation, a caspase-activated DNase inhibitor is cleaved by the caspases. Furthermore, galectin-1 operates by a mechanism involving the endonuclease G (EndoG)/apoptosis-inducing factor (AIF) complex, first released in a Bax-dependent manner from mitochondria and thereafter translocated to the nucleus. The abbreviations used are: ext, extracellular medium; cyt, cytosol; ncl, nucleus; ims, intermembrane mitochondrial space; mat, mitochondrial matrix.

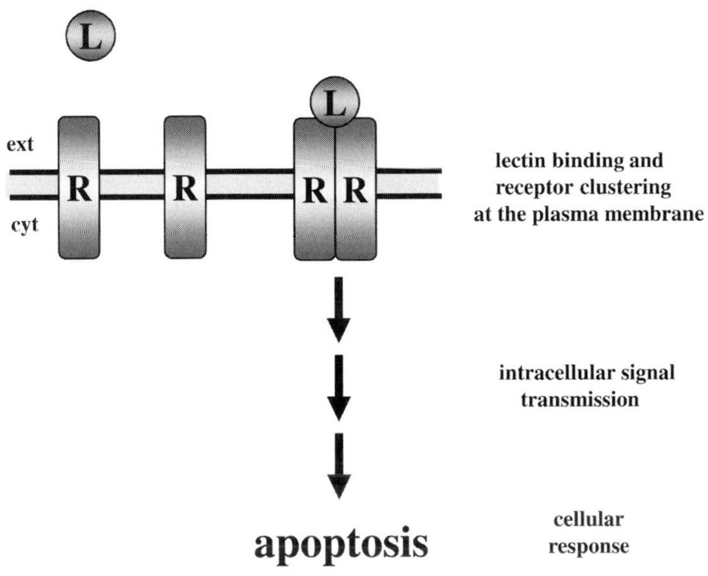
L
L
ext
R
R
R
R
cyt
lectin binding and
receptor clustering
at the plasma membrane
intracellular signal
transmission
apoptosis
cellular
response

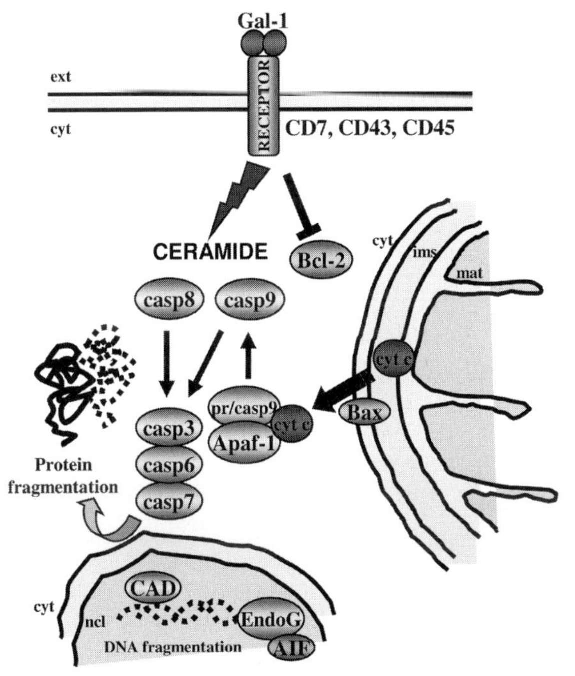
Gal-1
ext
RECEPTOR
cyt
CD7, CD43, CD45
CERAMIDE
Bcl-2
cyt
ims
mat
casp8
casp9
cyt c
pr/casp9
cyt c
Bax
casp3
Apaf-1
casp6
Protein
fragmentation
casp7
CAD
cyt
ncl
EndoG
DNA fragmentation
AIF

Table 27.5 Examples for galectin–carbohydrate interactions in immune processes.

Galectin type	Cellular expression	Biological activity	Function
Galectin-1	Vascular endothelium and stroma	Angiogenesis, apoptosis	Induction of microvessel formation, induction of apoptosis
Galectin-3 (Mac-2 IgE-binding protein)	Neutrophils, eosinophils	Inflammation, allergy	Proinflammatory effect, oxidative burst, induced interleukin-1 production, binding to IgE glycan amplifier of allergic reaction
Galectin-9	T lymphocytes	Allergy	Chemoattractant for eosinophils, induction of eosinophil aggregation

centrate on those that are known to be important for the inflammatory process and immune responses against pathogens. Compared to our knowledge about selectin functions, we are still in the early phase of understanding how galectins interfere with immune regulation. Twelve members of the galectin family have been identified in the human system, of these, galectins-1, -3 and -9 have apparent functions in the immune system. In general, galectins as multivalent, soluble molecules can cross-link their ligands on various cells and thus modulate immune responses in a paracrine or autocrine fashion by mediating cell–cell interactions and induce signal transduction (for compilation of ligands for galectins-1 and -3, please see Table 19.3). They act as adhesion molecules, and may alter the glycocalyx of cells by forming a lattice consisting of galectins and glycoconjugates. Through this latter function, the lateral mobility of surface receptors is possibly diminished and the threshold of receptor-mediated signaling increased. An important function of galectin-1 is connected to lymphocyte homeostasis. Contact with galectin-1 expressed by stromal cells in the thymus and lymph nodes can induce apoptosis of activated T lymphocytes, but not of resting T lymphocytes. This indicates a change in glycosylation of T lymphocytes during the process of activation, also relevant in communication between T effector and T regulatory cells [5] (see Chapter 25.2 for examples of orchestrated changes of galectin and glycan expression). Galectins-1, -2, -3, -7 and -9 are able to induce apoptosis of T lymphocytes when they bind to the endothelial layer followed by a cascade of differential intracellular signaling (Figure 27.2) [6]. Some examples of galectin interactions during immune reactions are summarized in Table 27.5.

There is evidence that galectins are also involved in certain steps of tumor progression, i.e., immune modulation, metastasis formation and tumor-induced angiogenesis (see also Figure 25.3 for impact on tumor cell proliferation) [7]. As tumorigenesis is often accompanied by inflammation, it is conceivable that the proinflammatory functions of galectin-3-promoted adhesion of neutrophils to vascular endothelium may also further the inflammation as a precancerous state (Table 27.4). Galectin-3 expression may additionally contribute to metastasis formation by supporting tumor cell contact to the endothelium. Specific homing of malignant lymphoma cells to bone marrow seems to be galectin mediated. A prominent tumor-specific glycan, the Thomsen-Friedenreich (TF) antigen (CD176), is a binding partner for galectin-3. Galectins-1 and -3 seem to have an additional function in angiogenesis (Table 27.4). Exogenously added galectin-3 induces tube formation of vascular endothelial cells.

One major function of galectins in the immune system is the recognition of microbial and parasitic glycoconjugates, which may then trigger further positive or negative immune responses [8]. For example, galectins-3 and -9 recognize surface lipophosphoglycans of *Leishmania major* and galectins-1 and -3 stimulate oxidative burst in neutrophils as a cytotoxic anti-pathogenic mechanism. All these functions depend on recognizing sugar-encoded information, especially from spatially accessible branch ends of glycan antennae. In addition to selectins and galectins, these sites are also targets for another lectin group, the siglecs (for molecular display, see Figure 19.1).

27.6 Siglecs

The siglecs are a group of sialic acid-binding I-type (belonging to the immunoglobulin supergene family) membrane-spanning lectins expressed not only in the cells of the immune system, but also in other cell types and tissues, e.g., the nervous system (Table 27.6). Siglecs can mediate heterotypic cell–cell interactions by virtue of their sialic acid-binding capacity and may also be involved in *cis–cis* interaction between siglecs expressed on the cell surface and neighboring sialic acid residues [9]. Although the structures and molecular prerequisites for sialic acid binding in the N-terminal V-set immunoglobulin-like domains have been well studied *in vitro*, the physiological role of most siglecs is still unclear. The best functionally studied siglec is CD22. This is a B lymphocyte-specific surface receptor, which, in contrast to most other siglecs, exclusively interacts with α2–6-sialylated glycans and participates in the fine regulation of antibody production of activated B lymphocytes (Figure 27.3) [10]. Strikingly, within the hematopoietic system, B lymphocytes carry the largest quantity of α2–6-sialylated surface carbohydrates as potential CD22 ligands, and the strongest binding of CD22 to α2–6-sialylated glycans is indeed between B cells. Interestingly, in the regulation of first-line-defense immunoglobulins, murine siglec-G seems to have a negative regulatory impact on the production of so-called natural antibodies of B1-type B lymphocytes.

While CD22 has a propensity towards α2–6-linked sialoglycans, most other siglecs tend to bind preferentially to α2–3-sialylated glycans (Table 27.6). It is striking that most siglecs can be found in cells of the myelomonocytic lineage of the hematopoietic system and on NK cells, and that they are absent from T cells. It is therefore most likely that siglecs play a role in the innate immune system and also during inflammatory reactions. However, this is still hypothetical. Of further interest is the interaction of siglecs on various immune cells with sialic acids

Table 27.6 Human siglecs and their properties.

Siglec	Synonyms	CD	Sialic acid specificity (linkage type)	Number of immunoglobulin domains	Cellular expression	Function
Siglec-1	Sialoadhesin	CD169	2,3 > 2,6 > 2,8[a]	17	Dendritic cells[b], monocytes macrophages	–
Siglec-2	–	CD22	2,6	6/7[c]	B lymphocytes	Fine modulation of B lymphocyte activation
Siglec-3	–	CD33	2,6 > 2,3	2	Monocytes, hematopoietic precursors	Growth regulation?
Siglec-4	MAG[d]	–	2,3 > 2,6	5	Schwann cells, oligodendrocytes	Maintenance of myeloid sheath
Siglec-5	–	CD170	2,3; 2,6 > 2,8	4	Monocytes, granulocytes, B lymphocytes	–
Siglec-6	OB-BP1	CDw327	2,6 (CD175s)	3	B lymphocytes, placenta	–
Siglec-7	AIRM1	CDw328	2,8 > 2,6 > 2,3	3	Monocytes, NK cells, T lymphocytes	Regulation of NK cell cytotoxicity?
Siglec-8	–	–	2,3>2,6	3	Eosinophils, mast cells	–
Siglec-9	–	CDw329	2,3; 2,6	3	Monocytes, NK cells, granulocytes	–
Siglec-10	–	–	2,3; 2,6	5	Monocytes, eosinophils, B lymphocytes	–
Siglec-11	–	–	2,8	5	Macrophages	–

[a] 2,3; 2,6, and 2,8 refer to the respective α-anomeric linkage of sialic acid to either galactose, *N*-acetylgalactosamine or, in the case of α2,8, to sialic acid in subsequent α2,3 linkage.
[b] DC = dendritic cells.
[c] CD22 may occur in a short (six Ig domains, 130 kDa) or longer form (seven Ig domains, 140 kDa) [10].
[d] Myelin-associated glycoprotein.

(a)

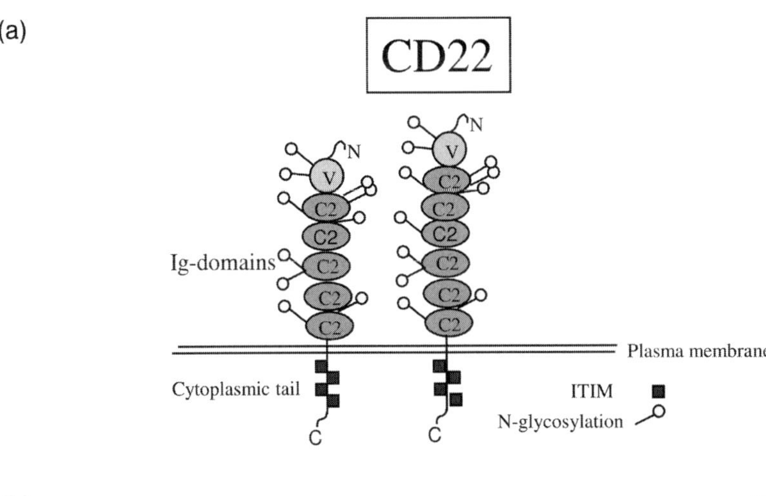

(b)

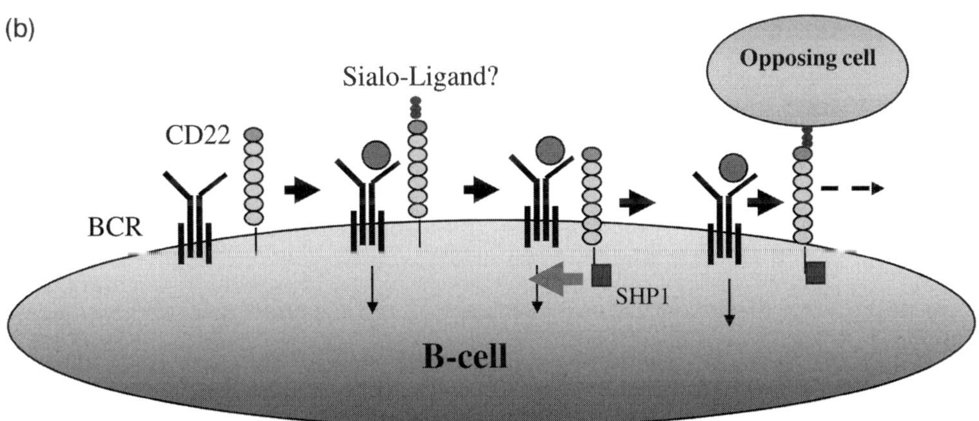

Figure 27.3 (a) Molecular structure of the B cell-specific lectin CD22. CD22 (siglec-2) is a single-chain type I *N*-glycosylated integral membrane sialoglycoprotein [10]. CD22 is a B-cell-surface expressed lectin in two forms: the major form is a 140-kDa sialoglycoprotein consisting of six immunoglobulin C2-like domains and one IgV domain, whereas the smaller form of 130 kDa lacks the fourth immunoglobulin C2 domain. The cytoplasmic tail contains four immunoreceptor tyrosine-based inhibitory motifs (ITIM) that are responsible for the inhibitory signaling pathway. (b) Involvement of CD22 in fine-tuning of B cell activation. CD22 is an inhibitory coreceptor of the B cell receptor (BCR) and fine regulates B lymphocyte activation. In some studies CD22 has also been associated with positive signaling. CD22 specifically recognizes α2,6-sialylated lactosaminyl sequences. Some cellular as well as soluble sialoglycoproteins have been identified as potential ligand carriers (CD45, IgM, haptoglobin); however, the physiologically active ligands and their carriers have not yet been described. For inhibitory activity, CD22 has to be in close proximity to the BCR. Upon cross-linking of the BCR induced by binding of an antigen, CD22 associated to the BCR is tyrosine phosphorylated by kinase Lyn. This provides docking sites for the tyrosine phosphatase SHP1, which dephosphorylates components of the BCR complex, thus dampening the BCR-mediated B lymphocyte activation. By this, CD22 helps to prevent an over-broadening humoral immune response. CD22-mediated immune regulation is controlled by at least two mechanisms: either by *cis* interaction with sialic acid on the same cell, thereby preventing interaction with exogenous sialoglycans or those on opposing cells, or by sialylation of CD22 itself in the terminal, carbohydrate-binding domain of the lectin. The latter phenomenon has also been described for sialoadhesin (CD169).

expressed on pathogenic microorganisms like *Neisseria meningitidis, Haemophilus influenzae* or *Corynebacterium diphtheriae* to name just a few. Apparently sialic acids are used by these microorganisms for host mimicry. These sialoglycans expressed on pathogens might also bind to CD22 on B lymphocytes, thereby modulating the antibody production of B lymphocytes. This is an example for the ability of pathogens to adapt recognition systems of the host for their own advantage. Having introduced the three predominant groups of lectins known to be involved in immune reactions, we now give some examples of other lectins taking part in inflammatory reactions and host defense against pathogens, in particular bacteria–host interactions.

27.7
Other Lectins Involved in Antigen Recognition and Inflammatory Processes

Another important example for lectins that play a role in pathogen recognition is the dendritic cell-specific ICAM-3 grabbing non-integrin (DC-SIGN), a prototype for type II C-type lectins (see also Chapter 19 and X-ray structure in Figure 16.1i). DC-SIGN may function as a pathogen recognition receptor as well as an adhesion coreceptor for host defense receptors like Toll-like receptors. As an adhesion receptor DC-SIGN is involved in the contact between the professional antigen-presenting dendritic cell and T lymphocytes by binding to intercellular adhesion molecule (ICAM)-3 on T cells. Additionally, and similar to selectins, it mediates rolling of dendritic cells on vascular endothelium by interaction with ICAM-2. As a pathogen receptor DC-SIGN recognizes several bacteria, fungi and viruses [11]. DC-SIGN and its liver homolog L-SIGN have within their N-terminal domain a high affinity site for mannose-containing and fucosylated oligosaccharides. In contrast to selectins, DC-SIGN also recognizes the non-sialylated form of Le^x (CD15) and, for instance, pseudo-Le^y of *Schistosoma mansoni*.

During uptake of pathogens by dendritic cells, which is a prerequisite for proper major histocompatibility complex (MHC)-mediated presentation of pathogen-derived peptides to T cells, DC-SIGN contributes to vesicular antigen uptake.

The C3b receptor CR3 (CD11b, CD18) is a member of the integrin family and is expressed on neutrophils, NK cells and minor subpopulations of T and B lymphocytes (see Chapter 29 for lectin functions of integrins). It functions as an adhesion molecule and as receptor for ICAM-1. Additionally, it can also bind β-glucans of bacteria, thereby bringing the receptor into an activated state, which permits neutrophil phagocytosis.

It was believed previously that T cells recognize exclusively peptides presented by the MHC for their activation (see also Chapter 10.4). This dogma has now changed and it is known that T cells are also able to recognize glycolipids via their MHC-like CD1 molecules and consequently glycolipids (e.g., α-galactosylceramide). These reactions may modify lymphocyte reactivities, and thereby the outcome of infectious, autoimmune and allergic diseases, and possibly of tumors.

CD1a–c molecules present microbial fatty acids, glycolipids, phospholipids and lipopeptides to the T cell receptor (TCR) [12].

27.8 Glycans Involved in Bacteria–Host Interactions

Humans are in constant contact with microbes, mainly in the gastrointestinal tract where the number of microorganisms exceeds by far the number of all cells of our body. The contact with the microbial flora ranges from symbiotic, beneficial, via commensal to pathogenic for the human host. However, most molecular details of these complex interactions are not well understood. This is in part due to the fact that the complexity of the human intestinal system is difficult to simulate with *in vitro* models. Several carbohydrate–lectin interactions have been modeled *in vitro* using cell lines, such as the adhesion of uropathogenic *Escherichia coli* strains by their FimH adhesin to mannosylated glycans of the superficial facet cells of the bladder urothelium (see also Chapter 17 for bacterial lectins). Only certain *Helicobacter pylori* strains found in the stomach of many humans are pathogenic, causing active gastritis. In some cases this may progress to chronic atrophic gastritis, which can be the pre-stage of gastric adenocarcinomas or mucosa-associated lymphoma. *H. pylori* is able to specifically interact with blood group Le^b glycans, which are expressed in approximately 70% of mankind. Why *H. pylori* colonization occurs preferentially in individuals with blood group O and not in those with blood groups A or B is not understood. Further, *H. pylori* is able to produce an LPS *O*-antigen that contains Le^x/Le^y immunodeterminants. Variations in these determinants may also explain why some hosts have a relatively benign response to *H. pylori* infections and others a pathological one. It has been hypothesized that the enormous complexity of mammalian glycan structures developed by evolutionary pressure. This complexity may have arisen 'in part from our need to both evade pathogenic relationships and to co-evolve symbiotic relationships with our non-pathogenic resident microbes' [13, 14]. For this interdependent process J. Gordon coined the term 'glycan legislation' [13]. Competition between invading microorganisms and the host defense against the attacks may have been the driving force to create the diversity of surface oligosaccharides on both sides – pathogens and host – and for the pathogens to continuously invent new mechanisms of mimicry to escape recognition by the immune system [15]. It may even be that today the selection process is accelerated due to modern anti-pathogenic treatment, which has led to the race between bacterial resistance against antibiotic attack and the development of new generations of antibiotics. Considering the complex interactions between host cells and bacteria that are constantly taking place in the intestinal system, it is possible to imagine that under certain circumstances bacteria–host interactions may lead to chronic inflammatory reactions. We now discuss the role of carbohydrates in these types of diseases.

27.9 Glycosylation in Inflammatory Bowel Diseases

As an example for chronic inflammatory diseases we here deal in more detail with colitis ulcerosa and Crohn's disease both known under the term inflammatory bowel diseases (IBD), which – being chronic inflammatory processes – have a high risk of turning into colorectal cancer [16]. Although the etiology of these two diseases is not quite clear, there are many indications that point to an autoimmune mechanism, possibly induced by an immune response of the host to the intestinal microflora. With regard to mucosal glycosylation, it is striking that similar changes can be observed in IBD as in carcinoma. One predominant change is the blockade of chain elongation to core 2 *O*-glycosylation by a quantitative change in the respective glycosyltransferases, and a concomitant increase in smaller carbohydrate sequences such as the TF structure (CD176) or the even shorter Tn/sialylated Tn glycan (CD175/CD175s) (Figure 27.4). CD176 is indeed a marker for malignancy, e.g., colon cancer, and its expression increases with malignancy. The unmasked CD176 surface structure is rarely found on normal cells. However, in its sialylated form, CD176 is expressed on almost all hematopoietic cells and many other somatic cells. One reason for the lack of CD176 expression on normal cells is that it may represent a signal for negative selection and can, upon linkage with respec-

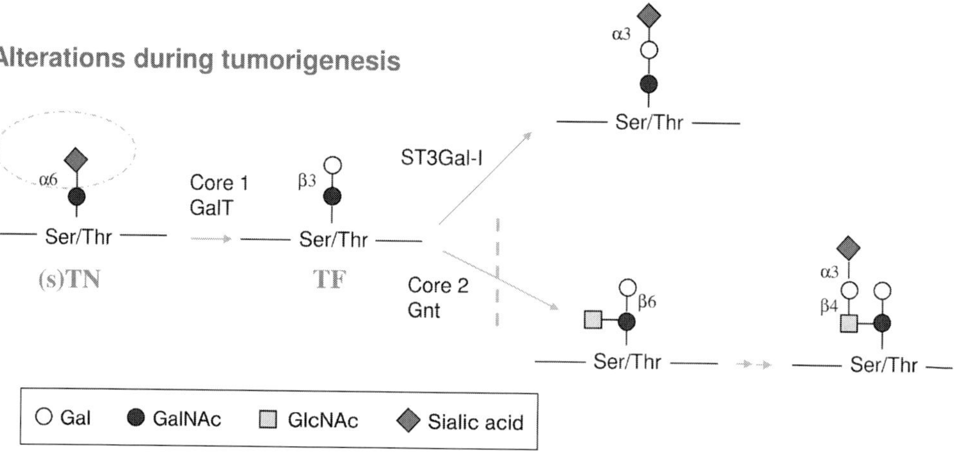

Figure 27.4 Synthesis pathway of *O*-glycans of mucin-type glycoproteins. Under certain circumstances, e.g., tumorigenesis, the synthesis towards core 2 glycans stops due to down-regulation of the respective β1,6-N-acetylglucosaminyl-transferase (C2GnT), with the consequence that glycoproteins carry shorter core 1 glycans like CD175 (Tn) or CD176 (TF). Expression of terminal CD176 structures under normal conditions is a rare event, possibly because it may transfer apoptosis signals. For tumor cells this may be advantageous, because CD176 may help metastasizing tumor cells to adhere to the asialoglycoprotein receptor of liver tissue.

tive receptors, lead to apoptosis of the cell. Other changes that occur in inflamed tissues and tumors are enhancement of CD15 and Le^y (CD174). Interestingly, CD174 is highly expressed not only on tumor cells, but also on activated endothelial cells at sites of inflammation (Figure 27.5). Recent studies have shown that CD174 surface expression on vascular endothelial cells is enhanced by inflammatory signals like TNF and IL-1β, and may also be involved in the homing process of leukocytes and lymphocytes to inflammatory sites [17]. Some carrier glycoproteins for these oligosaccharides (e.g., CD176) have been identified, such as CD44 on tumor cells and CD34 on distinct leukemia cells. Another alteration of glycans during chronic mucosal inflammation and colon cancer is a reduced sulfation of specific glycans, possibly due to a changed arrangement of the respective transferases in the Golgi.

A number of lectins have been detected in mucosa of IBD patients; however, their functional significance is not entirely clear. Galectin-3, mostly described as a proinflammatory factor, induces apoptosis in T cells as described above. In chronically inflamed mucosa of IBD patients, galectin-3 was found to be downregulated as compared to normal mucosa, serum levels were elevated [18]. Of note, this lectin is also a functional marker for cardiac inflammation [19]. Increased levels of CD62P have been found in IBD patients with acute-phase inflammation, which may further the recruitment of leukocytes to the inflammatory lesions. Obviously, the experimental and clinical information available to date points to a crucial function of carbohydrate–lectin interactions in chronic inflammatory diseases such as IBD, but does not yet suffice to gain a coherent mechanistic concept of their involvement.

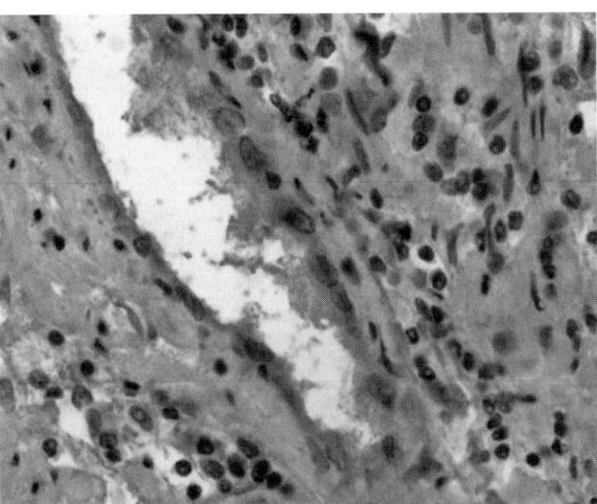

Figure 27.5 Immunohistological analysis of CD174 (Le^y) expression in capillaries of colon tumor-infiltrated tissue. Paraffin-embedded tumor tissues were stained with an anti-CD174 monoclonal antibody. Le^y is strongly expressed on vascular endothelial cells adjacent to highly inflamed tissue (right side of vessel, see massive leukocyte infiltration). (Immunohistology by kind permission of Dr M. Andrulis, Institute of Pathology, University of Heidelberg.)

27.10 Conclusions

In this chapter we have described distinct interactions of lectins and carbohydrates expressed on surfaces of host cells and pathogens. We have restricted ourselves to some of the many structures that are involved in the complicated network and cross-talk between the immune and vascular systems during inflammatory reactions as well as during pathogen recognition by immune cells. Although still at an early stage, glycoimmunology has already contributed many exciting new concepts to all sectors of immunology and provides explanations for mechanisms involved in early steps of the immune response, especially those of complex adaptive immune responses. Carbohydrate–lectin interactions also bridge innate and adaptive immunity, since many of the lectin reactivities fulfill tasks in both systems.

Summary Box

Inflammation is the response of the immune and the vascular system to injuries caused by pathogens, damaged cells or irritating agents. Communication of immune cells and vascular endothelial cells with inflammatory reagents is to a large extent mediated by carbohydrate–lectin interactions. These interactions regulate adhesion processes and organ-specific homing of immune cells, and also influence intracellular signaling. On the carbohydrate side, predominant structures comprise those of histo-blood groups or related sequences such as Le^x, Le^a and Le^y (CD15, CD173 and, CD174) carried by distinct glycoproteins. On the lectin side, three major families are involved: selectins, siglecs and galectins.

References

1 Rosen SD. Ligands for L-selectin: homing, inflammation, and beyond. *Annu Rev Immunol* 2004;22:129–56.
2 Gabius HJ. Cell surface glycans: the why and how of their functionality as biochemical signals in lectin-mediated information transfer. *Crit Rev Immunol* 2006;26:43–80.
3 Kannagi R. Regulatory roles of carbohydrate ligands for selectins in the homing of lymphocytes. *Curr Opin Struct Biol* 2002; 12:599–608.
4 Ley K. The role of selectins in inflammation and disease. *Trends Mol Med* 2003; 9:263–8.
5 Wang J *et al.* Cross-linking of GM1 ganglioside by galectin-1 mediates regulatory T cell activity involving TRPC5 channel activation: possible role in suppressing experimental autoimmune encephalomyelitis. *J Immunol* 2009;182:4036–45.
6 Villalobo A *et al.* A guide to signalling pathways connecting protein–glycan interaction with the emerging versatile effector functionality of mammalian lectins. *Trends Glycosci Glycotechnol* 2006;18:1–37.
7 Thijssen VLJL *et al.* Galectins in the tumor endothelium: opportunities for combined cancer therapy. *Blood* 2007;110:2819–27.

8 Young AR, Meeusen EN. Galectins in parasite infection and allergic inflammation. *Glycoconj J* 2004;19:601–6.
9 O'Reilly MK, Paulson JC. Siglecs as targets for therapy in immune-cell-mediated disease. *Trends Pharmacol Sci* 2009;30:240–8.
10 Schwartz-Albiez R *et al.* CD22 antigen: biosynthesis, glycosylation and surface expression of a B lymphocyte protein involved in B cell activation and adhesion. *Int Immunol* 1991;3:623–33.
11 Cambi A *et al.* How C-type lectins detect pathogens. *Cell Microbiol* 2005;7:481–8.
12 Moody DB. How a T cell sees sugar. *Nature* 2007;448:36–7.
13 Hooper LV, Gordon JI. Glycans as legislators of host–microbial interactions: spanning the spectrum from symbiosis to pathogenicity. *Glycobiology* 2001;11:1R–10R.
14 Xavier RJ, Podolsky DK. Unravelling the pathogenesis of inflammatory bowel disease. *Nature* 2007;448:427–34.
15 Gagneux P, Varki A. Evolutionary considerations in relating oligosaccharide diversity to biological function. *Glycobiology* 1999; 9:747–55.
16 Campbell BJ *et al.* Altered glycosylation in inflammatory bowel disease: a possible role in cancer development. *Glycoconj J* 2001; 18:851–8.
17 Moehler T *et al.* Involvement of α1,2-fucosyltransferase I (FUT1) and surface-expressed Lewis^y (CD174) in first endothelial cell–cell contacts during angiogenesis. *J Cell Physiol* 2008;215:27–36.
18 Frol'ová T *et al.* Detection of galectin-3 in patients with inflammatory bowel diseases: new serum marker of active forms of IBD? *Inflamm Res* 2009;58:503–12.
19 Liu YH *et al.* N-Acetyl-seryl-aspartyl-lysyl-proline prevents cardiac remodeling and dysfunction induced by galectin-3, a mammalian adhesion/growth-regulatory lectin. *Am J Physiol Heart Circ Physiol* 2009;296: H404–12.

28
Sugars as Pharmaceuticals

Helen M. I. Osborn and Andrea Turkson

From discussions in the previous chapters it is evident that sugar coding has considerable consequences on biological systems. Carbohydrates have, for example, been implicated in a wide range of disease processes. This has motivated both synthetic and medicinal chemists alike to investigate new treatments for these diseases using carbohydrate-based therapies. In this chapter a selection of carbohydrate-based pharmaceuticals that offer promise for treating specific diseases are discussed. The strategies discussed involve (i) selective blocking of carbohydrate–lectin interactions that are central to the disease pathway, (ii) immunization using monoclonal antibodies for carbohydrate antigens, and (iii) inhibition of enzymes that synthesize the disease-associated carbohydrates. The development of carbohydrate-based therapies is an area that is constantly evolving as more disease-associated carbohydrates are identified and methods for synthesizing carbohydrates continually improve (for refined synthetic procedures, please see Chapter 3).

28.1
Cancer Therapeutics

Cancer is a major health threat across the world with the latest statistics suggesting that more than one in three people will develop cancer at some point in their life (www.cancerhelp.org.uk). It is documented that of the secreted and cell-surface proteins associated with cancer, 80% are heavily glycosylated, that is, they contain a high sugar content which is often different to that found on healthy cells (see Chapters 6 and 7 and Table 25.2). Thus, formation of tumorous tissue is often associated with changes in glycosylation, i.e. an under- and/or overexpression of naturally occurring glycans compared to healthy cells. These tumor cells with altered surfaces are able to spread from a primary tumor site to other parts in the body, creating secondary tumors, by a process known as metastasis. Binding to the selectins (see Chapters 19, 25, 27 and 29) lining blood vessels increases the

The Sugar Code. Edited by Hans-Joachim Gabius

ISBN: 978-3-527-32089-9

aggressiveness and metastatic capacity of tumor cells as they migrate around the body [1].

The difference between surface carbohydrates in normal and tumor cells provides many possible strategies that can be exploited for the treatment of cancer and these are summarized in Table 28.1 (see also Chapter 25.2).

Therapies have been developed that use surface carbohydrates on tumor cells as markers to generate antibodies against the tumor cells. An example of this is provided by the study of globo H antigen – a surface carbohydrate found on prostate, colon and human breast tumor cells (see Figure 28.1 for the structure of globo-H antigen and Chapter 1 for common structural abbreviations). The challenges associated with preparing carbohydrates of biological and therapeutic interest are discussed in Info Box 1.

Recently, mice treated with globo-H antigen bound to a protein carrier have been shown to generate large amounts of antibodies that are capable of recognizing tumor cells [2]. Monoclonal antibodies generated against tumor-associated carbohydrate antigens should be able to deliver anticancer drugs to their specific site of action. Thus, it has been possible to increase the therapeutic indices of anticancer drugs via ligand-mediated targeting of liposomal anticancer drugs. These liposomes or phospholipid bilayer vesicles can be used to carry drugs that are either trapped inside their hydrophilic aqueous interiors or associated to their

Table 28.1 Possible carbohydrate-based approaches for the treatment of cancer.

Disease pathway	Therapeutic opportunity	Examples of carbohydrate-based therapies
Biosynthesis of unnatural carbohydrates on the tumor surface	Raise antibodies to the tumor-associated carbohydrate antigens to develop a vaccination strategy	Carbohydrate antigens on protein carrier for potential treatment of prostate, colon and breast cancer; theratope (sialyl-Tn antigen conjugate vaccine) for metastatic colorectal and breast cancer
Biosynthesis of unnatural carbohydrates on the tumor surface	Use antibodies generated from the tumor-associated carbohydrate antigens to deliver agents to cancer cells	Ligand-targeted liposomal therapeutics
Biosynthesis of unnatural carbohydrates on the tumor surface	Inhibit the carbohydrate processing enzymes	Naturally occurring aza and imino sugars, as well as synthetic derivatives – inhibition of metastatic tumors and tumor growth, as well as pulmonary and colon cancers
Biosynthesis of unnatural carbohydrates on the tumor surface	Inhibit the interactions of the tumor-associated carbohydrates with the lectin receptors, to minimize metastasis	Multivalent sialyl-Le^x derivatives, as potential antimetastatic agents

Info Box 1

It is often extremely difficult in the laboratory to make the carbohydrates required for therapies. This is due to the fact that carbohydrates can exist in several isomeric forms and selective methods have to be developed to access the isomers needed. This is discussed in more detail in Chapter 1. Enzymes provide useful alternatives to chemical methods for preparing carbohydrates and are finding widespread usage. Combined chemical–enzymatic approaches have also proved to be of value and these are illustrated in Chapter 3. A new technology that offers considerable promise for preparing carbohydrate targets is solid-supported carbohydrate synthesis. In recent years significant developments have been made so that an automated synthesis of many carbohydrate targets is now achievable, allowing easy access to disease-associated carbohydrates.

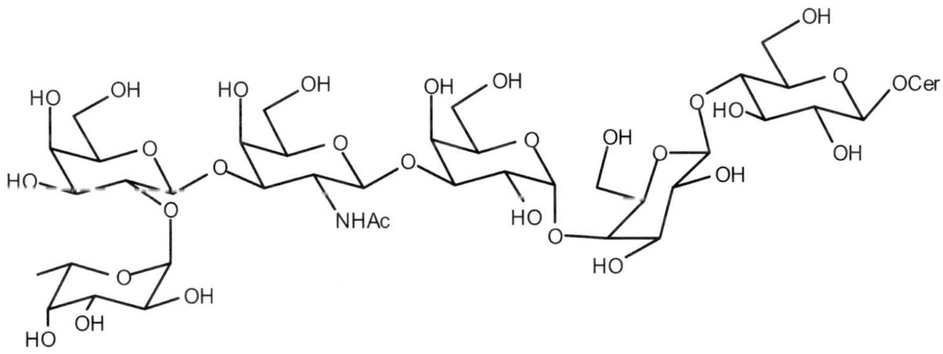

Figure 28.1 Structure of globo-H antigen.

hydrophobic bilayers. The drug package is delivered to its target cell using the homing antibody and then internalized, releasing the active drug through breakdown of the liposome–drug unit by lysosomal and endosomal enzymes. This internalization can be verified by confocal microscopy. Significantly, doxorubicin and vincristine have been delivered in this way. New advances in ligand-targeted liposomal therapeutics have been recently reviewed, highlighting the enormous advantages and the many problems encountered with this type of strategy [3].

An alternative anticancer therapy based on carbohydrates has examined the value of using carbohydrate-derived analogs as inhibitors of enzymes involved in the synthesis of disease-associated carbohydrates. Specifically, aza and imino sugars, which are able to mimic the shape and electronic nature of the transition state of substrates of specific enzymes involved in carbohydrate biosynthesis, have received considerable attention [4, 5]. Further details on the enzymatic process during glycan maturation can be found in Chapter 6 and further details on computer modeling for the design of enzyme inhibitors can be found in Chapter 2. Some examples of inhibitors of carbohydrate biosynthesis that are showing promise as therapeutic agents are highlighted in Figure 28.2. Many of these are derived from natural products, as discussed in Info Box 2.

In another line of cancer therapy research, efforts have been made to inhibit the interactions between the cancer-associated carbohydrates and the tissue receptors (lectins) (for examples, please see Chapter 25). This has proved particularly interesting for aborting metastatic processes, as interactions between tumor-associated carbohydrates and receptors on endothelial cells are widely implicated as being essential for the metastatic spread. Thus, considerable effort has been committed to synthesizing mimetics of tumor-associated carbohydrate antigens with the aim of producing derivatives that bind more strongly to the lectin receptors than the natural carbohydrate ligands [6, 7]. Indeed, recent clinical trials have reported that administration of sialyl-Lewis^x (Le^x) (Figure 28.3) mimetics offers such a potential. These mimetics can potentially block selectin–carbohydrate binding by occupying the selectin-binding sites (for the X-ray structure of sialyl-Le^x and P-selectin, please see Figure 16.1h, for further information on selectins, please see Chapter 27).

Although many anticancer therapies based on carbohydrates are still at the research and development stage, it is likely they will ultimately offer many advantages over conventional therapies. The potential of carbohydrate-based vaccines for treating cancer has been greatly improved by conjugating carbohydrates, which themselves display limited immunogenicity, with immunogenic proteins, or by presenting them as dendrimers (see Chapter 4 for structural aspects). Improved antibody-inducing capabilities have also been obtained by adding separate adjuvants to their formulations.

Info Box 2

Many leads for drug development are obtained from natural systems. Nature is able to prepare complex compounds through the use of enzymes. These can be isolated and their biological properties assessed through biological screens. Compounds that show the desired activity can then be further developed and new derivatives prepared as part of the drug development process. Imino sugars are examples of naturally occurring molecules that have been further developed by scientists to produce useful drugs. These sugar analogs are found in plants and microbes, and may inhibit enzymes associated with a range of diseases. The first natural imino sugar, nojirimycin, was discovered in Japan in 1966 from the broth containing a *Streptomyces* species. Castanospermine is another effective enzyme inhibitor that can be isolated from the Australian tree astanospermum. Research into modifying nojirimycin, castanospermine and related analogs has resulted in new inhibitors with improved selectivities, activities and solubilities. These analogs have been marketed for the treatment of many diseases including diabetes (miglitol was the first imino sugar drug to reach the market) and Gaucher's disease (see Chapter 10.14).

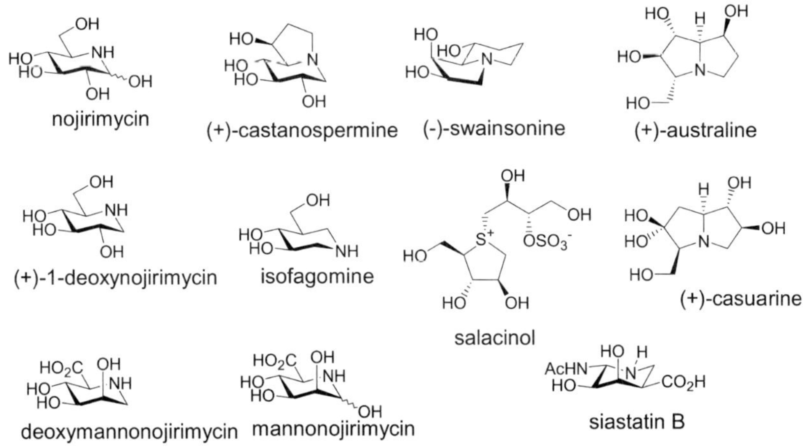

nojirimycin (+)-castanospermine (-)-swainsonine (+)-australine

(+)-1-deoxynojirimycin isofagomine salacinol (+)-casuarine

deoxymannonojirimycin mannonojirimycin siastatin B

Figure 28.2 Some naturally occurring glycosidase (carbohydrate-trimming enzyme) inhibitors.

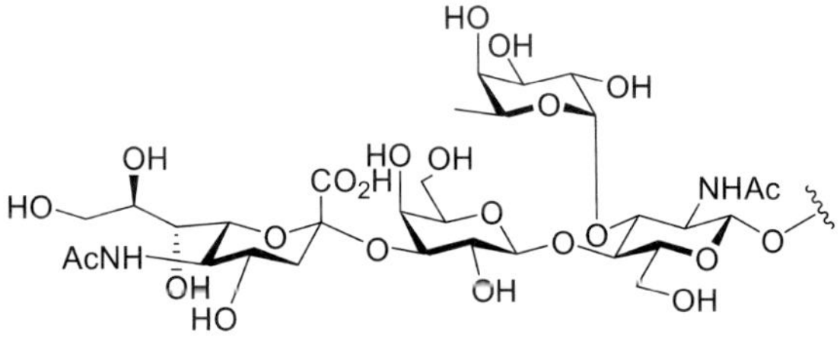

Figure 28.3 Structure of sialyl-Le^x.

28.2 Viral Infections: HIV-1 and Influenza

Carbohydrate-based therapies are being used to treat a range of viral infections, including AIDS and influenza. In these cases the infective pathways again rely on carbohydrate–protein interactions for their initiation, so methods that disrupt the synthesis and/or interaction of the carbohydrates are likely to offer hope for treatment of the diseases (see Chapter 17.2 for details of viral lectins and Chapter 4.1 for strategies to block viral adhesion). For example, with HIV-1, an unusual glycoprotein gp120 is displayed on the viral surface. More than 50% of the weight of gp120 has been ascribed to the carbohydrates and, of these carbohydrates, 24 have been identified as *N*-linked oligosaccharides, with 11 predominantly consisting of mannose glycans or hybrids thereof [8]. gp120 plays an important role in the infective process for the HIV as it is involved in the initial binding of the virus to lectins on the T cells. At present, clinical treatments for HIV-1 have been developed based upon inhibition of the HIV-1 reverse transcription enzyme and the HIV-1 protease enzyme that cleaves polyproteins into smaller functional protein units. Carbohydrate-based inhibitors have included the nucleoside analog reverse transcriptase

Table 28.2 Possible carbohydrate-based approaches for the treatment of viral infections[a].

Disease pathway	Therapeutic opportunity	Examples of carbohydrate-based therapies
Biosynthesis of unnatural carbohydrates on the AIDS virus	Inhibition of the HIV-1 reverse transcriptase enzyme and inhibition of HIV-1 protease enzyme which cleaves polyproteins into smaller functional protein units and hence inhibits adhesion	Nucleoside analog inhibitors such as zidovudine and 6-*O*-butanoylcastanospermine and *N*-butyl-deoxynojirimycin
Adhesion of the virus to the cell to initiate infection	Block the virus–cell adhesion mechanisms	Multivalent sialic acid-containing derivatives, e.g., to block influenza infections
Cleavage of sialic acid residues attached to cellular glycolipids and viral glycoproteins	Stop the virus from exiting the host cell, thus reducing the amount of virus that is released to infect other cells	Zanamivir and oseltamivir

[a] For further details on multivalent glycoclusters please see Chapter 4 and for further details on sialidase inhibitors see Chapter 17.3.

inhibitor zidovudine (3′-azido-3′-deoxythymidine) and others such as 6-*O*-butanoyl-castanospermine (BuCast; celgosivir) and *N*-butyl-deoxynojirimycin (Zavesca) (see Figure 28.4 and Table 28.2 below).

To act as competitive inhibitors of HIV-1, nucleoside analogs need to be phosphorylated within the cell to their triphosphate form. Once incorporated into the DNA chain, the nucleoside acts as a chain terminator, preventing chain elongation due to a lack of the 3′-hydroxyl required on the ribose moiety. Widespread use of antiretroviral drugs such as zidovudine has, however, led to the loss of therapeutic effect [9]. A study conducted on patients with advanced HIV-1 showed a desensitization of HIV-1 isolates to zidovudine during the course of treatment. Such resistance to reverse transcriptase inhibitors has been attributed to substitution of residues near the HIV-1 reverse transcriptase nucleotide binding site. Protease inhibitors, which bind via substrate-based compounds, acquire high-level resistance through mutation and amino acid substitution within the substrate-binding pocket [10]. A further carbohydrate-based drug, celgosivir, inhibits the gp120 protein glycosylation producing a different cell-surface carbohydrate that is unable to interact with the T cell lectins [11].

When considering new methods for treating influenza, it is worth remembering that within the human population, influenza A and B are the major pathogenic virus strains (please see Chapter 17.2). The influenza virus possesses surface proteins that interact with specific, membrane-bound oligosaccharides on human cells. Inhalation of these airborne viral particles is the main cause for the spread of influenza viruses. Influenza A can be subdivided depending on the surface antigens present, i.e., hemagglutinin or neuraminidase [12]. Multivalent binding

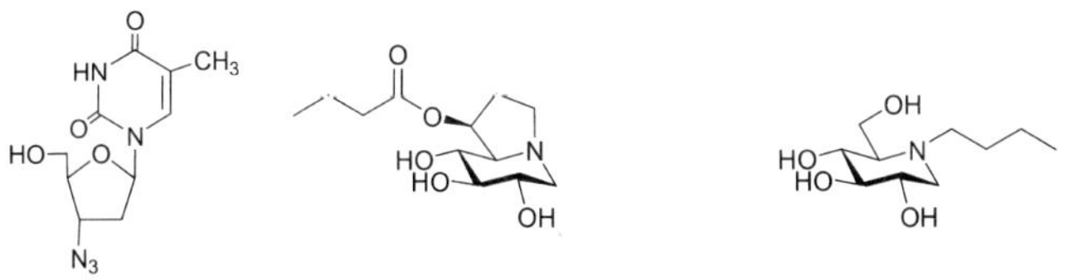

Zidovudine (AZT)
6-O-Butanoyl castanospermine (BuCast)
N-Butyl-deoxynojirimycin (Zavesca)

Figure 28.4 Structures of carbohydrate-based inhibitors.

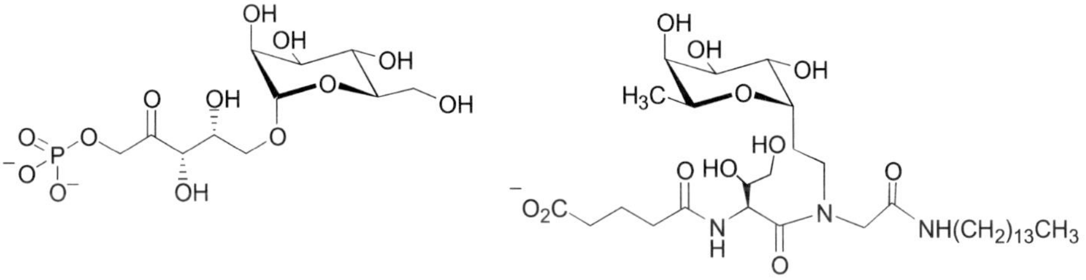

Figure 28.5 Sialic acid glycomimetics.

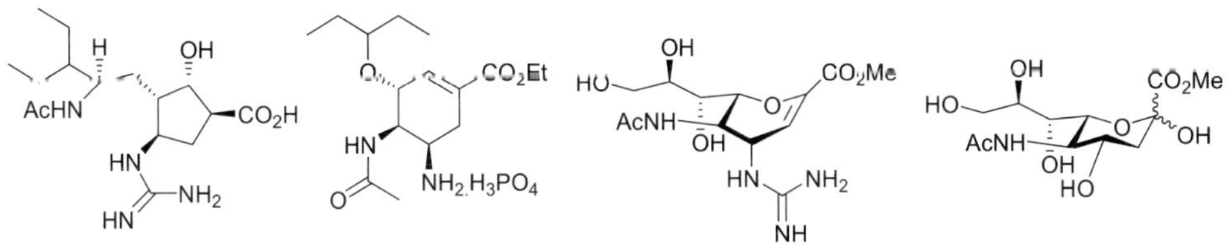

Peramivir
Oseltamivir
Zanamivir
N-Acetyl neuraminic acid (Sialic acid)

Figure 28.6 Structures of some anti-influenza drugs.

of hemagglutinin to glycoconjugates terminating in sialic acid initiates infection in host cells. Sialic acid-functionalized multivalent conjugates have been synthesized that are capable of blocking this virus–cell adhesion mechanism (see Chapter 4) [13]. Some sialic acid glycomimetics have shown 10^4-fold greater potency in inhibitory activity compared with the parent polymers that bear only sialic acid residues (Figure 28.5) [14].

Neuraminidase inhibitors are glycomimetic drugs that inhibit the cleavage of sialic acid residues attached to cellular glycolipids and viral glycoproteins. Drugs such as zanamivir and oseltamivir (Figure 28.6 and Figure 17.3) are able to bind to the neuraminidase that protrudes from the surface of the influenza virus – this action stops the virus from exiting the host cell, reducing the amount of virus that is released to infect other cells, and thus halting the spread of the virus throughout respiratory mucus. If treatment using zanamivir and oseltamivir is commenced within 48h of contracting laboratory-confirmed influenza, symptoms can be reduced in approximately 0.7–1.5 days.

28.3 Diabetes

As illustrated above, many new strategies for treating diseases have been developed that are based on the inhibition of enzymes involved in the biosynthesis of diseased-associated carbohydrates. In some cases the same drug can be used to treat different diseases. Thus, many of the carbohydrate-processing inhibitors that have attracted interest as anticancer agents have also been studied as antiviral compounds for the treatment of AIDS. An additional application for these glycosidase inhibitors is found in their treatment of diabetes. For example, acarbose is an α-glucosidase inhibitor that is effective in the intestine for slowing down the digestion of carbohydrates. This lengthens the time required for carbohydrates to be converted to glucose, thereby facilitating better blood glucose control. Carbasugar AO-128 (voglibose) is also an α-glucosidase inhibitor and has been shown to exert antiobesity activity in addition to antidiabetic action. Voglibose is well tolerated and may be used in combination with other drugs such as glibenclamide. Miglitol is an oral α-glucosidase inhibitor and is used in the treatment of non-insulin-dependent diabetes mellitus (Figure 28.7). Further details of this, and other carbohydrate-based therapies, can be found in a review article by Osborn [15].

The α-glucosidase inhibitors acarbose and miglitol offer advantages over other oral agents for type 2 diabetes in that they do not cause weight gain. They can therefore offer particular advantage for the treatment of type 2 diabetes in the clinically obese. They have also proved effective in combination with other oral hypoglycemics and insulin, and have been used in type 1 subjects to reduce glycemic fluctuations and reduce insulin dosage. They do, however, cause gas, bloating and diarrhea, but these side-effects can be minimized by commencing treatment with a relatively low dose that can be increased slowly.

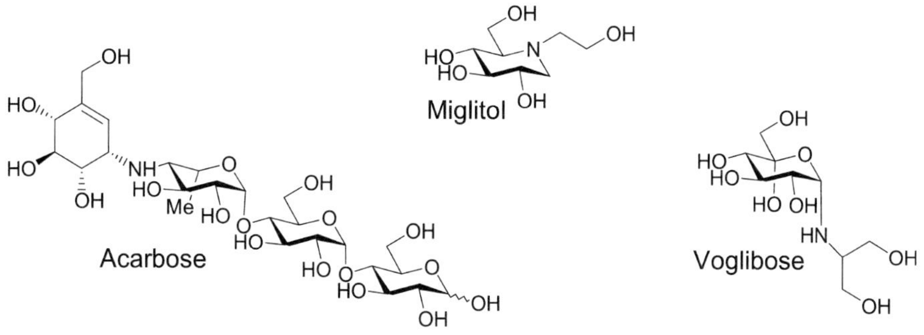

Acarbose

Miglitol

Voglibose

Figure 28.7 Structures of current antidiabetic drugs.

28.4 Carbohydrate-Based Antibacterial Agents

Due to the increase in bacterial resistance to a plethora of contemporary antimicrobial agents, antibiotics can actually produce major health problems. The rapid and widespread emergence of antibiotic-resistant bacteria has resulted in an urgent need to discover new antibiotics/antibacterial agents. As a result of the aging population and the increase in immunocompromised patients there is a dramatic increase in the incidence of life-threatening infections [15]. Again, a variety of approaches are being pursued to develop novel carbohydrate-based therapies that will be appropriate as antibacterial agents. These are summarized in Table 28.3.

Many lectins on the surface of bacteria show strong and specific binding for carbohydrates expressed on human cells (see Chapter 17.1 for details of bacterial lectins), and such interactions form an essential part of the infection pathway. Moreover, microbial enzymes can modify carbohydrate chains on host cells, resulting in an increased surface density of lectin receptors, which can enhance a bacterium's virulence. Evidence has shown that otherwise healthy people who suffer from recurrent infections often have an unusually high tissue expression of carbohydrate adhesin [19]. Promisingly, the lectin–oligosaccharide interactions (for an explanation of principles of this interaction, see Chapter 13; for methods to determine lectin specificity, see Chapter 14) can be competitively inhibited by

Table 28.3 Possible carbohydrate-based approaches for the treatment of bacterial infections.

Disease pathway	Therapeutic opportunity	Examples of carbohydrate-based therapies
Interaction of bacterial receptors with carbohydrates on the surface of human cells, necessary to initiate infection	Inhibit interaction via administration of multivalent derivatives of carbohydrates; the bacteria can no longer bind to the human cells and cannot initiate infection	Soluble forms of human cell-surface oligosaccharide components are being investigated and developed for rational anti-infective drug design, e.g., for prevention and treatment of infections caused by *E. coli* O157:H7 hemorrhagic colitis and *C. difficile*
Overcome resistance of bacteria to current antibiotics	Glycorandomization has been utilized to optimize antibiotic properties and this has resulted in the production of monoglycosylated vancomycins that rival vancomycin itself [16–18]	New derivatives of vancomycin
Vaccination	Generation of antibodies to specific bacterial saccharides	Vaccination for (i) *S. pneumoniae* 19F; (ii) *H. influenzae* type b, to combat childhood meningitis; (iii) pneumococcal infections

administration of submillimolar concentrations of synthetic or natural carbohydrate derivatives, provided that the administered derivatives have a high affinity for the bacterial lectins. In such cases, the bacteria are no longer able to interact with the host and therefore pass through the body without initiating infection (Figure 28.8).

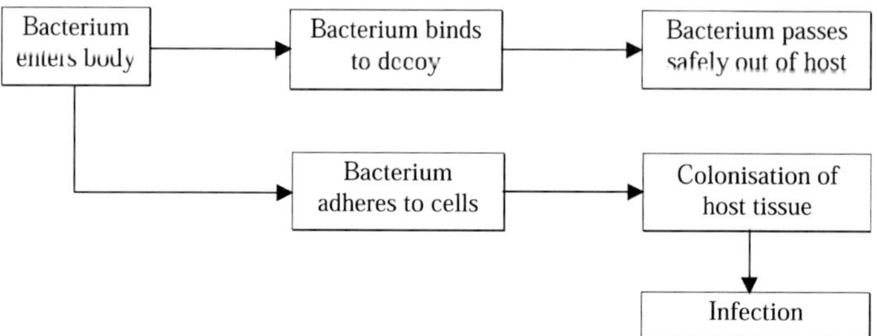

Figure 28.8 Administration of decoy carbohydrates.

The anti-infective carbohydrates and their biomimetics can be administered in monomeric or multivalent form in solution or presented immobilized on accessible surfaces, to block or arrest the targeted adhesion event. For example, Synsorb Biotech developed two anti-infective agents that consist of insoluble powders bearing carbohydrates that mimic the natural cell-bound carbohydrate targets of two bacterial toxins, Synsorb PK, effective against *Escherichia coli* O157 : H7 hemorrhagic colitis (Figure 28.9), and Synsorb Cd, effective against *Clostridium difficile*-associated diarrhea (see also Chapter 17.3). These agents underwent phase III clinical trials; however, these trials have now been discontinued.

Work of a more preliminary nature has also illustrated that treatment and/or prevention of gastrointestinal diseases can be effected by administration of carbohydrates recognized by such pathogens as enterotoxigenic *E. coli*, *Vibrio cholerae*, *Cryptosporidium* and *Helicobacter pylori* [20].

While medicinal chemists have sought to synthesize anti-infective agents in recent years, it is also worth noting that a number of anti-infective agents occur naturally, e.g., in human breast milk that contains numerous soluble oligosac-

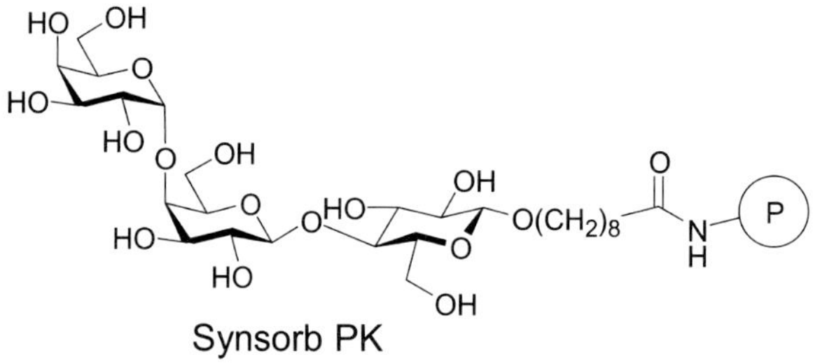

Figure 28.9 Structure of Synsorb PK.

charides that provide newborn babies with a mechanism for aborting infection processes (Figure 28.10). A prominent example is the ... Galβ1-4GlcNAcβ1-3Galβ1 ... trisaccharide that has been proposed as a receptor for adherence of *Streptococcus pneumoniae* to buccal epithelial cells. At corresponding concentrations, sialylated milk oligosaccharides strongly inhibit binding of influenza A virus and S-fimbriated enteropathogenic *E. coli* to their respective host cells (see [15] for further details).

An alternative strategy to prevent bacterial infections is to synthesize and administer capsular polysaccharides or fragments from bacterial cell surfaces that give rise to highly specific immune responses. An example of this has been reported, in which a spacer-containing nonasaccharide fragment of *S. pneumoniae* 19F, a common cause of respiratory infections in children, was synthesized (Figure 28.11) [21].

Although these synthetic structures have a lower relative molecular mass than the natural polysaccharides, they are often of sufficient size to function efficiently as immunogenic components of conjugate vaccines. Advantages of these synthetic

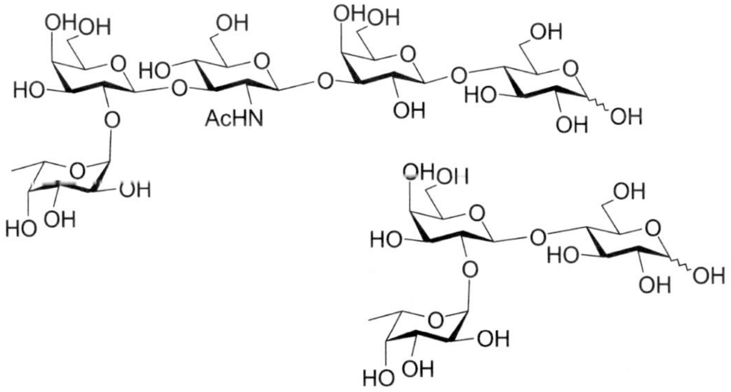

Figure 28.10 Structures of some human milk oligosaccharides.

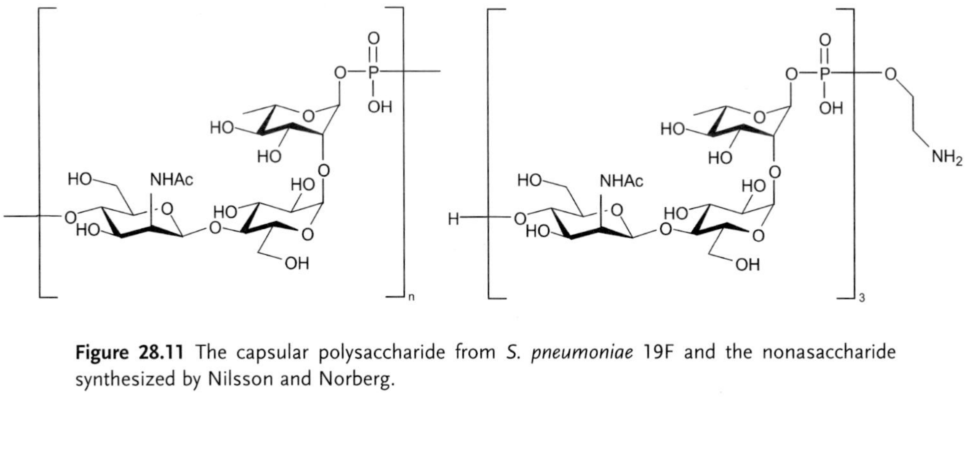

Figure 28.11 The capsular polysaccharide from *S. pneumoniae* 19F and the nonasaccharide synthesized by Nilsson and Norberg.

products are their defined structures and lack of bacterial contaminants, and, in some cases, the economy of production (typically less than 10 mg per dose of synthetic vaccine is required). Such polysaccharide vaccines, especially the modern conjugated vaccines, have proved to be very efficient in preventing a range of human disease. For example, ActHIB/OmniHIB is a conjugate vaccine used to prevent infection caused by *Haemophilus influenzae* type b to combat childhood meningitis. A number of new glycoconjugate vaccines have also been approved for use. Prevnar is used to prevent pneumococcal infections that can cause earaches, meningitis, blood poisoning and pneumonia. Typhim Vi is recommended for travelers to developing countries where the standards of hygiene and sanitation are poor [15].

28.5
Carbohydrate-Based Antithrombotic Agents

One further area in which carbohydrate-based therapies have seen considerable development is in antithrombotic agents, e.g., for the treatment of deep vein thrombosis. Heparin was the first polysaccharide-based drug to demonstrate powerful anticoagulant activity and find widespread use in humans (see also Figure 1.7 and Chapter 11.1 for general information on glycosaminglycan structure). It has been used in the clinic since 1937 to treat thrombosis. It is isolated from mammalian tissues as a complex mixture of glycosaminoglycan polysaccharides, and consists of repeat units of *N*-acetylglucosamine, D-glucuronic acid and L-iduronic acid, bearing *O*- and *N*-sulfate groups at varying points on the polymer chain (for structure of sugar units, please see Figure 1.6). It is administered intravenously and acts very rapidly–within minutes of receiving it, most patients have excellent anticoagulation that will prevent further clotting. Its mode of action is to reduce the action of antithrombin III (ATIII), a protease inhibitor involved in the blood clotting cascade. Thrombin is produced through the action of factor Xa, another protease, and can bind to ATIII to effect clotting of the blood. ATIII acts on factor Xa to prevent thrombin formation; hence, inhibition of factor Xa or thrombin by ATIII will control the blood clotting process [2]. Advantages of heparin over other anticoagulants such as warfarin are its low cost and fast action. Disadvantages, however, include the need for frequent blood tests to assess the level of coagulation, hospitalization to administer the drug intravenously, the risk of major bleeding in about 5% of patients and local reactions in the skin around the site of infusion.

A series of low-molecular-weight (LMW) heparins produced by enzymatic or chemical depolymerization of heparin to generate structures of chain length between 13 and 22 saccharides have been developed as alternative antithrombotic agents. Table 28.4 portrays LMW heparin-based drugs that have been approved for treatment of thrombosis, and Table 28.5 displays further drugs that are undergoing clinical trials. LMW heparins are salts of sulfated glucosaminoglycans with an average molecular mass of less than 8000 Da. Different methods of production

Table 28.4 Examples of LMW heparin-based drugs currently on the market.

Name	Target	Company
Dalteparin sodium (Fragmin)	Thrombosis, anticoagulant	Pfizer
Nadroparin calcium (Fraxiparin)	Thrombosis, anticoagulant	Sanofi-Synthelabo
Enoxaparin sodium (Clexane, Lovenox)	Thrombosis, anticoagulant	Aventis
Ardeparin (Normiflo)	Thrombosis, anticoagulant	Wyeth
Danaparoid (Orgaran)	Thrombosis, anticoagulant	Organon
Fondaparinux (Arixtra)	Thrombosis, anticoagulant	Organon/Sanofi-Synthelabo

Table 28.5 Further drugs that are undergoing clinical trials.

Name	Target	Company	Current phase
GH9001	Thrombosis	Inflazyme/Glycodesign	Phase I
Deligoparin (OP2000)	Antithrombosis, inflammatory bowel disease	Opocrin, Incara, Elan	Phase III
SR90107/ORG31540	Thrombosis	Sanofi/Organon	Phase I

give rise to different preparations of LMW heparins with altered molecular weight range and number of sulfation sites. Each commercially produced LMW heparin has a different recommended dose and reacts differently based on molecular weight.

LMW heparins have an advantage over heparin in that their activity is more predictable with a longer duration of action [2]. The medication can be given once or twice daily under the skin and patients can be treated at home. This eliminates or reduces the time patients need to spend in hospital. LMW heparins have been shown to be at least as effective as unfractionated heparin in treating patients with deep vein thrombosis. However, unfractionated heparin remains the anticoagulant of choice, rather than LMW heparins, for therapy for pregnant women. This is because it does not cross the placenta and therefore does not affect fetal coagulation.

Fondaparinux sodium is a synthetic pentasaccharide and the first synthetic agent to be a selective antithrombin-mediated inhibitor of factor Xa. Fondaparinux has complete bioavailability after subcutaneous injection and the peak plasma level is obtained after approximately about 2 h. Fondaparinux exhibits significantly better results than Enoxaparin in preventing venous thromboembolism after major orthopedic surgery and has recently been approved for use in thromboprophylaxis postsurgery. The clinical development of Fondaparinux for other thromboprophylactic indications is ongoing [22, 23].

28.6 Conclusions

This chapter has discussed a range of diseases selected to demonstrate the value of carbohydrate-based therapeutics. For many of these diseases no effective therapies are currently available, emphasizing the need for the development of new therapeutic strategies with new modes of actions. In many respects this area is still in its infancy and it is likely that, as our understanding of the roles of carbohydrates in biological systems improves, our ability to develop new carbohydrate-based therapies will progress. The promising results presented here portray many exciting therapies that are worthy of further study and many reports of carbohydrate-based therapeutics are likely to emerge in the forthcoming years.

Summary Box

Carbohydrates play many roles in biological systems, and therapies based on them are finding value for diseases in which carbohydrates are involved. This offers particular promise for treating cancer, viral and bacterial infections, diabetes, and thrombosis. Carbohydrate-based therapies are proposed to work by a range of mechanisms, including inhibition of enzymes that are used to make the disease-associated carbohydrates and blocking the interactions between the disease-associated carbohydrate and the specific receptor.

References

1 Aarnoudse CA *et al.* Recognition of tumor glycans by antigen-presenting cells. *Curr Opin Immunol* 2006;18:105–11.

2 Burger A. *Medicinal Chemistry and Drug Discovery*, 6th edn, pp. 203–48. Wiley Interscience, New York, 2003.

3 Sapra P, Allen TM. Ligand-targeted liposomal anticancer drugs. *Prog Lipid Res* 2003; 42:439–62.

4 Asano N *et al.* Sugar-mimic glycosidase inhibitors: natural occurrence, biological activity and prospects for therapeutic application. *Tetrahedron: Asymm* 2000;11: 1645–81.

5 Lillelund VH *et al.* Recent developments of transition-state analogue glycosidase inhibitors of non-natural product origin. *Chem Rev* 2002;102:515–53.

6 Gorelik E *et al.* On the role of cell surface carbohydrates and their binding proteins (lectins) in tumor metastasis. *Cancer Metastasis Rev* 2001;20:245–77.

7 McEver, RP. Selectin–carbohydrate interactions during inflammation and metastasis. *Glycoconj J* 1997;14:585–91.

8 Botos I *et al.* Structures of the complexes of a potent anti-HIV protein cyanovirin-N and high mannose oligosaccharides. *J Biol Chem* 2002;277:34336–42.

9 Larder BA *et al.* HIV with reduced sensitivity to zidovudine (AZT) isolated during prolonged therapy. *Science* 1989;243:1731–4.

10 Hansen J-E *et al.* *Complex Carbohydrates in Drug Research (Alfred Benzon Symposium 36)*, pp. 414–27. Munksgaard, Copenhagen, 1994.

11 Taylor DL *et al.* 6-*O*-Butanoylcastanospermine (MDL 28,574) inhibits glycoprotein processing and the growth of HIV. *AIDS* 1991;5:693–8.

12 Arvin AM, Greenberg HB. New viral vaccines. *Virology* 2006;344:240–9.

13 Choi S-K *et al.* Generation and *in situ* evaluation of libraries of poly(acrylic acid) presenting sialosides as side chains as polyvalent inhibitors of influenza-mediated hemagglutination. *J Am Chem Soc* 1997;119:4103–11.

14 Yarema KJ, Bertozzi CR. Chemical approaches to glycobiology and emerging carbohydrate-based therapeutic agents. *Curr Op Chem Biol* 1998;2:49–61.

15 Osborn HMI *et al.* Carbohydrate-based therapeutics. *J Pharm Pharmacol* 2004;56: 691–702.

16 Fu X *et al.* Antibiotic optimization via *in vitro* glycorandomization. *Nat Biotech* 2003;21:1467–9.

17 Griffith BR *et al.* Model for antibiotic optimization via neoglycosylation: synthesis of liponeoglycopeptides active against VRE. *J Am Chem Soc* 2007;129:8150–5.

18 Thibodeaux CJ, Liu HW. Manipulating nature's sugar biosynthetic machineries for glycodiversification of macrolides: recent advances and future prospects. *Pure Appl Chem* 2007;79:785–99.

19 Stapleton A *et al.* Binding of uropathogenic *Escherichia coli* R45 to glycolipids extracted from vaginal epithelial-cells is dependent on histoblood group secretor status. *J Clin Invest* 1992;90:965–72.

20 Lavelle EC. Targeted delivery of drugs to the gastrointestinal tract. *Crit Rev Ther Drug* 2001;18:341–86.

21 Nilsson M, Norberg T. Synthesis of a spacer-containing nonasaccharide fragment of streptococcus pneumoniae 19F capsular polysaccharide. *J Chem Soc Perkin Trans* 1998;1:1699–704.

22 Tan KT *et al.* Factor X inhibitors. *Expert Opin Invest Drugs* 2003;12:799–804.

23 Samama MM, Gerotziafas GT. Evaluation of the pharmacological properties, and clinical results of the synthetic pentasaccharide (fondaparinux). *Thromb Res* 2002;109:1–11.

29
Platelet Glycoproteins as Lectins in Hematology

Karin Hoffmeister and Hervé Falet

In response to hemorrhage, circulating platelets roll on the exposed subendothelium, adhere and form aggregates to seal a vascular leak. These processes involve platelet receptors such as von Willebrand factor (VWF) receptor complex, selectins and integrins. Platelet glycans have been investigated mainly in relation to platelet survival. Specifically, platelets exposed to temperatures below 37 °C are rapidly cleared by hepatic lectins and clustered platelet surface glycoproteins – a phenomenon that prohibits platelet refrigeration. Transfusion of human platelets, stored in blood banks as concentrates, remains the treatment of choice for thrombocytopenia and bleeding. However, platelets must be stored at room temperature, leading to major problems such as viral and bacterial growth, the loss of platelet functionality, and the complexity of managing supply with demand. A better understanding of the factors that mediate platelet clearance may lead to improved platelet storage. This chapter focuses on platelet adhesive receptors such as glycoprotein (GP)Ib and P-selectin (for illustration of the lectin site of P-selectin, please see Figure 16.1h; for its domain structure, see Figure 19.1; for its role in inflammation, see Chapter 27.4) and the P-selectin glycoprotein ligand (PSGL)-1. These glycoproteins interact with lectins or 'act' as lectins *in vitro* and *in vivo* (for definition of the term 'lectin', please see Chapter 15).

29.1
Platelet Physiology

Human platelets are anuclear discoid cell fragments that measure 2–4 μm in diameter. They circulate in the bloodstream at a concentration of $150–400 \times 10^3/\mu l$ with a half-life of 8–10 days. Platelets are formed by megakaryocytes in the bone marrow, and are eliminated primarily by the spleen and the liver. Platelets play an essential role in hemostasis and coagulation. Platelet dysfunction or low blood count predisposes to hemorrhage, while their hyperactivation increases the risk of thrombosis [1]. Once activated, platelets change shape and secrete their granular contents (Figure 29.1). Platelet adhesion, activation and aggregation are mediated

The Sugar Code. Edited by Hans-Joachim Gabius

ISBN: 978-3-527-32089-9

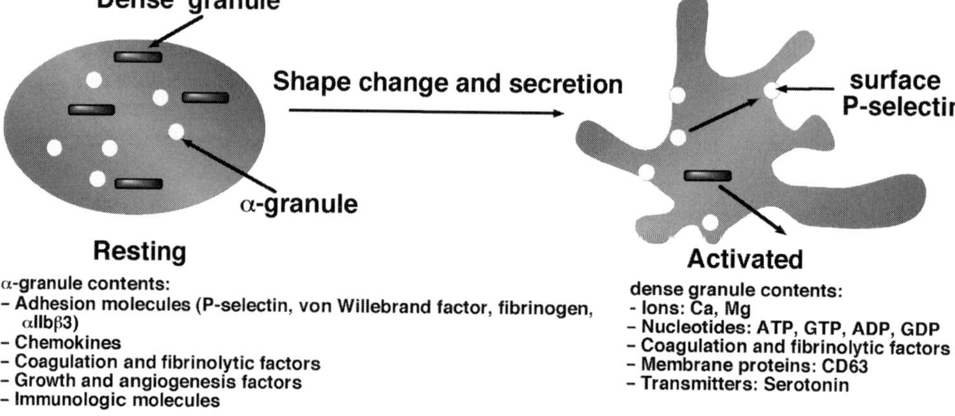

Figure 29.1 Following activation resting platelets rapidly change shape and secrete their granule contents.

by specific surface receptors that include integrins, such as the fibrinogen receptor (please see Figure 25.3 for a role of the $\alpha_5\beta_1$ integrin in growth regulation and its ligand properties to a tissue lectin via proper glycosylation) $\alpha_{IIb}\beta_3$, the leucine-rich von Willebrand factor (VWF) receptor GPIb-IX–V complex, immunoglobulin family receptors (GPVI) and P-selectin (Figure 29.2). Integrins are noncovalent heterodimers composed of α- and β-subunits, mediating cell–cell and cell–matrix interactions. $\alpha_{IIb}\beta_3$-Integrin (or GPIIb–IIIa) is a calcium-dependent receptor for fibrinogen and it is essential for platelet aggregation. $\alpha_{IIb}\beta_3$ has a very low affinity for fibrinogen on resting platelets. After platelet activation, $\alpha_{IIb}\beta_3$ receives signals from inside the cell to rapidly change its conformation to bind fibrinogen with

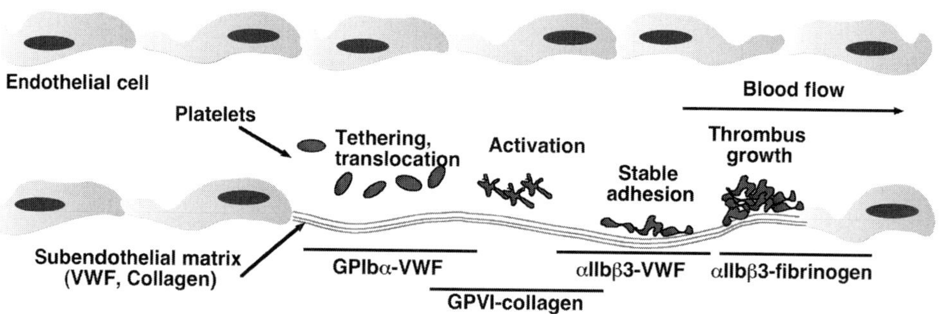

Figure 29.2 Multistep adhesive and signaling interactions of platelets with exposed subendothelium. Platelet tethering and rolling are initiated by binding of GPIbα to VWF. Platelet activation is then mediated by collagen binding to the immuno receptor GPVI. Platelets firmly adhere to the subendothelium and form a thrombus by binding through $\alpha_{IIb}\beta_3$ to VWF and fibrinogen.

high affinity. Its absence or deficiency leads to Glanzmann thrombocytopenia – a rare hematopoietic disorder associated with moderate to severe bleeding tendency and normal platelet morphology [2]. Although highly specialized in their role in hemostasis, platelets are also innate inflammatory cells. They have inflammatory and antimicrobial activities, playing an important role in wound healing, tumor growth and angiogenesis.

Bleeding due to thrombocytopenia is a major cause of morbidity in clinical disorders such as sepsis or cancer. Platelet transfusion remains the sole replacement therapy for patients having active bleeding. Unlike red blood cells or other transfused products, platelets for transfusion are stored at room temperature in gas-permeable bags for 5–7 days or less – a limit imposed by the risk of bacterial growth and the loss of platelet functionality. Platelet refrigeration can dramatically reduce the risk of bacterial growth and preserve platelet function [3], thereby permitting extended storage of platelets. However, almost 40 years ago, Murphy and Gardner demonstrated that refrigerated, transfused human platelets are rapidly cleared from the circulation [4].

29.2 GPIb-IX–V Complex

The GPIb-IX–V complex (VWF receptor) mediates platelet adhesion to collagen-bound VWF and activation under high shear rates. It belongs to the leucine-rich family of proteins and consists of four distinct transmembrane subunits: GPIbα, connected to GPIbβ by a disulfide bond, and the noncovalently associated GPIX and GPV. Absence or deficiency of GPIbα, GPIbβ or GPIX is responsible for Bernard–Soulier syndrome – a bleeding diathesis characterized by severe thrombocytopenia with abnormally large platelets and impaired platelet adhesion [5]. GPIbα is a type I membrane-spanning subunit containing an N-terminal, ligand-binding domain including leucine-rich loops and an anionic peptide sequence with three tyrosine residues, a sialomucin core with multiple *O*-glycans (for information on mucins and the typical forms of mucin-type glycosylation, please see Chapter 7) which projects the N-terminus above the platelet membrane, a transmembrane domain and a cytoplasmic tail (Figure 29.3) [5]. The extracellular N-terminal domain of GPIbα contains the binding site for VWF and interacts with various other ligands, such as thrombin, platelet P-selectin and neutrophil/macrophage β_2 integrin $\alpha_M\beta_2$ (Mac-1, CD11b/CD18) [5]. Activated leukocytes use integrin $\alpha_M\beta_2$ to adhere firmly to platelets. After vessel wall injury, platelet GPIb-IX–V mediates the initial adhesion contact to VWF bound to collagen within the exposed vascular subendothelium under arterial shear rates, allowing platelets to roll at much slower velocities. Typically, this glycoprotein contains up to four *N*-linked core-fucosylated glycans (for information on *N*-glycans and their core substitutions including their role as molecular switches for conformational changes, please see Chapters 2, 6 and 8), two located within the *N*-terminus [6] and *O*-linked glycans within the mucin-like region (Figure 29.3) [7]. The exposure of individual sugars

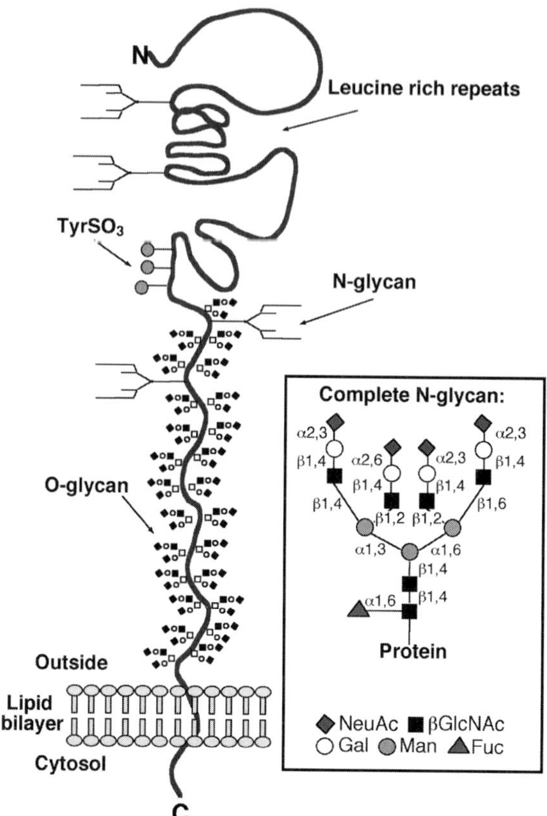

Figure 29.3 Schematic diagram of the human VWF receptor subunit GPIbα. A typical *N*-glycan structure identified on GPIbα is shown (inset box).

such as Gal or βGlcNAc is detectable by their binding to specific lectins, for example *Ricinus communis* agglutinin (RCA) I and succinylated wheat germ agglutinin (sWGA) (for a survey of specificities of plant lectins, please see Table 18.1).

29.3 Cold-Induced Platelet Clearance

An effort to address the clinically relevant problem of why refrigerated platelets fail to circulate led to the definition of a previously unsuspected, carbohydrate-dependent platelet clearance mechanism. The hepatic macrophage carbohydrate-binding integrin $\alpha_M\beta_2$ selectively recognizes clustered GPIbα subunits of the VWF receptor following short-term (4h) platelet refrigeration, resulting in the phagocytosis and clearance of platelets [8–10]. Experiments using α_M-deficient, but not

VWF-, complement- or P-selectin-deficient mice showed marked improvement in the survival of refrigerated platelets. Removal of GPIbα's N-terminal ligand binding-domain using the *O*-sialoglycoprotein endopeptidase restored the circulation of refrigerated murine wild-type platelets, indicating that the external domain of GPIbα initiates clearance [8]. Subsequent work narrowed carbohydrate recognition by integrin $\alpha_M\beta_2$ to exposed βGlcNAc residues on *N*-linked GPIbα glycans (Figure 29.4) [9, 10]. While resting platelets weakly bind sWGA, refrigerated platelets have markedly increased binding, suggesting that altered epitope presentation and/or clustering of exposed βGlcNAc on GPIbα can facilitate lectin binding to refrigerated platelets, representing another example for a physiologic modulation of lectin binding and for sugar-encoded information (for further examples, please see Chapters 19, 25 and 27) (Figure 29.4). What causes these alterations of specific carbohydrate epitopes on platelet glycoproteins during refrigeration? Actin rearrangement during refrigeration is likely to initiate surface VWF receptor redistribution from linear arrays into aggregates [8]. Platelet refrigeration can therefore cause profound changes in the presentation of exposed glycans on the platelet surface. Glycan clustering is detected early after refrigeration [8, 9], but may increase with long-term platelet storage and refrigeration in plasma (Figure 29.4). Integrin $\alpha_{IIb}\beta_3$, the most abundant platelet integrin with 80 000 copies per platelet, also contains incomplete glycans with exposed βGlcNAc and/or Gal moieties [11]. It remains to be determined if changes in the $\alpha_{IIb}\beta_3$ integrin and its glycans occur during platelet refrigeration.

A potential method for preventing the rapid clearance of refrigerated platelets for transfusion was envisioned to be enzymatic galactosylation of surface βGlcNAc residues on platelet glycoproteins using a β1–4-galactosyltransferase (β4GalT).

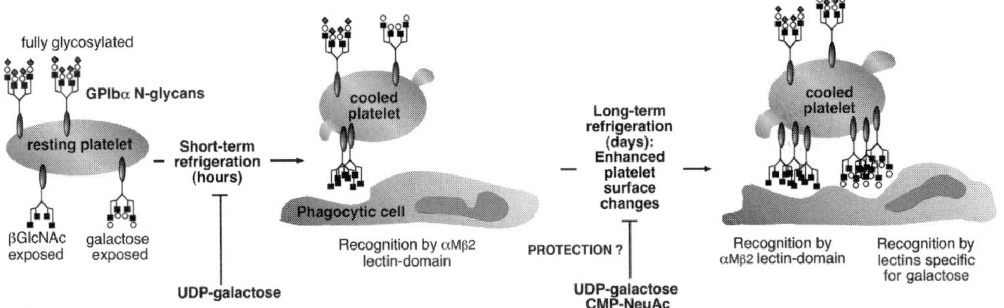

Figure 29.4 Proposed mechanisms for cold-induced platelet clearance. The VWF receptor complex, specifically GPIbα, has complete and incomplete *N*-linked glycans with exposed βGlcNAc and/or Gal residues. Clustering of VWF receptors and of exposed βGlcNAc initiates recognition and phagocytosis by the macrophage $\alpha_M\beta_2$ integrin. Coverage of exposed βGlcNAc residues by Gal prevents phagocytosis of short-term refrigerated platelets. Extended refrigeration may be followed by additional surface changes such as 'hyperclustering' of receptors and their associated glycans and/or carbohydrate residues, leading to recognition and phagocytosis by $\alpha_M\beta_2$ integrin-independent mechanisms, probably through galactose receptors and/or scavenger receptors.

Surprisingly, both human and murine platelets have functional platelet galactosyltransferase(s), especially the β4GalT1 (for further information on glycosyltransferases, please see Chapters 6 and 7), on their surface (for more information on platelet-associated glycosyltransferases, see Info Box). β4GalT catalyzes the coupling of galactose in a β1,4-linkage to exposed βGlcNac residues on the *N*-linked glycans of GPIbα. This reaction improved the circulation of short-term (4h) chilled murine platelets [9] (Figure 29.4). This was surprising since, while depriving the α_M-lectin domain of its βGlcNAc ligand on refrigerated platelets, galactosylation theoretically provides a new ligand for asialoglycoprotein (ASGP) receptors (for historic aspects of its detection, please see Chapter 15.4; for structural aspects Chapter 19). It is possible that the number of exposed βGlcNAc residues on GPIbα is small, such that even after clustering and galactosylation, the Gal density is insufficient to engage galactose-recognizing lectins [9]. However, this theory does not account for the recognition of long-term refrigerated human or murine platelets by the ASGP receptor. Subsequent experiments showed that galactosylation was not beneficial in improving the circulation of 48-h refrigerated platelets in humans or in mice [12]. Evidently, different mechanisms are involved in the clearance of short-term and long-term refrigerated platelets.

Of note, sialic acid in α2,3-linkage can function as an anticlearance signal for blood cells and serum proteins [13]. Platelets lose sialic acid from their membrane glycoproteins as they age and are removed from the circulation. Mice deficient in the sialyltransferase ST3Gal-IV [14] have low platelet counts, most likely caused by accelerated platelet clearance due to the high density of exposed Gal residues. Therefore, carbohydrate density and/or its presentation alter platelet survival.

29.4
Long-Term Platelet Refrigeration May Reveal New Insights into Platelet Clearance

Like short-term refrigerated platelets, long-term refrigerated platelets are rapidly removed in the liver following transfusion. Since prolonged storage in the cold increases the binding of the before mentioned plant lectin RCA-I [Hoffmeister, K.M., Rumjantseva, V. *et al.*, unpublished data], we postulate that clustering of Gal residues engages a lectin-based recognition system different from the previously defined α_M–βGlcNAc interaction, most likely through galactose-binding receptors on macrophages and/or hepatocytes. Since both galactose and βGlcNAc exposure would be expected to contribute to the clearance of refrigerated platelets, a combination of both galactosylation and sialylation would be required to rescue the circulation of long-term refrigerated platelets (Figure 29.4). Considering the complexity of the changes induced by long-term platelet refrigeration, whether glycosylation (here platelet sialylation) alone will be sufficient to overcome the rapid clearance of long-term refrigerated platelets awaits experimental testing. Having dealt with this recognition system, we next introduce a P-selectin activity in this respect.

Info Box

Over 30 years ago, G.A. Jamieson and A.J. Barber proposed that externally disposed glycosyltransferase activity mediates platelet adhesion, suggesting that Golgi enzymes may function as lectins during platelet activation [Jamieson GA *et al.*, Platelet collagen adhesion characterization of collagen glucosyltransferase of plasma membranes of human blood platelets. *Nature New Biology* 1971; **234**: 5–7]. Subsequent work ruled out ecto-glycosyltransferase activity in nucleated cells and established the Golgi apparatus as the sole site of such enzymes, although no further studies examined platelets (see Chapter 6 for classical glycosylation in the Golgi). We presented evidence for a β4GalT on the surface of platelets. Upon surface receptor activation, platelets excrete glycosyltransferases and donor substrates, accommodating posttranslational modifications of extracellular proteins in an environment that is devoid of direct endoplasmic reticulum or Golgi regulation. Labeling with granule markers reveals no correspondence of Golgi markers with dense- or α-granules [Wandall HH, Hoffmeister KM *et al.*, unpublished data]. Extracellular signal-dependent glycosylation may be a novel function of platelets that demonstrates previously unrecognized diversity regarding the functional roles of posttranslational modification in eukaryotic cells.

29.5 P-Selectin and PSGL-1

Selectins are cell adhesion molecules expressed on platelets, endothelial cells and lymphocytes. They belong to the C-type lectin family (for further general information, please see Chapters 19, 20 and 27). There are three types of selectins, that is P-, E- and L-selectins. Platelets and endothelial cells express P-selectin on the membrane of their α-granules and Weibel–Palade bodies, respectively. P-Selectin moves to the membrane surface during platelet and endothelial cell activation, and plays an essential role in the initial recruitment of white blood cells to the site of injury during inflammation. A well-characterized ligand for P-selectin is PSGL-1 –a mucin-type glycoprotein expressed on all white blood cells as well as on platelets. The affinity between the lectin-like domain of P-selectin and carbohydrate groups present on PSGL-1 mediates the adhesion and rolling of white blood cells along the blood vessel wall, and allows white blood cells to leave the blood vessel and enter the site of inflammation (Figure 29.5) [15] (for further information on the mechanics of tethering and the catch bonds, please see Chapter 19 and Chapter 27 and also Figure 27.1).

The requirement of platelets for experimental pulmonary metastasis and the ability of tumor cells to aggregate platelets was first recognized in 1968 by G.J. Gasic *et al.*, who demonstrated impaired experimental pulmonary metastasis in mice following platelet depletion. Others have proposed that platelets promote

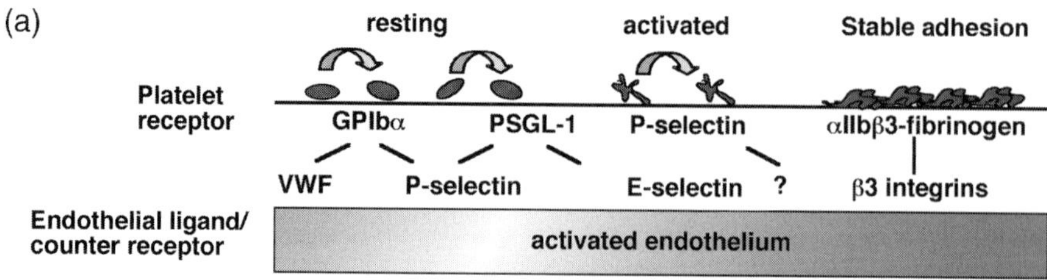

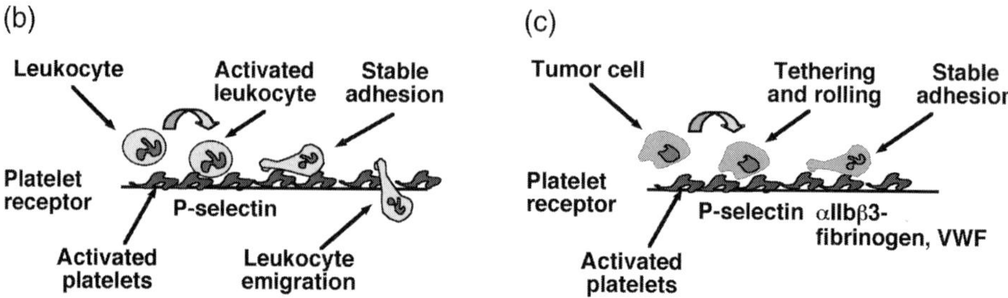

Figure 29.5 (a) Adhesion molecules involved in platelet rolling along the activated endothelium. The endothelial ligand-mediating rolling of activated platelets is unknown. (b) Platelet-mediated leukocyte recruitment at sites of vascular damage. Adhesion of leukocytes to activated platelets is mediated by P-selectin. Firm adhesion and emigration of leukocytes to the site of inflammation is mediated by β_2 integrins (not shown). (c) Platelet-mediated tumor cell recruitment at sites of vascular damage. Platelets activated at sites of vascular damage mediate tumor cell recruitment through P-selectin-dependent tethering and rolling, and β_3 integrin-dependent adhesion.

angiogenesis, specifically during tumor growth [16]. Platelets activated at the site of vascular damage mediate tumor cell recruitment through P-selectin–tumor cell tethering and rolling followed by stable adhesion mediated by $\alpha_{IIb}\beta_3$ and VWF (Figure 29.5). Some investigators have suggested that a specific tumor cell/endothelium/platelet interaction may contribute to tumor-induced angiogenesis, since activated platelets release vascular endothelial growth factor–a potent proangiogenic factor. Interestingly, pro- and antiangiogenic proteins are organized into separate platelet α-granules and differentially released. It is possible that dense granules containing Golgi enzymes and donor substrates are well organized and differentially released following platelet activation.

29.6 Conclusions

After vessel wall injury, platelets tether and roll on exposed subendothelial tissues through the interaction between the GPIbα subunit of their GPIb-IX–V receptor and collagen-bound VWF. Unlike leukocyte rolling mediated by selectins, GPIbα-dependent platelet rolling on VWF has never been reported to be glycan dependent. However, addressing a clinically relevant problem, that is how to refrigerate platelets for transfusion, led to two novel and unexpected observations: (i) that GPIbα glycans are incompletely assembled and are essential for platelet survival, and (ii) that the monocyte/macrophage integrins $\alpha_M\beta_2$ can act as lectins by recognizing incompletely assembled and clustered platelet glycans to phagocytize chilled platelets.

Summary Box

Platelets, as leukocytes, use a multistep process of tethering and reversible adhesion on sites of vascular damage–a process which is mediated through GPIbα interaction with collagen-bound VWF. The reversible binding of GPIbα to VWF is independent of carbohydrates. Platelet carbohydrates, specifically on GPIbα, are essential in dictating platelet survival. Platelets are unique in having functional glycosyltransferases expressed on their surfaces. It is tempting to speculate that reversible glycosylation of platelet surface glycans is a mechanism that alters platelet survival.

References

1 George J. Platelets. *Lancet* 2000;355:1531–9.
2 Shattil S, Newman P. Integrins: dynamic scaffolds for adhesion and signaling in platelets. *Blood* 2004;104:1606–15.
3 Blajchman M, Goldman M. Bacterial contamination of platelet concentrates: incidence, significance, and prevention. *Semin Hematol* 2001;38:20–6.
4 Murphy S, Gardner F. Effect of storage temperature on maintenance of platelet viability–deleterious effect of refrigerated storage. *N Eng J Med* 1969;280:1094–8.
5 Andrews R *et al.* The glycoprotein Ib–IX–V complex. In: *Platelets* (Ed.: Michelson MA), pp. 145–63. Elsevier, San Diego, CA, 2007.
6 Tsuji T, Osawa T. The carbohydrate moiety of human platelet glycocalicin: the structures of the major Asn-linked sugar chains. *J Biochem (Tokyo)* 1987;101:241–9.
7 Tsuji T *et al.* The carbohydrate moiety of human platelet glycocalicin. *J Biol Chem* 1983;258:6335–9.
8 Hoffmeister K *et al.* The clearance mechanism of chilled blood platelets. *Cell* 2003; 10:87–97.
9 Hoffmeister K *et al.* Glycosylation restores survival of chilled blood platelets. *Science* 2003;301:1531–4.
10 Josefsson E *et al.* The macrophage $\alpha_M\beta_2$ integrin α_M lectin domain mediates the phagocytosis of chilled platelets. *J Biol Chem* 2005;280:18025–32.
11 Bauvois B *et al.* Membrane glycoprotein IIb is the major endogenous acceptor for human platelet ectosialyltransferase. *FEBS Lett* 1981;125:277–81.
12 Wandall H *et al.* Galactosylation does not prevent the rapid clearance of long-term 4 °C stored platelets. *Blood* 2008;111:3249–56.
13 André S *et al.* From structural to functional glycomics: are substitutions as molecular switches for shape and lectin affinity of N-glycans. *Biol Chem* 2009;390:557–65.
14 Ellies L *et al.* Sialyltransferase ST3Gal-IV operates as a dominant modifier of hemostasis by concealing asialoglycoprotein receptor ligands. *Proc Natl Acad Sci USA* 2002;99:10042–7.
15 McEver R. P-Selectin/PSGL-1 and other interactions between platelets, leucocytes and endothelium. In: *Platelets* (Ed.: Michelson MA), pp. 231–49. Elsevier, San Diego, CA, 2007.
16 Folkman J. Angiogenesis: an organizing principle for drug discovery? *Nat Rev Drug Discov* 2007;6:273–86.

30
Neurobiology Meets Glycosciences

Robert W. Ledeen and Gusheng Wu

The initial encounter of neurobiology with glycosciences appears to have occurred in the nineteenth century, at roughly the same period in which lectinology had its beginning (for a survey of history of lectinology, please see Chapter 15). Much of the credit for those auspicious beginnings belongs to J.L.W. Thudichum, considered by many the father of neurochemistry, who was a physician/scientist of widely diverse talents. He became known among his peers as 'The chemist of the brain' for his pioneering achievements including isolation and chemical analysis of biomolecules from human and animal brain. For this he put to good use the combustion apparatus given him by his teacher and renowned analytical chemist Justus von Liebig. Limited to relying on solvent extraction, crystallization and hydrolytic procedures, Thudichum accomplished the remarkable feat of isolating in relative purity a number of brain lipids. These included cerebroside and its close relative, sulfatide–the first glycoconjugates to be characterized in brain (or possibly any tissue). He showed cerebroside (galactosylceramide) to consist of a long-chain base with a primary hydroxyl attached to a six-carbon sugar, which he liberated by acid hydrolysis and termed 'cerebrose'; this later acquired its present name, galactose. This galacto-form of cerebroside (Figure 30.1) is especially rich in the nervous system owing to its prominence in the myelin sheaths encasing the larger axons. Thudichum's investigations were carried out in a climate of intense controversy, his preparations being described as 'impure smeary masses', the true nature of which no chemist would be able to decipher (it was alleged). Prominent neuroscientists of his day, and even somewhat later, spoke disparagingly of the chemical approach that produced what was described as 'cerebral hash' from which little could be learned. Thudichum's seminal contributions, which laid the groundwork for subsequent revelations on the rich diversity of sphingolipids [including notably glycosphingolipids (GSLs)] in the nervous system, became more fully appreciated following his death in 1901. The Biochemical Society today awards a Thudichum Medal to those who have made outstanding contributions to neurochemistry. Sphingosine itself was first isolated and analyzed by Thudichum who named it 'in

The Sugar Code. Edited by Hans-Joachim Gabius

ISBN: 978-3-527-32089-9

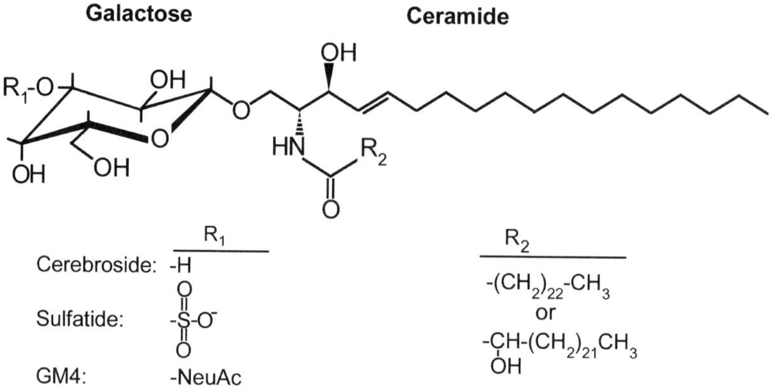

Figure 30.1 Structures of GSLs based on galactosylceramide. All three structures (based on R_1) are found in myelin and myelin-forming cells of the CNS, while cerebroside and sulfatide are also in PNS myelin. R_2 represents two of the major fatty acids, 24:0 and 2-OH 24:0, that are found in all indicated structures. These glycolipids (exclusive of GM4) were discovered and at least partially characterized as to chemical composition by Thudichum over 120 years ago. It is noteworthy that the term 'cerebroside', assigned by Thudichum, is retained as standard nomenclature to this day.

commemoration of the many enigmas which it presented to the inquirer'. Failure to solve the riddle of the sphinx, according to legend, incurred the penalty of death (which has been jokingly invoked to explain the notable diligence of Thudichum and subsequent sphingolipid researchers; please see also Info Box in Chapter 10).

What might be called the 'Thudichum approach' of analyzing whole-brain samples has given way to more meaningful study of specific tissue regions and cell types within the nervous system. These include central nervous system (CNS) gray matter, the primary locus of neurons as the information processing units of brain, and white matter that is rich in neuron-associated glia. Macroglial elements include astrocytes, so named because of their star-like processes that extend to neuronal junctures and blood vessels, and oligodendrocytes that ensheath a significant number of CNS axons with multilamellar lipid-rich myelin wrappings that promote saltatory conduction. Schwann cells fulfill similar functions in the peripheral nervous system (PNS). Microglia comprise approximately 7–15% of brain cells in both gray and white matter, and subsume functions involving brain neuropathology. A fifth glial type in the CNS, the oligodendrocyte precursor cell, has special functions pertaining to repair that come into play after brain injury. Studies of individual cell types have revealed that they all possess unique glycan profiles corresponding to specific cellular functions and in regard to glycophenotyping with plant and mammalian lectins (for further information on these lectins as tools, please see Chapters 18, 19, 24 and 25). The following sections highlight some of the best-known glycan features of these neural cells, and their elucidated roles in normal and pathological conditions.

30.1
Glucose and Glycogen as Energy Sources

Glucose is both the primary energy source for neural cells and a major building block for a number of CNS and PNS glycoconjugates. Glucose enters the brain by means of the Glut-1 transporter in endothelial cells that comprise the blood–brain barrier. Unlike many other cells that convert glucose into glycogen as an energy reserve, neurons do not catalyze this reaction under normal conditions even though they possess the glycogen synthase enzyme. Paradoxically, failure to keep this enzyme silent results in glycogen synthesis with concomitant damage to neurons by apoptotic signaling [1]. Neurons nevertheless depend on glycogen indirectly through metabolism of this polysaccharide by astrocytes, its primary locus. These cells carry out active glycogenolysis with release of lactate that is utilized, along with glucose itself, as a major energy substrate by neurons. Glucose also modifies neural cell activities in several ways independent of energy provision, for example glycosaminoglycan (GAG) production, insulin signaling, c-Jun N-terminal kinase/mitogen-activated protein kinase activation and UDP-GlcNAc formation–the substrate for an important posttranslational modification (see below).

30.2
Gangliosides as Primary Glycans of the Nervous System

Gangliosides are defined as GSLs containing at least one sialic acid in the oligosaccharide chain; the vast majority possess glucose as the first carbohydrate attached to ceramide (Figure 30.2; please see also Chapter 10 for survey on glycolipids). The first identification and structural characterization of specific gangliosides resulted from the pioneering studies of Ernst Klenk and coworkers on lipid storage diseases of brain during the 1930s and 1940s [2]. Klenk assigned the term 'ganglioside' in relation to their special abundance in brain gray matter, known to be the primary locus of neurons or 'Ganglionzellen'. It was later recognized, however, that gangliosides are not unique to neurons and occur, albeit at lower concentrations, in all neural cell types and indeed in virtually all vertebrate (and even some invertebrate) tissues. The nervous system is unique among mammalian tissues in possessing gangliosides (rather than glycoproteins) as the predominant sialoglycoconjugates (for the different forms of glycoprotein sialylation, please see Chapters 6 and 7); gangliosides represent around 75% of total bound sialic acid. Neurons contain the highest concentration–the majority of these belonging to the gangliotetraose family characterized by a four-carbohydrate backbone structure to which one or more sialic acids are attached (Figure 30.2 and also Chapter 10.7). Gangliosides, as with GSLs in general, are named and classified according to the oligosaccharide structure, irrespective of the frequent heterogeneity in the lipophilic components of ceramide. A recent review [3] has listed 188 different gangliosides that have been characterized in vertebrate tissues to date, of which perhaps about

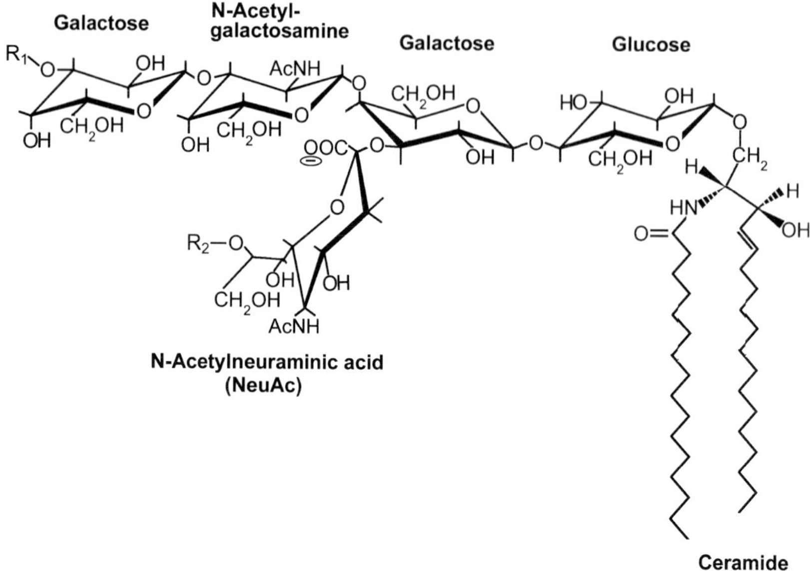

Figure 30.2 Structures of the five major gangliosides of mammalian brain. GM1, $R_1 = R_2 = H$; GD1a, R_1 = Neu5Ac, R_2 = H; GD1b, R_1 = H, R_2 = Neu5Ac; GT1b, $R_1 = R_2$ = Neu5Ac; GQ1b, R_1 = Neu5Ac(2-8)Neu5Ac, R_2 = Neu5Ac. The zig-zag lines of ceramide represent hydrocarbon chains. [Reproduced with permission from R. Ledeen. Gangliosides of the neuron. *Trends Neurosci*, 1985; **8**, 169–174.]

a quarter occur as major or minor gangliosides of the nervous system. An approximately equal number of neutral GSLs have been characterized in vertebrate tissues, although these are relatively sparse in the nervous system. These numerical counts would expand considerably with inclusion of invertebrates. It seems likely that more GSLs await discovery, in keeping with improved methodologies such as high-performance liquid chromatography, two-dimensional thin-layer chromatography and mass spectrometry (for glycan analysis using these methods, please see Chapter 5). GSLs can be designated according to International Union of Pure and Applied Chemistry–International Union of Biochemistry systematic nomenclature, although many investigators have found it convenient for the most common gangliosides to use the system proposed by Svennerholm following his landmark resolution of these substances by thin-layer chromatography [4]. Both systems were therefore listed in Table 10.3.

The five structures shown in Figure 30.2 comprise the major gangliosides of the mammalian nervous system, with neurons as their principal locus. These are based on GM1 as the monosialo prototype, the di-, tri- and tetrasialo analogs (GD1a, GD1b, GT1b and GQ1b) containing additional *N*-acetylneuraminic acid (Neu5Ac) (for abbreviations and structures of monosaccharides, please see Chapter 1) attached to the hydroxyls indicated. The much larger number of quantitatively minor gangliosides can vary significantly among the myriad neuronal and glial cell subtypes. The most abundant of these (that is, GM3, GM2 and GD3) are often major forms in extraneural tissues. Minor gangliosides of the nervous system include structures containing GlcNAc in place of GalNAc (lacto series) and some with NeuAc attached to the 6-hydroxyl of GalNAc within the gangliotetraose core (α-series). Sulfated gangliosides have also been observed along with gangliotetraose forms containing fucose attached to the 2-hydroxyl of terminal galactose [3]. Oligodendrocytes contain the five gangliosides of Figure 30.2 as their major species, with proportionately more GM1 in their extended membrane that becomes myelin in the CNS. Sialosylgalactosylceramide (GM4) is uniquely found in vertebrate myelin, occurring there as a major ganglioside along with GM1 in primates and avian species. A notable departure from the glucosylceramide motif of most GSLs, GM4 contains sialic acid attached to galactosylceramide (Figure 30.1) and is also unique in possessing ceramide with long-chain (for example C-24) unsubstituted or 2-hydroxy-substituted fatty acids – analogous to galactosylceramide from which it is derived. Most brain gangliosides contain ceramide with stearate (18 : 0) as the major fatty acid and sphingosine or its C-20 homolog as the major long chain bases. Astrocytes isolated from mammalian brain contain a ganglioside pattern similar to that of neurons, but it is not clear whether these are synthesized *in situ* or transferred from neurons since astrocytes in culture express mainly GM3 and GD3. Relatively little is known about gangliosides of the other neural cell types. The main sialic acid in these various mammalian CNS gangliosides is Neu5Ac, which is sometimes acetylated at the 8- or 9 hydroxyl group and can also be substituted in other ways [5]. *N*-Glycolylneuraminic acid (-NH-$COCH_2OH$ on Neu5Gc in place of -NH-$COCH_3$) occurs as a minor component in some brain gangliosides of certain mammalian species.

30.3 Ganglioside Metabolism

With the exception of GM4 (see above), gangliosides of the nervous system (and other tissues) are derived from glucosylceramide by stepwise addition of two to several carbohydrates. These can include two to five sialic acids. The biosynthetic pathways begin with ceramide and glucosylceramide synthesis in the endoplasmic reticulum, followed by translocation to the Golgi apparatus where glycosyltransferases sequentially attach additional carbohydrates (please see also Chapter 10). This process was elegantly explored in the pioneering studies of Roseman *et al.* on biosynthesis of such gangliosides as GM3, GM2 and GM1 [6], leading to what became known as 'The Roseman Pathway'. The sugar-donating unit in most reactions is UDP-hexose or UDP-*N*-acetylhexosamine, whereas CMP-Neu5Ac fulfills that function for sialic acid. Further work by many investigators revealed a number of GSL families based on distinct biosynthetic pathways, four of which are depicted in Figure 30.3. Structure GM1a in the a-series refers to the normally designated GM1 to distinguish it from the isomeric GM1b in the o-series. The

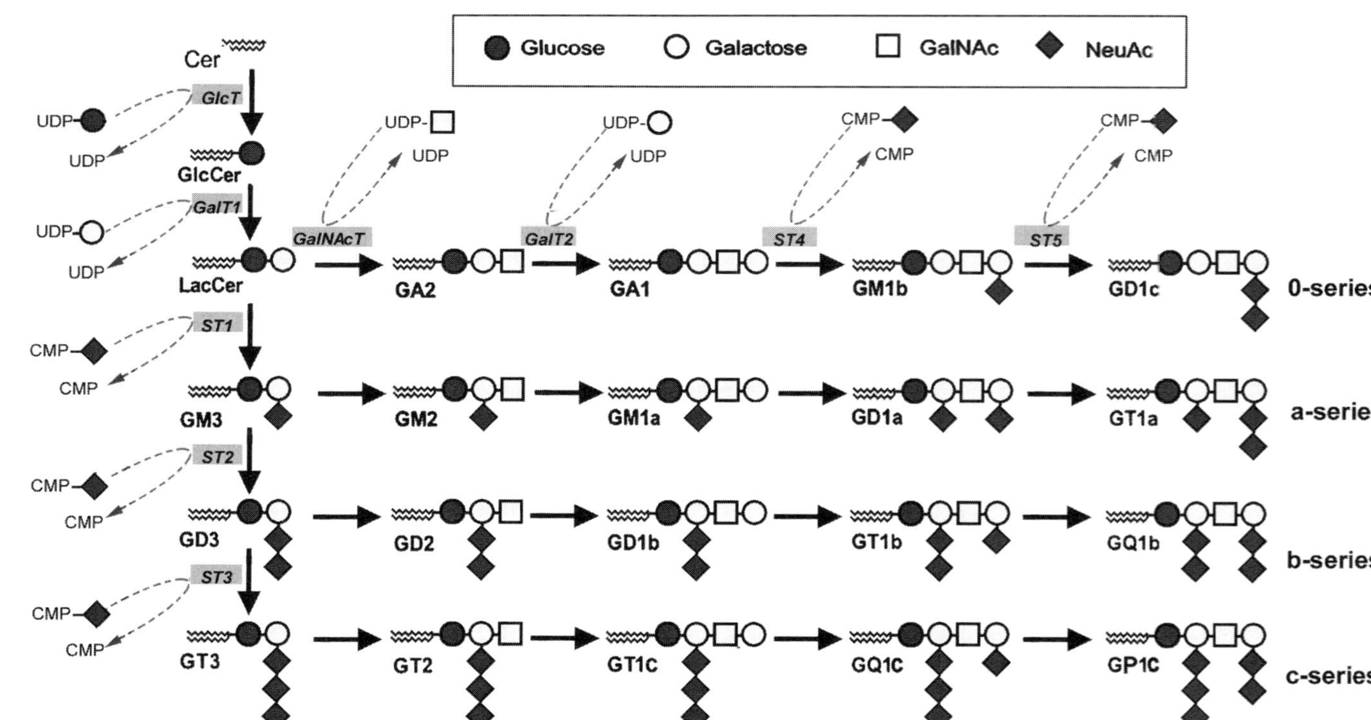

Figure 30.3 Biosynthetic pathways of gangliotetraose gangliosides, showing glycosylation sequences for synthesis of the a-, b-, c- and o-series. [Adapted with modifications from S. Ngamukote, *et al.* Developmental changes of GSLs and expression of glycogenes in mouse brain. *J Neurochem*, 2007; **103**, 2327–2341.]

latter is found to a limited extent in the nervous system and elsewhere, and is distinguished from GM1a in its susceptibility to sialidase and nonreactivity toward cholera toxin B-subunit (see below). In contrast to the numerous defects in GSL catabolic enzymes (see below), there is a paucity of reports on inborn errors of synthetic enzymes. One such report [7] describes an infantile-onset symptomatic epilepsy syndrome due to mutation of GM3 synthase with resultant absence of gangliosides of the a-, b- and c-series (Figure 30.3; for more information on this disorder, please see Chapter 22.7).

Glycohydrolases responsible for catabolic metabolism also progress one sugar at a time, in a manner formally analogous to reversal of the synthetic reactions of Figure 30.3; however, the relevant catabolic enzymes are quite distinct from those in the synthetic pathways and reside largely in lysosomes (please see also Chapter 10). Failure of specific glycosidase activities due to genetic mutations gives rise to lysosomal storage disorders such as the gangliosidoses [8]. The first of these to be biochemically characterized was Tay–Sachs disease (GM2 gangliosidosis), eventually shown to result from mutation of the gene for the α-subunit of β-hexosaminidase A that cleaves GalNAc from GM2 in normal brain (Table 30.1). GM2 is a convergent product in metabolism of the a-, b- and c-series gangliotetraose gangliosides (please see also Chapter 10), and several known mutations result in GM2 accumulation in lysosomal storage vesicles leading to neuronal destruction and usually early death. The O variant (Sandhoff's disease) results from mutation in the gene for the β-subunit, thereby inactivating both β-hexosaminidase A (αβ) and B (ββ). Finally, the AB variant arises from defects in the gene for the GM2 activator, a necessary cofactor for β-hexosaminidase activity; the latter enzyme is normal in these patients when supplemented with the activator. GM1 gangliosidosis results from neuronal accumulation of GM1 due to an inherited defect of GM1 β-galactosidase. There are clinically distinct forms of this disorder (Table 30.1), some due to mutation of the galactosidase itself and another to mutation in the protective protein that protects β-galactosidase and an associated sialidase from premature proteolytic degradation. Many mutations have been found in both β-hexosaminidase and β-galactosidase, the form and severity of the disease correlating with residual enzyme activity.

Sialidase (neuraminidase) catalyzes removal of terminal sialic acid residues from gangliosides and glycoproteins, often with resultant influence on cellular activity. Four genetically distinct forms of mammalian sialidase have been cloned and characterized, each with a predominant cellular localization and substrate specificity [5]. Neu3 is associated with the plasma membrane and shows preferential reactivity toward gangliosides [9]. This enzyme converts oligosialogangliosides of the a-, b- and c-series to GM1, whose sialic acid resists hydrolysis, and this exerts a regulatory role on neuronal differentiation and transformation in neuroblastoma cells [10] (please see Chapter 25 for the functional role of the GM1–lectin interaction) as well as primary neurons [11]. Such observations point up the significance of GM1 in neuronal process outgrowth, related at least in part to its influence on calcium regulatory mechanisms in addition to its effect on neurotrophic receptors (see below); GM2 and GM3 also influence Ca^{2+} regulation in some cells [12]. Neu1

Table 30.1 Forms of gangliosidosis and corresponding pathology.

Type	Defective gene	Pathology
GM2 gangliosidosis		
B variant (Tay–Sachs)	α-subunit of HexA (αβ) and S (αα); more than 50 different mutations	Massive storage of GM2 and GA2 in grey and white matter; severity correlates inversely with residual activity of HexA; occurs as infantile (classical Tay–Sachs, fatal), juvenile and adult forms
B1 variant	α-subunit of HexA (αβ) and S(αα) 3 mutations at Arg178	Mutated HexA active toward uncharged substrates (GA2) but not toward GM2, which is stored; a juvenile disease progression
O variant (Sandhoff)	β-subunit of HexA (αβ) and Hex B (ββ)	Storage of GM2, GA2 and globoside throughout the CNS; also in visceral organs; pathology and symptoms similar to B-variant
AB variant	GM2 activator	Comparable to B variant; symptom appearances slightly delayed
GM1 gangliosidosis		
Type 1 (infantile)	GM1-β-galactosidase	Storage of GM1 and GA1 in visceral organs and brain, progressive motor and mental retardation, short lifespan.
Type 2 (late infantile)	GM1-β-galactosidase	Similar to type 1, but less severe, with lifespan about 10 years.
Type 3 (adult)	GM1-β-galactosidase	Mild and slowly progressive neurological symptoms.
Morquio type B	GM1-β-galactosidase	Skeletal deformation without involvement of the CNS; oligosaccharides with terminal galactose accumulate in viscera

Lysosomal storage diseases that result from genetic defects in β-hexosaminidase (HexA and B) and β-galactosidase are indicated with corresponding pathological and clinical symptoms. Many mutations have been discovered in some, with varying effects on enzyme activities. β-Hexosaminidase S is involved in degradation of GAGs. GM2 activator is a glycoprotein that is a necessary cofactor for β-hexosaminidase. GA1 and GA2 refer to uncharged asialo-GM1 and asialo-GM2, respectively. Please see also Info Box in Chapter 11.

is the primary sialidase of lysosomes, but can also appear on the cell surface in some situations, while Neu2 is cytosolic. The locus and function of Neu4 are not yet known. In addition to converting gangliotetraose gangliosides to GM1, these enzymes also hydrolyze such structures as GM4, GM3, GD3 and o-series gangliosides to neutral GSL products.

30.4 Gangliosides of the Peripheral Nervous System

The PNS has not been studied as intensively as the CNS in relation to GSL composition, but both sensory and motor nerves are known to contain the four gangliotetraose gangliosides (Figure 30.2) characteristic of brain [13]. However, these are not the major gangliosides of the PNS, that being the *N*-acetylglucosamine-containing species 3′LM1 (NeuAcα2,3Galβ1,4GlcNAcβ1,3Galβ1,4Glcβ1,1′ Cer). This ganglioside in peripheral nerves occurs primarily in myelin and has not been detected in CNS myelin. Interestingly, the most abundant acidic GSL in the human PNS is not a ganglioside, but rather inositolphosphoryl galactoslylceramide, whose acidity is due to phosphate rather than Neu5Ac. Anti-GSL antibodies are directly involved in the pathogenesis of Guillain–Barré syndrome and related peripheral neuropathies [14]. These are autoantibodies directed against some 20 or more endogenous GSLs, most of which are gangliosides of the PNS. The high antibody titers in these patients are closely associated with specific antecedent infections, such as *Campylobacter jejuni* and cytomegalovirus. The mechanism has been proposed to involve molecular mimicry, based on the fact that lipopolysaccharides from these microorganisms contain structural components that closely resemble the terminal carbohydrate structures of the peripheral GSLs. Clinical symptoms appear to require the lipopolysaccharide entity, since anti-GSL antibody formation via direct injection of gangliosides into humans or animals is usually not accompanied by Guillain–Barré syndrome-like symptoms. In addition to gangliosides, peripheral nerves also contain sulfated GSLs that can become targets of molecular mimicry, the primary example being sulfoglucuronosyl paragloboside containing sulfated glucuronic acid: $^{-}SO_3$-GlcAβ1,3Galβ1,4GlcNAcβ1,3 Galβ1,4Glcβ1,1′Cer. The terminal carbohydrates comprise the human natural killer (HNK)-1 epitope (for structural depiction, please see Figure 1.7c, see also below). Having reviewed the structural and metabolic features of these GSLs, we now survey what is known of their functional behavior in the nervous system.

30.5 Ganglioside Functional Activities

Numerous studies have documented a wide variety of neurotrophic, neuroprotective properties of GM1 and other gangliosides, many of these involving application of exogenous ganglioside *in vivo* or *in vitro* [15]. Although in some cases these turned out to represent nonspecific perturbations rather than reflections of true physiological function, the biological results were often significant (see Info Box 1). Among several examples was GM1 rescue of substantia nigra dopaminergic neurons in mice subjected to 1-methyl-4-phenyl-1,2,3,6-tetrahydropyridine–a model of Parkinson's disease [16]. A clearer picture of physiological function is coming into focus with the newer emphasis on specific molecular mechanisms of endogenous gangliosides, these often involving direct association between ganglioside and protein. The main locus of most gangliosides is the plasma membrane where, in the case of neurons, they make a major contribution to the carbohydrate-rich glycocalyx that forms the interface for cellular interactions. Within that matrix the oligosaccharide chains of gangliosides (and GSLs generally) are relatively close to the cell surface and thus well situated for *cis*-interactions with membrane proteins. This contrasts with other glycoconjugates whose glycan chains can extend a considerable distance from the plasma membrane, thereby facilitating *trans*-interactions with other cells or extracellular components. However, there is evidence that gangliosides can undergo *trans*-interactions in some instances (for example myelin-associated glycoprotein (MAG) or siglec-4; for further information on siglecs, a lectin family, please see Chapters 19 and 27.6; see also below) just as some glycoproteins show *cis*-reactivity. GSLs of the membrane often reside within microdomains termed lipid rafts, which are heterogeneous, highly dynamic, cholesterol- and sphingolipid-enriched aggregations that compartmentalize cellular processes; in addition to GSLs, they sequester numerous glycoprotein receptors in addition to the raft-specific glycosylphosphatidylinositol (GPI)-anchored proteins (for further information on GPI-anchored proteins, please see Chapter 9). GM1 often serves as marker for lipid rafts and caveoli, although its absence from such microdomains has been noted (please see also Chapter 10.9).

GM1 has been perhaps the most widely studied ganglioside in relation to protein modulatory activities (see Info Box 1). It has the distinction of being one of the few sialoglycoconjugates whose sialic acid resists hydrolysis by most sialidases in the intact molecule; this moiety becomes susceptible to sialidase following removal of terminal galactose and *N*-acetylgalactosamine. The negative charge carried by the carboxyl of sialic acid is generally required for its modulatory functions. Such regulatory actions are not limited to the plasma membrane, as seen in the GM1 activation of a sodium–calcium exchanger in the nuclear membrane (see Info Box 1). The copresence of GD1a and sialidase in the same membrane provides a mechanism for regulatory maintenance of GM1 expression at that locus. In addition to modulation through direct association, GM1 can also influence protein activity 'from a distance', so to speak, through cross-linking in a manner that triggers downstream signaling. In this way GM1 cross-linking resulted in co-cross-linking of associated integrin, which in turn caused protein tyrosine kinase activation with subsequent TRPC5 Ca^{2+} channel opening and neurite induction (see Info Box 1). Cross-linking can also result from carbohydrate recognition by galectins (for further information on structural and functional aspects of this interaction, please see Chapters 19 and 25). Other gangliosides were shown to exert modulatory activities to other proteins. These can be inhibitory as well as excitatory, as seen in GM3-mediated inhibition of autophosphorylation, and hence activity, of the insulin receptor. GD3 interacts with mitochondrial proteins causing cytochrome *c* release and caspase activation in ceramide-induced apoptosis. Such findings *en toto* have led to the speculation that the great variety of ganglioside (and other GSL) structures is nature's way of creating modulators that are tailor-made to react stereospecifically with particular membrane proteins.

GSL–protein interaction is well illustrated by the ability of gangliosides to serve as opportunistic receptors for various bacteria, viruses and toxins. High-affinity association of bacterial toxins to certain membrane gangliosides is illustrated in the tight binding to GM1 by the B-subunits of cholera toxin and *Escherichia coli* enterotoxin (please see Chapter 17.1.3.2) [17]. Such toxins use this attachment mechanism as a prelude to insertion of their cyclic AMP-elevating A-subunits into

Info Box 1

Gangliosides modulate receptors, enzymes, ion channels and other proteins through direct association or indirectly through signaling sequences – triggered by cross-linking. GM1 has been the most studied from this standpoint, several examples of which follow:

- The tropomyosin-related kinase A receptor for nerve growth factor is tightly associated with GM1 and fails to fulfill its functional role – stimulation of neuronal differentiation – in its absence [T. Mutoh *et al.* Stable transfection of GM1 synthase gene into GM1-deficient NG108-15 cells, CR-72 cells, rescues the responsiveness of Trk-neurotrophin receptor to its ligand, NGF. *Neurochem Res* 2002; **27**, 801–806].
- Opioid receptors are converted by GM1 from inhibitory to excitatory mode, leading to opioid tolerance and dependence [G. Wu *et al.* Interaction of the δ-opioid receptor with GM1 ganglioside: conversion from inhibitory to excitatory mode. *Mol Brain Res* 1997; **44**, 341–346].
- A sodium–calcium exchanger in the nuclear envelope is tightly associated with and potentiated by GM1 [X. Xie *et al.* Potentiation of a sodium-calcium exchanger in the nuclear envelope by nuclear GM1 ganglioside. *J Neurochem* 2002; **81**, 1185–1195].
- Cross-linking of GM1 leads to TRPC5 Ca^{2+} channel activation – an example of 'modulation at a distance' [G. Wu *et al.* Induction of calcium influx through TRPC5 channels by cross-linking of GM1 ganglioside associated with $\alpha_5\beta_1$ integrin initiates neurite outgrowth. *J Neurosci* 2007; **27**, 7447–7458]. A human lectin is capable to act as natural GM1 cross-linking agent [J. Wang *et al.* Cross-linking of GM1 ganglioside by galectin-1 mediates regulatory T cell activity involving TRPC5 channel activation: possible role in suppressing experimental autoimmune encephalomyelitis. *J Immunol* 2009; **182**, 4036–4045]. The roles of galectins-1 and -3 as GM1 receptors in neuroblastoma cells are described in Chapter 25.2.
- GM1 in the endoplasmic reticulum induces Ca^{2+} release via the endoplasmic reticulum stress response system [A. Tessitore *et al.* GM1-ganglioside-mediated activation of the unfolded protein response causes neuronal death in a neurodegenerative gangliosidosis. *Mol Cell* 2004; **15**, 753–766].
- GM1 binds to calmodulin and inhibits calmodulin-dependent enzymes [K. Higashi, T. Yamagata. Mechanism for ganglioside-mediated modulation of calmodulin-dependent cyclic nucleotide phosphodiesterase activity through binding of gangliosides to calmodulin and the enzyme. *J Biol Chem* 1992; **267**, 9839–9843]. Inhibition of nitric oxide synthase, a calmodulin-dependent enzyme, was proposed as a mechanism for the neuroprotective effect of GM1 (and other gangliosides) [T.M. Dawson *et al.* Neuroprotective effects of gangliosides may involve inhibition of nitric oxide synthase. *Ann Neurol* 1995; **37**, 115–118].

the cell, with resultant pathophysiological consequences. Gangliosides of the b-series (for example GD1b and/or GT1b) were proposed to serve as the shared coreceptor(s) by which botulinum neurotoxin enters axon terminals. GD1b has been proposed as a receptor (or coreceptor) for tetanus toxin. Microorganisms themselves are able in some instances to exploit this pathway to gain entry into cells. The B-subunit of cholera toxin, owing to its relative specificity and high affinity toward GM1, has found wide use for histochemical localization of this ganglioside as well as functional studies. The latter can involve removal of GM1 from its associated protein as well as cross-linking of GM1 as described above.

Genetically altered mice lacking specific gangliosides have provided important clues regarding ganglioside function [18]. Disruption of the GalNAcT gene that synthesizes GM2, GD2 and GT2 (Figure 30.3) caused depletion of those and all gangliotetraose gangliosides, resulting in significant neurological abnormalities over time. These appeared to result from impaired Ca^{2+} regulatory mechanism(s), as seen in apoptotic destruction of CNS neurons stemming from failure to restore Ca^{2+} homeostasis [19]. Surprisingly, elimination of b-series gangliosides by disruption of the GD3S gene produced relatively few neurological abnormalities. Such findings suggest plasticity in which surviving gangliosides are at least partially able to functionally replace depleted ones. However, the GalNAcT and GD3S double-knockout mice, which express only GM3, showed a spontaneous lethal phenotype in adults. Depletion of the GlcT gene, which encodes glucosylceramide synthase, eliminated the vast majority of GSLs and proved embryonic lethal. Such studies have revealed GSLs as essential for normal development of the nervous system as well as for survival of the organism (for additional information on these and other animal models, please see Chapter 23).

To summarize neural glycolipids, we can see that nature has been bountiful in providing neural cells with an abundance of low-molecular-weight glycans – the gangliosides. Molecular biology together with classical biochemistry is providing a detailed understanding of the means by which these glycoconjugates modulate neural cell behavior. Myelin is another nervous system structure vitally dependent on glycolipids – the cerebrosides and sulfatides. We turn now to another type of neural glycoconjugate, the glycoproteins, which are macromolecules of considerable complexity that function as key determinants in nervous system development and behavior. As with gangliosides, glycan interactions exert a primary role in these processes.

30.6 Neural Glycoproteins: Overview

The importance of glycosylation as a posttranslational modification in the nervous system is evident in the enormous variety of glycoproteins that are expressed in the CNS and PNS. These include many of the well-characterized neural enzymes, such as acetylcholinesterase, dopamine β-hydroxylase and Na^+/K^+-ATPase. Most membrane proteins are glycosylated, as is the case for the majority of ion channels which provide the physiological basis of neuronal excitability. The nicotinic cholinergic receptor channel, for example, consists of five subunits that are glycosylated with a total of about 75 carbohydrate residues. Virtually all the monosaccharides found in GSLs are expressed in the glycans attached to proteins. As with GSLs, this diversity is cell-type specific and developmentally regulated. Glycoproteins of the nervous system, as for other tissues, contain either *N*-glycan or *O*-glycan oligosaccharides depending on the nature of their linkage to the protein backbone. Biosynthetic pathways, involving coordinate processing by the rough endoplasmic reticulum and Golgi apparatus, differ significantly for these two groups (please see Chapters 6 and 7). The opportunities for inherited defects in glycan biosynthesis leading to neurological disorders are thus manifold, and were shown to result in epilepsy, psychomotor disturbance and mental retardation, among others (please see Chapter 22). In general, the more upstream the blockade of synthesis, the more severe the pathology. Glycan structures in all their complexity are undoubtedly a necessary feature of the enormously complex array of neuronal, glial, and other cell types and subtypes that interact in multiple ways in the developing and mature nervous system. As indicated below, neural recognition molecules are vital to developmental processes, while myelin glycoproteins have received special attention in relation to their role in blocking regeneration.

A dynamic posttranslational modification found in neural (and virtually all metazoan) cells involves transfer of GlcNAc from UDP-GlcNAc to Ser/Thr residues without further elongation (please see Chapter 7.4.2 and Table 7.8). Prominent in, but not limited to nuclei, this reaction is necessary for cell viability. Evidence suggests that *O*-GlcNAc and *O*-phosphate are reciprocal and compete for the same Ser/Thr residues. It was recently found that phosphorylation of tau protein is inversely regulated by *O*-GlcNAcylation and that abnormal hyperphosphorylation of tau, which leads to neurofibrillary degeneration in Alzheimer's disease, can be caused by the decreased glucose metabolism and *O*-GlcNAcylation that characterize this disorder [20]. This is one of many theories pointing to glycoconjugate involvement in the etiology of Alzheimer's disease.

30.7 Neural Recognition Glycoproteins

Neural recognition molecules in general have a major role in signaling mechanisms and cell–cell interactions that regulate neural development. The polysialic acid (PSA; please note: *Pisum sativum* agglutinin in Table 18.1 is also abbreviated as PSA) glycan is a linear homopolymer of variable length amounting to 50 or more sialic acid residues, all joined by 2,8 linkages (see gangliosides GD1b and GT1b of Figure 30.2 for depiction of this linkage). These chains are linked to two *N*-glycans located in the fifth immunoglobulin-like domain of the neural cell adhesion molecule (NCAM), forming one of the most abundant glycan structures in the developing mammalian nervous system. The three known isoforms of NCAM vary in terms of molecular weight, mode of membrane attachment and

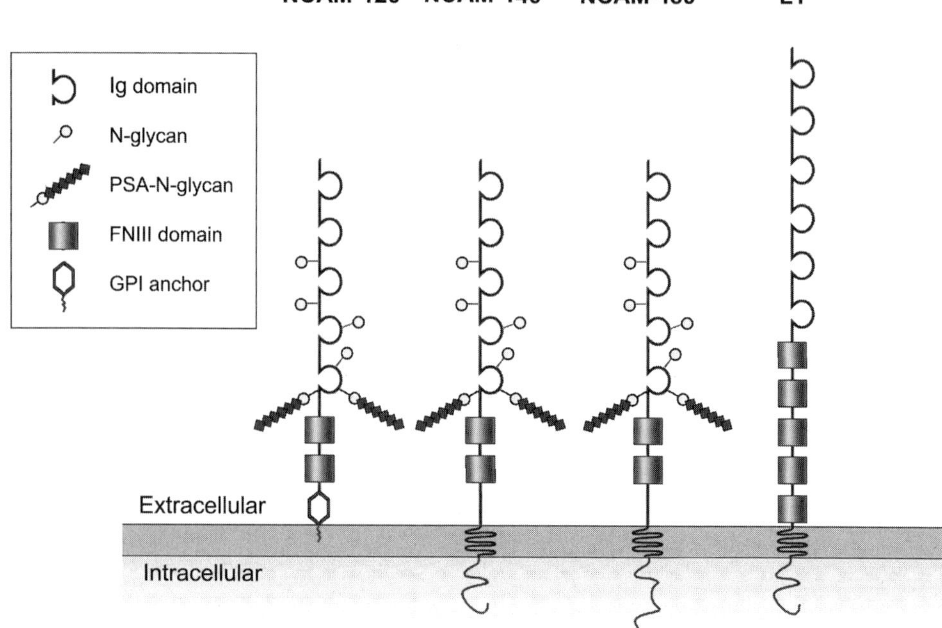

Figure 30.4 Neural cell adhesion molecules of the immunoglobulin superfamily. Glycoproteins of this family contain immunoglobulin-like domains and fibronectin type III (FNIII) repeats. Of the proteins encoded by the NCAM1 gene, many isoforms are produced by splicing, but NCAM-120, -140, and -180 are the major ones. NCAM members possess five Ig-like domains and two FNIII domains, while L1 contains six Ig-like domains and five fibronectin type III repeats. Shown are the structures of the three major isoforms of NCAM from mouse brain (the glycans of NCAM exhibit structural diversity). These each contain six *N*-glycan chains with two PSA chains attached to the fifth Ig-like domain; HNK-1 is also present on some glycans (please see Chapter 6 for more detail on NCAM structures). L1 has many *N*-glycosylation sites (not shown), about which less is known; it contains HNK-1, but no PSA. [Adapted with modifications from (a) [22] and (b) C. Albach *et al.* Identification of *N*-glycosylation sites of the murine neural cell adhesion molecule NCAM by MALDI-TOF and MALDI-FTICR mass spectrometry. *Anal Bioanal Chem*, 2004; **378**, 1129–1135.]

cytoplasmic unit (Figure 30.4). Two polysialyltransferases, ST8Sia-II and ST8Sia-IV, that synthesize PSA on NCAM have been identified [21]. The former is more important during embryogenesis, while the latter has a role in those few areas of the mature brain with ongoing neurogenesis or plasticity (for example olfactory interneuron precursors, dentate gyrus of the hippocampus). Other regions of the mature brain contain relatively little PSA. Surprisingly, complete genetic ablation of NCAM resulted in a fairly mild phenotype–the brain showing normal cytoarchitecture and only small decrease in size. In contrast, complete ablation of PSA produced animals with severely retarded postnatal development and early death. From this and related findings it was concluded that a major function of PSA is to mask NCAM and thus prevent inappropriate NCAM interactions during development. This is in accord with the prevailing model of PSA function as that of sterically inhibiting cell–cell apposition during neuronal differentiation. PSA–NCAM has an important role in regeneration of axons and dendrites, thereby accounting for its upregulation after a lesion. Immature synapses express more PSA–NCAM than mature ones, suggesting a role in learning and memory. L1 is another member of the immunoglobulin superfamily, structurally similar to NCAM but lacking PSA (Figure 30.4). A donor of high oligomannosidic glycans, L1 can associate with NCAM, an oligomannoside receptor, and disruption of this interaction interferes with neurite outgrowth [22]. Mutations in the gene for L1 lead to human genetic diseases including mental retardation.

The HNK-1 glycan, so named as the epitope recognized by the monoclonal antibody that binds to human natural killer cells, consists of 3′-sulfated glucuronic acid attached to lactosamine (for structure, please see Chapter 1 and Figure 1.7c). It is present on *N*- and *O*-glycans associated with a large variety of recognition molecules, including NCAM, L1, chondroitin sulfate proteoglycans (CSPGs), integrins, tenascin and various myelin glycoproteins (see below) [22]. Receptors that bind HNK-1 have been identified, such as the extracellular matrix (ECM) glycoproteins laminin and merosin and the cell-surface lectins L- and P-selectins (for further information on selectins, please see Chapters 19 and 27.3 and Figure 29.5; for an X-ray structure of P-selectin, see Figure 16.1h). Interestingly, in myelinating mice, HNK-1 is expressed selectively by Schwann cells associated with motor axons, but not by those associated with sensory axons. HNK-1 also occurs on glycolipids in Schwann cells which show similar preferential expression on motor nerves [22].

Oligomannosidic glycans as components of neural recognition glycoproteins are uniquely abundant in the nervous system. These glycans contain variable numbers of α- and β-mannosyl residues attached to the peptide backbone via the GlcNAc core [22]. They are normally transient structures that appear on glycoproteins during biosynthesis, further processing of the oligosaccharide in the Golgi apparatus having been blocked where they persist. In brain, as opposed to most other tissues, oligomannosides are carried to the cell surface as part of such molecules as L1. Oligomannosidic glycans are particularly abundant in the rodent brain and in the adult are concentrated in synapses. Glutamatergic synapses also contain lectins that bind to these glycans. An adhesion molecule on glia known as AMOG, which is also the β-subunit of Na^+/K^+-ATPase, contains 80% of its glycans as oligomannosides.

The role of recognition glycoproteins in neurite outgrowth includes both inhibitory and conducive reactions. Tenascins are ECM glycoproteins that interact with such proteoglycans as heparin and heparan sulfate (please see Chapter 11 for information on these proteoglycans). Tenascin-C and -R promote neuron migration or neurite extension when uniformly distributed as substrate in brain or the PNS, but inhibit such activities when expressed at discontinuous boundaries. They are highly enriched at nodes of Ranvier of both CNS and PNS axons. The specific molecular domains in tenascin-C and -R that are conducive and inhibitory have been structurally identified [23]. Semaphorins and ephrins also regulate axonal guidance in the developing nervous system through a combination of attractive and repulsive signals. Finally, proteoglycans, often in conjunction with recognition glycoproteins, have major functional roles in developmental and repair processes (see below). Glycoconjugates of all types figure prominently in neural stem cell functioning during the earliest stages of development (see Info Box 2; for further information on stem cell glycans as markers, please see Table 24.3).

Info Box 2

Growing interest is being directed to glycoconjugate composition and function in neural stem cells (NSCs). [M. Yanagisawa, R.K. Yu. The expression and functions of glycoconjugates in neural stem cells, *Glycobiology* 2007; **17**, 57R–74R]. They serve as excellent biomarkers due to their occurrence on the cell surface and dramatic changes in expression pattern at various stages of differentiation. They also have important functional roles in determining cell fate. As development proceeds, NSCs give rise to neural precursor cells (NPCs) which in turn develop into neurons, astrocytes or oligodendrocytes. As one of many examples, stage-specific embryonic antigen (SSEA)-1 is recognized by a monoclonal antibody to the Lewis^x structure [Galβ1,4(Fucα1,3)GlcNAcβ-]. Initially considered a specific marker for NSCs/NPCs, it is now known that Lewis^x is not restricted to these although it is still considered to represent an immature status of cells. Carrier molecules for the SSEA-1 epitope in mouse embryonic NPCs include CSPGs, β_1 integrin and a glycolipid. Among the many other glycans that could be cited, the Notch receptor containing an *O*-fucose glycan within the EGF domains provides an example of inhibitory influence on cell fate and/or maintenance of a progenitor pool; it is now seen as promoting astrocyte development while inhibiting that of neurons and oligodendrocytes. The above-mentioned *O*-GlcNAcylation has been described as necessary for viability in embryonic stem cells and mouse embryogenesis (for further information on *O*-GlcNAcylation, please see Chapter 7).

30.8 Glycoproteins of the Synapse

Carbohydrates were initially considered as only of secondary importance in understanding the synapse, but it has gradually become clear that glycans are integral to the formation and functioning of these linkages that mediate communication among the 10^{11} CNS neurons with their estimated 10^{14} synaptic connections [24]. As mentioned, synapses contain a high proportion of oligomannosidic glycans and also NCAM, with or without attached PSA. The latter has important roles in synaptogenesis and synaptic plasticity, both in the brain and neuromuscular junction. Mice lacking ST8Sia-IV (see above) were defective in both long-term potentiation and long-term depression in the Schaffer collateral-CA1 projections of the hippocampus. A role for PSA in localizing trophic factors via GAGs at the synapse has been suggested. The pre- and postsynaptic membranes of the synapse are held together by adhesion complexes, and the search for membrane-specific glycoconjugates revealed neuroligin to be localized in the postsynaptic membrane, whereas neurexin 1β is largely confined to the presynaptic membrane [25]. Both of these are type I membrane proteins with a glycosylated extracellular sequence, a single transmembrane region and a relatively short intracellular segment. Significantly, they bind to each other only in the presence of Ca^{2+}, which is known to occur in the synaptic gap. Proteoglycans also occur prominently at the synapse and neuromuscular junction (see below).

30.9 Glycoproteins of Myelin

Myelin is formed as a spiraling, multilamellar, lipid-rich extension of the plasma membrane of Schwann cells in the PNS and oligodendrocytes in the CNS. It serves to insulate the axon and promote rapid saltatory conduction. As with the gangliosides, glycoproteins of CNS and PNS myelin differ significantly [26]. Glycoproteins comprise a majority of total protein in PNS myelin, but a rather small minority in CNS myelin. Po is a 30-kDa type 1 transmembrane glycoprotein that comprises over half the total protein in compact PNS myelin. It contains a single extracellular Ig-like domain and one *N*-linked oligosaccharide that is very heterogeneous; many of these are terminated in sialic acid and sulfated glucuronic acid, the latter comprising the HNK-1 epitope. Po is viewed as stabilizing the compact lamellar structure through homophilic interactions that may occur in both *cis*- and *trans* modes. The other glycoprotein of PNS myelin is PMP-22, a 22-kDa protein that accounts for less than 5% and appears to be a tetraspanin protein with a glycosylation pattern similar to that of Po. The first glycoprotein to be identified in CNS myelin was MAG (see above), a 100-kDa molecule that contains about 30% by weight carbohydrate [26]. The oligosaccharides are heterogeneous and negatively charged due to sialic acid and/or sulfate, the latter reflecting HNK-1. It has five extracellular Ig-like domains (it is an I-type lectin of the siglec group, that is siglec-4; please see Table 27.6, Chapter 19 and below for further information), a single transmembrane region and a cytoplasmic domain that occurs in two developmentally regulated forms. It is a minor component of both CNS and PNS myelin, localized mainly in the periaxonal and other specialized glial membranes; this contrasts with the major CNS myelin proteins (proteolipid protein, myelin basic protein) which are not glycosylated and occur in compacted myelin. Studies with MAG null mice indicated MAG is not essential for myelination, although such mice exhibited subtle structural abnormalities in the periaxonal region of CNS myelin sheaths. MAG is now seen as a bifunctional protein, able to stimulate axon regrowth from young neurons while inhibiting regrowth of adult axons after injury (see below). Myelin oligodendrocyte glycoprotein (MOG) is localized on the outer surface of CNS myelin sheaths and oligodendrocytes. Like MAG, it is a member of the Ig superfamily but with a single Ig-like variable domain. It contains one site for *N*-linked glycosylation and two hydrophobic, potential transmembrane domains. It is highly immunogenic and its possible implication in multiple sclerosis(MS) was suggested by the finding that autoantibodies and T cells to MOG occur in a subset of MS-patients. Experimental autoimmune encephalomyelitis, a widely used model of MS, is generated by immunization of rodents, marmosets, and other animals with MOG, a property that is shared with other CNS myelin proteins. The similarly named oligodendrocyte-myelin glycoprotein (OMgp) is also a minor component of CNS myelin with an important role in regulating formation of nodes of Ranvier.

Glycoproteins of CNS and PNS myelin differ in another important respect: the former, but not the latter, are strongly inhibitory toward axon growth. This is one factor contributing to failure of regeneration following axonal injury in the CNS, in contrast to PNS capability. This role of MAG has been well documented in relation to its ability to interact with the GPI-linked Nogo-66 receptor (NgR) on the axon – a property shared with NogoA, a nonglycosylated protein of CNS myelin, and the above-mentioned OMgp [27]. The small G-protein RhoA and Rho kinase (ROCK) act downstream of the NgR complex, providing a link to cytoskeletal regulation and outgrowth inhibition (Figure 30.5). MAG is a member of the siglec family of sialic acid-binding lectins (siglec-4a) specific for NeuAcα2,3Gal (Table 27.6), which explains its selective binding to GD1a and GT1b (see Figure 30.2) of mammalian axons. Cross-linking of GT1b by MAG is considered part of the inhibitory mechanism involving NgR and possibly $p75^{NTR}$ as coreceptor [27]. Resulting activation of Rho-ROCK represents a downstream convergent reaction mediating neuritogenic inhibition [28]. Detailed knowledge of these inhibitory participants and pathways presents opportunities for therapeutic intervention. CNS myelin is not the only cause of axon outgrowth inhibition, components of the glial scar also contributing (see below).

30.10
Proteoglycans and Extracellular Matrix of the Nervous System

Finally, we consider the proteoglycans, another prominent macromolecular glycan type in the nervous system of considerable complexity in terms of structure and function. Proteoglycans are prominent components of the ECM of all multicellular organisms and contain a protein core to which long repeating linear polymers of disaccharides termed GAGs are linked (for more information on proteoglycans, please see Chapter 11). Proteoglycans can also occur as integral membrane components on the cell surface. All known GAGs have been detected in brain, the main ones being chondroitin sulfates and heparan sulfates. The neural ECM contains a large hyaluronin scaffold to which glycoproteins and proteoglycans are associated. The lectican family of CSPGs makes up the predominant hyaluronan-binding proteins, and includes neurocan, brevican, versican and aggrecan (please see Figure 19.1 for molecular display, Chapter 11.3 and Table 19.2 for functions of lecticans). The latter two are widely distributed, while neurocan and brevican are abundant in the adult CNS. They function as spatial organizers within the matrix by virtue of their pericellular localization and interactions with both ECM

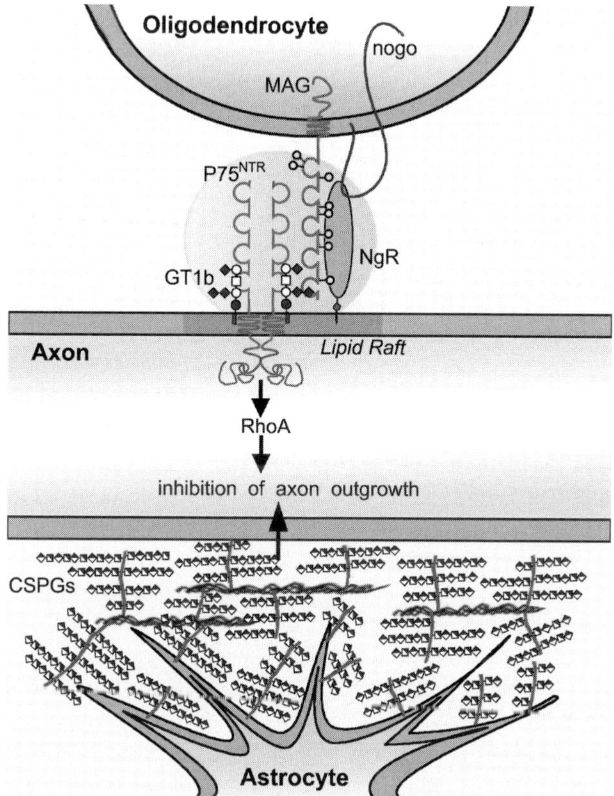

Figure 30.5 Proposed mechanism for inhibition of axon outgrowth by glycans of myelin and ECM. MAG expressed by oligodendrocytes contains eight *N*-linked oligosaccharides (blank circles) in its extracellular region. MAG binds to ganglioside GT1b and GPI-linked NgR of the axonal membrane, recruiting these into a lipid raft microdomain. A complex is formed with neurotrophin receptor $P75^{NTR}$ and the latter transmits an inhibitory signal by activating small GTPase, RhoA. NgR alone also interacts with Nogo ligand expressed in myelin. Astrocytes inhibit axon outgrowth by producing CSPGs after injury. GAG side-chains are shown. [Adapted with revisions from (a) M. Vinson *et al.* Lipid rafts mediate the interaction between myelin-associated glycoprotein (MAG) on myelin and MAG-receptors on neurons. *Mol Cell Neurosci* 2003; **22**, 344–352 and (b) [27].]

and the cell-surface glycocalyx. Such assemblies have been described as perineuronal nets, which are believed to contain a hyaluronan–lectican–tenascin-R complex that constitutes the core assembly of the mature brain ECM; the glycoprotein tenascin-R (see above) is proposed to function as a molecular cross-linker within the complex. In the nervous system, as elsewhere, proteoglycans are characterized by enormous structural diversity based on sugars/proteins and the degree of sulfation along the glycan chains of 50–150 disaccharide repeat units.

In relation to development, the neural ECM has essential roles in cell migration, neurite outgrowth, synaptogenesis, and regulation of synaptic plasticity. As with tenascins (see above), CSPGs promote neuritogenesis when present as a uniform substrate, but have the opposite effect when present as a sharp substrate boundary. Lectican structures encompass a major globular domain at each end, joined by a stretched domain to which chondroitin sulfate chains are attached [29]. They exist as splice variants that, by virtue of their sulfate-induced negative polarity, constitute the inhibitory signals toward navigating axons or motile cells. The reactive gliosis resulting from CNS injury involves dense deposits of lecticans, principally versican and neurocan, in the glial scar. This area is a physical and biochemical barrier to axon regeneration, explaining why removal of chondroitin sulfate from the core proteins with the enzyme chondroitinase ABC increases the regenerative potential of damaged axons. Matrix metalloproteinases also promote regeneration by degrading the core protein of some CSPGs in addition to other inhibitory molecules such as Nogo and tenascin-C. Versican, of around 600 kDa size, is produced by oligodendroglia as well as oligodendroglial precursor cells and exists as two splice variants with different chondroitin sulfate chains: V1 versican predominates during prenatal development and promotes neurite outgrowth, whereas V2 occurs in myelinated tracts and inhibits neuritogenesis. Neurocan, of around 250 kDa size, is predominantly neuronal, but is made and processed by reactive astrocytes after injury [29]. The role of proteoglycans is also becoming of interest in relation to chronic neurodegenerative diseases, such as amyloidopathies. The well known extracellular plaques with aggregated amyloid β in Alzheimer's disease are characterized by dense accumulations of CSPGs and heparan sulfate proteoglycans. These were proposed to increase the stability of tau and amyloid β proteins, thereby enhancing degenerative plaque formation. Another point of view, however, is that CSPGs might protect against neuronal loss, consistent with the finding that regions with perineuronal net-containing neurons seem to be the least affected in Alzheimer's disease.

An additional proteoglycan component of the glial scar that inhibits axonal growth is NG2, expressed on the surface of oligodendrocyte precursor cells for which it is a marker [30]. NG2 is a high-molecular-weight CSPG that is a type 1 transmembrane protein with a short cytoplasmic tail and large extracellular domain containing a single GAG chain. Enzymatic removal of the latter did not alter the inhibitory effect, whereas polyclonal antibodies against NG2 did reduce growth inhibition. A considerable portion of NG2 is shed or secreted from the cells and becomes incorporated into the ECM via binding to collagen (a minor component of the ECM). Considering the manifold interactions of proteoglycans of neural cells with each other and the ECM, what once appeared as an impenetrable enigma is now giving way, molecule by molecule, to a gradual understanding of the conducive and inhibitory properties of these highly complex neural glycoconjugates.

30.11 Conclusions

As the most complex of vertebrate tissues, the nervous system is highly dependent on the information-dense glycoconjugates for developmental and functional processes. The nervous system contains the same general categories of glycoconjugates as other tissues, but with many unique features. These arise to a large extent from unique structural elements, including excitable cells (neurons), required to communicate rapidly, and supporting cells (glia) that produce specialized elements (myelin, astrocytic processes) to aid neuronal function. It is noteworthy that some glycoconjugates, such as CSPGs, tenascin-C and -R, and the MAG protein of myelin, that aid in development can prove inhibitory at later stages involving damage repair and regeneration.

Summary Box

Glycolipids are prominent glycan carriers of the nervous system. They consist primarily of gangliosides, the major sialoglycoconjugates of brain present in all neural cells, and galactocerebrosides plus sulfatides, key components of CNS and PNS myelin. Molecular mechanisms by which GM1 and other gangliosides modulate neural protein activities involve direct protein–ganglioside association and also cross-linking of GM1 leading to signaling and ion channel activation. Glycoproteins in the form of recognition molecules have major roles in the various phases of neural differentiation. These are characterized by the presence of polysialic acid, HNK-1 and/or oligomannosidic glycans. The myelin sheath, which insulates CNS axons and aids saltatory conduction, contains glycoproteins such as MAG and OMgp which promote myelination but also block axonal regeneration. These inhibitory mechanisms are known to involve the GPI-linked NgR and ganglioside GT1b on the axonal membrane. Glycoproteins of PNS myelin are more abundant, but much less inhibitory than those of the CNS. The ECM is rich in proteoglycans which have dual roles in both promoting and blocking neuronal differentiation. CSPGs are especially prominent in the glial scar and contribute to the failure of damaged CNS neurons to regenerate. PNS neurons, lacking these inhibitory elements, are able to regenerate damaged axons. Glycoconjugates of neural stem cells are subjects of interest and will likely have a prominent role in this rapidly growing field.

References

1 Vilchez D *et al.* Mechanism suppressing glycogen synthesis in neurons and its demise in progressive myoclonus epilepsy. *Nat Neurosci* 2007;10:1407–13.

2 Klenk E. On the discovery and chemistry of neuraminic acid and gangliosides. *Chem Phys Lipids* 1970;5:193–7.

3 Yu RK *et al.* Glycosphingolipid structures. In: *Comprehensive Glycoscience* (Ed.: JP Kamerling), pp. 73–122. Elsevier, Oxford, 2007.

4 Svennerholm L. Chromatographic separation of human brain gangliosides. *J Neurochem* 1963;10:612–23.

5 Schauer R. Sialic acids: fascinating sugars in higher animals and man. *Zoology* 2004; 107:49–64.

6 Roseman S. The synthesis of complex carbohydrates by multiglycosyltransferase systems and their potential function in intercellular adhesion. *Chem Phys Lipids* 1970;5:270–97.

7 Simpson MA *et al.* Infantile-onset symptomatic epilepsy syndrome caused by a homozygous loss-of-function mutation of GM3 synthase. *Nat Genet* 2004;36:1225–9.
8 Kolter T, Sandhoff K. Sphingolipids – their metabolic pathways and the pathobiochemistry of neurodegenerative diseases. *Angew Chem Int Ed Engl* 1999;38:1532–68.
9 Miyagi T *et al.* Molecular cloning and characterization of a plasma membrane-associated sialidase specific for gangliosides. *J Biol Chem* 1999;274:5004–11.
10 Kopitz J *et al.* Effects of cell surface ganglioside sialidase inhibition on growth control and differentiation of human neuroblastoma cells. *Eur J Cell Biol* 1997;73:1–9.
11 Rodriguez JA *et al.* Plasma membrane ganglioside sialidase regulates axonal growth and regeneration in hippocampal neurons in culture. *J Neurosci* 2001;21:8387–95.
12 Ledeen RW, Wu G. Ganglioside function in calcium homeostasis and signaling. *Neurochem Res* 2002;27:637–46.
13 Svennerholm L *et al.* Gangliosides and allied glycosphingolipids in human peripheral nerve and spinal cord. *Biochim Biophys Acta* 1994;1214:115–23.
14 Ariga T, Yu RK. Antiglycolipid antibodies in Guillain–Barré syndrome and related diseases: review of clinical features and antibody specificities. *J Neurosci Res* 2005; 80:1–17.
15 Skaper SD, Leon A. Monosialogangliosides, neuroprotection, and neuronal repair processes. *J Neurotrauma* 1992;9:S506–16.
16 Schneider JS *et al.* GM1 ganglioside rescues substantia nigra pars compacta neurons and increases dopamine synthesis in residual nigrostriatal dopaminergic neurons in MPTP-treated mice. *J Neurosci Res* 1995;42:117–23.
17 Fishman PH. Role of membrane gangliosides in the binding action of bacterial toxins. *J Membr Biol* 1982;69:85–97.
18 Allende ML, Proia RL. Lubricating cell signaling pathways with gangliosides. *Curr Opin Struct Biol* 2002;12:587–92.
19 Wu G *et al.* Cerebellar neurons lacking complex gangliosides degenerate in the presence of depolarizing levels of potassium. *Proc Natl Acad Sci USA* 2001;98:307–12.
20 Gong C-X *et al.* Impaired brain glucose metabolism leads to Alzheimer neurofibrillary degeneration through a decrease in tau *O*-GlcNAcylation. *J Alzheimers Dis* 2006;9:1–12.
21 Mühlenhoff H *et al.* Brain development needs sugar: the role of polysialic acid in controlling NCAM functions. *Biol Chem* 2009;390:567–74.
22 Kleene R, Schachner M. Glycans and neural cell interactions. *Nat Rev Neurosci* 2004;5: 195–208.
23 Loers G, Schachner M. Recognition molecules and neural repair. *J Neurochem* 2007; 101:865–82.
24 Martin PT. Glycobiology of the synapse. *Glycobiology* 2002;12:1R–7R.
25 Berninghausen O *et al.* Neurexin Iβ and neuroligin are localized on opposite membranes in mature central synapses. *J Neurochem* 2007;103:1855–63.
26 Quarles RH. Myelin sheaths: glycoproteins involved in their formation, maintenance and degeneration. *Cell Mol Life Sci* 2002;59: 1851–71.
27 Liu BP *et al.* Extracellular regulators of axonal growth in the adult central nervous system. *Phil Trans Roy Soc B* 2006;361:1593–610.
28 Guan KL, Rao Y. Signaling mechanisms mediating neuronal responses to guidance cues. *Nat Rev Neurosci* 2003;4:941–56.
29 Viapiano MS, Matthews RT. From barriers to bridges: chondroitin sulfate proteoglycans in neuropathology. *Trends Mol Med* 2006;12:488–96.
30 Tan A *et al.* NG2: a component of the glial scar that inhibits axon growth. *J Anat* 2005; 207:717–25.

Glossary

a

ab initio calculations

In quantum chemistry (QC) *ab initio* (Latin: from beginning) means that the calculations are solely based on first principles, no empirical data are included.

acrosome

A Golgi-apparatus-derived cell organelle that is a part of the sperm head. It covers about half of the sperm nucleus and contains enzymes to assist in penetration of the egg and its surrounding cell layers at fertilization.

acrosome reaction

The release of enzymes and other proteins from the acrosome of a spermatozoon, which occurs after the spermatozoon has bound to the zona pellucida.

activated (nucleotide) sugars

Sugars that are linked either to nucleotides or to lipids (*see* dolichol) to be available in a high-energy state for glycosyl transfer. They therefore serve as donor substrates for glycosyltransferase reactions.

affinity chromatography

A method exploiting the specific interaction between a ligand and its binding partner. Usually the ligand is immobilized on a resin packed into a column. On applying a crude extract to the column, the immobilized ligand retains its binding partner, thus separating it from the bulk of non-binding components. It can subsequently be removed from the column by changing the buffer composition. To purify lectins, carbohydrates (mono-, oligo-, and polysaccharides, glycosides, glycopeptides or glycoproteins) are attached to an insoluble hydrophilic matrix, and the lectin-containing crude extract is then passed over a column filled with this material. Desorption of the lectin can be achieved by a buffer containing the free carbohydrate, by changing the pH value or by adding borate or chaotropic reagents to the elution buffer (*see also* chromatography).

The Sugar Code. Edited by Hans-Joachim Gabius

ISBN: 978-3-527-32089-9

agglutinin
See lectin.

aglycon
The non-sugar part of a glycoside.

angiogenesis
Tightly regulated formation of new blood vessels from pre-existing ones or even subsets of tumor cells. Angiogenesis takes place during normal growth and development as well as during wound healing, inflammation and tumor growth. Cellular components involved in angiogenesis are vascular endothelial cells, pericytes and in larger vessels smooth muscle cells. Angiogenesis is regulated by the concerted action of several growth factors (such as vascular endothelial growth factor) and chemokines.

anoikis
Programmed cell death (*see* apoptosis) of anchorage-dependent cells induced by defects in appropriate cell-matrix interaction [Greek: *anoikos* (ἄνοικος) = homeless].

anomeric (Edward-Lemieux) effect
A stereoelectronic effect that relates to the distance between the electronegative substituent on the anomeric carbon in a pyranoside and the dipole of the ring oxygen. In accord with the anomeric effect, axial substitution is preferred over the sterically less hindered equatorial one due to maximal spatial separation.

anomery
A special form of isomery specific for carbohydrates. It concerns the hydroxy group that originates from the aldehyde or keto group of a monosaccharide after ring closure (*see* pyranose and furanose). As long as this anomeric hydroxy group bears no substituent, it can assume either an equatorial or an axial position in equilibrium (*see* Fig. 1.2). Conventionally, the α-form of D-monosaccharides (e.g., D-glucose, D-mannose, D-galactose) presents the anomeric hydroxy group axially, the β-form equatorially. In L-monosaccharides (e.g., L-fucose or L-rhamnose), the inverse is true (*see* Fig. 1.6). As soon as the anomeric hydroxy group is substituted, for instance by a small organic substituent or by a further monosaccharide, mutarotation (dynamic equilibrium of anomers via the open-chain form, monitored by measuring optical rotation of polarized light) is no longer possible, and the anomeric configuration is fixed (*see* Fig. 1.4).

antibody
A Y-shaped glycoprotein generated by the body's immune system to interact with an antigen present on a foreign molecule via contact sites generated by disulfide bridge-linked chains. Binding can then initiate rejection of the foreign molecule.

antithrombotic
An agent that prevents or interferes with the formation of thrombi.

apoptosis
Term coined in 1972 [Greek: *apoptosis* (ἀπόπτωσις) = falling off] as a result of analyzing ultrastructural changes characteristic of dying cells, such as chromatin degradation and cell shrinkage, synonymous to programmed cell death (or suicide).

atomic force microscope
Type of scanning probe microscope that enables a surface to be imaged at high resolution by rastering a sharp tip in close, but not direct, contact over the surface.

azurophil
An old-fashioned word, introduced by Paul Ehrlich, to describe cellular structures with an affinity for basic stains, such as methylene blue or methylene azure.

b

bacillosamine
2,4-Diamino-2,4,6-trideoxy-D-glucose; in its di-*N*-acetyl form, it is a component of bacterial protein-linked glycans, such as the *N*-glycans of *Campylobacter*.

basement membrane
Sheet-like 40–120-nm-thick structure of the extracellular matrix, underlying endothelial and epithelial cells and surrounding other types of cells. Made up of collagen IV, perlecan, agrin and collagen XVIII. Other constituents are laminins-8, -9 and -10 and nidogen. Acts as mechanical barrier and is essential for angiogenesis.

biopolymer
Class of polymers produced by living organisms.

blastocyst
A stage of development of a mammalian embryo that follows the morula. It consists of a hollow sphere of trophoblastic cells, a central fluid-filled cavity (blastocoel) and a cluster of cells in the interior, called inner cell mass or embryoblast.

blood group
Serologically detectable difference on blood cells, classically in ABH epitopes. These are defined by glycan structures (*see* Fig. 1.5).

brefeldin A
A fungal compound that blocks translocation of proteins from the endoplasmic reticulum to the Golgi apparatus. It inhibits the GTP exchange factor of the protein

study. Different NMR-active nuclei (^{1}H, ^{13}C, ^{15}N, etc.) strongly differ in their chemical shifts. Moreover, most of the ^{1}H atoms within a molecule also differ in their chemical shifts dependent on their individual environment, providing a powerful tool for structure elucidation.

chitin
Linear *N*-acetylated aminopolysaccharide made of β1,4-linked *N*-acetylglucosamine units. Single chitin chains aggregate to form microfibrils with different physicochemical properties depending on the orientation of the chains.

chitin synthase
UDP-*N*-acetyl-D-glucosamine:chitin β4-*N*-acetyl-D-glucosaminyl-transferase (EC 2.4.1.16); membrane-integral family 2 glycosyltransferase catalyzing the polymerization of *N*-acetylglucosamine.

chitinase
Poly[β1,4-(*N*-acetyl-β-D-glucosaminide)] glycanohydrolase; collective term for glycolytic enzymes that degrade chitin by hydrolyzing *O*-glycosidic bonds. It includes endo- and exochitinases (EC 3.2.1.14), *N*-acetylglucosaminidases (EC 3.2.1.52) and chitobiases (EC 3.2.1.29).

chitosan
Linear aminopolysaccharide made of β1,4-linked glucosamine units, naturally containing *N*-acetylglucosamine to various extents.

chondroitin AC lyase
Enzyme (EC 4.2.2.5) performing eliminative degradation of polysaccharides containing β1,4-D-hexosaminyl and β1,3-D-glucuronosyl linkages to disaccharides containing 4-deoxy-β-D-gluc-4-enuronosyl groups.

chondroitin-sulfate-ABC endolyase
Enzyme (EC 4.2.2.20) performing endolytic cleavage of β1,4-galactosaminic bonds involving *N*-acetylgalactosamine. Its reaction results in a mixture of Δ^4-unsaturated oligosaccharides of different sizes that are ultimately degraded to Δ^4-unsaturated tetra- and disaccharides.

chromatography
First described in 1906 but developed to a highly efficient separation method much later. Many variations are in use based on differences (in electrical charge, molecular size, hydrophobicity, etc.) in the substances to be separated (*see also* affinity chromatography). The origin of the term dates back to the first experiments in which colored plant pigments were separated [Greek: *chroma* (χρῶμα) = color, graphein (γράφειν) = to write].

circular permutation
Phenomenon of sequence inversion on the level of the protein first detected by comparison of the amino acid sequence of the lectin concanavalin A with those of other related legume lectins (for details: *see* Info Box 2 in chapter 18).

circumforaneous
A word to be looked up by the reader, along with sesquipedalian. It requires sesquipedalian tendencies to use circumforaneous.

clearance (survival) signal
Structural determinant that regulates the rate at which blood cells or (glyco)proteins are removed from the body, for example the status of sialylation in glycans. Respective glycoengineering is a means to prolong serum circulation of therapeutic glycoproteins such as erythropoietin.

collectin
A member of a group of pattern recognition receptors in innate immunity. They share a cysteine-rich N-terminal section, a collagen-like region and an α-helical neck domain for oligomerization (stalk section) to position the C-type lectin domains (head part). This topology enables sensing foreign or aberrant glycan signatures (variation in genomic representation of individual group members across animal kingdom).

comparative genomics
Comparison of results from whole-genome analysis allowing study of relationships between genomes, such as the similarities and differences in the evolution of proteins. An example is comparison of genes for lectins from invertebrates, such as worm and fly, to those of vertebrates, such as fish and mouse. This provides insights into how evolution has differentially recruited and amplified particular domains, the 'building blocks' of proteins, into the functional repertoire of genomes from different branches of life. For example, the C-type lectin-like domain is very abundant in worm, less abundant in mouse and human, much less common in fly.

competitive inhibitor
An inhibitor that competes with the normal substrate for an enzyme's active site. Binding is reversible.

configuration
The three-dimensional arrangement (or sequence) in space of atoms or functional groups that characterizes a stereoisomer. To change a configuration, bond breaking and bond reforming in a different sequence of events must occur. Configuration should be contrasted with conformation, in which changes are brought about only by bond rotation (for example chair and boat conformers of pyranose rings).

conformation
Three-dimensional structure of a molecule. Conformers are different forms of a molecule resulting from rotations around covalent bonds (*see also* configuration).

conformational analysis
Analysis aims at determining the three-dimensional structures of molecules. NMR spectroscopy, X-ray analysis and circular dichroism measurements belong to the most important methods used for conformational studies. All computational methods that can predict the energy hypersurface of the molecules give structure to the energy landscape. Conformations adopted by a molecule are represented by minima on the potential energy surface, computational conformational analysis thus means searching for stable minima on energy surfaces.

congenital disorder of glycosylation (CDG)
Inherited human disease based on defects in the biosynthesis of glycan chains of glycoconjugates, leading for example to hypoglycosylation of glycoproteins. Abnormally low levels of sialylation can readily be detected by isoelectric focusing.

copy number polymorphism
The presence of different numbers of a given gene in different individuals. For example, the number of genes for HNP-1, a human α-defensin, can vary from 4 to 11, or more.

cortical reaction
The release of the contents of the oocyte cortical granules into the perivitelline space after contact with the fertilizing spermatozoon; the cortical reaction prevents polyspermy in mammalian species.

C-type lectin
Member of a large group of extracellular Metazoan proteins with diverse functions containing one or more C-type lectin-like domains (CTLDs), a type of lectin domain with distinct folding which commonly involves a Ca^{2+} ion for ligand binding by coordination bonds (*see* Fig. 16.1).

C-type lectin-like domain (CTLD)
The carbohydrate recognition domain (CRD) of group C of animal lectins. A protein domain 110–140 residues long with a characteristic conserved sequence signature allowing identification from analysis of the protein sequence. The main characteristics of the CTLD fold are two antiparallel β-sheets and two α-helices, two disulfide bridges and a hydrophobic core stabilizing the long-loop region containing the sugar-binding site involving a Ca^{2+} ion (*see* Fig. 20.1).

cuticle
Multi-layered structure covering predominantly epidermal cells. Arthropod cuticles are complex biocomposites containing chitin, cuticular proteins and lipids.

cyanovirin-N
An antiviral lectin from the cyanobacterium *Nostoc ellipsosporum* that neutralizes virus infectivity by binding to specific high-mannose-type oligosaccharides in HIV-1 or influenza A.

d

defensin
Member of a family of multifunctional host-defense peptides represented in vertebrates by three subfamilies called α-, β-, and θ-defensins.

dendrimer
Repeatedly branched molecule; the term comes from the Greek *dendron* (δενδρον) = tree. Synonymous terms are arborols and cascade-molecules. Hence, a glycodendrimer is a dendrimer containing multiple copies of carbohydrates. A dendron is a partial fragment of a dendrimer, it is referred to as a wedge of the dendrimer.

density functional theory (DFT)
A quantum mechanical theory, used to calculate electronic properties of atoms and molecules.

docking
A computational technique to accommodate ligand molecules within a receptor's binding site. The binding affinity can be thus predicted and sets of ligand molecules scored accordingly.

dolichol
Generic name for terpenoids, lipids with 13–24 repeats of isoprene units terminated by an alcohol group. Dolichol can be phosphorylated, and is then involved in initial steps of *N*-glycosylation.

dystroglycan
A two-subunit glycoprotein component of mammalian cell membranes acting as a transmembrane 'bridge' between the cytoskeleton and the extracellular matrix; the glycosylation of dystroglycan, especially its *O*-mannosylation, is defective in some congenital muscular dystrophies.

e

electron-density map
Three-dimensional representation of the electron clouds of a molecule based on X-ray diffraction data (*see* Fig. 13.4). A 2Fo-Fc map shows the position where the model is and where the model should be built.

β-elimination
A method for release of glycans, especially *O*-linked oligosaccharides, from glycoproteins, involving use of an alkaline solution such as sodium hydroxide and,

often, a reducing agent such as sodium cyanoborohydride for following product stabilization.

embryoblast
Cells of the blastocyst that give rise to the embryo.

enantiopeptide
A peptide or peptide analog synthesized exclusively with D-amino acids.

endoplasmic reticulum-associated degradation (ERAD)
An activity that regulates (glyco)protein availability for the extracellular space. It is the final effector phase of quality control within the secretory pathway. It routes inactive, dead-end folding states and excess of non-assembled subunits of oligomeric proteins to proteasomal annihilation after their marking with ubiquitin, a process preceded by sensing the deviation to initiate disposal. Herein, *N*-glycans effectively signal protein-folding states.

enthalpy of binding
The enthalpy change (ΔH) accompanying the binding process. It is equal to the heat absorbed or released in the process at constant pressure. Protein-carbohydrate interactions are typically exothermic.

entropy of binding
The entropy change (ΔS) in a binding process. Entropy is a measure of the disorder of a system. Binding of carbohydrates by proteins typically implies a decrease in entropy that opposes complex formation.

enzyme inhibitor
An agent that binds to an enzyme and inhibits its activity.

epididymis
A complex tubular structure attached to the testis that provides sites for sperm maturation and storage.

epimery
A special form of isomery for monosaccharides. They have more than one asymmetric centre. The change of configuration at only one site results in isomers called epimers. For instance, the pair D-glucose and D-galactose, which differ only in the configuration of the hydroxy group at C4, consists of epimers. Change at the anomeric centre at C1 is excluded, this form of isomery being referred to as anomery.

epitope
Antigenic determinant specifically recognized by antibodies or antigen receptors (*see also* hapten)

equilibrium dialysis
A classic method used to determine the association constants (K_a) of sugar-protein interactions. A lectin solution is placed inside a semi-permeable membrane, and then brought to equilibrium with a sugar solution of known concentration. A sugar can freely pass through the membrane. Therefore, if the sugar binds to the lectin, the concentration of the sugar inside the membrane is increased, depending on the binding affinity.

evanescent-field
An optic near-field standing wave exhibiting decay exponentially with distance from the interface at which it is formed. The principle is applied to the sensitive monitoring of lectin microarrays, where bound glycans or glycoconjugates, appropriately labeled by fluorescent dyes, are detected *in situ*, i.e., without washing procedures.

EXT1/EXT2 protein
Copolymerase involved in heparan sulfate (HS) biosynthesis. After assembly of the core tetrasaccharide and a GlcNAc (chain initiation), chain polymerization takes place by the alternating addition of GlcA and GlcNAc residues catalysed by EXT1/EXT2 (EC 2.4.1.225 and EC 2.4.1.224). Modification of the resulting HS chain involves epimerization (GlcA→IdoA) and sulfation of GlcNAc and IdoA residues. EXT genes encode tumor suppressors. Mutations cause hereditary multiple exostoses.

extracellular matrix (ECM)
Matrix comprising the interstitial matrix and the basement membrane. The interstitial matrix is present between cells, is produced by resident cells and secreted into the ECM. ECM provides support and anchorage for cells, acts as compressive buffer against stress and storage depot for growth factors and cytokines. It adsorbs large volumes of water.

f

fimbriae/pili
Long polymers on bacterial surfaces built up of various numbers of protein subunits.

free oligosaccharide (FOS)
Oligosaccharide cleaved from the dolichol donor and from glycoconjugates. FOS are generated in the lumen of the ER and in the cytosol, and then enter the non-lysosomal section of their metabolic fate. Here, former *N*-glycans reach the state of a $Man_5GlcNAc_1$ hexasaccharide, further processed in lysosomes.

frontal affinity chromatography (FAC)
A quantitative affinity chromatographic method used to determine the association constants (K_a) of sugar-protein interactions. An excess volume of diluted sugar

solution is continuously applied onto the lectin-bearing column and the volume representing the elution front is measured. If the sugar binds to the lectin, the elution front is delayed, depending on the affinity. K_a can be calculated from the elution front.

furanose
Monosaccharides consisting of a five-atom ring. The ring is closed by reaction between the aldehyde or keto group with a hydroxy group from the same molecule, forming an inner hemiacetal or hemiketal. The term follows the name of the simplest five-membered O-containing heterocyclus, furan.

g

galactosylceramide
A glycosphingolipid containing galactose attached glycosidically to the 1-hydroxyl of the sphingosine moiety within ceramide. Also termed cerebroside, it is a major component of the myelin sheath. Its 3'-sulfation produces a sulfatide.

galectin
A member of a family of lectins with conserved carbohydrate recognition domain of about 130 amino acids present in protists (fungi, sponges), invertebrates and vertebrates, initially detected by lactose-inhibitable hemagglutination (formerly called electrolectin, galaptin or S-type lectin).

ganglioside
A glycosphingolipid with one or more sialic acids as part of the oligosaccharide chain, especially abundant in the central nervous system.

gene targeting
A genetic technique to inactivate an endogenous gene. This method can be used to delete a gene, remove exons, and introduce point mutations.

Gibbs free energy of binding
The change in Gibbs free energy (ΔG) coupled to the binding process. It is equal to the energy released upon binding and determines the binding affinity ($\Delta G = -RT \ln K$, where R is the gas constant, T the absolute temperature and K the binding equilibrium constant). ΔG is a function of the enthalpy (ΔH) and entropy (ΔS) changes associated to the process ($\Delta G = \Delta H - T\Delta S$).

glycan
Oligo- or polysaccharide. Glycan may also be used to refer to the carbohydrate portion of a glycoconjugate such as a glycoprotein, glycolipid, or a proteoglycan.

glycan antenna
Linear glycan chain on proteins or lipids. Branching reactions of glycosyltransferases generate new antennae, hence multivalency.

glycan branch
Glycan chains, newly extending from a formerly linear chain or core. For example, the lipid-linked oligosaccharide precursor in *N*-glycan synthesis has three different branches. These branches are processed during remodeling to mature complex-type *N*-glycans. They can present up to five branches.

glycan profiling
Determination of the diversity of glycan structures expressed on glycoconjugates or cells. In general, memory-matching approaches are taken for either free glycans, intact glycoproteins or cell surface glycans. Multi-dimensional liquid chromatography, capillary electrophoresis, mass spectrometry, lectin affinity profiling, or their strategic combinations are used for this purpose.

glycocalyx
The outer surface coat of a cell consisting predominantly of glycosylated macromolecules such as glycoproteins, glycosphingolipids and proteoglycans. The glycocalyx has protective functions, its constituents are receptors transmitting signals into the cell or promoting adhesion to the extracellular matrix or other cells.

glycoconjugate
General term for a compound in which carbohydrates are covalently linked with other chemical species such as a protein or a sphingolipid (*see also* neoglycoconjugate).

glycodendrimer
Dendritic glycoconjugate in which saccharide portions (generally accessible at the surface) are conjugated according to the principles of dendritic growth or ligated to pre-existing, highly functionalised and repetitive dendritic scaffolds (*see also* dendrimer).

glycoform
A distinct form of a glycoprotein with a distinct glycan structure per glycosylation site. Natural glycoproteins consist of a heterogeneous population with a number of possible glycans at each glycosylation site and, thus, consist of a population of glycoforms.

glycogene
Gene encoding an enzyme/functional protein for glycan synthesis including glycosyltransferases, sugar-nucleotide synthases, sugar-nucleotide transporters and sulfotransferases.

glycolipid
Biomolecule containing one or more sugar units bound by a glycosidic linkage to a hydrophobic membrane-anchoring compound, such as an acylglycerol, a sphingoid, a ceramide (*N*-acylsphingoid) or a prenylphosphate.

glycome
The complete range of glycan structures synthesized by a species or cell type; compare this to the words 'genome', 'proteome', 'transcriptome', etc.

glycomics
An analogous term to genomics and proteomics meaning comprehensive analysis of glycans produced by a single organism or individual.

glycomimetic
A molecule that can imitate the structure or function of a glycan or one of its fragments.

glycosaminoglycan (GAG)
Long linear, highly charged polysaccharide composed of a repeating pair of monosaccharides (hexuronic acid, HexA, and *N*-acetylated or substitution-free amino sugar HexN/HexNAc). Mainly found covalently linked to a protein core. Examples are chondroitin, dermatan, heparan and keratan sulfates (CS, DS, HS, KS), heparin, and hyaluronic acid.

glycosidase
Enzyme that hydrolyzes glycosidic linkages. Exoglycosidases remove terminal non-reducing-end sugars. Endoglycosidases hydrolyze internal glycosidic linkages. Glycosides containing an aglycon can also be cleaved.

glycoside cluster effect
Natural synergistic and cooperative effect encountered when multiple ligand copies are simultaneously presented to complementarily clustered receptors. This phenomenon is responsible for the exponential affinity enhancement of certain protein-carbohydrate interactions, which are rather weak on a per-saccharide basis in terms of selectivity and specificity (*see also* multivalency and polyvalency).

glycosidic hydroxy group
Used synonymously to anomeric hydroxy group.

glycosphingolipid
Glycosylated lipid detivative (sphingolipid). The carbohydrate residue is attached by a glycosidic linkage to *O*1 of the sphingoid. A variety of 'core' structures containing various linkages are known, including ganglio-type (based on Galβ1-3GalNAcβ1-4Galβ1-4Glcβ-Cer) in mammals and arthro-type (based on GalNAcβ1-4GlcNAcβ1-3Manβ1-4Glcβ-Cer) in insects and nematodes.

glycosphingolipid microdomain
Cluster of glycosphingolipids in biomembranes. Cholesterol preferentially colocalizes with glycosphingolipids in these clusters, also rich in sphingomyelin. Such domains are often referred to as 'lipid rafts'. A functional interaction between 'lipid

rafts' and membrane-associated proteins has been implicated in a number of physiological and pathological processes, including cell signaling, molecular trafficking and the function of the immune, vascular, digestive and reproductive systems. In order to emphasize the functional roles of these domains in cell adhesion and signal transduction, they have also been termed 'glycosynapse'.

glycosphingolipid storage disorder
Inherited human disorder resulting from defects in enzymes involved in glycosphingolipid degradation, also referred to as sphingolipidoses. These diseases cause severe central nervous damage, which is often lethal in early infancy.

glycosyl donor and acceptor
In joining two sugar residues through the formation of a glycosidic bond in a glycosylation reaction, the glycosyl donor (a saccharide with a leaving group at the anomeric centre, which donates a glycosyl moiety, i.e., the sugar ring without the anomeric oxygen) is the sugar residue in which the anomeric carbon forms part of the linkage. The glycosyl acceptor (a saccharide with at least one free hydroxyl group) is the sugar residue in which a carbon other than the anomeric centre forms part of that linkage.

glycosylphosphatidyl inositol (GPI)
The basic structural unit of GPI anchors (*see* GPI anchor).

glycosyltransferases
Enzymes (EC 2.4) transferring monosaccharide units from a nucleotide di(mono)phosphate-activated sugar to an acceptor molecule, resulting in the formation of di-, oligo- or polysaccharides. Transfer can also occur to Ser, Thr, or Tyr residues of proteins (*O*-linked) or to Asn in the consensus sequence (*N*-linked).

glycosynapse
See glycosphingolipid microdomain.

glycotype
The unique glycosylation potential or capacity of a species or a specific cell type.

glypiation
Proteoglycans presented on the exocytoplasmic surface of the plasma membrane through a GPI anchor are 'glypiated'.

glypican
Member of a family of six (in vertebrates) heparan-sulfate-containing proteoglycans anchored to the cell surface via a GPI anchor. Function similar to syndecans.

GPI anchor
Glycosylphosphatidyl inositol as a molecular means to anchor proteins in membranes; found in all eukaryotic cells. The membrane-residing lipid part is covalently linked through a glycoinositol chain with its terminal ethanolamine to the protein's C terminus. Its basic function is the attachment of membrane proteins to the external surface of the plasma membrane. GPI anchors enhance the mobility of proteins and support intracellular signaling and targeted cellular transport.

GPI transamidase
A multimeric enzyme complex mediating post-translational transfer of pre-formed GPI to proteins bearing a C-terminal GPI attachment signal. Variations of subunits in different species (protozoa) may be employed for drug design.

h

hapten
A small molecule that can elicit an immune response only when attached to a large carrier such as a protein. Its presence blocks antibody-antigen binding (*see also* epitope).

hemagglutination assay
The classic method for determining lectin activity by adding serially diluted lectin-containing solutions to aliquots of erythrocyte suspensions. Lectin-mediated cell association is measured.

hemagglutinin
See lectin.

hemostasis
Arrest of bleeding from an injured blood vessel.

heparin lyase
Enzyme (EC 4.2.2.7) performing eliminative cleavage of polysaccharides containing β1,4-linked D-glucuronate or L-iduronate residues and α1,4-linked 2-sulfoamino-2-deoxy-6-sulfo-D-glucose residues to give oligosaccharides with terminal 4-deoxy-α-D-gluc-4-enuronosyl groups at their non-reducing ends.

heparitin sulfate lyase (heparitinase I/II)
Enzyme (EC 4.2.2.8) acting on *N*-acetylated or *N*-sulfated glucosaminido-glucuronic acid linkages of HS. Sulfate groups at the 6-position of the glucosamine moiety are impeditive. Heparitinase II acts on HS producing disulfated, *N*-sulfated and *N*-acetylated-6-sulfated disaccharides and small amounts of *N*-acetylated disaccharides. Total degradation of HS is only achieved by the combined action of both heparitinases.

heptose
A seven-carbon sugar.

heteronuclear single quantum correlation (HSQC)
NMR pulse sequence that allows the correlation of directly attached NMR-active nuclei. It is particularly used for identifying ^{1}H-^{13}C or ^{1}H-^{15}N nuclei pairs. In this manner, most of the amide N-H pairs within a polypeptide can be distinguished as well as the C-H resonances belonging to a given pyranose moiety.

hexose
A six-carbon sugar.

homozygosity mapping
Strategy aimed at identifying disease loci based on the assumption that consanguineous affected individuals are likely to have two recessive copies of the same disease allele.

host tropism
The way in which a virus or bacterium preferentially target specific host species or specific cell types within those species.

hyalectan (lectican)
A member of a group of chondroitin-sulfate-containing proteoglycans combining two functional sites in their modular arrangement at opposite ends: a hyaluronan-binding region and a C-type lectin domain (found in aggrecan, brevican, neurocan and versican).

hydrogen bond
Interaction between a hydrogen atom covalently bound to a strongly electronegative atom (such as oxygen or nitrogen), which attracts the electron cloud of the hydrogen atom and leaves it positively polarized, and a lone pair of electrons on a second electronegative atom, which acts as hydrogen bond acceptor. In hydroxy groups, the sp^3-hybridised oxygen atom can participate in two hydrogen bonds as acceptor, while the proton can act as donor.

hypermannosylation
The tendency of yeast species to add many mannose residues, in the Golgi, to their *N*-glycans. The result can be, depending on species, *N*-glycans containing one hundred or more mannose units.

hypotonia
Clinical condition of low muscle tone, often equivalent to reduced muscle strength.

i

inflammation

The cooperative response of the immune and vascular systems to injury caused by invading pathogens, damaged or malignant cells, irritating chemicals or physical agents.

integrin

A member of a family of dimeric plasma membrane glycoproteins that mediate cell attachment to the extracellular matrix and that are implicated in cell adhesion and migration. They bind (glyco)proteins such as fibronectin and act as ligands for lectins via their glycan part.

isoelectric focusing

Electrophoresis technique that separates molecules by their difference in their isoelectric points, i.e., the pH at which a specific protein carries no net charge. Practical for detecting low-level sialylation in congenital disorders of glycosylation.

isothermal titration calorimetry (ITC)

A technique used to directly measure the heat released during a chemical reaction, triggered when mixing the reaction partners. ITC is frequently used to characterize the thermodynamics of protein-ligand interactions, to obtain the Gibbs free energy (ΔG) and enthalpic/entropic contributions.

isozyme

Enzymes encoded by different genes but catalyzing the same chemical reaction. These enzymes display different kinetic parameters, different regulatory properties, or different expression patterns.

l

L1

A member of the immunoglobulin superfamily similar to the neural cell adhesion molecule but lacking polysialic acid chains.

lectin

A protein or glycoprotein that binds to carbohydrates (mono-, oligo-, or polysaccharides or glycosides). Carbohydrate-binding antibodies, enzymes that act on carbohydrates and sensor or carrier proteins for free mono- or oligosaccharides are excluded from the class of lectins by definition. Many lectins agglutinate cells. Therefore, the terms agglutinin, hemagglutinin (if red blood cells are agglutinated) or phytohemagglutinin (if the lectin originates from plant material) are often used.

lectin domain

Protein domain capable of binding sugar without enzymatic activity. There are several different types of lectin domains, corresponding to the different lectin subtypes, e.g., C-type lectin-like, I-type or β-trefoil domains. Lectin domains have different, characteristic protein folds that bind sugars by different mechanisms.

lectin microarray
A method for glycan profiling, by spotting a series of lectins with distinct sugar-binding specificity on a microarray platform ready for interaction with labeled glycans. Signal read-out aids structural characterization of the studied glycan or cells.

leukocyte
Large white cell of the immune system that defends the body against both infectious disease and foreign materials.

ligand
A substance that is able to bind to and form a complex with a biomolecule to serve a biological function by the interaction (Latin: *ligare* = to bind).

lipid raft
See glycosphingolipid microdomain.

lipopolysaccharide
Lipid-containing polysaccharides of the cell wall of Gram-negative bacteria. The basic units are a diglucosamine-diphosphate with amide- and ester-bound β-hydroxy-myristic acid (lipid A), an inner core glycan including 2-keto-3-deoxy-D-manno-octulosonic acid (Kdo) and heptose, as well as a head part of about 40 sugar moieties. It has a direct role in pathogenesis of infection by Gram-negative bacteria (septicemia) and is a potent and pleiotropic activator of immune cells. The complex with proteins on the cell is referred to as endotoxin.

lysosomal storage disease
A group of human genetic disorders characterized by excessive lysosomal accumulation and excretion of incompletely degraded GAG (mucopolysaccharidoses) or glycosphingolipids (sphingolipidoses; *see also* glycosphingolipid storage disorders), resulting from a deficiency of one or more lysosomal glycosidases responsible for the degradation of dermatan and/or heparan sulfates or glycosphingolipids.

m

major histocompatibility complex (MHC)
Glycoproteins expressed on the cell surface, encoded by a highly polymorphic gene cluster, that function in T lymphocyte-mediated immune responses and present peptides either of self antigens or foreign antigens to T lymphocytes. Three classes of MHC molecules exist: MHC class I proteins present peptides to cytotoxic $CD8^+$ T lymphocytes, MHC class II proteins present peptides to $CD4^+$ T helper lympho-

cytes. In contrast, MHC class III proteins consist of complement factors such as C2 and C4 and are plasma proteins involved in unspecific immune responses.

malaria

A disease caused by protozoan blood parasites of the genus *Plasmodium* transmitted by infected female *Anopheles* mosquito. Four species are known to infect humans: *P. falciparum*, *P. vivax*, *P. ovale* and *P. malariae*. The parasites undergo asexual multiplication in the erythrocytes (erythrocytic schizogony or merogony). The ring-stage trophozoites (newly infected erythrocytes) mature into schizonts, which rupture releasing merozoites. The released merozoites infect new erythrocytes. This so-called blood stage is responsible for the clinical manifestations of malaria. Some parasites differentiate into sexual erythrocytic stages (gametocytes), which are taken up during a blood meal by a female *Anopheles* mosquito. Inoculation of the sporozoites into a new human host (after the multiplication of the parasites in the mosquito, known as the sporogonic cycle) perpetuates the malaria life cycle.

metastasis

The breaking away of individual cancer cells from an established tumor that results in them entering the blood supply. This allows secondary tumors, the metastatic lesions, to be established elsewhere in the body.

mitogenicity

The term describes the ability of a host of macromolecules, among them many lectins, to induce eukaryotic diploid cells to undergo mitosis, that means to divide or to proliferate. One of the frequent applications of plant lectins is to enlarge the number of blood lymphocytes for further analyses.

molecular dynamics (MD)

A computer simulation method into which Newton's law of motion is integrated for each atom of the molecular system in order to model system parameters over time periods (*see* molecular mechanics for energy terms).

molecular mechanics (MM)

A computational method to model molecular structure (conformation), assuming atoms being balls and bonds being strings. The force field is composed of several deformation (stretching, bending, torsion) terms, and terms corresponding to calculation of non-bonded energy as well as electrostatic repulsion or attraction. Additional terms may be added describing distinct features characteristic for carbohydrates that is the anomeric effect.

morula

Berry-like cluster of about 16–64 cells without a central cavity formed by the first mitotic cleavage divisions of the fertilized oocyte.

mucin
Highly glycosylated glycoproteins coded by a family of genes (MUC genes). Includes secreted, gel-forming and one non-gel-forming, membrane-associated member. Secreted and membrane-associated forms show characteristic peptide domains. All mucins contain serine/threonine-rich variable number of tandem repeat (VNTR) sequences that carry the *O*-GalNAc-linked, mucin-type glycans. They give the mucins their characteristic physicochemical and rheological properties. Protozoa can also recruit GlcNAc for *O*-glycan anchoring.

multivalency/valency
The number of times a given ligand (or receptor site) appears on a molecule (*see also* polyvalency).

multivalent interaction
The interaction between multiple receptors on one biological entity (for example a bacterium) with multiple ligands on another entity (for example a human cell or a glycodendrimer).

myelin-associated glycoprotein (MAG)
This glycoprotein with five immunoglobulin-like domains (4× C2-set, 1× V-set) is present in myelin of both the central and peripheral nervous system. Promotes myelination, inhibits axonal regeneration. It belongs to the siglec group of I-type lectins (members of the immunoglobulin superfamily) and is also termed siglec 4.

myxobacterium
A large group of slime-producing Gram-negative bacteria with unusual gliding mobility, social behavior, and the ability to form fruiting bodies under starvation conditions.

n

neighboring group participation
The direct interaction of the reaction centre with a lone pair of electrons of an atom, or with the electrons of a σ or π bond. They are present within the same molecule but not conjugated with the reaction centre.

neoglycoconjugate
General term for synthetic glycoconjugate analogues resulting from chemical or chemo-enzymatic synthesis. They present carbohydrates covalently linked on unnatural scaffolds and are useful in detecting, studying, characterizing, and exploiting protein-carbohydrate interactions. Neoglycoconjugates are useful tools to provide large quantities of various architectures developed as effectors or inhibitors of biological mechanisms. Potent multivalent neoglycoconjugates are based on a wide range of carriers (polymeric backbones) including polymers (glycopolymers), polyamino acids (neoglycoproteins), lipids and liposomes (neoglycolipids and glycoliposomes) and dendrimers (glycodendrimers).

neoglycoprotein
Protein to which a defined glycan structure has been chemically conjugated (*see also* neoglycoconjugate)

neural cell adhesion molecule (NCAM)
A glycoprotein with five immunoglobulin-like domains containing polysialic acid chains. Occurs in three isoforms, important in calcium-independent cell adhesion during neural development.

***N*-glycosylation**
Frequent cotranslational protein modification by attachment of the reducing end of a GlcNAc residue of a glycan. The nitrogen (N) atom of the amide of the Asn residue in the glycosylation sequon Asn-X-Ser/Thr acts as acceptor.

Notch
A transmembrane glycoprotein with a single pass peptide and multiple epidermal growth factor (EGF) domains. The EGF domains may carry *O*-fucose or *O*-glucose glycans that are recognized by the Notch ligands Delta and Jagged (in mammals). Notch is coded for by a single gene in *Drosophila melanogaster* and four genes in man. The glycoprotein mediates signaling leading to cell fate decisions such as lateral inhibition or cell lineage resolutions.

nuclear magnetic resonance (NMR) spectroscopy
Powerful technique for determining the structure, conformation and binding features of small, medium-size and large molecules, both in solution and in the solid state, based on the quantum mechanical magnetic properties of an atom's nucleus.

nuclear Overhauser effect (NOE)
Transfer of spin polarization between nuclear spins, used as molecular ruler in NMR spectroscopy to determine interproton distances, that is structural (conformational) features.

o

***O*-α-*N*-acetylgalactosamine, mucin-type glycosylation (*O*-GalNAc)**
Glycan chains linked through an α-glycosidic linkage of *N*-acetyl-D-galactosamine to serine and threonine residues on a polypeptide backbone.

***O*-β-*N*-acetylglucosamine glycosylation (*O*-GlcNAc)**
Addition of a single *N*-acetyl-D-glucosamine unit to serine or threonine residues in β-linkage. The process is called O-GlcNAcylation of a protein.

***O*-fucose glycosylation**
Addition of glycan chains linked through an α-glycosidic linkage of L-fucose to serine and threonine residues on a polypeptide backbone, for example Notch.

O-glucose glycosylation
Addition of glycan chains linked through a β-glycosidic linkage of D-glucose to serine and threonine residues on a polypeptide backbone.

oligomannosidic N-glycan
N-Linked oligosaccharide consisting of the standard chitobiose core and at least five mannose residues, for example $Man_9GlcNAc_2$. They are components of neural recognition glycoproteins uniquely abundant in the mammalian nervous system.

O-mannose glycosylation
Addition of glycan chains attached through an α-glycosidic linkage of D-mannose to serine and threonine residues on a polypeptide backbone.

opsonized
Bearing a surface that has been modified in a manner that facilitates subsequent ingestion by a phagocyte. Among the host defense molecules that opsonize bacteria are antibodies, complement components, and certain lectins, including some defensins and the collectins.

ortholog
A type of homologous sequence, homolog. Orthologs are genes in different species that evolved from a common ancestral gene by speciation. Usually, orthologs retain the same function

p

pandemic
An epidemic of infectious disease that spreads across a large region or even worldwide.

parasite
An organism that lives on or in other organisms, from which it obtains nutrients to live, and causes harm in the process. The term originates from the Greek words *para* that means beside, and *sitos,* which means food.

paucimannosidic N-glycan
N-Linked oligosaccharide consisting of the standard chitobiose core and between two and four mannose residues; the core may be modified by fucose residues. These structures are typical of plants and invertebrates.

pentose
A five-carbon sugar.

pentraxin
Phylogenetically conserved protein in innate immunity characterized by a disc-like arrangement of five non-covalently associated protomers (present in mammals,

fish, amphibians and horseshoe crab). The first member of this group (C-reactive protein in human serum) was discovered in 1930 due to its reactivity with phosphocholine in the capsular C-polysaccharide of *Streptococcus pneumoniae*, leading to precipitation.

peptidoglycan
Bacterial cell wall component maintaining the shape of the cell and counteracting osmotic pressure. It consists of polysaccharide chains [disaccharide unit: *N*-acetylglucosamine in β1,4-linkage to *N*-acetylmuramic acid (MurNAc); the following β1,4-linkage is cleaved by lysozyme] linked to a peptide network via lactyl groups of MurNAc. These cross-linking peptides typically contain four amino acids, alternating in L- and D-configuration. In Gram-negative bacteria and Gram-positive bacilli, meso-diaminopimelic acid resides in position 3 for introducing direct cross-links. Potent stimulator of the immune system, induces fever, target for host defense lectins.

peritrophic matrix
Pseudo-membrane lining and protecting midgut epithelia of many invertebrates.

phase variation
Expression of fimbriae on a bacterial cell surface is subject to changes with time.

phospho(ryl)choline
(2-Hydroxyethyl)trimethylammonium phosphate, the trimethyl form of phospho (ryl)ethanolamine; a non-reducing terminal component of glycans from nematodes and of lipopolysaccharide from some bacterial species, also present in phosphoglycerides.

phytohemagglutinin
See lectin.

PI-PLC
Phosphatidylinositol-specific phospholipase C, an enzyme belonging to the group of hydrolases. It cleaves between the glycerol and the phosphate group of a GPI or GPI-protein, only if the 2-position of the inositol is not substituted. Also known for its generation of inositol-1,4,5-triphosphate (IP_3).

PI-PLD
Phosphatidylinositol-specific phospholipase D, an enzyme belonging to the group of hydrolases. It cleaves between the inositol ring and the phosphate of the phosphoglycerol group of a GPI or GPI-protein. It is capable to cleave a GPI with substitution on the 2-position of the inositol ring.

Plasmodium falciparum
Causative agent of malaria.

platelet
Non-nucleated blood cell involved in primary hemostasis.

polysialic acid (PSA or polySia)
A linear homopolymer of variable length of 50 or more sialic residues, joined by α2,8 or α2,9 linkages. It is found in mammalian tissues of neuroectodermal and mesodermal origin, most commonly known when attached to two *N*-glycans in the fifth immunoglobulin-like domain of the neural cell adhesion molecule. Its presence in bacteria, e.g., as capsular polysaccharide (colominic acid) in *Escherichia coli* and in *Neisseria meningitidis*, when presenting lactones or de-*N*-acetylated sites, makes it a tool for vaccination. Further substitutions by sulfation are known from sea urchin (here flagellasialin). Please note: when men are concerned about PSA, they mean prostate-specific antigen, for lectinologists it can mean *Pisum sativum* agglutinin.

polyvalency
Clustering of (carbohydrate) epitopes that can raise binding affinity to neighboring partners by orders of magnitude (*see also* glycoside cluster effect and multivalency).

prepropeptide
A biosynthetic precursor of a peptide typically containing three domains: an N-terminal signal sequence, a domain that is removed during processing (the pro-domain), and the domain containing what will become the mature peptide.

protegrin
A family of unusually potent antimicrobial peptides originally isolated from porcine leukocytes. Protegrins contain 16–18 amino acid residues and have a β-hairpin structure stabilized by two intramolecular disulfide bridges.

protein body
Storage organelle formed from the central vacuole of a plant cell during its development. Protein bodies are lined by a single membrane and contain the typical storage proteins, lectins, several hydrolases and *myo*-inositolhexaphosphates (phytin). They are also called storage vacuoles.

protein domain
Many proteins, including lectins, contain more than one component protein module, called a domain. A domain has a compact three-dimensional structure, which may fold independently of the rest of the protein, and may function and evolve semi-independently. The sequence of a protein domain is typically 100–200 amino acids long.

protein domain architecture
The number and order of different protein domains in a multidomain protein. Many lectins, notably most C-type lectins, are multidomain proteins containing several other non-lectin domains; these complement the sugar-binding function of the lectin domain(s) and contribute to the overall function of the lectin. The C-type lectin-like domain is arranged in a large number of different protein architectures in C-type lectins in a particular organism, and these architectures vary greatly among different major branches of life such as worm, fly and mouse.

protein xylosyltransferase
Pacemaker enzyme (EC 2.4.2.26) for the biosynthesis of CS-, DS- and HS-containing proteoglycan (PG).Transfers xylose post-translationally to the protein core of PG, and is secreted in soluble form into ECM and blood plasma where it establishes a marker for the rate of PG synthesis in a multicellular organism.

proteoglycan (PG)
A family of high-molecular-weight polyanionic substances consisting of a protein core to which one or more different unbranched glycosaminoglycan (GAG) chains are linked in a post-translational process. Main component of extracellular matrix (ECM) or ECM-rich organs.

protohominid
Your grandfather's grandfather's grandfather's ... grandfather. Also, your grandmother's grandmother's grandmother's ... grandmother, going back to evolutionary antecessors of hominids.

pseudogene
A mutated gene, no longer capable of producing a functional protein. Expressed pseudogenes are transcribed into mRNA, but their transcripts contain a premature termination codon, deletion or frameshift mutation that prevents effective translation. Human DEFT genes are expressed pseudogenes.

pyranose
Monosaccharides consisting of a six-atom ring. The ring is closed by reaction between the aldehyde or keto group with a hydroxy group from the same molecule, forming an inner hemiacetal or hemiketal. The term follows the name of the simplest six-membered O-containing heterocyclus, pyran.

q

quantum chemical (QC)/molecular mechanics (MM) method
A combined QC and MM approach applied to solve structural issues of large molecular systems of biological importance. The active site of the protein-ligand complex (where bond breaking and formation of new bonds occurs) is modeled on the QC level, whereas the bulk environment within the protein and the surrounding solvent are calculated on the MM level.

quartz crystal microbalance
Mass sensing technique based on the piezoelectric frequency change due to mass attached to the quartz crystal.

r

receptor tyrosine kinase
Receptor present at the plasma membrane that harbors intrinsic tyrosine kinase activity in its intracellular region. Members of this group are capable to trigger outside-inside signaling upon binding of a ligand to its extracellular region.

retro-peptide
A peptide analog synthesized with its component L-amino acids present in reverse order along the peptide's backbone. It can also be called an inverso-peptide.

retrosynthesis
A method for purposefully arranging organic chemical synthesis into consecutive steps. The synthetic strategy is planned in reverse, beginning with the final product, then in backward steps moving to the starting materials. This procedure fragments the target molecule into subtargets, which are then fragmented further, all the way to the starting materials.

s

saccharide
A general synonym for carbohydrate or sugar.

saposin
Sphingolipid activator protein. It extracts membrane-bound glycosphingolipids for lysosomal degradation or CD1 loading (*see* CD1 protein).

saturation transfer difference (STD) spectroscopy
NMR technique that allows the binding of a particular ligand (or mixture of ligands) to a large receptor to be monitored, by observing NMR resonance signals of the ligand following a transfer of energy from the fully saturated receptor. In particular cases, it can also be used to deduce the ligand-binding epitope and to estimate binding affinities and dissociation kinetics.

schistosomiasis (bilharziosis or bilharzia, named after the German physician Theodor Bilharz, who described it in 1852)
Disease caused by *Schistosoma* spp., commonly known as blood flukes and bilharzia, the most significant infection of humans by flatworms. It is considered by the World Health Organization as second in importance only to malaria, with hundreds of millions infected worldwide. Adult worms parasitize mesenteric blood vessels. Eggs are passed through urine or feces to fresh water, where larval stages can infect a new host by penetrating the skin.

sclerotization
Hardening of cuticles due to chemical cross-linking of proteins and chitin.

secretory pathway
The series of organelles facilitating classical secretion of proteins that have been synthesized by ribosomes attached to the ER. A variety of post-translational modifications, such as glycosylation, occur during a protein's 'journey' through the secretory pathway (including ER and Golgi apparatus). Defective products are removed after ubiquitin attachment by proteasomal degradation (*see* endoplasmic reticulum-associated degradation).

selectin
A group of mammalian surface glycoproteins establishing a subgroup of C-type lectins. They share the following modular display: N-terminal surface-exposed C-type lectin domain, an epidermal-growth-factor-like module and two to nine complement-binding consensus repeats followed by the transmembrane region and a short cytoplasmic tail. The three group members are E-, L-, and P-selectin (CD62E, L, P). They are involved in interactions between endothelial cells/platelets and leukocytes and in lymphocyte homing to lymph nodes [formerly referred to as LEC-CAMs (lectin cell adhesion molecules) and individually as ELAM-1, homing receptor and GMP140/PADGEM]. The three genes are tightly clustered on a syntenic region of human and murine chromosome 1.

sialic acid
An acidic (carboxyl) nine-carbon sugar with a pyranose, biosynthetically consisting of pyruvate (phosphoenolpyruvate) joined to *N*-acetylmannosamine. An obligatory component of gangliosides and frequently present in glycoproteins at branch ends.

sialidase
An enzyme, present in four isoforms, that removes terminal sialic acid from oligosaccharide chains, but not the internal sialic acid of ganglioside GM1. Also termed neuraminidase, here known for its role in influenza.

siglec
A member of a subgroup of lectins with cell surface-exposed immunoglobulin (Ig)-like domains (I-type lectins) of the C2- and V-set types. The latter module is reactive with sialic-acid-containing glycans. Siglecs are sialic acid-binding Ig superfamily lectins (present in mammals, birds and fish).

single-molecule force spectroscopy
Application of optical tweezers or the atomic force microscope for measuring binding forces (piconewtons) in single ligand-receptor systems or unfolding forces in single biomolecules.

sleeping sickness: African trypanosomiasis
Disease caused by the protozoal parasite *Trypanosoma brucei* and transmitted by tsetse flies. Approximately 10^7 molecules of the variable surface glycoprotein (VSG), which is inserted into the plasma membrane via a GPI anchor, shield the parasite from the immune system.

solvation
Association of solvent molecules with the solute molecules or ions. Solvation energy can be approximated by calculation of energy contributions, corresponding to cavity, dispersion, electrostatic and specific interaction terms between the solute (carbohydrate) and solvent (water) molecules.

sphingoid
Class of long-chain aliphatic amines, containing two or three hydroxyl groups. Sphingoids are a primary part of sphingolipids.

sphingolipid
A lipid containing sphingosine as component within the hydrophobic part of an amphipathic structure. In most cases, this unit is ceramide, which usually has a polar moiety (e.g., an oligosaccharide chain or phosphocholine) attached to the first (primary) hydroxyl of sphingosine. Sphingomyelin and gangliosides are characteristic sphingolipids, the latter a glycosphingolipid.

sphingosine
An 18-carbon amino alcohol (2-amino-4-octadecene-1,3-diol). The most abundant sphingoid in animal tissues.

stem cell
Cell with self-renewing capabilities that can differentiate into multiple cell lineages.

stereoselectivity
The preferential formation of one stereoisomer over another in a chemical reaction. When the stereoisomers are enantiomers, the phenomenon is called enantioselectivity; when they are diastereoisomers, it is called diastereoselectivity.

sub-telomeric
Occupying a location just below the end of a chromosome. The telomere structure located at a chromosome's end plays important roles in its stability and resistance to ageing.

sugar
A term used almost synonymously to carbohydrate.

sulfotransferases
Enzymes (EC 2.8.2) performing acceptor-specific transfer of sulfate groups (often) from the coenzyme 3`-phosphoadenosine-5`-phosphosulfate (PAPS) to different substrates (protein, lipid or a monosaccharide residue of mono-, oligo- or polysaccharides).

surface plasmon resonance (SPR)
A quantitative technique used to determine the association constants (K_a) of ligand-receptor (here sugar-protein) interactions. An analyte solution is passed over a sensor chip surface containing the immobilized ligand. Binding is measured by the change of refractive index and is recorded in resonance units (RU).

Svennerholm nomenclature
A short-hand nomenclature system for gangliosides frequently encountered in the literature. For example, the gangliotetraosylceramide with α2,3-(mono)sialylation of the central Gal moiety (at position II) II^3Neu5Ac-$GgOse_4$Cer is termed GM1.

syndecan
Family of four transmembranous proteoglycans (PG) carrying three to four HS and one to two CS chains at the extracellular domain. They interact with a large variety of proteins including fibroblast growth factors, fibronectin, antithrombin and others. A short intracellular domain is involved in intracellular signaling.

t

γ/δ T cells
A small subset of T lymphocytes, most prominent among the intraepithelial lymphocytes of the intestinal mucosa. Their T cell receptor is composed of one γ chain and one δ chain.

tenascin
Extracellular matrix glycoprotein that interacts with glycosaminoglycans such as heparin and heparan sulfate.

thrombus
An aggregate of blood factors that can result in vascular obstruction at the site of its formation.

toxin
A chemical, often a protein, with a poisonous effect on biological entities such as cells, plants, animals and humans. Bacterial AB_5 or plant AB toxins are lectins, examples given by cholera toxin or ricin.

toxoplasmosis
Disease caused by *Toxoplasma gondii*, a protozoan (unicellular) parasite infecting most species of warm-blooded animals, including humans. The definitive hosts

are *felidae* (cat family) allowing the multiplication via the sexual stages. Human infection may be acquired by ingestion of undercooked infected meat containing *Toxoplasma* cysts, ingestion of oocysts (from cat faeces), organ transplantation or blood transfusion, transplacental transmission and accidental inoculation of tachyzoites (the aggressive stage of the parasite). Healthy people who become infected with *Toxoplasma gondii* rarely show symptoms. However, the parasite forms cysts that are reactivated many years later; immunosuppression is a key factor for this process. Infection during pregnancy poses a serious risk to the fetus.

transfusion
Process of transferring whole blood or plasma products from one person to another. Detection of incompatibility led to the concept of blood groups.

trophoblast
Cells of the blastocyst, excluding those which form the embryo; the cells of the trophoblast attach the early embryo to the endometrium and contribute to the formation of the fetal membranes and placenta.

Trypanosoma brucei gambiense
Causative agent of sleeping sickness.

Trypanosoma brucei rhodesiense
Causative agent of sleeping sickness.

Trypanosoma cruzi
Causative agent of Chagas disease.

tumor suppressor
Protein whose normal function is to suppress cell proliferation. Members of this family are frequently absent or inactivated by mutations in tumor cells.

v

virus-like particle (VLP)
Non-infectious particle derived from the structural proteins of a virus, with or without lipid layer.

w

whole-genome analysis
Analysis of the sequences of whole genomes of model organisms, such as fly, worm, fish, frog, mouse and human, allows systematic classification of their proteins, and their component protein domains and protein-domain architecture. This provides a powerful means to assess the abundance of particular domains, including lectin domains, in the genome.

x

X-ray crystallography
A technique for determining the three-dimensional structure of molecules that have been crystallized through the analysis of the scattering of X-ray beams by the electrons in the crystal.

X-ray diffraction pattern
The diffraction pattern consists of reflections of different intensity produced by interference of the atoms in crystals with X-ray waves. It is the fingerprint of the periodic arrangement of atoms in the crystal.

z

zona pellucida
Transparent, non-cellular layer surrounding the oocyte mainly formed by glycoproteins.

Index

The Sugar Code. Edited by Hans-Joachim Gabius

ISBN: 978-3-527-32089-9

d

q

r